Springer Geology

The book series Springer Geology comprises a broad portfolio of scientific books, aiming at researchers, students, and everyone interested in geology. The series includes peer-reviewed monographs, edited volumes, textbooks, and conference proceedings. It covers the entire research area of geology including, but not limited to, economic geology, mineral resources, historical geology, quantitative geology, structural geology, geomorphology, paleontology, and sedimentology.

More information about this series at http://www.springer.com/series/10172

Sreepat Jain

Fundamentals of Invertebrate Palaeontology

Macrofossils

Springer

Sreepat Jain
New Delhi
India

ISSN 2197-9545 ISSN 2197-9553 (electronic)
Springer Geology
ISBN 978-81-322-3880-5 ISBN 978-81-322-3658-0 (eBook)
DOI 10.1007/978-81-322-3658-0

Printed on acid-free paper

This Springer imprint is published by Springer Nature
The registered company is Springer (India) Pvt. Ltd.
The registered company address is: 7th Floor, Vijaya Building, 17 Barakhamba Road, New Delhi 110 001, India

For my son Parth

Contents

About the Author

Dr. Sreepat Jain is Professor at the Department of Applied Geology, Adama Science and Technological University, Adama, Ethiopia. Professor Jain has 16 years of teaching and over two and a half decades of research experience. He has two doctorates, one from India and the other from United States. He was awarded the "TA Excellence in Teaching" for meritorious teaching in the United States. In addition, he has received the prestigious "Prof. S. K. Singh Memorial Gold Medal" of the Paleontological Society of India for best paper. He has travelled extensively in US and Europe in the pursuit of higher academics. Also he is the recipient of the DST-DAAD Fellowship. He has sound teaching knowledge in disciplines of Geoscience and Petroleum engineering, alike and has an experience of establishing a new Petroleum engineering department. His areas of research include Jurassic ammonites and nannofossils, Jurassic to Recent foraminifers, and the use of biotic and abiotic proxies for deciphering paleoenvironments. He has published several research articles in national and international peer-reviewed journals and has authored two books of which The Fundamental of Physical Geology (Springer) has been recommended for the undergraduate Earth Science semester course in several American universities. It has also been downloaded over 30,000 times. Professor Jain is also a reviewer for several national and international geoscience journals.

Chapter 1
Life Through Ages: An Overview

The idea of this brief chapter is to introduce students to the concept of fossil occurrences and their use, through geological time. An approach that defines the geological time scale, a method of relating the timing and relationship between events that have occurred during Earth's history. The Geologic time is divisible into four Eons, Hadean, Archean, Proterozoic, and Phanerozoic (Fig. 1.1). The Eons are further divided into Eras; the Phanerozoic is divided into Paleozoic, Mesozoic, and Cenozoic (Fig. 1.1). The Eras are based on major changes in fossils record (such as extinction and origination; the latter is the appearance of new forms). The Eras are further divided into smaller units called Periods, in which a single type of rock system is formed; these are further divided into Epochs like the Paleogene is divisible into Paleocene, Eocene, and Oligocene (Fig. 1.1). However, in spite of a long impressive fossil record, it is also interrupted by major and minor extinction events (Fig. 1.1; see also Hart 1996; Hallam and Wignall 1997; Koeberl and MacLeod 2002; Taylor 2009). The Phanerozoic record of marine invertebrates, in particular, is interrupted by numerous, geologically short-term intervals (generally <3 Ma) during which biotic diversity and abundance declined significantly (<40 % at the familial level and <63 % at the generic level) (Raup and Sepkoski 1986) (see Fig. 1.1). Extinctions (major and minor) do not fall within the preview of this book but have only been mentioned to allude to the interruptions within the fossil record.

S. Jain, *Fundamentals of Invertebrate Palaeontology*, Springer Geology,
DOI 10.1007/978-81-322-3658-0_1

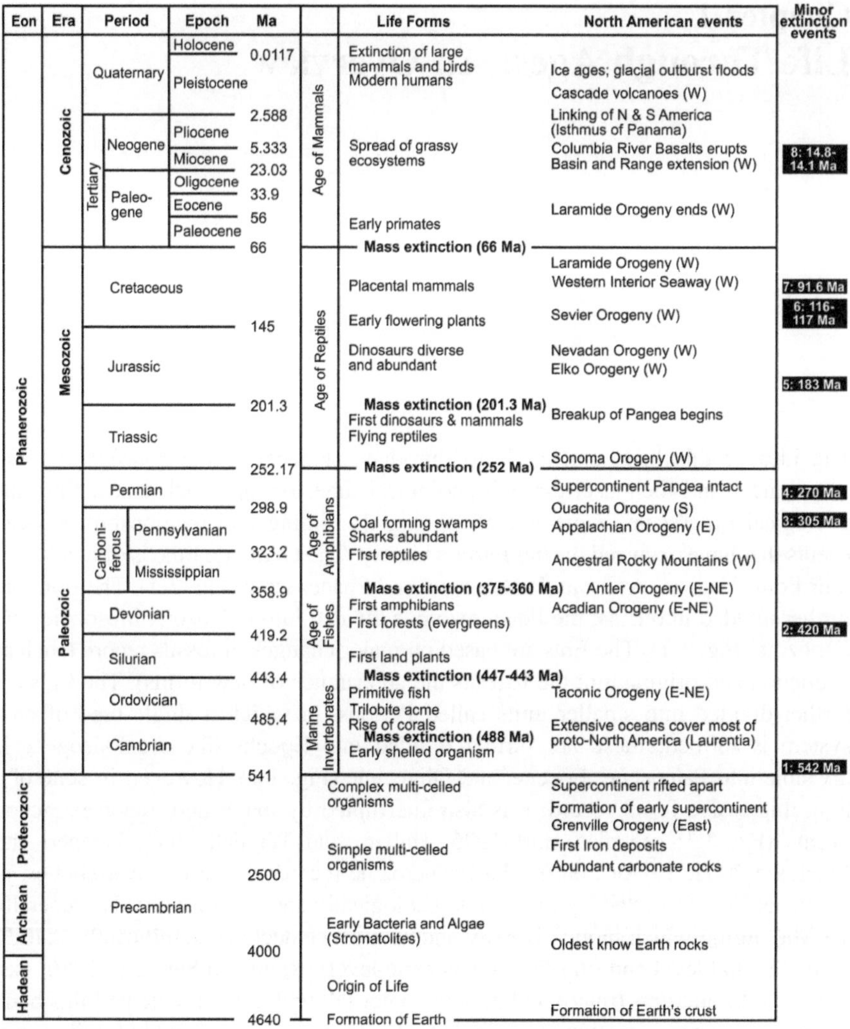

Fig. 1.1 The Geological timescale (all age used here and throughout the book, are after Gradstein et al. 2012) mentioning major advent of life forms, corresponding events in North America, and occurrence of major and minor extinction events

Here, for this book, only major invertebrate faunal groups, whenever, they assume important age or duration marker characteristics, are mentioned (see Figs. 1.2 and 1.3). Figure 1.3 shows marine family-level diversity with some distinctive invertebrate forms discussed in this book, through time. Thus, this chapter lays the foundation of the book where major invertebrate groups are mentioned, and detailed later in subsequent chapters.

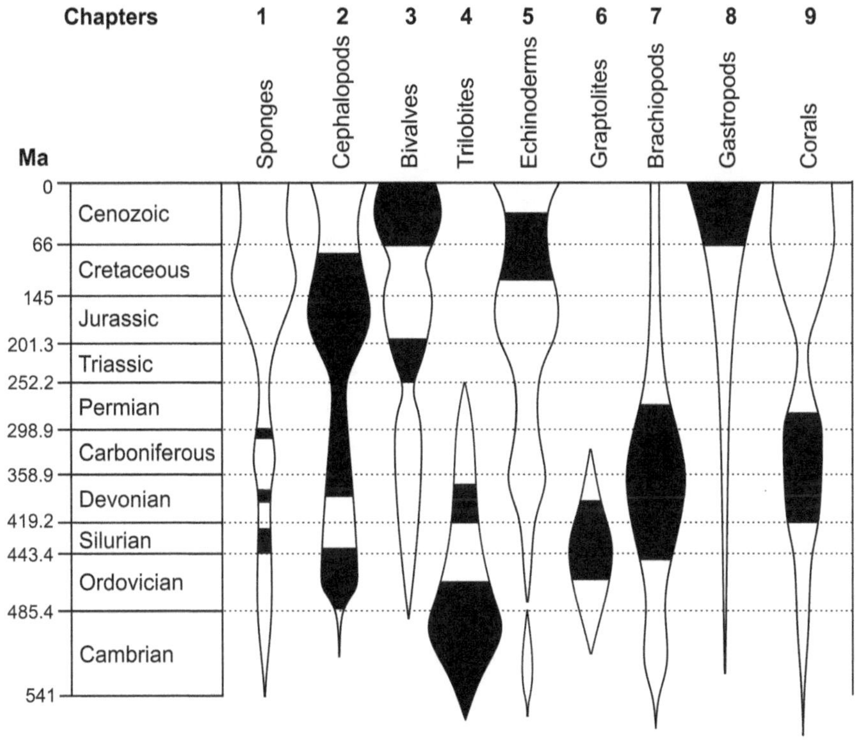

Fig. 1.2 Invertebrate groups discussed in the book. The *black shaded* portions indicate when an organism assumed important age or duration marker characteristics, i.e., they are stratigraphically useful and used for finer biostratigraphic divisions. The width of bands indicate the approximate abundance of each group through time

Interesting to note is that life (as a single-celled simple organism like Stromatolites, Bacteria, etc.) actually evolved very early (Fig. 1.4) and multi-cellularity (such as Sponges; see Chap. 2), came quite late during the Cryogenian (last ~760 Ma; Neoproterozoic; see Figs. 1.1 and 1.4). All ages mentioned in this book are after Gradstein et al. (2012).

The book contains illustrations of around 1200 species through time in over 3000 well-labeled hand-drawn classroom-friendly diagrams. The illustrations are also indexed (Appendix A, at the end of the book) mentioning the chapter number,

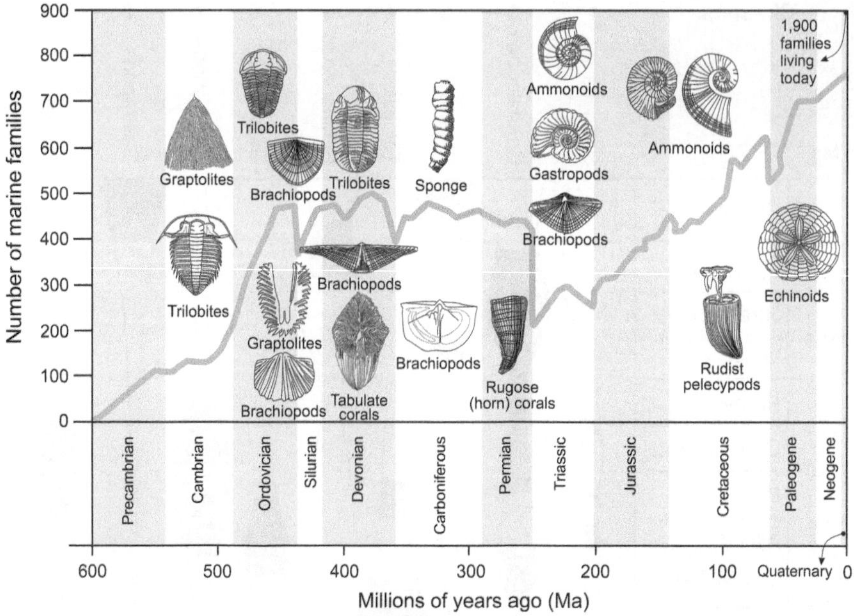

Fig. 1.3 Stratigraphic distribution of major invertebrate forms discussed in the book

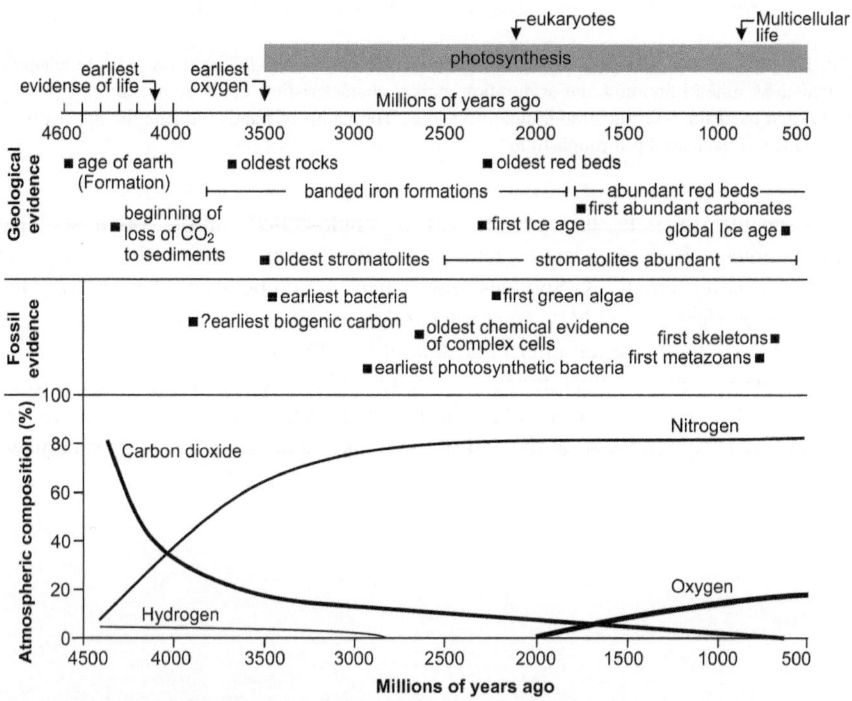

Fig. 1.4 Earliest events in Earth's history and the timing of the birth of life on Earth. All ages are after Gradstein et al. (2012)

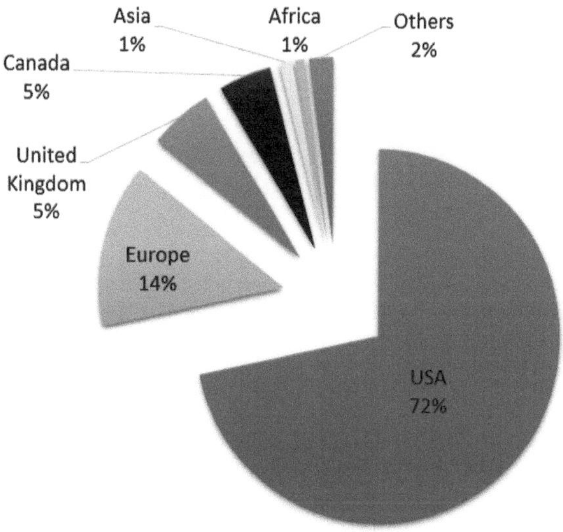

Fig. 1.5 The distribution of illustrated specimens based on geographic regions

species name, age, and locality of the illustrated specimen along with its figure number within the said chapter, to make the book more user-friendly. Of the 1200 illustrated species, 72 % are from the United States of America, 14 % from Europe, 5 % from Canada, 5 % from United Kingdom (including Scotland, Ireland, and Wales), 1 % from Asia and Africa (Fig. 1.5). Others (2 %) include a host of countries with single or less than 4 samples.

References

Gradstein, F.M., Ogg, J.G., Schmitz, M., and Ogg, G. 2012. The Geologic Time Scale 2012. doi:10.1016/B978-0-444-59425-9.00026-3.

Hallam, A., and P.B. Wignall. eds. 1997. Mass extinctions and their aftermath. Oxford University Press, Oxford, 328 p.

Hart, M.B. ed. 1996. Biotic recovery from mass extinction events. Geological Society Special Publication No. 102, 388 p.

Koeberl, C., and K.G. MacLeod. eds. 2002. Catastrophic events and mass extinctions: impacts and beyond. Geological Society of America Special paper 356, 746 p.

Raup, D.M., and J.J. Sepkoski. Jr. 1986. Periodic extinctions of families and genera. Science 231: 833–836.

Taylor, P. 2009. Extinctions in the history of life. Cambridge University Press, 204 p.

Chapter 2
Sponges

2.1 Introduction

The recent discovery of a 750 million years old sponge-like organism, *Otavia antiqua* [Fig. 2.1(1–5)], a calcareous sponge with Ca-based skeleton, from the Cryogenian–Ediacaran successions of Namibia (South Africa) makes sponges the oldest living animal (Brain et al. 2012; Maloof et al. 2010) [Fig. 2.1(6, 7)]. Their phosphatised body fossils also demonstrate a complex rigid structure [Fig. 2.1(1–5)], indicating the presence of a high level of organization, and supporting the results based on genetic sequencing and biomarkers that the first animals were sponges (Love et al. 2009; Sperling et al. 2010; Brain et al. 2012). In fact, with this discovery, the sponges are now the most basal metazoan taxon. They are also the most diverse and successful of the extant phyla, known so far (Gehling and Rigby 1996; Borchiellini et al. 2002; Philippe et al. 2009; Pick et al. 2010) (Fig. 2.1).

Sponges are invariably sessile in habit, being attached either by means of a stem or a bundle of anchoring spicules, or simply encrusting at the base (substrate). There are about 9000 living species of sponges: most are marine but few ($\sim$200 species) also inhabit freshwaters. Sponges of Class Calcarea largely inhabit shallow waters (<100 m) and are most common in intertidal habitats. Class Demospongiae that contains about 95 % of all sponge species are found at almost all depths ranging from intertidal to abyssal zones (Rigby et al. 1993).

Sponges are simple or primitive multicellular sedentary organisms that show remarkable variability in form (Fig. 2.2), size (from 1 mm to >1 m), and shape. Even among individuals of the same species, shapes vary depending largely on environmental factors such as hydrodynamics, light, and turbidity.

S. Jain, *Fundamentals of Invertebrate Palaeontology*, Springer Geology,
DOI 10.1007/978-81-322-3658-0_2

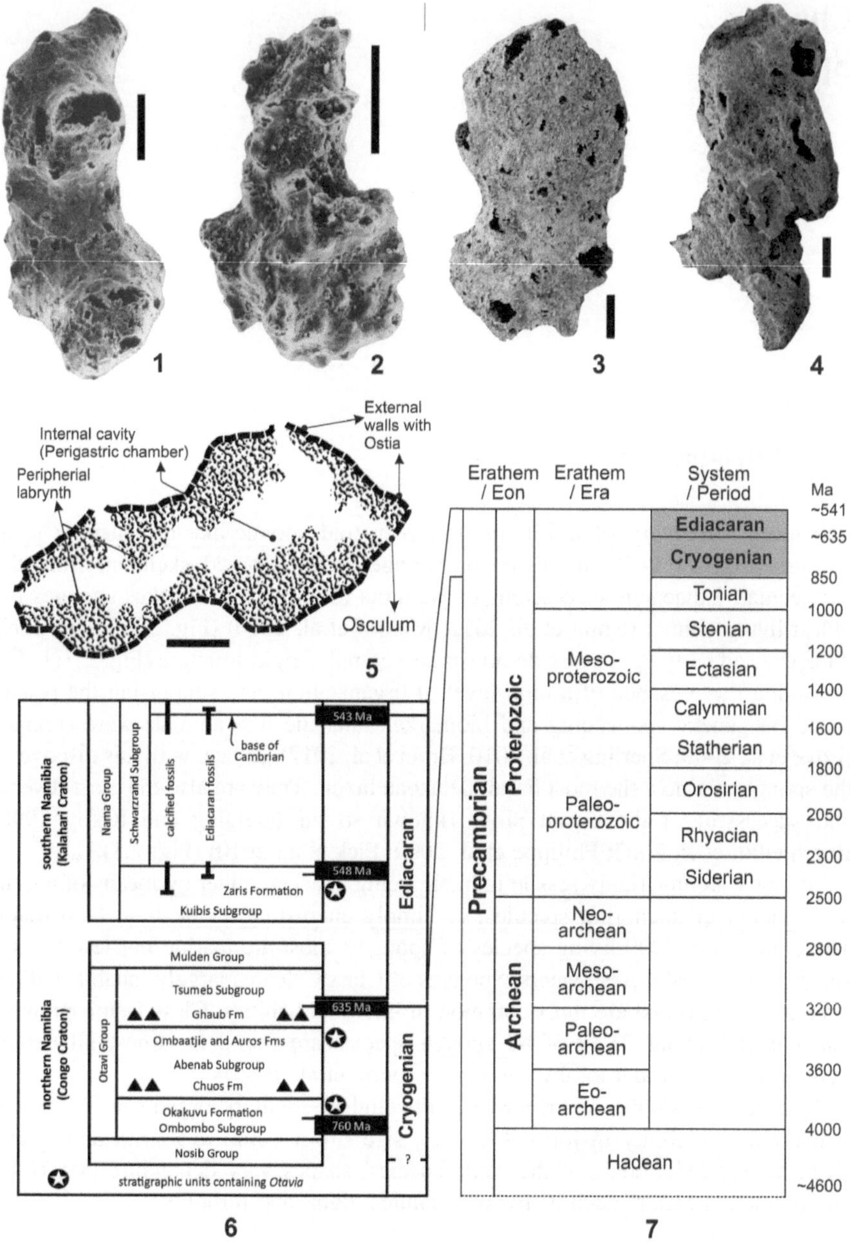
1
2
3
4
Internal cavity (Perigastric chamber)
Peripherial labrynth
External walls with Ostia
Osculum
5
543 Ma
base of Cambrian
southern Namibia (Kalahari Craton)
Nama Group
Schwarzrand Subgroup
calcified fossils
Ediacaran fossils
548 Ma
Zaris Formation
Kuibis Subgroup
Ediacaran
Mulden Group
Tsumeb Subgroup
635 Ma
Ghaub Fm
northern Namibia (Congo Craton)
Otavi Group
Ombaatjie and Auros Fms
Abenab Subgroup
Chuos Fm
Okakuvu Formation
Ombombo Subgroup
760 Ma
Cryogenian
Nosib Group
?
stratigraphic units containing Otavia
6
Erathem / Eon
Erathem / Era
System / Period
Ma
~541
Ediacaran
~635
Cryogenian
850
Tonian
1000
Stenian
1200
Ectasian
1400
Calymmian
1600
Statherian
1800
Orosirian
2050
Rhyacian
2300
Siderian
2500
Precambrian
Proterozoic
Meso-proterozoic
Paleo-proterozoic
Archean
Neo-archean
2800
Meso-archean
3200
Paleo-archean
3600
Eo-archean
4000
Hadean
~4600
7

◀ **Fig. 2.1** *Otavia antiqua*. Vertical bars measure 100 μm. *1–4* Scanning electron microscopy images of *Otavia*. Note the presence of a consistent globular to ovoid shape, external bounding surface pierced by numerous small pores and larger openings commonly forming raised mounds (particularly in Figs. 2.1 and 2.2). *5* Features of *Otavia*. Note the presence of an overall ovoid to globular shape, the external wall pierced with Ostia and the interior peripheral labyrinth surrounding an irregularly shaped internal void connected to the outside by large Oscula. *6–7* Cyrogenian and Ediacaran stratigraphy of the Neoproterozoic Otavi and Nama Groups of northern and southern Namibia (South Africa) yielding *Otavia antiqua*. Figures illustrated with permission from the South African Journal of Science (these are also free illustrations under the Creative Commons Attribution license)

2.2 Structural Features

Sponges are filter feeders where the water is pumped in through Ostia [small inhalant openings; singular: Ostium; see Fig. 2.3(1, 2)], or by the irregular beating of flagella of Choanocytes or Collar cells that bear a mobile, whip-like flagellum guarded by a cylindrical wall, the collar [Fig. 2.3(3, 4)]. The interior chambers of the sponge houses the Choanocytes [Fig. 2.3(4)]. These are responsible for the circulation of the fluid through numerous canals within the sponge body; the Osculum, the larger exhalant opening, allows for the exit of fluids [Fig. 2.3(3, 4)].

The sponges are broadly characterized by three body plans types: Ascon, Syon, and Leucon [Fig. 2.3(5–10)]; these are briefly described below.

2.2.1 Ascon Type

This is the most basic type [Fig. 2.3(5, 8)]. Such simple sponges, belonging to Class Calcarea, are usually smaller and largely radially symmetrical. They possess a typical central spongocoel lined by choanocytes, with single osculum where water exits from the spongocoel [the pseudogastric cavity; Fig. 2.3(5)]. The spongocoele opens to the outside through an excurrent pore called the Osculum. The surface has numerous pores (incurrent pores or Prosopores) that allows water to enter the sponge [Fig. 2.3 (8)]. This type of sponge represents a "flagellate chamber" and *Leucosolenia*, a living calcareous sponge, is a good example of this type of body plan.

2.2.2 Syon Type

This is made up of a group of several flagellate chambers of the Ascon type around an Excurrent canal [or Apochete; the pseudogastric cavity; see Fig. 2.3(6)]. The excurrent pores lead into this and it empties to the outside through an Apopore (or Osculum). In simple terms, the body wall folds to form secondary choanocyte chambers, which then empties into the spongocoel through a system of canals

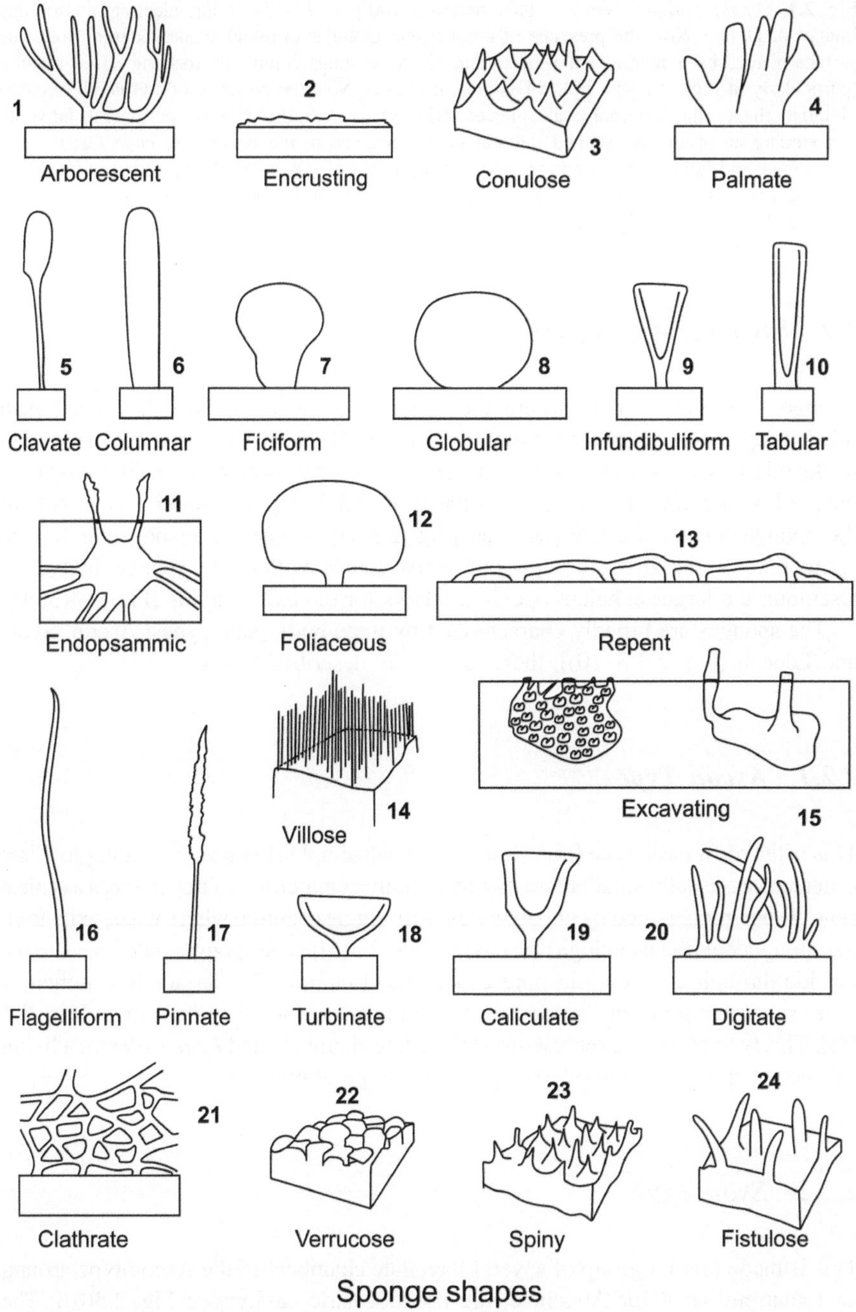

Fig. 2.2 Shape variability in sponges

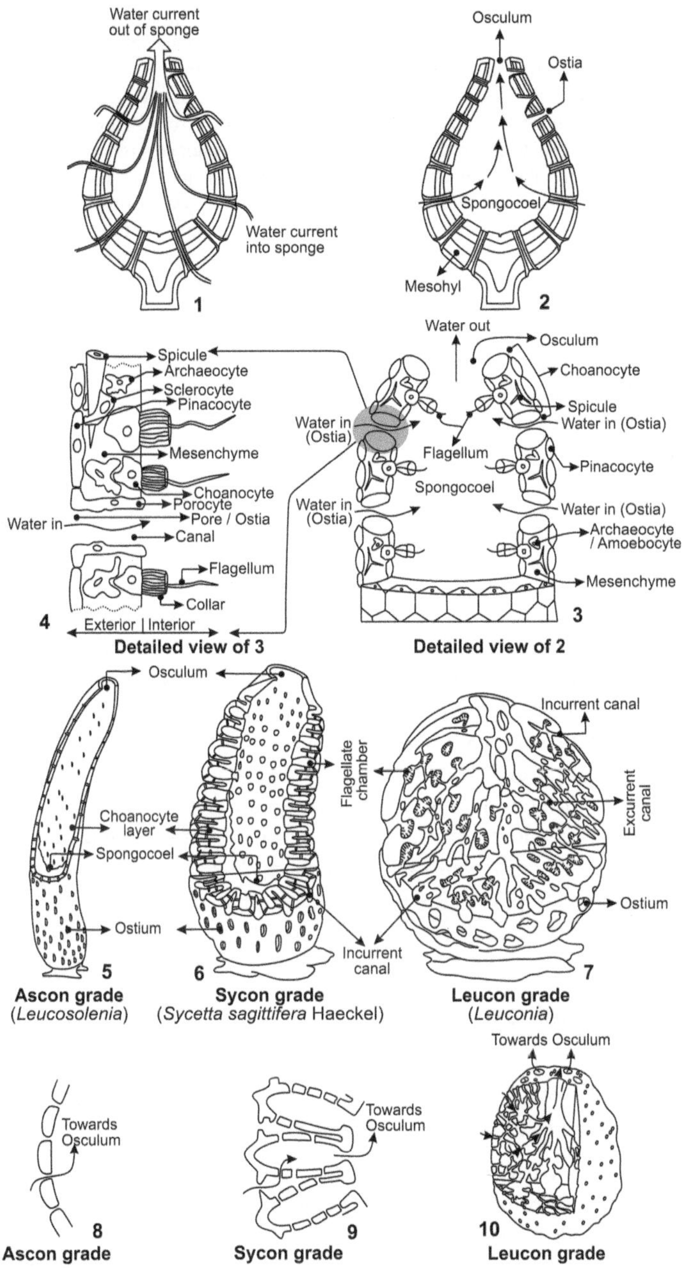
Water current out of sponge
Water current into sponge
1
Osculum
Ostia
Spongocoel
Mesohyl
2
Spicule
Archaeocyte
Sclerocyte
Pinacocyte
Mesenchyme
Choanocyte
Porocyte
Water in
Pore / Ostia
Canal
Flagellum
Collar
4
Exterior | Interior
Detailed view of 3
Water out
Osculum
Choanocyte
Spicule
Water in (Ostia)
Flagellum
Pinacocyte
Spongocoel
Water in (Ostia)
Archaeocyte / Amoebocyte
Mesenchyme
3
Detailed view of 2
Osculum
Choanocyte layer
Spongocoel
Ostium
Flagellate chamber
Incurrent canal
Excurrent canal
Ostium
Incurrent canal
5
6
7
Ascon grade (Leucosolenia)
Sycon grade (Sycetta sagittifera Haeckel)
Leucon grade (Leuconia)
Towards Osculum
Towards Osculum
Towards Osculum
8
9
10
Ascon grade
Sycon grade
Leucon grade

◀ **Fig. 2.3** Sponge canal systems. *1–2* Water enters the Spongocoel through the incurrent pores or Ostia and passes out through the Osculum. *3* The flagellated collar has withdrawn to small, radial chambers, each of which communicates to the Spongocoel. *3* The small chambers lined with collar cells are deeply embedded in Mesenchyme and connected by intricately branched incurrent and excurrent canals. *5–10* Major structural types of sponges (modified from Boardman et al. 1992). *5* Ascon type—a single flagellate chamber lined by a layer of choanocytes (choanosome). *6* Sycon type—several independent flagellate chambers opening directly into the Pseudogastric cavity (spongocoele). *7* Leucon type—several sycon structures with a system of pores connected by canals, and with incurrent canals leading toward flagellate chambers and then through excurrent canals and pores emptying into the cloaca. *Arrows* indicate the direction of water circulation. *8–10* Cross sections of body walls of three structural types. *5* and *8 Leucosolenia* Bowerbank; *6* and *9 Sycetta sagittifera* Haeckel; *7* and *10 Leuconia* Grant

[Fig. 2.3(9)]. Many calcareous sponges have such a plan of construction such as the recent *Sycetta sagittifera* Haeckel.

2.2.3 *Leucon Type (or Rhagon Type)*

This emerges from the merging of several Sycon units. The excurrent canals or apopores open into a pseudogastric cavity (or Cloaca) which empties to the outside, through a true osculum [Fig. 2.3(7, 10)]. The presence of a dermis and/or cortex in certain syconoid forms and in all leuconoid forms consequently entails the development of a complex network of incurrent canals (or prosochetes) between the incurrent pores (prosopores), which are open to the outside and that empty into the flagellate chambers through the prosopyles. These are homologous to the incurrent pores of the basic asconoid forms. All species of Demospongiae, and most of Calcarea, have a Leuconoid plan of construction. The Recent calcareous sponge *Leuconia* is a good example of this type of body plan.

2.3 Cell Terminology

The cell terminology used to describe the sponge body plan (see also Boury-Esnault and Rützler 1997) is briefly given below and illustrated Fig. 2.3.

2.3.1 **Archaeocyte (Amoebocyte)**: These are cells in the Mesenchyme (=Mesohyl). They possess pseudopods that are used for processing food and distributing it to other cells

2.3.2 **Choanocyte (Collar Cell)**: These cells line the inner cavity of the sponge. The flagellum enables the organism to obtain nutrients and oxygen by processing the flowing water

2.3.3 **Flagellum**: A whip-like structure of the choanocyte cell that moves, pushing water (containing nutrients) through the sponge

2.3.4 Mesenchyme (Mesohyl): A gelatinous layer between the outer body of the organism and the spongocoel (the inner cavity = pseudogastric cavity)

2.3.5 Osculum (Plural = Oscula): Large openings that allow the water to flow out of the organism

2.3.6 Pinacocyte (Epidermis): These are the thin, flattened cells of the epidermis (a layer of cells that covers the outer surface of the sponge)

2.3.7 Porocytes: These are cells with pores located all over the sponge body; the water flows into the sponge through them

2.3.8 Spicule: Located in the mesenchyme, these sharp spikes made of calcium carbonate, form the "skeleton" of most sponges

2.3.9 Spongocoel (Cloaca): It is the central, open cavity through which water flows

2.4 Skeleton

The internal skeleton of sponge which supports its soft parts (tissues) is either made of organic fibers or mineralized needle-like or multirayed spicules, or a combination of both (Bergquist 1998) (see Fig. 2.4). The spicules are distributed throughout the sponge's soft tissue or intertwined, and sometimes fused, into a rigid skeleton that facilitates their fossilization, sometimes even preserving the original shape (Uriz et al. 2003). The group's excellent geological record is largely due to its mineralized skeleton (both calcareous and siliceous). Hence, the discussion below on spicule is largely about the calcareous and siliceous types and of Class Desmospongea.

2.4.1 Spicules

The composition spicules is opaline silica or crystalline to microgranular calcium carbonate, although, no sponge will secrete both materials at the same time. Thus, the presence of a skeleton and its corresponding structure provides a first basic subdivision—sponges without a skeleton and sponges with a calcareous, collagenous, or siliceous skeleton.

Hence, sponge classification is based on the nature, shape, and the interrelationships of spicules (see also Butler 1962; De Vos et al. 1992; Uriz et al. 2003; Dohrmann et al. 2012). Their composition further enables differentiation wherein Calcarea, Archaeocyatha, and Sclerospongiae contain calcium carbonate as layered, granular to crystalline aragonite or calcite. In Sclerospongiae and the hypercalcified sponges in two subclasses of the Demospongea have intermixed siliceous and carbonate skeletal elements.

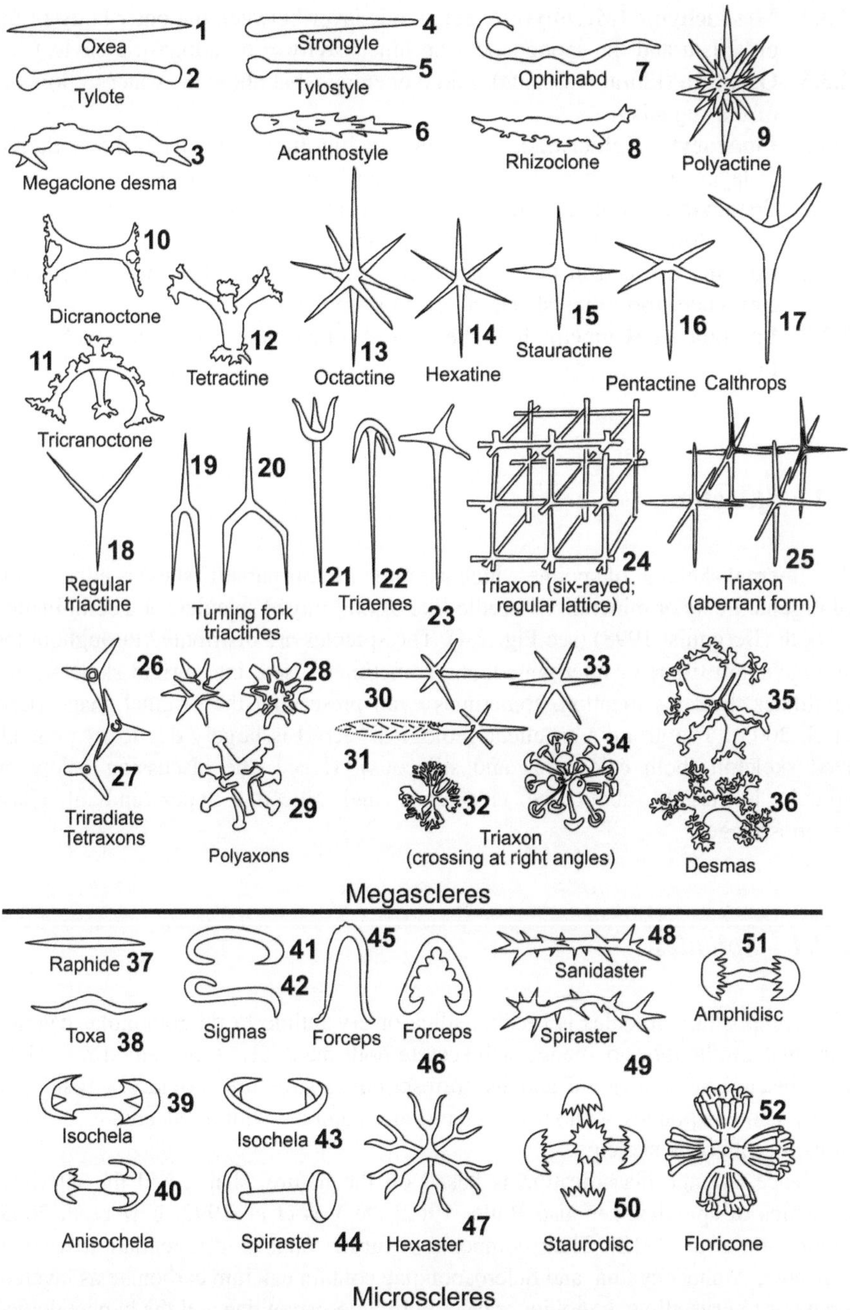

Fig. 2.4 Spicules. The sponge soft parts are supported by internal skeletons made of organic fibers, mineralized needle-like or multirayed spicules, or a combination of fibers and spicules. These are distributed throughout the soft tissue or intertwined, and sometimes fused, into a truly rigid skeleton which facilitates their fossilization. This mineralized skeleton makes up most of the geological record for the group

2.4.2 *Spicule Size and Nomenclature*

Both the spicule size and numbers of rays (axes) enable categorization. However, a single sponge may possess different kinds of spicules, and the same kind of spicule may also occur in several sponges.

Broadly, spicules are of two types: the large ones that make the skeletal framework are called the Megascleres [Fig. 2.4(1–36)] and the smaller ones, the Microscleres [Fig. 2.4(37–52)]. These are irregularly distributed and always form only an accessory element of the skeleton. The microscleres measure <100 μm and megascleres >300 μm (see also Wang et al. 2009). Therefore, the size of megascleres extends over 4–5 orders of magnitude; the microscleres are much more uniform in size. The largest known natural silica structure, the 3 m long Giant Basal spicule is of the recent hexactinellid sponge *Monorhaphis chuni* (Schulze). The diameter of megascleres commonly varies between 3 and 30 μm (De Laubenfels 1955) but can grow up to 12 mm (Levi et al. 1989). The spicules are secreted by specialized mobile cells (the archaeocyte/amoebocytes) within the mesenchyme [Fig. 2.3(4)].

Names for the general categories of spicules are formed by adding a numerical prefix, mono- (=one), di- (=two), tri- (=three), and tetra- (=four), to the word "axon" when the number of axes composing the spicule is referred to, or to the word "actine" when the number of rays is referred to. In the latter case additional prefixes pent- (=five) or hex- (=six) may occur. A rod-shaped spicule pointed at both ends is called a monaxonid diactine; if pointed at one end and rounded at the other, it is monactine. Tetraxonid and triaxonid spicules also occur, as do tetractine and hexactine spicules. Among calcitic spicules (as in class Calcarea), triactines and tetractines have three or four rays, respectively. The calcareous tetractines with eight rays (=Octactines) are characteristic of Heteractinida, a Palaeozoic class. The polyactines or polyaxons, or sphaeractines are those that possess multiple rays and axes of growth. All these terms refer to the larger spicules of sponges, that is, the megascleres that make up the primary framework of the skeleton [Fig. 2.4(1–36)].

2.4.2.1 Megascleres

Broadly, these siliceous skeletal elements are resolvable into a few fundamental types (Fig. 2.4), such as the following:

2.4.2.1.1 **Uniaxial spicules or Monaxons**. Straight or bent, smooth, prickly or knotty, bevelled, sharpened or truncated needles, rods, hooks, clasps, pins, and anchors (amphidisc). They almost always contain an axial canal, which may be either entirely sealed up, or open at one or at both ends

2.4.2.1.2 **Tetraxial spicules or Tetraxons**. The normal form is characterized by four equal rays intersecting like the bisectrices of the plane angles of a regular tetrahedron. Triaxial forms result from the occasional abortion of one of the rays. One of the rays may become elongated or otherwise modified so as to form anchors (triaens) with three simple or furcate hooks. Three of the rays may be numerously divided or foliately expanded so as to produce forms resembling thumb-tacks (trichotriaens, phyllotriaens; atrophy of the fourth ray in the last-named form reduces the spicule to a delicate silicious disk. A peculiar forking of the shaft gives rise to candelabras or amphitriaens, while other modifications may produce umbel-like spicules, etc. Certain skeletal elements of the Lithistids may be regarded as irregular tetraxons (desmas), in which the extremities of the four rays are prolonged in knotty, root-like excrescences

2.4.2.1.3 **Hexactinellid spicules (Hexactins or Triaxons)**. The groundform is an axial cross with six equal arms intersecting at right angles like the axes of a regular octahedron. Atrophy of one or more of the rays may result in pentaxial, tetraxial, triaxial, or even nail-shaped forms, without their real character becoming entirely obliterated. Bifurcation or other modifications of a number or all of the rays produce those exquisite siliceous structures so characteristic of the group Hexadinellida, which resemble candelabras, double-headed anchors, fir trees, pitchforks, rosettes, etc. The fusion of juxtaposed hexactins produces more or less symmetrical latticeworks with cubical interstices. Anaxile or polyaxile bodies of spherical, cylindrical, stellate, or discoidal shape, which are not derivable from either of the three ground forms, occur in only a few varieties of recent and fossil siliceous sponges

2.4.2.2 Microscleres

The Microscleres [Fig. 2.4(37–52)] are small-sized spicules that provide a dermal armor at the surface, strengthen the ground substance of the cortex or mesenchyme, or may reinforce the pinacoderm that line the canals. Many kinds of microscleres occur in Demosponges and Hexactinellids and are given names of Latin or Greek origin to describe their shapes [Fig. 2.4(37–52)]. The Microscleres, due to their low preservational potential through geological time, are of minor use in paleontology.

2.5 Classification

The fossil sponges are traditionally classified on the following three parameters: composition and forms of spicules, canal systems, and structural grades. Recently, a more balanced multicharacter approach is taken in which spicules, skeletal structure, soft parts, and life history characteristics are included resulting in three unchallenged classes: Calcarea, Hexactinellida, and Demospongea (see also Hooper

1991; Van Soest 1991; Clarkson 1993; Van Soest et al. 1994; Hooper and Wiedenmayer 1994; Hooper and Van Soest 2002) and the fourth Archaeocyatha (contentious) (Table 2.1; Fig. 2.5). To the established three, Chaetetids and Stromatoporoids (Sclerospongia) have recently been included, making them five (see Table 2.1).

Table 2.1 Traditionally, five classes have been recognized in the phylum, including the Calcarea, Demospongea, and Hexactinellida

Class	Subclass	Order	Age range
Calcarea Bowerbank 1864			**Cambrian-Holocene**
	Calcinea Bidder, 1898		?Precambrian, Cambrian-Holocene
		Clathrinida Hartman, 1958	Holocene
		Murrayonida Vacelet, 1981	?Precambrian, Cambrian-Holocene
	Calcaronea Bidder, 1898		?Cambrian,? Triassic, Jurassic-Holocene
		Leucosoleniida Hartman, 1958	Holocene
		Sycettida Bidder, 1898	Holocene
		Sphaerocoeliida Vacelet, 1977	Cretaceous
		Lithonida Doederlein, 1892	Jurassic-Holocene
Demospongea Sollas, 1875			**Precambrian-Holocene**
	Clavaxinellida Lévi, 1956		Precambrian-Holocene
	Choristida Sollas, 1880		Late Ordovician-Holocene
	Tetractinomorpha, Lévi, 1953		Middle Ordovician-Holocene
	Ceractinomorpha Lévi, 1953		Middle Cambrian, Middle-Late Ordovician, ? Pennsylvanian, Holocene
	Lithistida Schmidt, 1870		Cambrian-Holocene
Hexactinellida Schmidt, 1870			**Precambrian-Holocene**
	Amphidiscophora Schulze, 1887		Precambrian-Holocene
		Amphidiscosa Schrammen, 1924	Ordovician-Holocene
		Reticulosa Reid, 1958	Precambrian-Upper Permian

(continued)

Table 2.1 (continued)

Class	Subclass	Order	Age range
		Hemidiscosa Schrammen, 1924	Late Pennsylvanian-Cretaceous
		Hexasterophora Schulze, 1887	Ordovician-Holocene
		Lyssacinosa Zittel, 1877	Ordovician-Holocene
		Hexactinosa Schrammen, 1903	Permian-Holocene
		Lychniscosa Schrammen, 1903	Upper Triassic-Holocene
Heteractinida de Laubenfels, 1955			**Lower Cambrian-Lower Permian**
		Octactinellida Hinde, 1887	Early Cambrian-Lower Permian
		?Hetairacyathida Bedford and Bedford, 1937	Early Cambrian
?Sclerospongiae			**Cambrian-Holocene**
	Chaetetids and Stromatoporoids		

To these have recently been added Chaetetids and Stromatoporoids (Sclerospongia) (see text above for further explanation)

However, as simple as it seems, the classification of the Porifera is still largely based on morphological characters, spicules, and fibers. But, with recent molecular studies, discrepancies between the results of morphological and molecular analysis are increasingly becoming common and hence, new tools are needed to weigh the competing results.

Representative examples of fossil sponges through time are illustrated in Figs. 2.6, 2.7, 2.8, 2.9, 2.10, and 2.11, and the classes are briefly discussed below (Figs. 2.12 and 2.13).

2.5.1 Class Calcarea Bowerbank (Calcispongia or Calcareous Sponges)

This class includes sponges with calcareous skeletal elements that lack both silica and spongin. Skeleton is characterized by the range from three-rayed spicules of calcite or aragonite, to those with rigid skeletons of fused polygonal elements or imbricate calcitic plates. Their present day analog is exclusively marine and commonly found in shallow tropical environments.

Calcarea have a worldwide distribution. Their structure is of the leucon type but, in contrast to the other classes, of the sycon and ascon types (see Fig. 2.3 for types). Three types of spicules are found in virtually all species: monaxon diactinal; triaxon

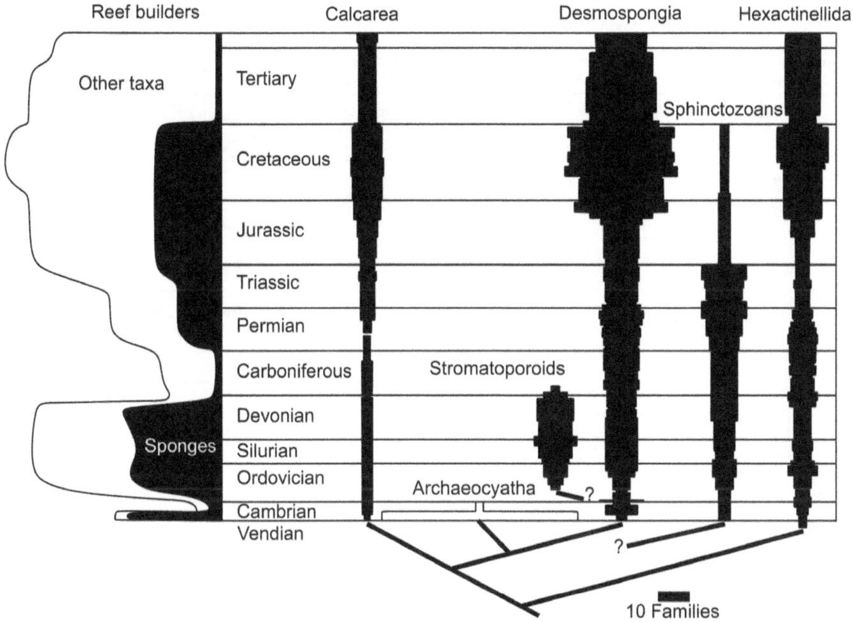

Fig. 2.5 Simplified classification of sponges. Traditionally, five classes have been recognized in the phylum, including the Calcarea, Hexactinellida, and Demospongea (see also Table 2.1). To these have recently been added Chaetetids and Stromatoporoids (Sclerospongia). Fossil "sphinctozoan" ("chambered sponges") families for which calcarean or demosponge affinities cannot be determined are known from the Cambrian to the Cretaceous. Calcarea is the smallest class with 860 species (9 % of the total sponges) (Fig. 2.5). This has exclusively calcareous skeletal material and are predominantly whitish and small, and of fragile consistency. The second class (600 species or 6 %) is the Hexactinellida ("glass sponges"), have siliceous skeletons built of hexaradiate spicules. These occur predominantly in deep oceanic habitats. The third and by far the most diverse class (8400 species or 85 %) are the Demospongiae. Their skeletons are built of siliceous spicules of various forms (but not hexaradiate) and are often cemented together with a keratinous protein called Spongin

triactinal with the rays arranged at an angle of 120° in the same plane of symmetry (triod) or in different planes (tripod); and a particular type in the shape of a tuning fork called "pharetron" (see also Fig. 2.4).

Some consider Heteractinida (Early Cambrian to Early Permian) as a class but most assign it as an order of Calcarea. Heteractinida is characterized by a skeleton made of large calcareous octactine spicules [Fig. 2.4(13)]. This is one of the only two classes of sponges to have become extinct. Examples of Class Calcarea illustrated here include the following: *Girtyocoelia typica* King, *Girtyocoelia beedei* (Girty), *Maeandrostia kansasensis* Girty, *Girtyocoelia dunbari* King, *Cotyliscus ewersi* King, *Amblysiphonella prosseri* Clarke, *Cystauletes mammilosus* King, *Barroisia anastomans* (Mantell), *Stellispongia glomerata* (Quenstedt), and *Corynella quenstedti* Zittel.

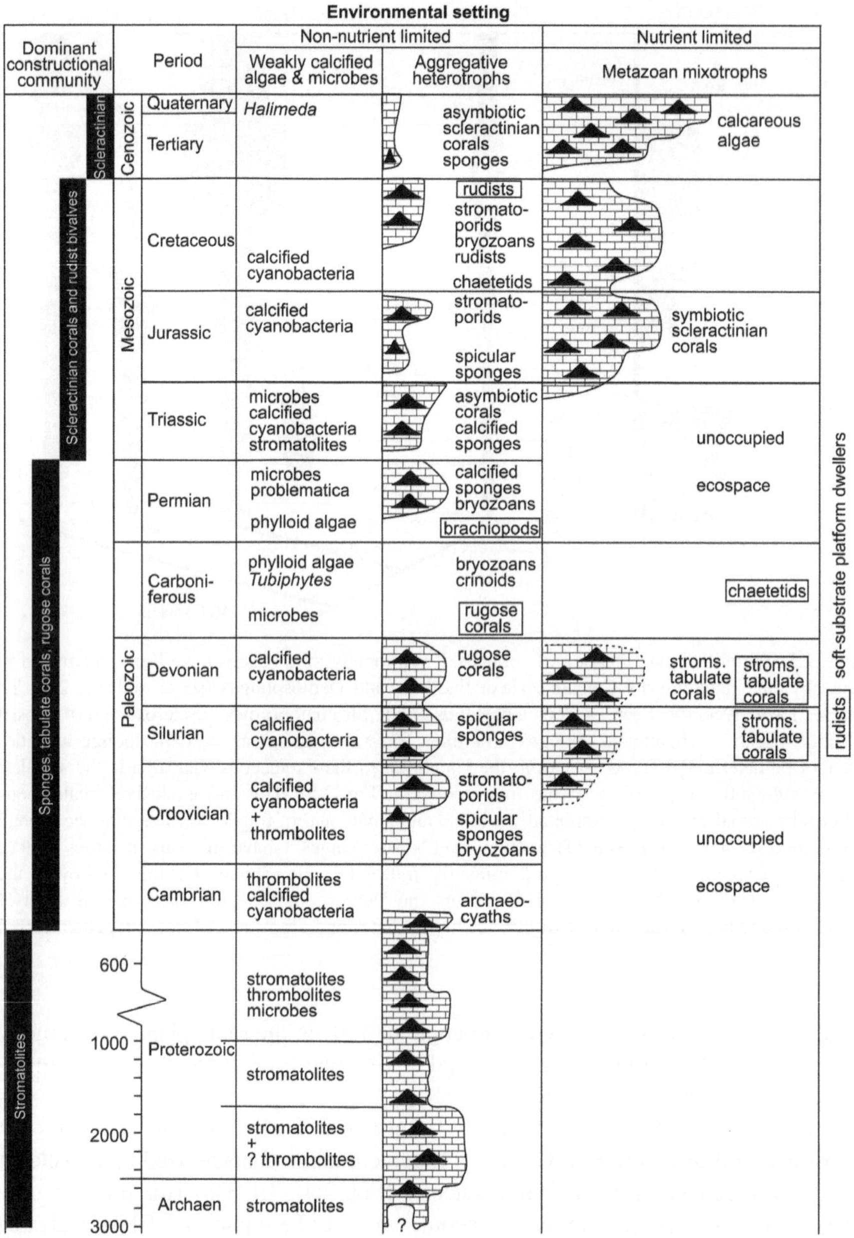

Fig. 2.6 Sponge reef builders across ages. A simplified diagram of the contribution of sponges in reef building through time is shown in Fig. 2.5

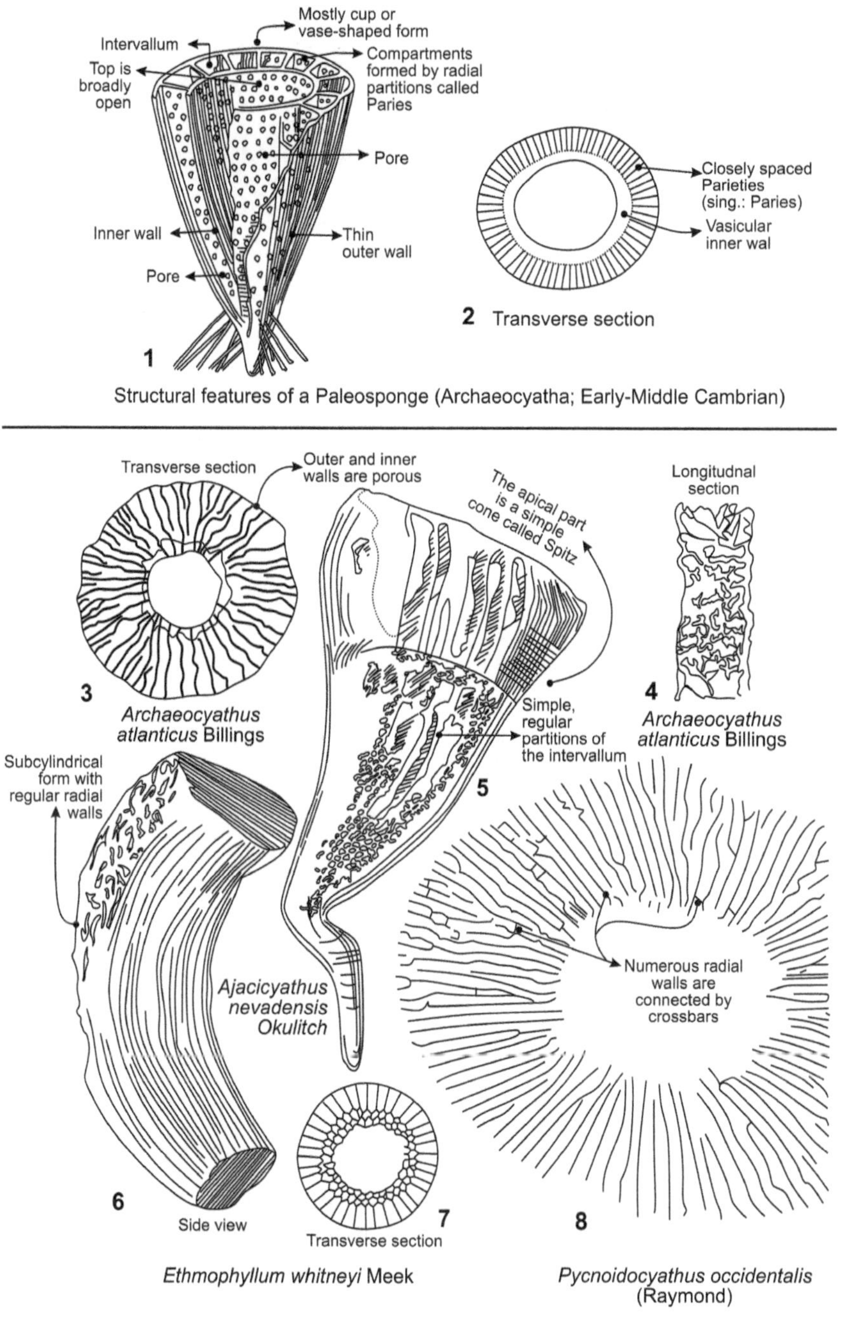

Fig. 2.7 Paleosponge (Archaeocyaths) and their major distinguishing characters. *1–2*: Structural features of a Paleosponge (Early-Middle Cambrian)

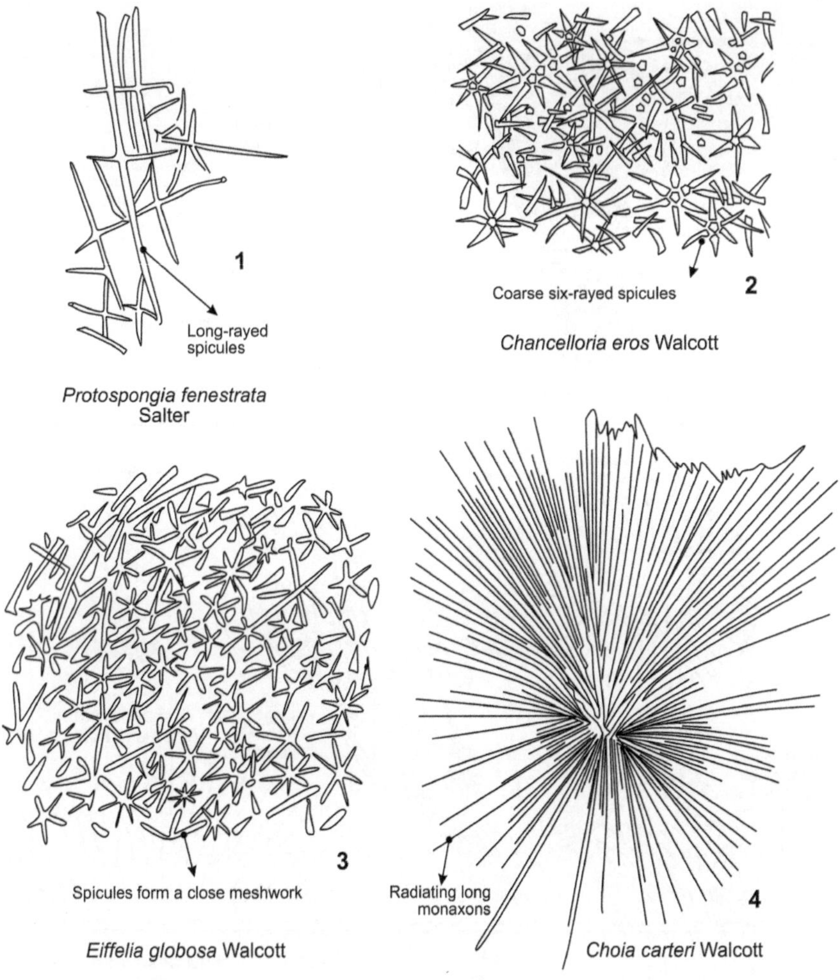

Fig. 2.8 Representative Middle Cambrian (Burgess Shales) sponges, British Columbia, Canada, and their major distinguishing characters

2.5.2 *Class Demospongea Sollas*

The Demosponges are a dominant, heterogeneous group, characterized by varied shapes, canal patterns, and spicule shapes (and their inter-relationships). There are several thousand known living species and about 500 fossil genera. They mainly inhabit shallow marine niches and are the only known living and fossil freshwater sponges. Their structure is Leucon type [Fig. 2.3(7, 10)]. They usually possess a skeleton made up entirely or partly of organic spongin fibers (with a poor paleontological record), siliceous spicules, or mixed spongin, and siliceous spicules (with a much better preservational geological record). Diversity of shapes and

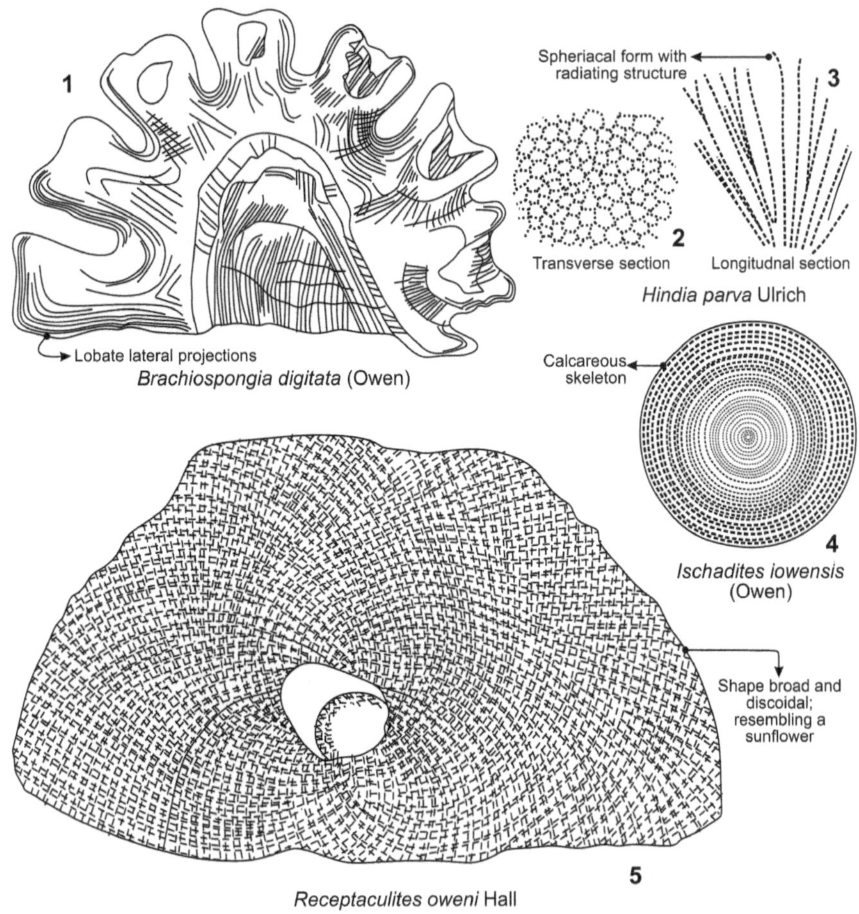

Fig. 2.9 Representative Middle Ordovician sponges and their major distinguishing characters

skeletal structures is huge (Fig. 2.2). The skeletons are composed either of spongin fibers or with siliceous spicules. These are usually divided into larger megascleres and smaller microscleres. Spicules of the class range from loose monaxons to tetractines whose rays do not join at right angles or to irregular root-like forms (Fig. 2.4). Their skeleton is composed of particular spicules or desmas, which are formed by the complex deposition of silica on an original megasclere (as "mortar") [Fig. 2.4(35, 36)]. The spicules are diactinal or tetractinal and are recognizable from the axial canals. Among Demosponges, Order Lithistida is the dominant fossil group, due to their higher preservational potential; their skeleton is made up of spicules fused together to form a rigid network, making them "as hard as rock," as their name implies.

Examples of Class Demospongea illustrated here include the following: *Protospongia fenestrata* Salter, *Chancelloria eros* Walcott, *Eiffelia globosa*

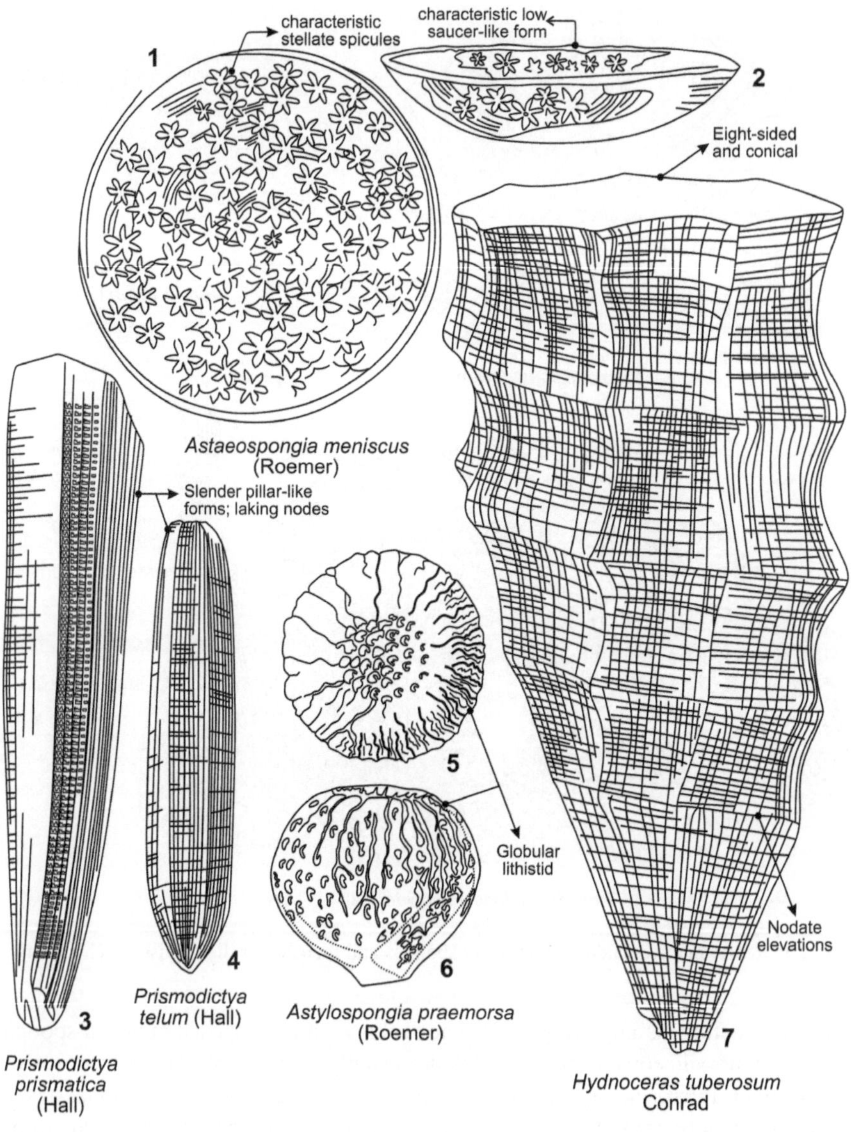

Fig. 2.10 Representative Late Ordovician-Late Devonian sponges and their major distinguishing characters

Walcott, *Brachiospongia digitata* (Owen), *Astaeospongia meniscus* (Roemer), *Prismodictya telum* (Hall), *Prismodictya prismatica* (Hall), *Hydnoceras tuberosum* Conrad, *Titusvillia drakei* Caster, *Ventriculites striatus* Smith, *Coscinopora infundibuliformis* Goldfuss, and *Coeleptychium agaricoides* Goldfuss.

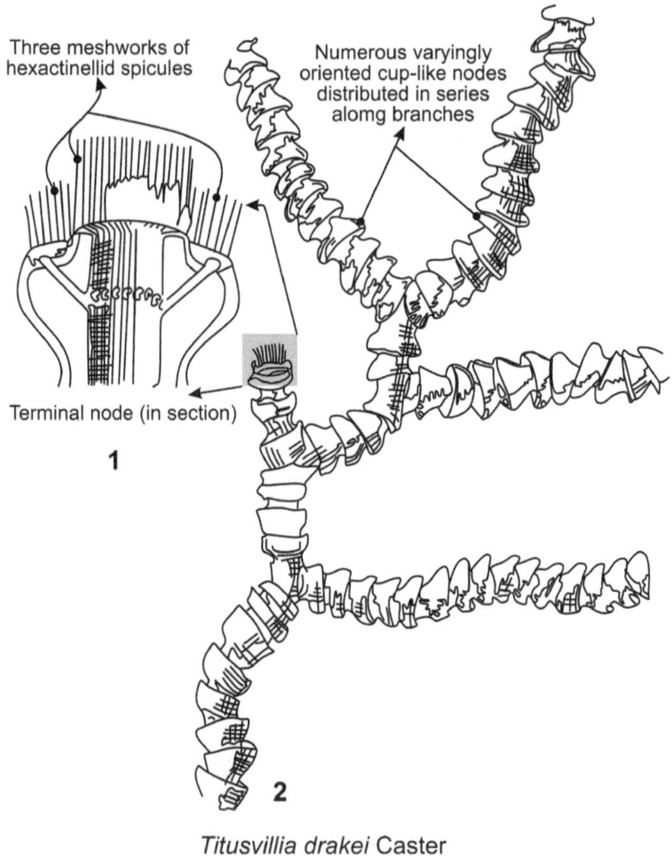

Fig. 2.11 Representative Early Mississippian sponge and their major distinguishing characters

2.5.3 Class Hexactinellida Schmidt (Hyalosponges or "Glass Sponges")

The siliceous hexactine spicules [Fig. 2.4(14)] characterize the skeleton of this exclusively marine class. In modern day seas, the Hexactinellida commonly inhabit on seafloors, from depths ranging from 200 to 2000 m, although many species have been reported from lower bathyal depths also (up to 4000 m).

In contrast to the demosponges, the hexactinellids form a homogeneous group of a simple leucon type with a large pseudogastric cavity of the sycon type [Fig. 2.3 (7–10)]. Their megascleres are characterized by rays arranged at right angles from a point of divergence. They build triaxon—hexaradiate spicules (i.e., the three axes of

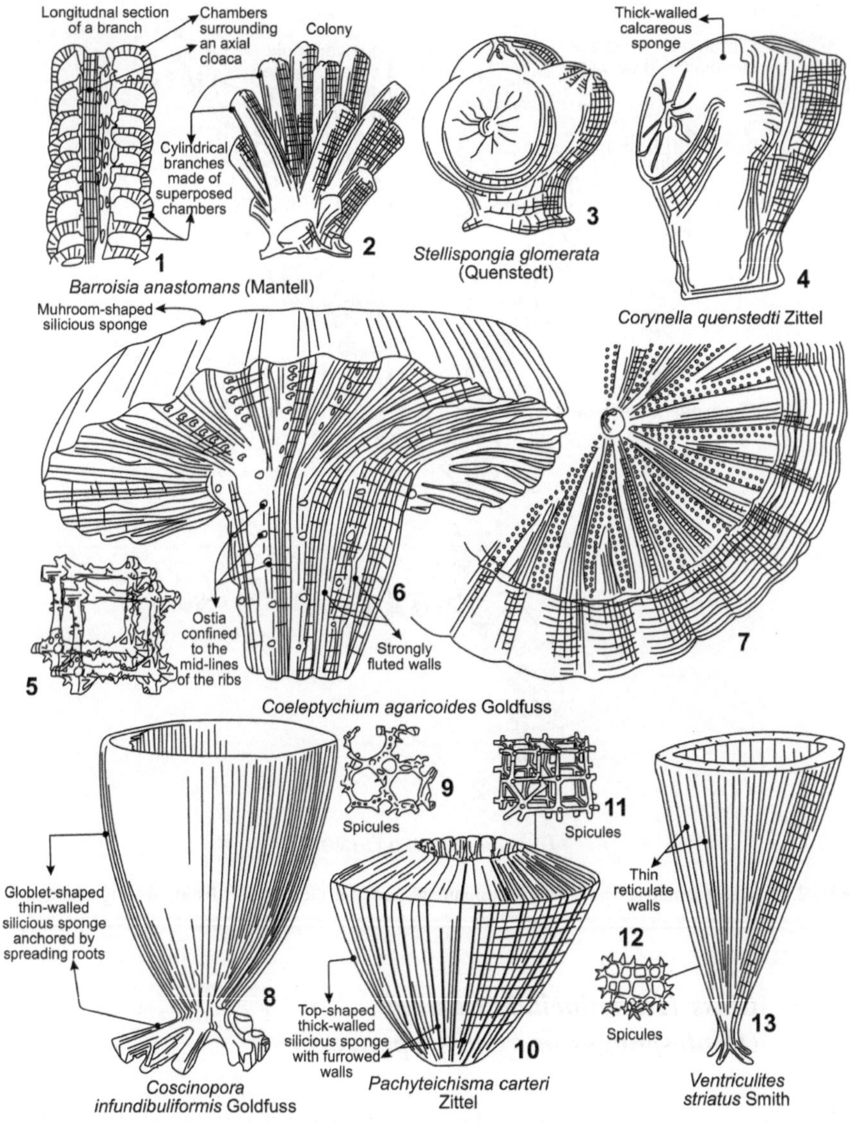

Fig. 2.12 Representative Early Cretaceous-Late Jurassic sponges and their major distinguishing characters

the spicules intersect at right angles) (Fig. 2.4). The triaxon hexactinal spicule, sometimes considered as the basic type, is characteristic only of the dictyids (with the hexactinellid family), and other types may be dominant in other species. A tetraxon spicule with the four rays lying in the same plane (also known as stauractinal) is a common type. There are also pentactinal (five), hexactinal

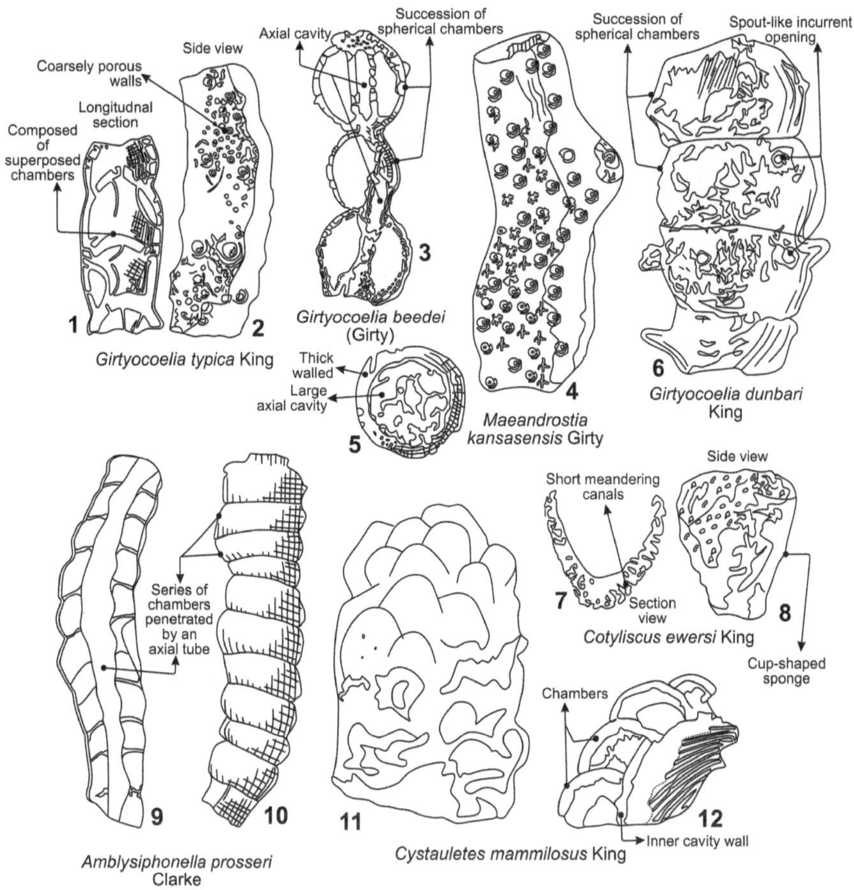

Fig. 2.13 Representative Late Pennsylvanian-Permian sponges and their major distinguishing characters

(six) and octactinal (eight) types, and even polyactinal spicules with more than eight rays. The heteractinids, with a skeleton made up essentially from polyactinal spicules assimilated into megascleres (Fig. 2.4). Hexactinellids lack any calcareous or spongin components in the skeleton.

2.5.4 *Class Sclerospongiae*

Class Sclerospongiae was proposed to include a few living sponges. It was introduced fairly recently to cover some recent "relict" sponges, and two fossil groups usually placed with the Coelenterata: the Stromatopora, historically compared to the hydrozoans, and the Chaetetids, sometimes also placed among the hydrozoans or, more frequently, with Tabulata (Corals; Chap. 9).

Stromatoporoids possess a laminated calcareous skeleton that is perforated by astrorhizal canals, and the living animal is thought to have occupied essentially only the upper surface. They have domal, tabular, branching, or bulbous growth forms.

The Chaetetids were shallow-water marine organisms. Their growth form, skeletal structure, and the rare occurrence of spicules indicate a strong relationship to other Sclerosponges. They have compound skeletons that range from plate-like or domed to columnar and are composed of clustered narrow tubes, or calicles, that are less than 1 mm in diameter and commonly polygonal to irregular in cross-section.

The Stromatoporoida exhibit greater diversity, and during the Devonian, and were among the main builders of the earliest bioherms (see Fig. 2.6) (see also Maldonado et al. 1999; Mueller 2003). Chaetetids, on the other hand, were important and abundant reef builders during the Paleozoic (Suchy and West 2001; May 2008).

2.5.5 Archaeocyatha

The Archaeocyaths (Fig. 2.12) are an extinct class of the phylum Porifera, close to the living Demospongiae (Debrenne 2007; Debrenne and Vacelet 1984). They are sessile, benthic, and filter feeders. They lived only in environments with restricted temperatures (stenothermal), salinity (stenohaline), and depth (stenobathic; 20–30 m). The archaeocyathans have been recorded from low latitudes during the Cambrian such as in Antarctica, Australia, China, Kazakhstan, Siberia, and North America. This latitudinal distribution is similar to that of modern colonial corals that inhabit warm shallow seas.

Typically, the archaeocyathan skeleton is solitary conical to branching [Fig. 2.12 (1)]. The central cavity of the organism is surrounded by a porous inner wall followed by a cavity called Intervallum and then outside it, a thin porous outer wall [Fig. 2.12(1)]. The Intervallum is divided into longitudinal openings by radial septa that acts as a bridge between the two porous walls [Fig. 2.12(1)]. At the base, root-like attachment structures occur that also act as holdfasts [Fig. 2.12(1)]. Various body forms have been reported from simple (single-walled) to more complex thalamid chambered forms. Rare are branched colonial forms with moderately complex walls (fan-shaped and bowl-shaped genera).

The Archaeocyaths were the oldest of the calcified sponges, and the first metazoans to build reefs (in association with calcimicrobes; see Fig. 2.6), are characteristic fossils used for the biozonation of the first, pre-trilobitic Cambrian stage (Tommotian) (Kerner et al. 2012).

The Archaeocyaths appeared close to the base of the Cambrian and became extinct by Middle Cambrian. The Cambrian Archaeocyathan occurrences can be broadly grouped into two associations—the Early Cambrian sponges from China, and the Middle Cambrian sponges in North America (mainly from the Canadian British Columbia and American Utah) (see also Carrera and Botting 2008). These first metazoan reef formers were relatively abundant during the Early Cambrian, from where over 300 genera have been described. The first subdivision of a stage based on archaeocyaths was established on the Siberian Platform (Zhuravleva 1960). Morocco, western Europe, Australia, and Canada have since provided regional scales, which allow stratigraphic comparisons that parallel trilobites biozones or replace them when necessary.

Appendix A gives the list of illustrated specimens mentioning the chapter number, species name, age, and locality along with its figure number within the chapter.

References

Bergquist, P. 1998. Porifera. In *Invertebrate zoology*, ed. D. Anderson, 10–27. USA: Oxford University Press.

Boardman, R.S., Cheetham, A.H. and A.J. Rowell. eds. 1992. Fossil invertebrates. Wiley, New York.

Borchiellini, C., M. Manuel, E. Alivon, N. Boury-Esnault, J. Vacelet, and Y. Le Parco. 2002. Sponge paraphyly and the origin of Metazoa. *Journal of Evolutionary Biology* 14(1): 171–179.

Boury-Esnault, N., and K. Rützler. 1997. Thesaurus of sponge morphology. *Smithsonian Contributions to Zoology* 596: 1–55.

Brain, C.K., Prave, A.R., Hoffmann, K.H. et al. 2012. The first animals: ca. 760-million-year-old sponge-like fossils from Namibia. South Africa Journal of Science 108(1/2), Art. #658. 10. 4102/sajs.v108i1/2.658.

Butler, P. 1962. Morphologic classification of sponge spicules, with descriptions of siliceous spicules from the Lower Ordovician Bellefonte Dolomite in central Pennsylvania. Journal of Paleontology, 191–200.

Carrera, M.G., and J.P. Botting. 2008. Evolutionary history of Cambrian spiculate sponges: implications for the Cambrian evolutionary fauna. *Palaios* 23: 124–138.

Clarkson, E.N.K. 1993. *Invertebrate palaeontology and evolution*. London: Chapman and Hall. 434 p.

De Laubenfels, M. 1955. Porifera. *Treatise on Invertebrate paleontology*, 21–112. Lawrence: Geological Society of America and University of Kansas Press.

De Vos, L., Ruetzler, K., Boury Esnault, N., Donadey, C. and J. Vacelet. 1992. Atlas of sponge morphology. Atlas de morphologie des Èponges. Washington and London: Smithsonian Institution Press.

Debrenne, F. and J. Vacelet. 1984. Archaeocyatha: is the sponge model consistent with their structural organization? In 4th international symposium on fossil Cnidaria, New York, 358–369.

Debrenne, F. 2007. Lower Cambrian archaeocyathan bioconstructions. *Comptes Rendus Palevol, Paris* 6(1–2): 5–19.

Dohrmann, M., C. Göcke, D. Jannussen, J. Reitner, C. Lüter, et al. 2012. Systematics and spicule evolution in dictyonal sponges (Hexactinellida: Sceptrulophora) with description of two new species. *Zoological Journal of the Linnean Society* 163: 1003–1025.

Gehling, J.G., and J.K. Rigby. 1996. Long expected sponges from the neoproterozoic ediacara fauna of South Australia. *Journal of Paleontology* 2: 185–195.

Hooper, J.N.A. 1991. Revision of the family Raspailiidae (Porifera: Demospongiae), with description of Australian species. Invertebr. Taxon. 5(6): 1179–1418.

Hooper, J.N.A. and R.W.M. Van Soest. 2002. Systema Porifera. A Guide to the Classification of Sponges. Kluwer Academic/Plenum Publishers, New York, (2 vols.) 1708 p.

Hooper, J.N.A., and F. Wiedenmayer. 1994. Porifera. In Zoological catalogue of Australia, edited by Wells, A., Vol. 12, 1–620.

Kerner, A., Vignes-Lebbe, R. and F. Debrenne. 2012. Computer-aided identification of the Archaeocyatha genera now available online. Carnets de Géologie / Notebooks on Geology, Brest, Letter 2011/02 (CG2011_L02), pp. 99–102.

Levi, C., Barton, J., Guillemet, C., Bras, E., Lehuede, P. 1989. A remarkably strong natural glassy rod: the anchoring spicule of theMonorhaphis sponge. Journal of Materials Science Letters 8: 337–339.

Love, G. D., Grosjean, E., Stalvies, C., et al. 2009. Fossil steroids record the appearance of Demospongiae during the Cryogenian period. Nature 457: 718–722.

Maldonado, M., Carmona, M.G., Uriz, M.J. and A. Cruzado. 1999. Decline in Mesozoic reef-building sponges explained by silicon limitation. Nature 401: 785–788.

Maloof, A., Rose, C., Beach, R., Samuels, B., Calmet, C., Erwin, D., Poirier, G., Yao, N., and F. Simons. 2010. Possible animal-body fossils in pre-Marinoan limestones from South Australia. Nature Geoscience. doi:10.1038/ngeo934.

May, A. 2008. Corals (Anthozoa, Tabulata and Rugosa) and chaetetids (Porifera) from the Devonian of the Semara area (Morocco) at the Museo Geominero (Madrid, Spain) and their biogeographic significance: Bulletin dell'Institut Scientifique Rabat, Section Sciences de la Terre 30: 1–12.

Muller, W.E.G. 2003. Sponges (Porifera), Springer, Berlin, 258 p.

Philippe, H., Derelle, R., Lopez, P., et al. 2009. Phylogenomics revives traditional views on deep animal relationships. Current Biology 19: 706–712.

Pick, K.S., Philippe, H., Schreiber, F., et al. 2010. Improved phylogenomic taxon sampling noticeably affects nonbilaterian relationships. Molecular Biology Evolution 27: 1983–1987.

Rigby, J.K., Budd, G.E., Wood, R.A., and F. Debrenne. 1993. Porifera. In The fossil, record edited by Benton, M.J., Vol. 2, 71–99. Chapman and Hall, London.

Sperling, E. A., Robinson, J. M., Pisani, D., and K.J. Peterson. 2010. Where's the glass? Biomarkers, molecular clocks, and microRNAs suggest a 200-Mry missing Precambrian fossil record of siliceous-sponge spicules. Geobiology 8: 24–36.

Suchy, D.R., and R.R. West. 2001. Chaetetid buildups in a Westphalian (Desmoinesian) cyclothem in Southeastern Kansas: Palaios 16(5): 425–443.

Uriz, M., Turon, X., Becerro, M., and G. Agell. 2003. Siliceous spicules and skeleton frameworks in sponges: origin, diversity, ultrastructural patterns, and biological functions. Microscopy Research and Technique 62: 279–378.

Van Soest, R.W.M. 1991. Demosponge Higher Taxa Classification Re-Examined. In J. Reitner, H. Keupp (eds), Fossil and Recent Sponges. Springer-Verlag, Berlin, Heidelberg: 54–71.

Van Soest, R.W.M., Van Kempen, T.M.G., and J.-C. Braekman, eds. 1994. Sponges in time and space. In Proceedings of the 4th international sponge symposium, 515 p. Balkema: Rotterdam.

Wang, X., Schroder, H., and W. Muller. 2009. Giant siliceous spicules from the deep-sea glass sponge *Monorhaphis chuni*. International Review of Cell and Molecular Biology 273: 69–184.

Zhuravleva, I.T. 1960. Arkheotsiaty Sibirskoy Platformy (Archaeocyaths of the Siberian Platform). Akademiya Nauk SSSR, Moscow, 344 p.

Chapter 3
Cephalopods

3.1 Introduction

Cephalopods are bilaterally symmetrical swimming marine carnivore molluscs that include modern day cuttlefish, octopus, squid, pearly *Nautilus,* and a large number of and mostly Paleozoic and Mesozoic forms. At present there are 800 living cephalopod species ($\sim$175 genera) and more than 17,000 ($\sim$600 genera) fossil forms; these are the largest group after Arthropods.

Cephalopods are active marine predators that were able to swim swiftly and compete with fishes within the marine habitat (although the early forms were most likely drifters). They are also the most advanced, intelligent, mobile, and largest of all the molluscs (Pojeta and Gordon 1987). As compared to Pelecypods (bivalves) and Gastropods, their brain is much more developed, with highly specialized sense organs. The cephalopods, especially the nautiloids in the Paleozoic and ammonoids in the Mesozoic are of enormous stratigraphic importance and are amongst the best known index fossils (i.e., organisms that are short-ranging in time but widespread in geographic distribution) (see also Tables 3.1, 3.2, 3.3, 3.4, 3.5, 3.6 and 3.7).

The earliest cephalopod, *Plectronoceras cambria* Walcott [a gastropod-like molluscs; Fig. 3.1(1–3)], comes from the Late Cambrian of NE China of the earliest Fengshanian stage (492.5–488.3 ± 1.7 Ma) from the basal Fengshan Formation (see also Dzik 1981; Clarke and Trueman 1985). The latest Cambrian of North America has also yielded *Palaeoceras mutabile* Flower [Fig. 3.1(4–9)] and *Ectenolites primus* Flower (Flower 1954).

The Nautiloids diversified into many different orders, and some of them were huge predators like the North American and European *Cameroceras* (Ordovician; 485.4 ± 1.9–443.4 ± 1.5 Ma) that grew as much as 10–11 m ($\sim$32–36 ft). However, most became extinct by the end of Devonian (358.9 ± 2.5 Ma). Only one order survived, the Nautilida (which peaked abundance between 515 and 250 Ma); from this only five species of *Nautilus* are alive today. During mid-Devonian, the ammonoids began to evolve as an offshoot of one of these

S. Jain, *Fundamentals of Invertebrate Palaeontology*, Springer Geology,
DOI 10.1007/978-81-322-3658-0_3

Table 3.1 Early-Middle Triassic and Permian index ammonites

Period	Epoch	Age/Stage	Substage	Tethyan Ammonoids
Triassic	Middle	Ladinian	Longobardian	Daxatina canadensis Frankites regolendanus
				Protrachyceras archelaus
				Protrachyceras gredleri
			Fassanian	Eoprotrachyceras curionii
		Anisian	Illtrian	Nevadites secedensis
				Reitziites reitzi
				Kellnerites felsoeoersensis
				Paraceratites trinodosus
				Schreterites binodosus
				Balatonites balatonicus
				Aghdarbandites ismidicus
				Nicomedites osmani
			Pelsoian	Lenotropites caurus
				Silberlingites mulleri
				Pseudokeyserlingites guexi
			Birthnian	Japonites welteri
				Neopopanoceras haugi
				Prohungarites-Subcolumbites
			Aegean	Procolumbites
				Columbites parisians
	Early	Olenekian	Spathian	Tirolites cassianus
				Anasiirites kingianus
				Meekoceras gracilitatis
				Flemingites flemingianus
			Sithian	Rohillites rohilla
				Gyronites frequense
				"Pleutogyronites" planidorsatus-Disciphiceras
		Induan	Dienerian	
			Late Griesbachian (Gangetian)	Otoceras tibeticum
				Otoceras woodwardi
				Otoceras fissisellatum
Permian	Lopingian	Changhsingian		Hypophiceras changxingense
				Pleuronodoceras occidentale
				Paratirolites kittli
				Shevyevites shevyrevi-Dzhulfites spinosus
				Iranites transcaucasicus
		Wuchiapingian		Phisonites triangulum

Table 3.2 Late Triassic index ammonites

Ma	Period	Epoch	Age/Stage	Substage	Tethyan Ammonoids
200, 201	Triassic	Late	Rhaetian		Choristoceras marshi
202					Choristoceras haueri
203					Cochloceras suessi
204			Norian	Sevatian	Sagenites reticulatus
205					Sagenites quinquepunctatus
206, 207					Halorites macer
208, 209, 210				Alaunian	Mesohimavatites columbianus
211					Cyrtopleurites bicrenatus
212				Lacian	Juvavites magnus
213, 214					Malayites paulckei
215, 216					Guembelites jandianus
217, 218			Carnian	Tuvalian	Anatropites spinosus
219, 220					Tropites subbullatus
221, 222					Tropites dilleri
223, 224, 225				Julian	Austrotrachyceras austriacum
226					Trachyceras aonoides
227, 228					Trachyceras aon
					Daxatina canadensis

Table 3.3 Early Jurassic index ammonites

Ma	Period	Epoch	Age/Stage	Substage	Tethyan Ammonoids
176	Jurassic	Early	Toarcian	Late	Pleydellia aalensis
177					Dumortieria pseudoradiosa
178					Phlyseogrammoceras dispansum
179, 180					Grammoceras thouarsense
					Haugia variabilis
181				Middle	Hildoceras bifrons
182				Early	Harpoceras serpentinum
					Dactyloceras tenuicostatum
183			Pliensbachian	Late Domerian	Emaciaticeras emciatum
184, 185					Arieticeras algovianum
186					Fuciniceras lavinianum
187				Early Carixian	Prodactylioceras davoei
188					Tragophylloceras ibex
189					Uptonia jamesoni
190, 191			Sinemurian	Late	Echioceras raricostatum
192					Oxynoticeras oxynotum
193					Asteroceras obtusum
194				Early	Caenisites turneri
195					Arnioceras semicostatum
196					Arietites bucklandi
197			Hettangian		Schlotheimia angulata
198					Alsatites liasicus
199					Psiloceras planorbis

Table 3.4 Middle-Late Jurassic index ammonites

Ma	Period	Epoch	Age/Stage	Substage	Tethyan Ammonoids
146	Jurassic	Late	Tithonian	Late	Durangites
147					Micracanthoceras microcanthum
148				Early	Micracanthoceras ponti/ Burckhardticeras peroni
					Semiformiceras fallauxi
149					Semiformiceras semiforme
150					Semiformiceras darwini
					Hybonoticeras hybonotum
151			Kimmeridgian	Late	Hybonoticeras beckeri
152					Aulacostephanus eudoxus
153					Aspidoceras acanthicum
				Early	Crussoliceras divisum
154					Ataxioceras hypselocyclum
					Sutneria platynota
155					Idoceras planula
156			Oxfordian	Late	Epipeltoceras bimammatum
157					Perisphinctes bifurcatus
158				Middle	Gregoryceras transversarium
159					Perisphinctes plicatilis
160				Early	Cardioceras cordatum
					Quenstedtoceras mariae
161		Middle	Callovian	Late	Quenstedtoceras lamberti
162					Peltoceras athleta
				Middle	Erymnoceras coronatum
163					Reineckeia anceps
				Early	Macrocephalites gracilis
					Bullatimorphites bullatus
164					Clydoniceras discus
165			Bathonian	Late	Hecticoceras retrocostatum
					Cadomites bremeri
166				Middle	Morrisiceras morrisi
					Tulites subcontractus
167					Procerites progracilis
				Early	Procerites aurigerus
168			Bajocian	Late	Zigzagiceras zigzag
					Parkinsonia parkinsoni
169					Garantiana garantiana
					Strenoceras niortense
170				Early	Stephanoceras humphriesianum
					Sonninia propinquans
171					Witchellia laeviuscula
					Hyperlioceras discites
172			Aalenian	Late	Graphoceras concavum
173				Middle	Brasilia bradfordensis
174					Ludwigia murchisonae
175				Early	Leioceras opalinum

Table 3.5 Early Cretaceous index ammonites

Ma: 100, 101, 102, 103, 104, 105, 106, 107, 108, 109, 110, 111, 112, 113, 114, 115, 116, 117, 118, 119, 120, 121, 122, 123, 124, 125, 126, 127, 128, 129, 130, 131, 132, 133, 134, 135, 136, 137, 138, 139, 140, 141, 142, 143, 144, 145

Period	Epoch	Age/Stage	Substage	Tethyan Ammonoids
Cretaceous	Early	Albian	Late Middle Early	Same as Boreal (NW Europe)
		Aptian	Late Clansayesian	Hypacanthoplites jacobi
				Nolaniceras nolani
			Middle Gargasian	Perahoplites melchioris
				Epicheloniceras subnodosocostatum
			Early Bedoulian	Dufrenoyi furcata
				Deshayesites deshayesi
				Deshayesites weissi
				Deshayesites oglanlensis
				Pseudocrioceras waagenoides
		Barremian	Late	Colchidites sarasini
				Imerites giraudi
				Hemihoplites feraudianus
				Gerhardtia sartousi
				Ancycloceras vandenheckii
			Early	Coronites darsi
				Kotetishvilia compressissima
				Nicklesia pulchella
				Nicklesia nicklesi
				Spitidiscus hugii
		Hauterivian	Late	Pseudothurmannia ohmi
				Balearites balearis
				Pleisiospitidiscus ligatus
				Subsaynella sayni
			Early	Lyticoceras nodosoplicatus
				Crioceratites loryi
				Acanthodiscus radiatus
		Valanginian	Late	Criosarasinella furcillata
				Neocomites peregrinus
				Saynoceras verrucosum
			Early	Busnardoites campylotoxus
				Timovella pertransiens
				Thurmanniceras otopeta
		Berriasian	Late	Subthrumannia boissieri
			Early	Subthrumannia occitanica
				Berriasella jacobi
				Durangites

Table 3.6 Late Cretaceous index ammonites

Ma	Period	Epoch	Age/Stage	Substage	Tethyan Ammonoids
66–99 (scale: 66, 67, 68, 69, 70, 71, 72, 73, 74, 75, 76, 77, 78, 79, 80, 81, 82, 83, 84, 85, 86, 87, 88, 89, 90, 91, 92, 93, 94, 95, 96, 97, 98, 99)	Cretaceous	Late	Maastrichtian	Late	Anapachydiscus terminus
					Anapachydiscus fresvillensis
				Early	Pachydiscus neubergicus to Pachy. epiplectus
			Campanian	Late	Nostroceras hyatti
					Didymoceras cheyennense
				Late / Middle	Bostrychoceras polyplocum
				Early	Hoplitoplacenticeras marroti / vari
					Delawarella delawarensis
					Placenticeras bidorsatum
			Santonian	Late / Middle / Early	Placenticeras polyopsis
			Coniacian	Late	Paratexanites serratomarginatus
				Early	Gauthiericeras margae
					Peroniceras tridorsatum
			Turonian	Late / Middle / Early	Same as Boreal (NW Europe)
			Cenomanian	Late / Middle / Early	

Table 3.7 The geological history of cephalopods

Geological history			
Age		**Events**	**Extinction**
Present		One Nautiloid genera with 5 species survived and is restricted to the south-west Pacific	
Tertiary		Rare Nautiloid records	
Cretaceous-Tertiary boundary (K/T)		Only 5 ammonoid families are known by the K/T boundary. However, the Nautiloids survived the crash; genera like *Aturia* and *Hercoglossa* are recorded in the Danian, Paleogene; similarly the belemnites also crashed although few Eocene records are noted but they were soon heavily outcompeted by the modern Coleoidea (octopus, squid, cuttlefish, etc.).	
Cretaceous	Late	Only 11 ammonoid families remained; widespread Heteromorphs	
Cretaceous / Jurassic		The remaining ammonoid genera radiated again with 90 families in the Jurassic and 85 in the Cretaceous; first confirmed fossil record of squid, octopi and *belemnites* from the Jurassic	
Triassic	Late (**Age of *Ceratites***)	Most *Ceratites* were wiped out by end-Triassic extinction leaving only few genera to survive; demise of straight shelled Nautiloids (the *Pseudorthocerids*)	
		The *Ceratites* dominated the Triassic with 80 families and over 500 genera	
	Early	Ammonoids order Phylloceratids arose that gave rise to all post-Triassic ammonoids	
Permian	Late	One ammonoid genera survived the Permian catastrophe; extinction of Bactritids and *Goniatites*	
		Decline in ammonoid diversity	
	Middle	27 ammonoid families	
Carboniferous	Pennsylvanian	Improvement in Late Pennsylvanian with 30 ammonoid families	
		Great reduction in the ammonoids diversity due to the Mississippian / Pennsylvanian extinction; only 9 ammonoid families survived	
	Mississippian	2 genera of ammonoids remained; the Actinoceratids struggled to survive	
Devonian	Famennian	Radiation of ammonoids-80 genera present	
	Fransnian / Famennian Event	Only 3 ammonoid genera survived the Fransnian / Famennian extinction event	
	Late	Rise of Bactritids, Ammonoids and Coleoids; 30 genera of ammonoids present	
	Early	Ammonoids diversified	
Silurian		Disappearance of Endoceratoids	
Ordovician (**Age of Nautiloids**)	Late	Decline of Endoceratoids and Actinoceratoids	
	Middle	20 genera of Actinoceratoids	
	Early	40 genera of Endoceratoids	
Cambrian	Late	The earliest recorded Nautiloid is *Plectronoceras* [Fig. 3.1(1–3)]	
		Cephalopods arose from a Monoplacophoran ancestor with a tall, conical, slightly curved shell such as that of Late Cambrian *Knightoconus* (*Knightoconus antarcticus*; Figs. 3.1 (10–11)	

extinct groups of straight-shelled nautiloids, called the Bactritida; the evolution of coiled shells began during this time (Table 3.7). The Bactritids are an obscure group, and are regarded as a transitional stock between nautiloids and ammonoids.

Gradually, three forms of ammonoids (*Goniatites*, *Ceratites,* and *Ammonites*) evolved, over time. The *Goniatites* in the mid-Devonian (393.3 ± 2.7–382.7 ± 2.8 Ma) through Permian (252.2 ± 0.5 Ma), the *Ceratites* during the Triassic (252.2 ± 0.5–201.3 ± 0.2 Ma) and following their extinction, the *Ammonites* in the Jurassic (201.3 ± 0.6–145 ± 4 Ma). Thus, their evolutionary history spans an impressive ~500 Ma, recording repeated speciation and extinction events (Table 3.7).

3.2 General Morphology

The name cephalopoda comes from the Greek words *cephale* meaning "head" and *podos* meaning "feet." The foot consists of a cluster of tentacles (arm-like appendages) that surrounds the mouth [Fig. 3.2(1)] which is used to capture prey. Part of the foot is modified into a muscular organ called the Hyponome [Fig. 3.2(1)] that jets water when it is compressed, facilitating swimming. Thus, the cephalopod is jet-propelled in the opposite direction at a velocity controlled by the force with which the water is expelled. Cephalopods are streamlined and the most streamlined living squid can reach speeds up to 70 km/h. In the chambered *Nautilus*, the ejection of water through the hyponome results from the retraction of the body into the conch with simultaneous contraction of the hyponome muscles rather than by the mantle muscles alone, as done by squids (Fig. 3.2).

At the onset, it must be kept in mind that no ammonite has yet been found that possesses preserved soft tissues. Hence, our best estimate of its physiology is based on the living *Nautilus*; a suitable comparison despite anatomical differences. Here, preserved hard parts and their observed main morphological features are briefly described with corresponding figures.

3.2.1 **Chamber (Camera; pl. Camerae)**: The living chambered *Nautilus* (the only cephalopod that has retained its exterior shell) is planispirally coiled and is divided into gas-filled chambers called Camerae [Fig. 3.2(1)]. Camera, excluding the siphuncle, is the space enclosed between two adjacent septa (Fig. 3.2)

3.2.2 **Septa**: These are curved calcareous walls, invariably concave adorally, and are pierced by the siphuncle. The chambers are partitioned by the Septa [Fig. 3.2(2)]

3.2.3 **Aperture**: This is the opening at the anterior end of the shell through which the head–foot protrudes [Fig. 3.2(1)]

3.2.4 **Peristome**: This is the edge of the aperture [Fig. 3.3(1)]

3.2.5 **Hyponomic sinus**: At the mid-ventral margin of the aperture there is an indentation (a large concave sinus in the middle of the aperture) called the Hyponomic Sinus [Fig. 3.3(4, 5)]. This depression accommodates the Hyponome and is invariably ventral in position. The side which has the hyponome is the Ventral side and the opposite is the Dorsal side [Fig. 3.3(4–7)]

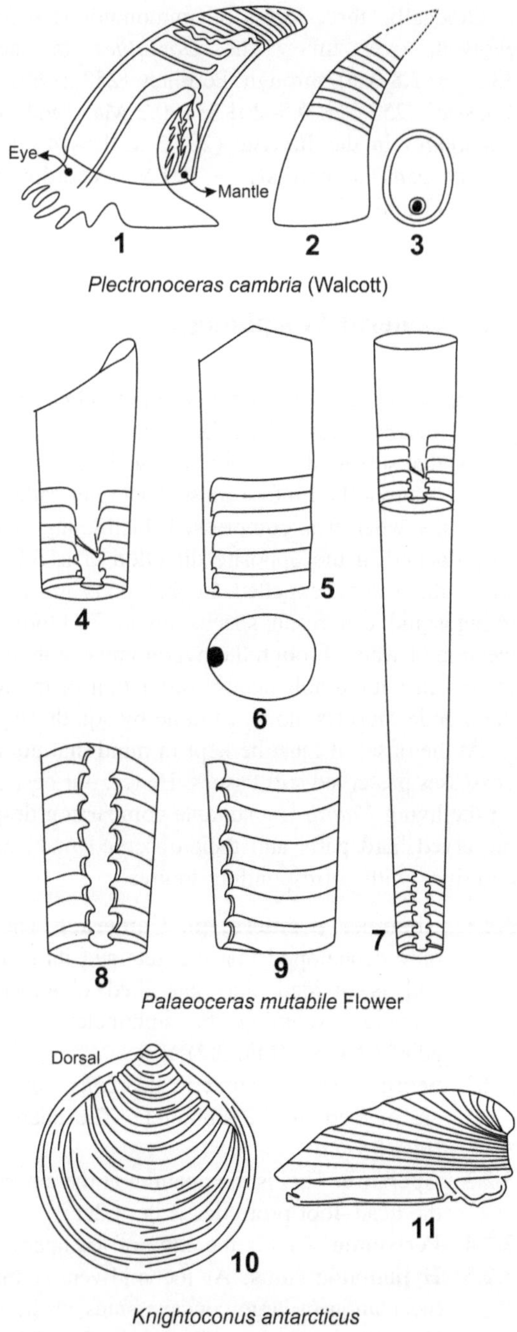

Fig. 3.1 Earliest known cephalopods. *1–3 Plectronoceras cambria* Walcott. It emerged in the Cambrian, 540 Million years in the Fengshan formation of NE China. It was a nautiloid. Its shell enclosed an air space which provided buoyancy. This is evidenced by the observation of septa with a siphuncle or tube for the extraction of water from its chambers, Figs. 3.4, 3.5, 3.6, 3.7, 3.8 and 3.9: *Palaeoceras mutabile* flower, *d–i*; Paratype, *d–f* (modified from Flower 1954). *4* Ventral view. *5* Lateral view, venter is on the *left*. *6* Cross-section of *5*. *7* Reconstructed shell, ventral view. *8–9* Holotype and ventral and lateral views, respectively, the venter is on the *left*. *10–11 Knightoconus antarcticus* recorded from Western Antarctica

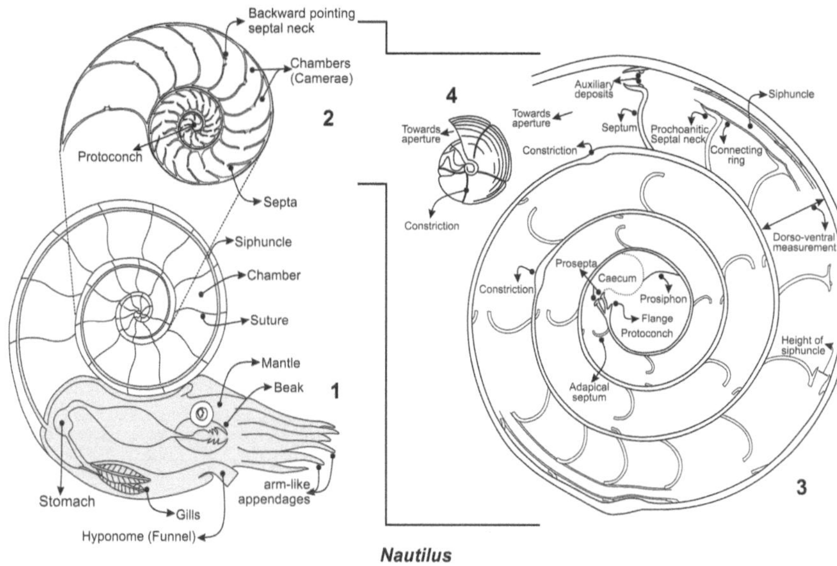

Fig. 3.2 *1–3* Structural details of a chambered Nautilus (adapted from Treatise on Invertebrate Paleontology, Part K. Mollusca 3: Cephalopoda General Features). *4* External expression of a constriction on an ammonite

3.2.6 **Venter**: When the shell is planispirally coiled, the ventral side that forms the circumference is the Venter [Fig. 3.3(2)]

3.2.7 **Phragmocone**: The remaining part of the chambered shell where the organism does not live is called the Phragmocone [Fig. 3.3(1)]

3.2.8 **Body chamber**: This is where the organism lives and is the last chamber of the shell [Figs. 3.2(1) and 3.3(1)]. This moves forward each time a new chamber is secreted. The length of the body chamber varies from half a volution (whorl) (as in Brevidome ammonoids) to more than a complete whorl (as in Longidome ammonoids) (see Fig. 3.4). The length of the living chamber is a function of the static stability and of the position of the aperture (see also Trueman 1941; Saunders and Shapiro 1986; Westermann 1996). These types are briefly mentioned below

3.2.8.1 **Brevidome**: These have a body chamber length of 160°–180° [Fig. 3.4(1)]. These are short rapidly expanding conchs. Example: *Dactylioceras*

3.2.8.2 **Mesodome**: These have a body chamber length of about 260° [Fig. 3.4(2)]. Example: Hildoceratides

3.2.8.3 **Longidome**: These have a body chamber length of more than 360° [Fig. 3.4(3)]. Example: Perisphinctids.

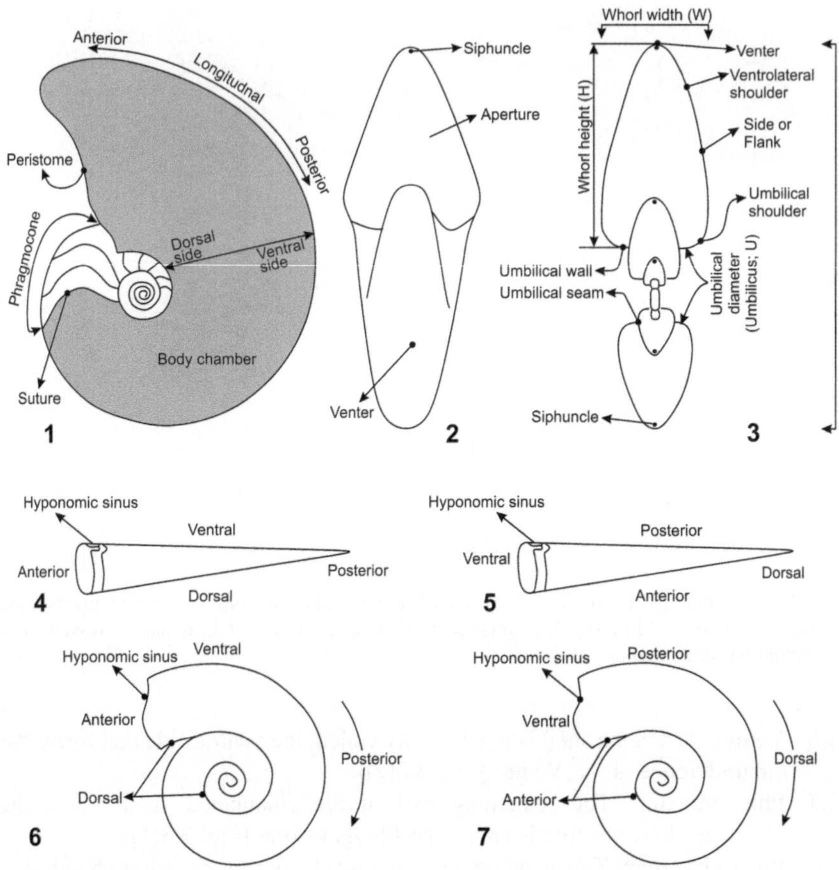

Fig. 3.3 *1–3* Shell terminology. *4–7* Position of the hyponomic sinus

3.3 Modifications of the Shell

The main morphological features are briefly described below and their corresponding figures mentioned therein.

3.3.1 Apertural Modifications

The presence of apertural modifications (Lappets; Lateral and Ventral) suggests that ammonites were highly visual as such a feature would be virtually invisible at darker depths. Hence, the ammonites were inferred to be fairly shallow water creatures (Saunders and Shapiro 1986; Westermann 1996). Some have suggested the use of Lappets for sexual display.

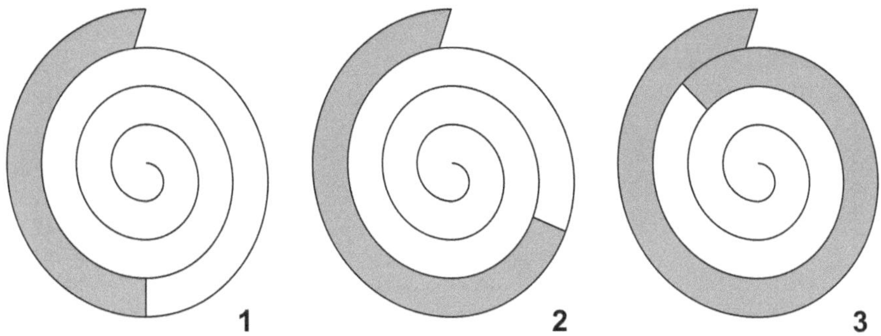

Fig. 3.4 Length of body chamber. *1* Brevidome ammonoids have a body chamber length between 160° and 180°. *2* Mesodome about 260°. *3* Longidome ammonoids with more than 360° body chamber length

A mature shell shows several modifications which are indications of maturity instead of being transient features of the growing shell (Makowski 1962, 1963; Callomon 1963, 1969) (Figs. 3.5, 3.6, 3.7 and 3.8). The presence or absence of Lateral lappets enables to assign dimorphism within an organism—Microconch (adult male; without lappets) and Macroconch (adult female; with lappets) (see Fig. 3.8(9–10); respectively). The smaller form, the microconch, looks very similar to the early whorls of the larger macroconch [Fig. 3.8(9–10)]. Sexual dimorphism has considerable implications for ammonite taxonomy, stratigraphy, paleoecology, and evolutionary reconstructions (see also Callomon 1963). However, in spite of numerous variations in the types of aperture (Figs. 3.5, 3.6, 3.7, and 3.8), these can broadly be grouped into five major types (Fig. 3.9), besides the Lateral and Ventral lappets, as enumerated below.

3.3.1.1 **Lateral Lappets**: These are projections from the lateral part of the peristome [Fig. 3.9(1–3)]

3.3.1.2 **Ventral Lapetts (Rostrum)**: The Ventral Lappet or the Rostrum projects from the venter or the ventral part of the peristome. Example: Late Callovian *Quesntedoceras* [Fig. 3.9(4–6)]

3.3.1.3 **Contracted**: Contraction is closing off [Fig. 3.9(7–9)],

3.3.1.4 **Constricted**: Constriction is necking down [Fig. 3.9(10–12)]

3.3.1.5 **Expanded**: Sometimes the later part of the whole body chamber is expanded [Fig. 3.9(13–15)].

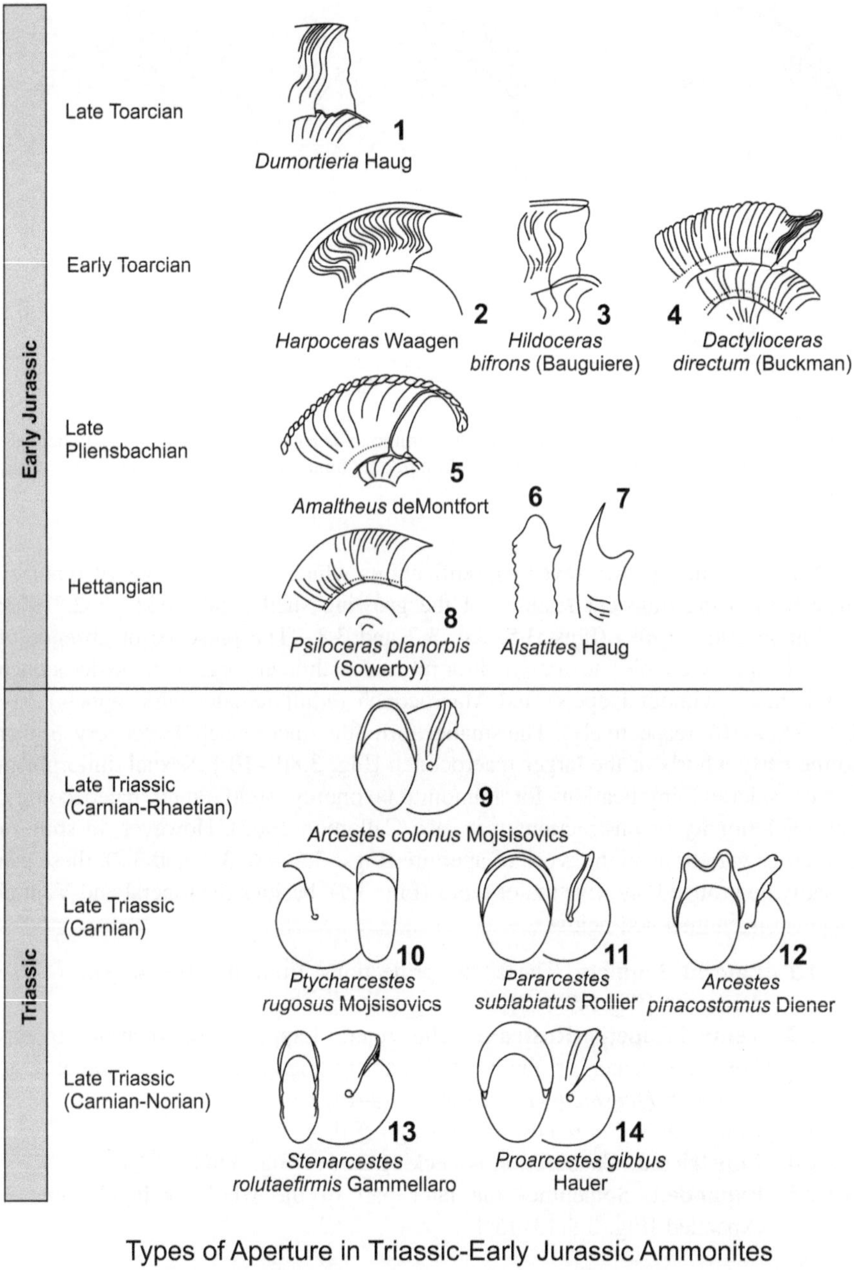

Fig. 3.5 Apertural modifications (lappets) of Triassic and Early Jurassic ammonites

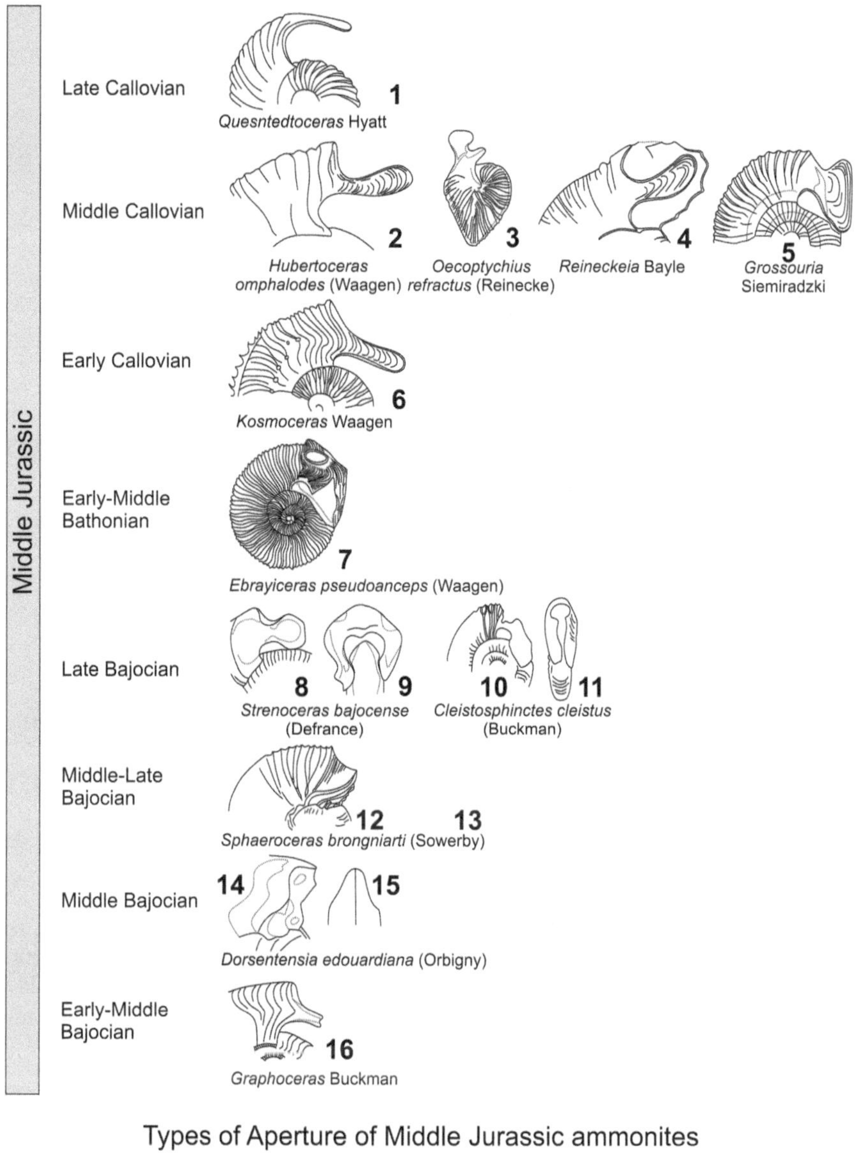

Fig. 3.6 Apertural modifications (lappets) of Middle Jurassic ammonites

3.3.2 Constrictions and Growth Lines

Cephalopod shells are either unornamented (total absence of constrictions) or could be varied with as much as nine constrictions (Fig. 3.10). Additionally, all cephalopod shells at maturity are also ornamented with Growth Lines (Fig. 3.11),

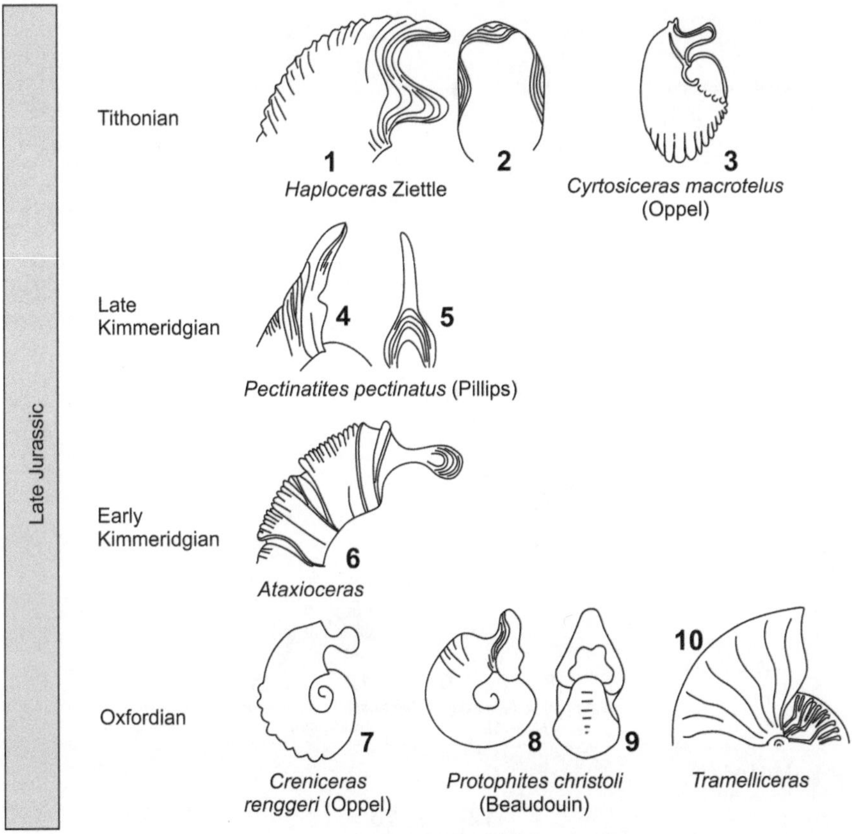

Fig. 3.7 Apertural modifications (lappets) of Late Jurassic ammonites

representing former growth positions of a chamber. As each chamber is added to the shell, it leaves a scar on the shell, visible as growth lines; there is a periodic increase in size and hence, represents the former position of the aperture (Fig. 3.11).

3.3.3 *Phragmocone*

The main morphological features of a phragmocone are briefly described below:

3.3.3.1 **Siphuncle**: It is a thin porous tube that connects cephalopod chambers up to the anterior part of the shell (Fig. 3.12). The siphuncle consists of soft and shelly parts, including septal necks, connecting rings, calcareous deposits, and siphuncular cord. These are briefly enumerated below and illustrated in Fig. 3.12

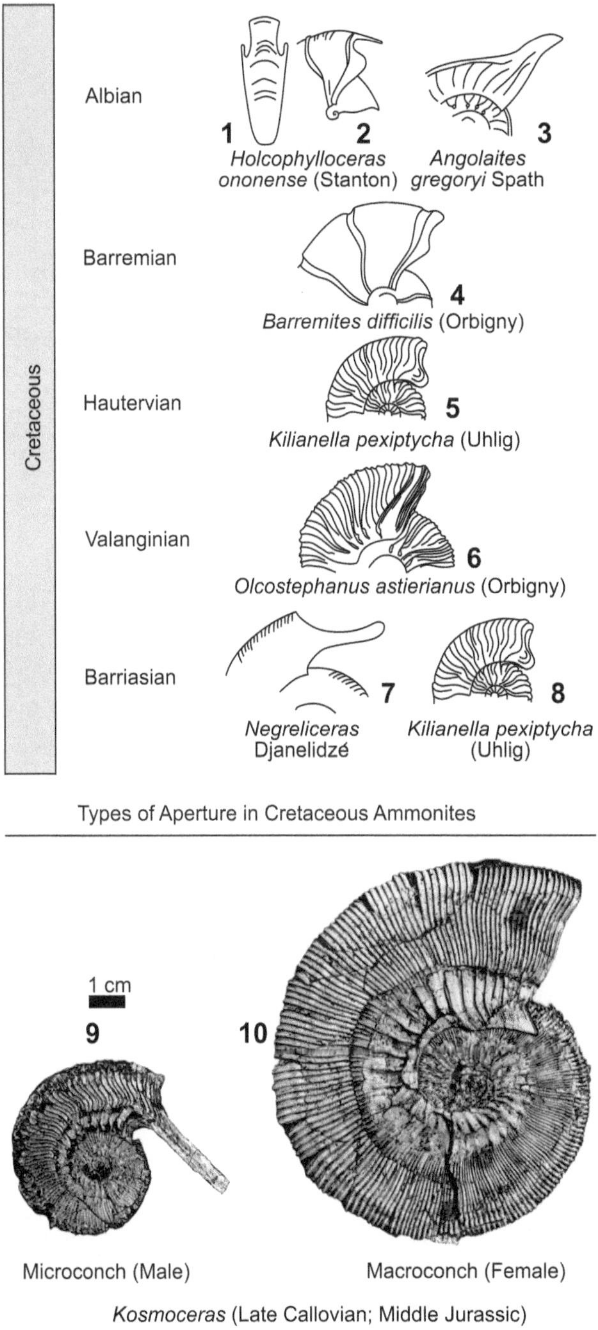

Fig. 3.8 *1–8* Apertural modifications (lappets) of Cretaceous ammonites. *9–10* The dimorphic pair. Microconch (adult male) and macroconch (adult female) of the Late Callovian (Middle Jurassic) *Kosmoceras* (courtesy Late John H. Callomon)

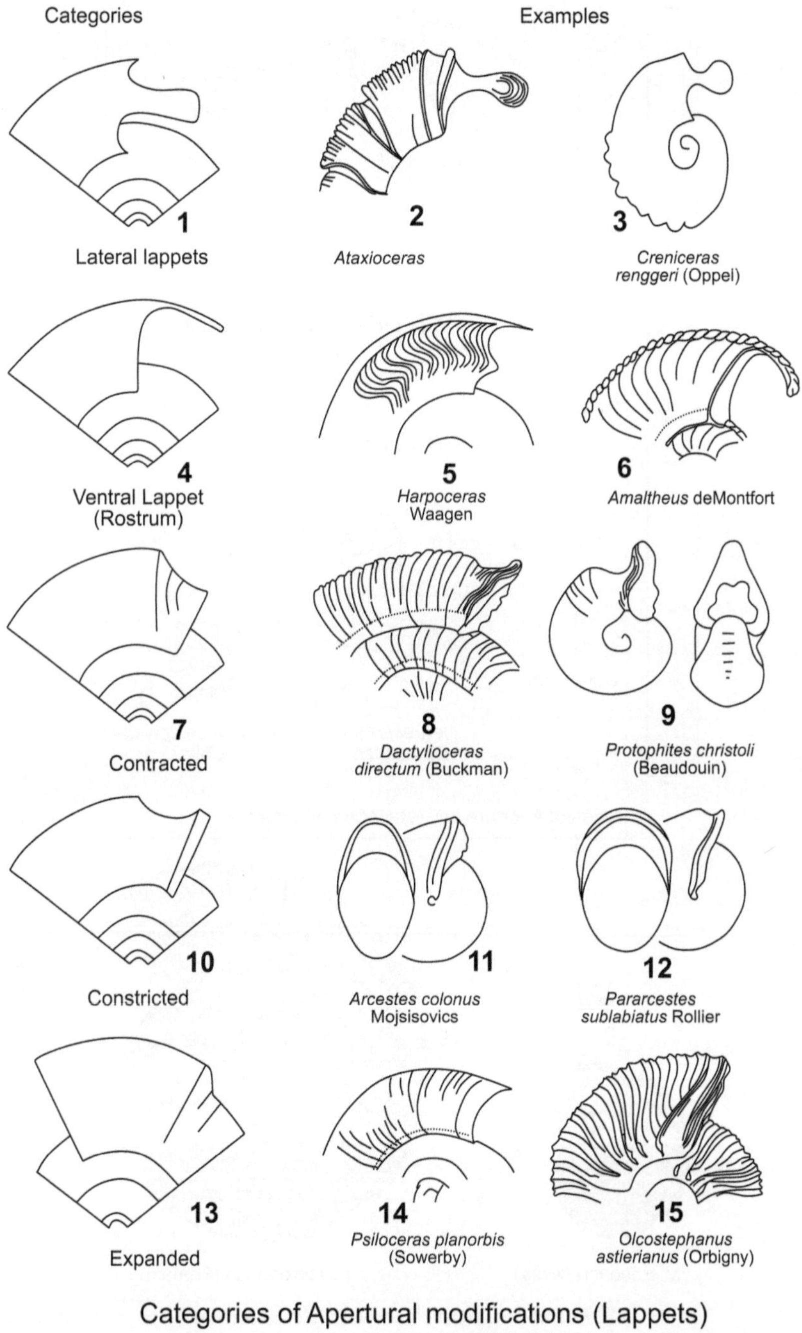

Fig. 3.9 *1–15* Broad categories of apertural modifications (lappets) noted in Triassic–Cretaceous ammonites with examples (wherever possible specific examples are given)

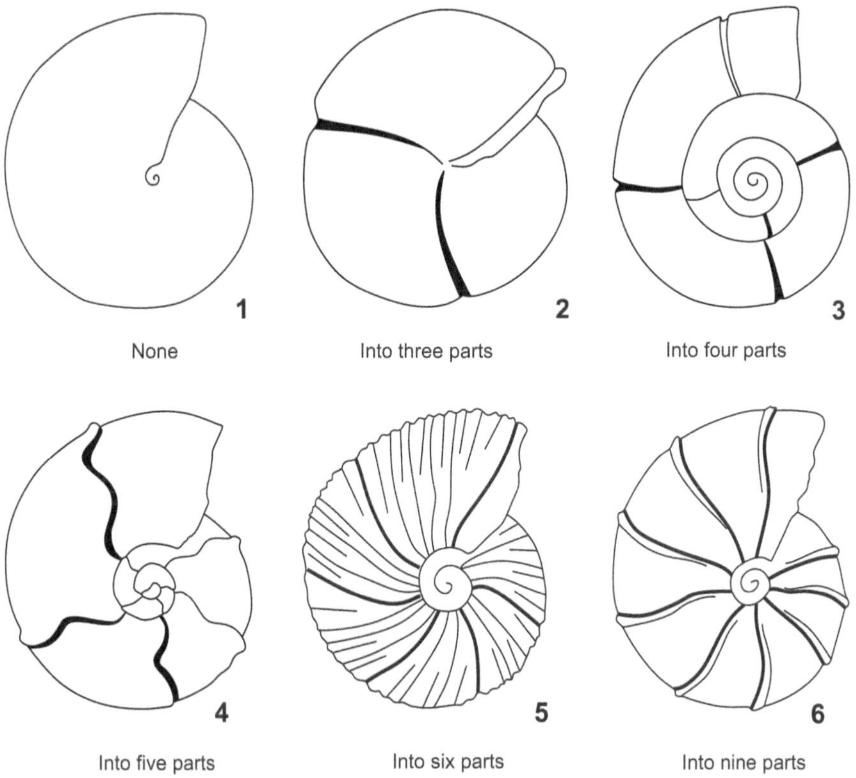

Fig. 3.10 Constrictions

3.3.3.2 **Septal neck**: It is the portion of septum that is bent adapically (or backwards) around the septal foramen and is reinforced by Septal Necks and Connecting Rings [Fig. 3.12(2–17)]. A Septum is the partition dividing the phragmocone into chambers; attached to the inside of shell wall [see Fig. 3.11(1)]. There are various types of septal neck orientations [Fig. 3.12(2–12)]. These are:

- **3.3.3.2.1** **Prochoanitic**: These are directed adorally, i.e., toward the aperture [Fig. 3.12(2)]
- **3.3.3.2.2** **Retrochoanitic**: These are directed adapically, i.e., away from the aperture [Fig. 3.12(3)]
- **3.3.3.2.3** **Achoanitic**: These are barely developed or are extremely short [Fig. 3.12(4)]
- **3.3.3.2.4** **Laxochoanitic**: They point inward at moderate lengths [Fig. 3.12(5)]
- **3.3.3.2.5** **Orthochoanitic**: These necks are directed adapically and are less than half the chamber length [Fig. 3.12(6)]

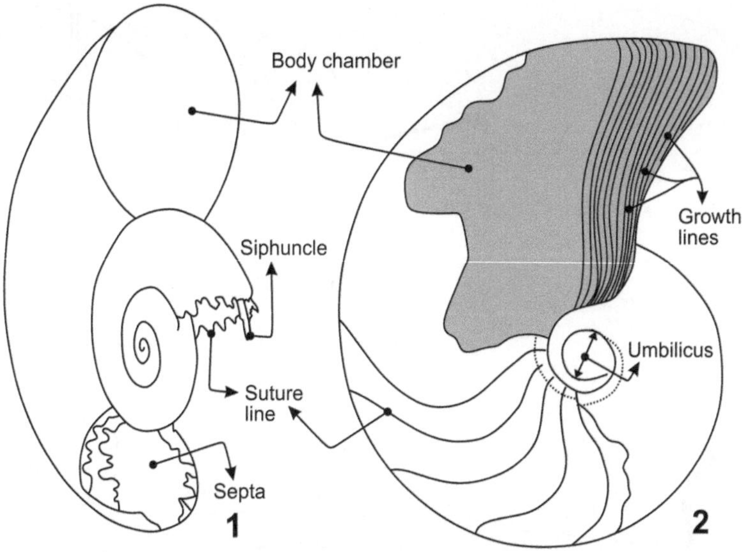

Fig. 3.11 Growth lines, the former growth positions of a chamber

3.3.3.2.6 **Hemichoanitic**: These necks extend half to three quarters of the chamber length [Fig. 3.12(7)]

3.3.3.2.7 **Subholochoanitic**: These curve inwards, just before reaching the next septum [Fig. 3.12(8)]

3.3.3.2.8 **Holochoanitic**: These necks reach to the next septum or slightly beyond it [Fig. 3.12(9)]

3.3.3.2.9 **Macrochoanitic**: These necks extend longer than the distance to the next septum [Fig. 3.12(10)]

3.3.3.2.10 **Suborthochoanitic**: These necks are barely recurved [Fig. 3.12(1)]

3.3.3.2.11 **Crytochoanitic**: These necks are recurved; some even touch the free part of the septum [Fig. 3.12(12)]. They are comparatively short retrochoanitic septal necks

3.3.3.3 **Connecting Rings**: Connecting Rings [Fig. 3.12(13)] are in part calcareous and conchiolinous tubular structures. They connect the septa or septal necks. Some rings are thin and simple; others are thick and composed of two or three layers of deposits. Connecting rings come in varied styles; they can be straight, concave, convex, or bulbous [Fig. 3.12(14–17)].

3.3.3.4 **Ectosiphuncle, Endosiphuncle, and Endosiphuncular Deposits**

3.3.3.4.1 **Endosiphuncular Deposits**: The nonliving part of the siphuncle that covers the living siphuncular chord is the Ectosiphuncle. The ectosiphuncle is composed of septal necks and connecting rings. The area inside the Ectosiphuncle is the Endosiphuncle [Fig. 3.12(18–27)] which is the space within the Ectosiphuncle (the walls of the siphuncle) and includes all organic tissues and calcareous structures. In *Nautilus,* the Ectosiphuncle is composed of a shelly material, conchiolin, and spicules of calcium carbonate. There are no shelly deposits in the Endosiphuncle in *Nautilus.*

3.3.3.4.1.1 **Endocones**: These are cone-shaped calcareous deposits [Fig. 3.12(18, 23)] formed in the adapical portion of the siphuncle. These are noted mainly in endocerid and discosorid shells

3.3.3.4.1.2 **Diaphrams**: These are transverse partitions [Fig. 3.12(19, 24)]

3.3.3.4.1.3 **Lamellae**: These are longitudinal partitions [Fig. 3.12(20, 25)]

3.3.3.4.1.4 **Rods**: These are round structures lying on the ventral wall of the siphuncle [Fig. 3.12(21, 26)]

3.3.3.4.1.5 **Annulosiphonate deposits**: These are donut shaped deposits inside the siphuncle [Fig. 3.12 (22, 27)]

3.3.3.4.2 **Cameral Deposits**: These are calcareous deposits secreted against the walls of the chamber during the organism's life. These are named according to their depositional position on the wall of the chamber [Fig. 3.12(28)].

3.3.3.4.2.1 **Episeptal**: These are deposited on the anterior wall of a septum [Fig. 3.12(28)], i.e., the concave (or adapertural) side of the septum

3.3.3.4.2.2 **Hyposeptal**: These are deposited on the posterior wall [Fig. 3.12(28)]; the convex (adapical) side of the septum

3.3.3.4.2.3 **Mural**: These line the outer wall of the chamber [Fig. 3.12 (28)] or along the mural parts of septa (i.e., parts of the septum attached to wall of shell).

The shapes and forms of cameral deposits are important in the classification of some cephalopods. Functionally, these aid in counteracting the positive buoyancy of the gas-filled phragmocones of orthoconic or cyrtoconic longicones.

3.3.3.5 **Suture**: The cephalopod body chamber during its secretion leaves a distinctive pattern at its edge called the Suture (Fig. 3.11). It is a line of attachment of the septum to the shell's interior that is not exposed on the exterior surface of the shell. Hence, fossil cephalopods commonly display sutures preserved as internal molds, with the outer shell being dissolved away. It is a character immensely useful for species level taxonomic identification. Although, particular types of sutures characterize distinct ammonoid families (Fig. 3.13) but if the sutures are not well preserved,

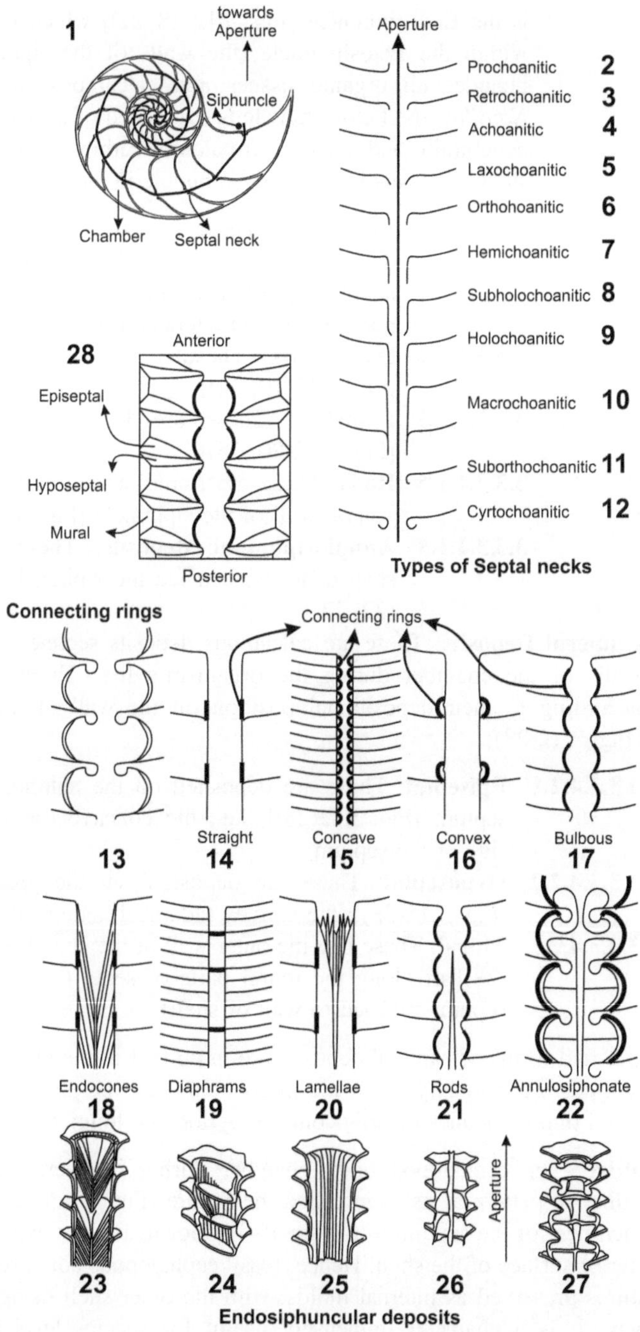

Fig. 3.12 *1–27* Internal structure of a shell. *1* Siphuncle. *2–12* Septal necks. *13–22* Connecting rings. *23–27* Endosiphuncular. *28* Cameral deposits (*bottom* panel illustrates the actual; *23–27*)

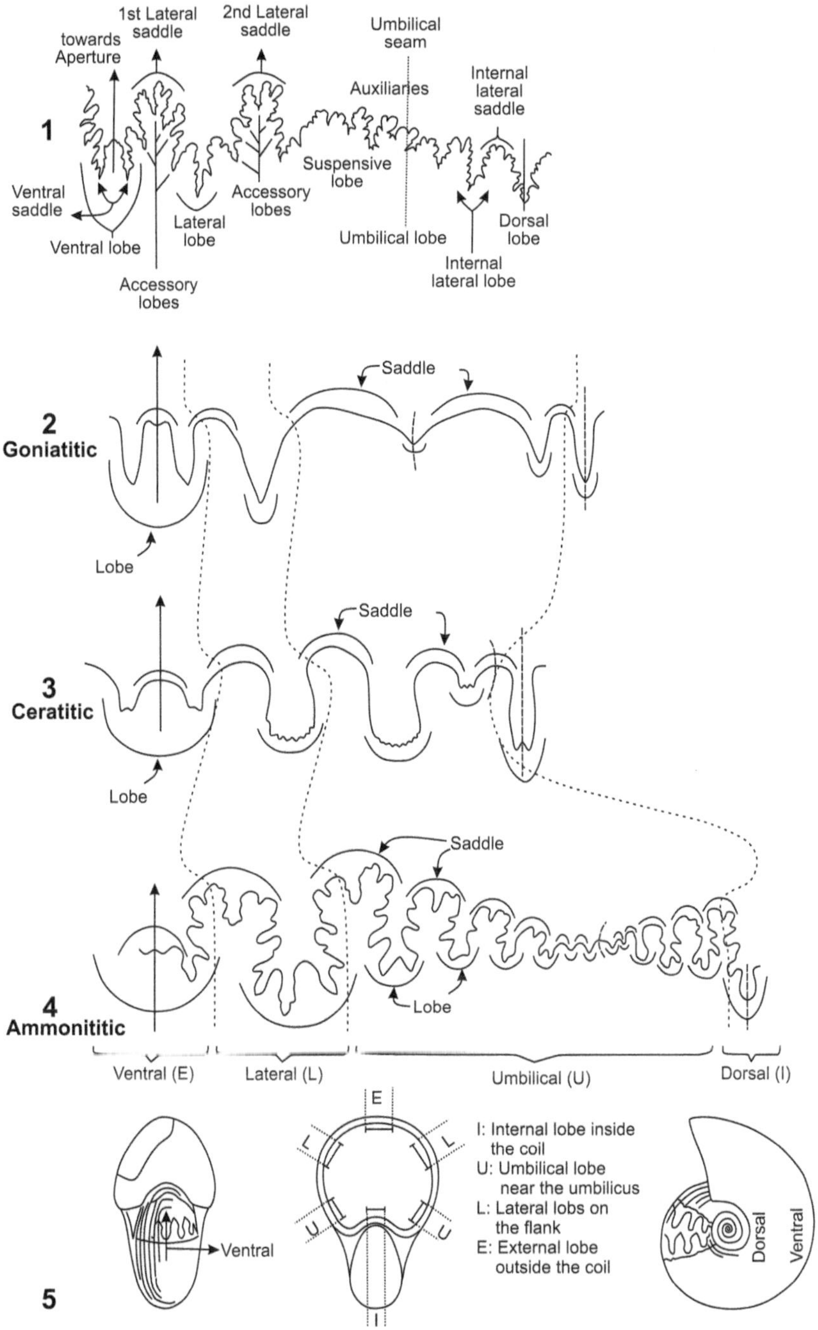

Fig. 3.13 Suture line. *1–4* Elements of a suture line. *2–4* Types of suture line. *5* and *7* External expression of suture line. *6* Position of lobes

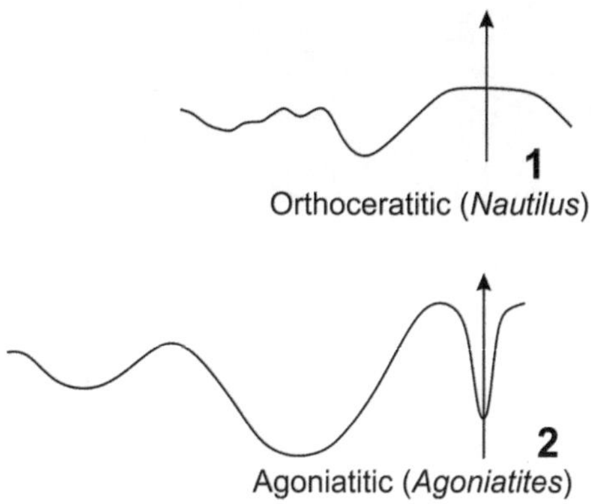

Fig. 3.14 *1* Orthoceratitic. *2* Agoniatitic sutures

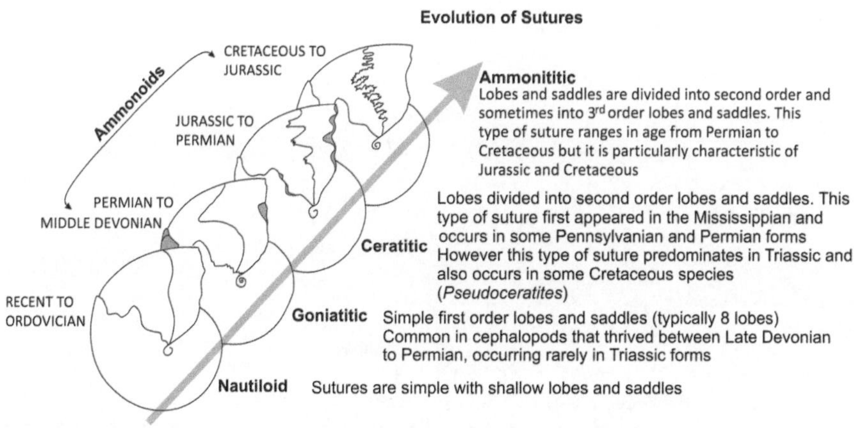

Fig. 3.15 Evolution of suture through time, suture types, and their major characteristics. Overall an increase in sutural complexity is noted; from simple sutures with shallow lobes and saddles (goniatites) to those where lobes and saddles are divided into second-order and sometimes into third-order lobes and saddles

homeomorphy in shell form and ornament can make specific identification very difficult.

3.3.3.5.1 **Lobes and Saddles**: Suture terminology commonly used is given in Fig. 3.13(1) and its outward expression (particularly the position of lobes) is given in Fig. 3.13(5). Parts of the suture line that are directed adorally (i.e., toward the aperture) are called Saddles, and those directed adapically (i.e., away from the aperture) are termed Lobes [Fig. 3.13(1)]. Ordinarily, only half of a suture, from midventer to middorsum is illustrated as sutures are usually bilaterally symmetrical; the midventral point is marked by an arrow pointing adaperturally (adorally or toward the aperture; Fig. 3.13(1–4). Generally, three principal types of sutures are recognized (Fig. 3.13). These are:

3.3.3.5.1.1 **Goniatitic** (named after the Mississippian genus *Goniatites*): These have strong, mostly angular lobes, and angular to rounded saddles [Fig. 3.13 (2)]. Lobes are narrowly rounded to pointed; the ventral one is commonly divided into two prongs by a median saddle. Saddles are typically but not invariably rounded

3.3.3.5.1.2 **Ceratitic**: These have strong rounded saddles and serrated lobes [Fig. 3.13(3)]

3.3.3.5.1.3 **Ammonititic**: Ammonitic sutures have complex lobes and saddles [Fig. 3.13(4)]
Although, there are two more types. These are quite rare and include

3.3.3.5.1.4 **Orthoceratitic**: These are relatively simple sutures with shallow (broadly undulating and gently rounded) lobes and saddles [Fig. 3.14(1)]

3.3.3.5.1.5 **Agoniatitic**: These have broad lateral lobes and saddles with a characteristic narrow mid ventral lobe [Fig. 3.14(2)].

Overall an increasing sutural complexity is noted from simple sutures with shallow lobes and saddles (*Goniatites*) to those where lobes and saddles are divided into second-order and third-order lobes and saddles in *Ammonites* (Fig. 3.15).

3.3.4 Shell Shape

The cephalopod shell shapes are varied, ranging from being coiled in one plane (planispirally coiled), to open spirals, called Heteromorphs (Fig. 3.16). Various shell shapes are illustrated in Fig. 3.16 and briefly described below

3.3.4.1 **Orthocone**: These are straight shells [Fig. 3.16(1)]

3.3.4.2 **Cyrtocone**: These are curved shells that complete less than one whorl [Fig. 3.16(2)]
For both Orthocones and Cyrtocones, the longer ones are called Longicones [Fig. 3.16(1–2)] and the shorter ones are Brevicones [Fig. 3.16(3–4)]

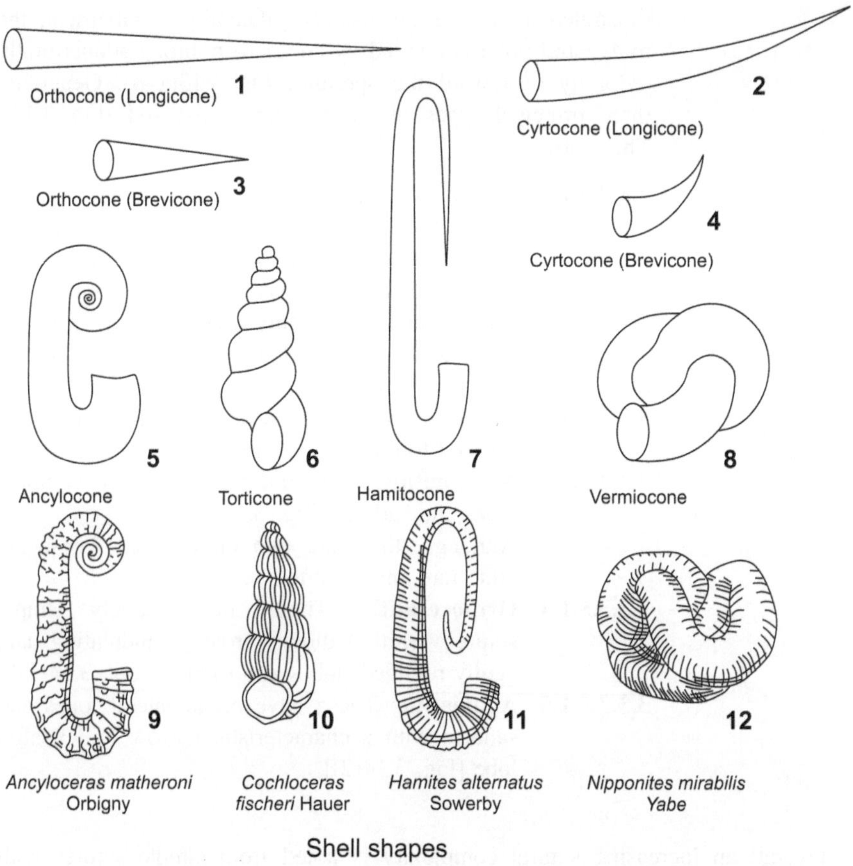

Fig. 3.16 Shell types along with relevant examples of open spiral forms (the Heteromorphs)

3.3.4.3 **Ancylocone**: They possess an open or closed, planispiral or helical coiled early whorls, followed by a hook [Fig. 3.16(5, 9)]

3.3.4.4 **Torticone**: They have helical whorls [Fig. 3.16(6, 10)]

3.3.4.5 **Hamitocone**: These shells form two or more straight shafts [Fig. 3.16(7, 11)]

3.3.4.6 **Vermicone**: These are irregular, worm-like shells [Fig. 3.16(8, 12)].

3.3.5 Shell Outline

Shell outlines vary considerably (Fig. 3.17). These include

3.3.5.1 **Platycone**: Spiral with flat whorl sides (flanks). Example: *Meekoceras* [Fig. 3.17(1)]

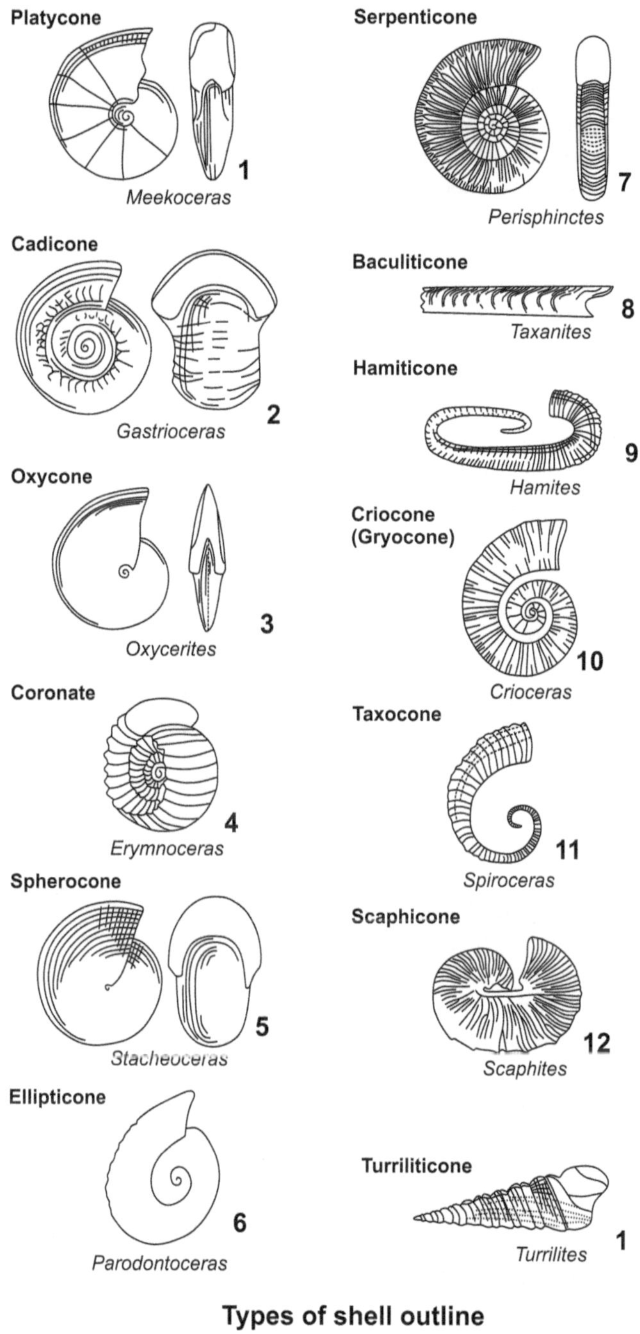

Fig. 3.17 Types of shell outline and examples

3.3.5.2 **Cadicone**: Thick shell with low whorls. Example: *Gastrioceras* [Fig. 3.17(2)]

3.3.5.3 **Oxycone**: Disc-like shape. Example: *Oxycerites* [Fig. 3.17(3)]

3.3.5.4 **Coronate**: The whorls are vertically flattened to give a crow-like look. Example: *Erymnoceras* [Fig. 3.17(4)]

3.3.5.5 **Spherocone**: Shell has globular outline. Example: *Stacheoceras* [Fig. 3.17(5)]

3.3.5.6 **Ellipticone**: Oval or egg-shaped outline caused by the bending of the last part of the outer most whorl. Example: *Parodontoceras* [Fig. 3.17(6)]

3.3.5.7 **Serpenticone**: These are closed spiral forms whose whorls touch each other. Example: *Perisphinctes* [Fig. 3.17(7)]

3.3.5.8 **Baculiticone**: These are straight forms. Example: *Taxanites* [Fig. 3.17 (8)]

3.3.5.9 **Hamiticone**: These are single or double hairpin forms. Example: *Hamites* [Fig. 3.17(9)]

3.3.5.10 **Criocone (Gryocone of old literature)**: These are open spiral, i.e., a loosley coiled shell whose successive whorls are not in contact with each other. Example: *Crioceras* [Fig. 3.17(10)]

3.3.5.11 **Taxocone**: Very open spiral. Example: *Spiroceras* [Fig. 3.17(11)]

3.3.5.12 **Scaphicone**: Flat spiral forms with a separated body chamber (the straight part of the hook). Example: *Scaphites* [Fig. 3.17(12)]

3.3.5.13 **Turriticone**: 3D spiral forms sometimes open or ending in a separated body chamber. Example: *Turrilites* [Fig. 3.17(13)].

3.3.6 Shell Form

3.3.6.1 **Heteromorphic**: These non-planispirally coiled shells are characterized by the detachment of the body chamber from the rest of the spiral (Pojeta 1987) [see Fig. 3.18(1, 3)]. This detachment led to the formation of very aberrant forms such as the Cretaceous *Nipponites* [=a very long tubular shell coiled in a series of U-bends into a tangle; Fig. 3.16(2)]. Such aberrant forms (=Hetermorphism) was first noted in the Triassic, with diversification in the Jurassic and acme during the Cretaceous. In general, the Triassic is characterized by simpler forms, with more complex ones in the Jurassic and Cretaceous (Lehmann 1981; Westermann 1996)

3.3.6.1.1 **Heteromorph**: Example: *Nipponites mirabilis* Yabe [Figs. 3.16(12), 3.18(2)]

3.3.6.1.2 **Open heteromorph**: Example: *Hamites alternatus* Sowerby [Fig. 3.18(5)]

3.3.6.1.3 **Part open-plan heteromorph**: Example: *Ancycloceras matheroni* Sowerby [Fig. 3.16(9)]

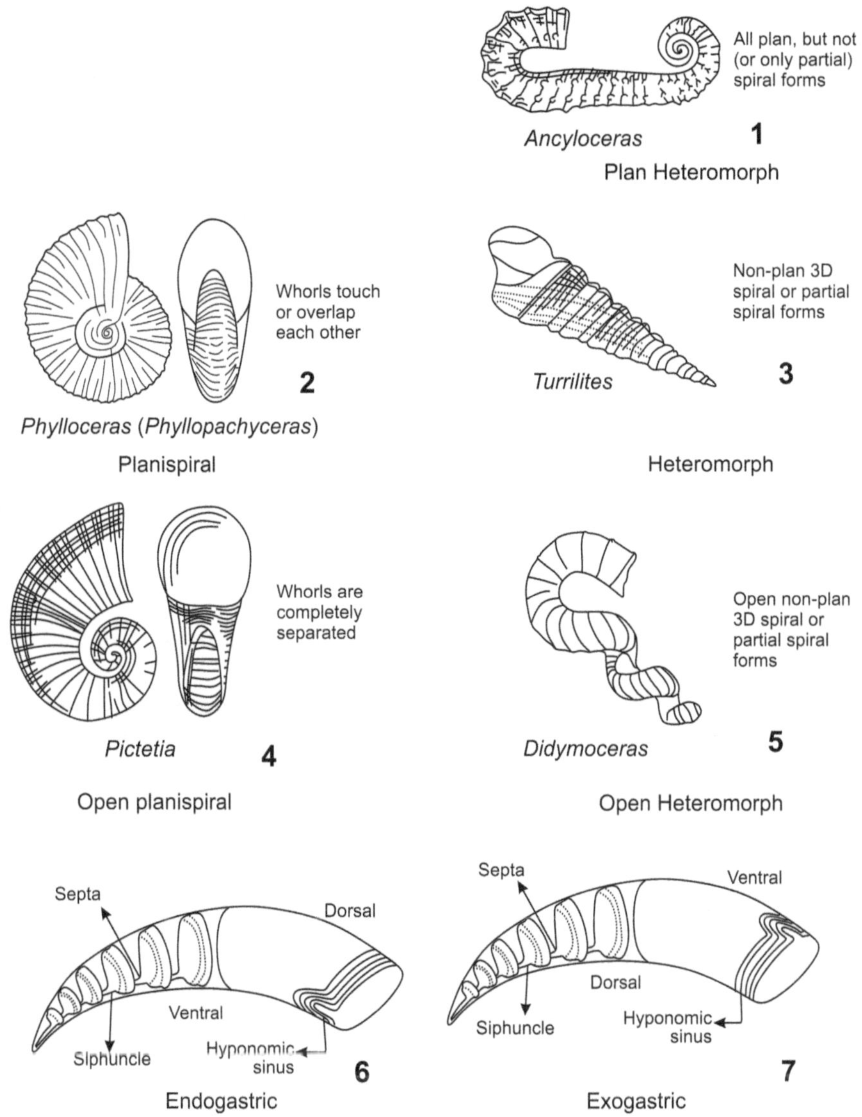

Fig. 3.18 *1–5* Shell forms. *6* Endogastic shell. *7* Exogastic shell

3.3.6.2 Planispiral: This category includes ammonites ranging from Cadiconic (Fig. 3.17(2); *Meekoceras gracilitate* White) to Serpenticonic (Fig. 3.17 (5); *Perisphinctes tiziani* Oppel)

3.3.6.2.1 Planispiral: Example: *Meekoceras gracilitate* White [Fig. 3.18(2)]

3.3.6.2.2 Open-plan spiral: Example: *Crioceras* [Fig. 3.18(7)]

3.3.6.2.1.1 Endogastric: These are coiled or curved shells whose dorsal side, or Dorsum, is convex and on the outer side [Fig. 3.18(6)]

3.3.6.2.1.2 Exogastric: These are coiled or curved shells whose ventral side, or Venter, is convex [Fig. 3.18 (4)] or whose venter is on or near the outer convex side.

3.3.7 Whorl

One complete volution (360°) of a coiled shell [Fig. 3.19(1)].

3.3.7.1 Umbilicus (U): It is the space that is enclosed on either sides by the last whorl [Fig. 3.19(1)]

3.3.7.1.1 Involute: These are shells with narrow umbilicus [Fig. 3.19(4, 5)]

3.3.7.1.2 Mesovolute: In these, the whorls overlap each other partially [Fig. 3.19(6, 7)]

3.3.7.1.3 Evolute: These are shells with wide umbilicus [Fig. 3.19(8, 9)]

3.3.7.1.4 Open: These shells have whorls that are completely open [Fig. 3.19(10–13)]

3.3.7.2 Umbilical seam: Attachment of the shell wall with the preceding whorl [Fig. 3.19(3)]

3.3.7.3 Umbilical wall: It is the portion between umbilical shoulder and umbilical seam [Fig. 3.19(1–3)]

3.3.7.4 Umbilical shoulder: The shell wall (high or low) bends toward the preceding whorl [Fig. 3.19(1–3)]

3.3.7.5 Ventrolateral shoulder: The shell bends (steep or sloping) toward the venter [Fig. 3.19(1–3)]

3.3.7.6 Venter: The underside of an organism's shell, marked by the hyponomic sinus and often by the conchal furrow [Fig. 3.19(1, 2)]

3.3.7.7 Flank or Side: It is the area between the ventrolateral shoulder and the umbilical shoulder [Fig. 3.19(1)]

3.3.7.8 Diameter (D): It is the maximum length of the shell [Fig. 3.19(2)]

3.3.7.9 Whorl thickness (T): It is the maximum width of the shell [Fig. 3.19(3)]

3.3.7.10 Whorl height (H): It is the maximum height of the whorl [Fig. 3.19(3)].

While describing the cephalopod shell, the parameters commonly used are: Diameter (D), Whorl width (T), Whorl height (H), Umbilical diameter (U), and their ratios—coiling (U/D) and thickness (T/H). An arrow or a cross marking at the last septum, if visible, indicates the end of the phragmocone or the beginning of the body chamber [Fig. 3.19(1–3)].

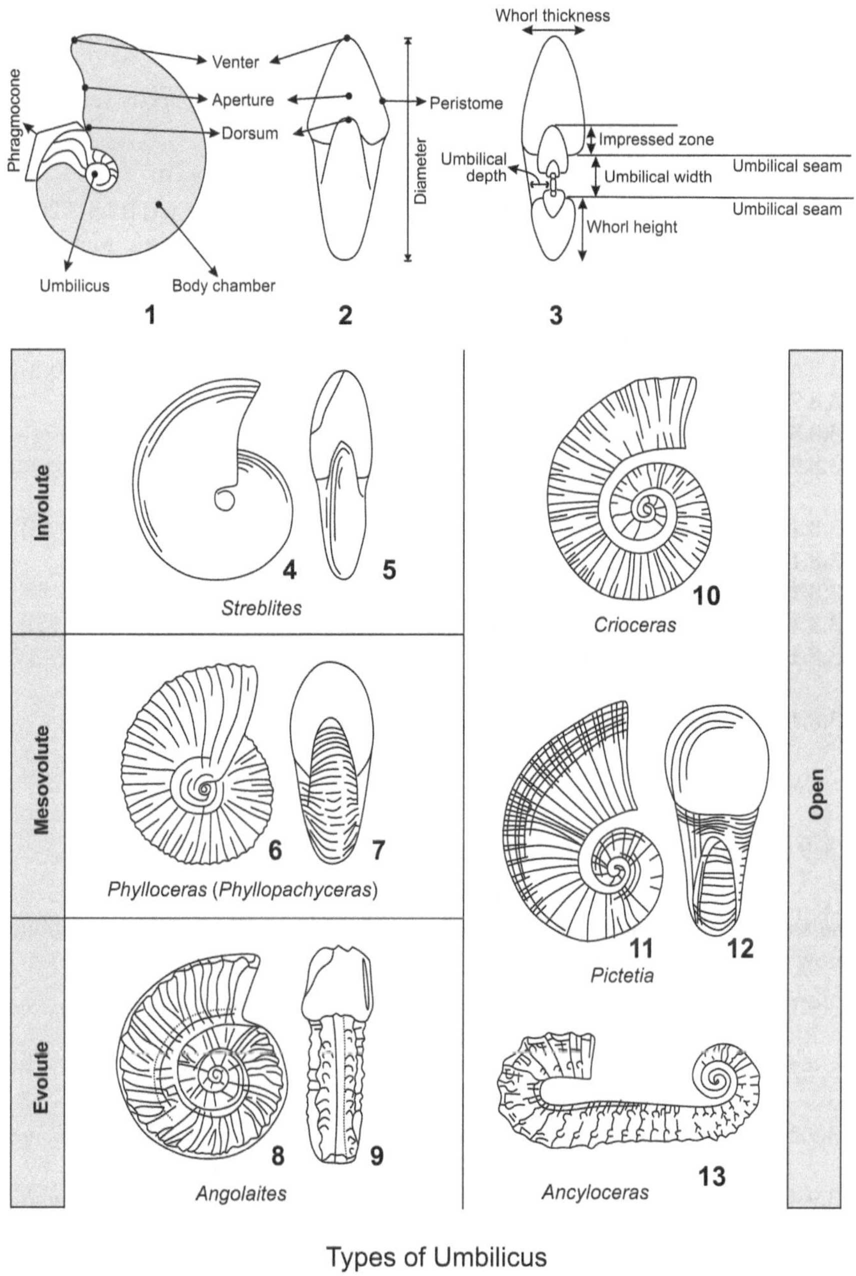

Fig. 3.19 *1–3* Terminology used to describe shell. *4–13* Types of Umbilicus and examples

3.3.8 Types of Whorl Section

A whorl section is the cross sectional shape of the whorl. This shape is very varied [Fig. 3.20(1, 2)] and includes

3.3.8.1 **Round**: Example: *Liroceras liratum* Girty [Fig. 3.20(3)]
3.3.8.2 **Ellipse**: Example: *Sporadoceras milleri* (Flower and Caster) [Fig. 3.20(4)]
3.3.8.3 **Wide-ellipse**: Example: *Parajaubertell kawakitana* Matsumoto [Fig. 3.20(5)]
3.3.8.4 **Square**: Example: *Keyserlingites subrobustus* (Mojsisovics) [Fig. 3.20(6)]
3.3.8.5 **High-square**: Example: *Prouddenites primus* Miller [Fig. 3.20(7)]
3.3.8.6 **Wide-square**: Example: *Bihenduloceras gregoryi* Spath [Fig. 3.20(8)]
3.3.8.7 **Octagonal**: Example: *Saghalinites cala* Forbes [Fig. 3.20(9)]
3.3.8.8 **Trapezoid**: Example: *Tropites subbulatus* (Hauer) [Fig. 3.20(10)]
3.3.8.9 **Reverse trapezoid**: Example: *Plesiovascoceras santoni* (Reeside) [Fig. 3.20(11)]
3.3.8.10 **Triangular**: Example: *Girthiceras pernodosum* Diener [Fig. 3.20(12)]
3.3.8.11 **Peak**: Example: *Prodromites primus* Miller [Fig. 3.20(13)]
3.3.8.12 **Egg**: Example: *Epengonoceras dumbli* (Cragin) [Fig. 3.20(14)]
3.3.8.13 **Reverse egg**: Example: *Cooperoceras texanum* Miller [Fig. 3.20(15)]
3.3.8.14 **Lancet (Lanceolate)**: Example: *Hudlestonia affinis* (Seebach) [Fig. 3.20 (16)]
3.3.8.15 **Gothic**: Example: *Inyoites oweni* Hyatt and Smith [Fig. 3.20(17)].

3.3.9 Types of Venter

The shape of the ventral side (the Venter) is also varied. The types are mentioned below and illustrated in Fig. 3.21.

3.3.9.1 **Carinate**: Shell has a marked keel. Example: *Tropites subbulatus* (Hauer) [Fig. 3.21(1)]
3.3.9.2 **Fastigate**: Shell has a triangular keel. Example: *Sublunuloceras lariense* Waagen [Fig. 3.21(2)]
3.3.9.3 **Lanceolate**: Shell has a flat or no keel at all. Example: *Dipoloceras* (*Oxytropidoceras*) *roissyi* D'Orbigny [Fig. 3.21(3)]
3.3.9.4 **Tabulate**: Shell has a tabulate keel. Example: *Stenopoceras dumblei* (Hyatt) [Fig. 3.21(4)]
3.3.9.5 **Sulcate**: Shell has a sulcus in the middle of the keel. Example: *Pseudacompsoceras vectense* Spath [Fig. 3.21(5)]
3.3.9.6 **Tabulate-Sulcate**: Shell has a flat ventral side with a single groove. Example: *Sharpeiceras schluteri* Hyatt [Fig. 3.21(6)]
3.3.9.7 **Tricarinate-bisulcate**: Shell has a flat ventral side with three keels and two grooves. Example: *Delecticeras delectum* Arkell [Fig. 3.21(7)]

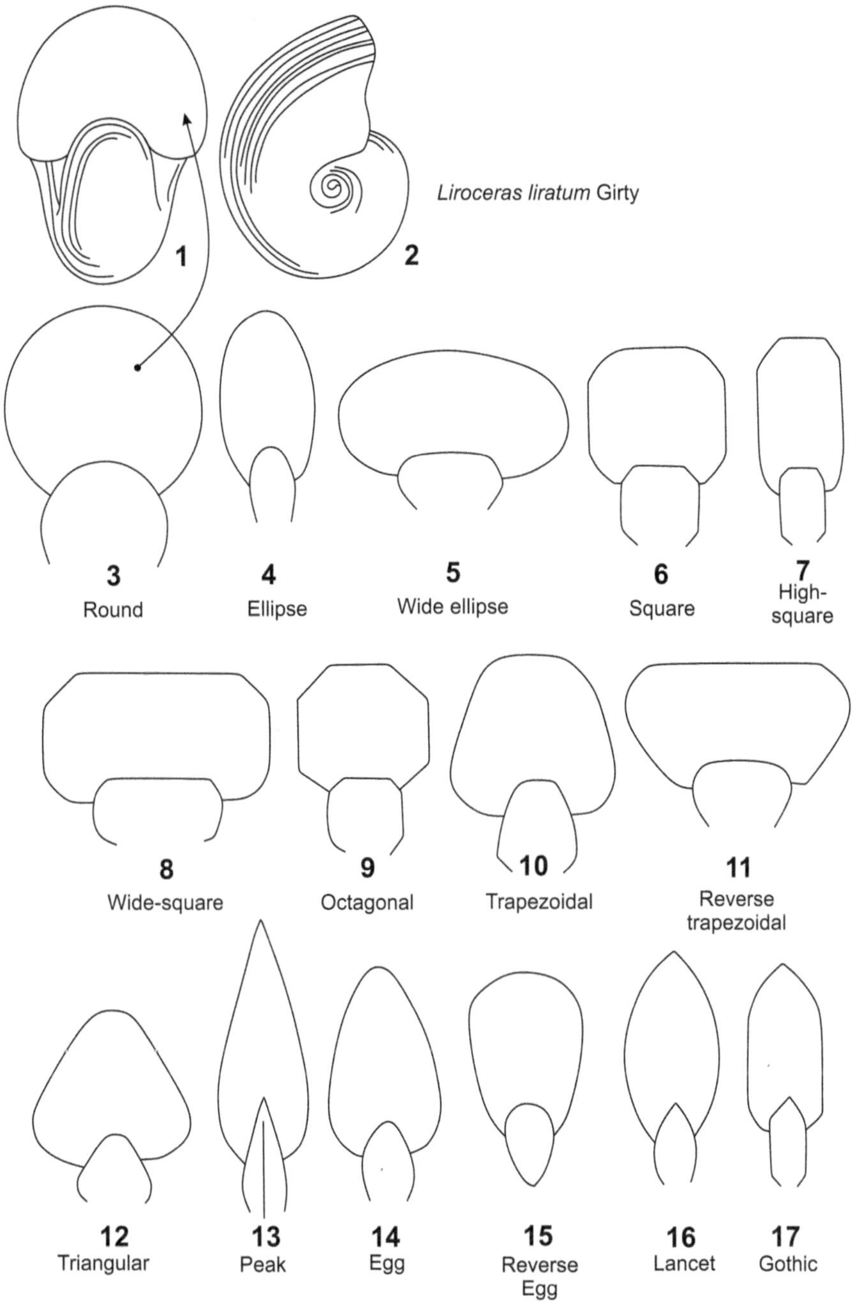

Fig. 3.20 Types of whorl section

3.3.9.8 **Concave-bicarinate**: Shell has two keels with a wide hollow groove. Example: *Otohoplites raulinianus* D'Orbigny [Fig. 3.21(8)]

3.3.9.9 **Septicarinate**: Shell has a hollow keel with a floor. Example: *Haugia variabilis* (D'Orbigny) [Fig. 3.21(9)]

3.3.9.10 **Crossing ribs**: Shell has uninterrupted ribs crossing the ventral side. Example: *Erymnoceras coronatum* (Brugière) [Fig. 3.21(10)]

3.3.9.11 **Crenulate**: Shell has a keel with small teeth. Example: *Ochetoceras (Cubaochetoceras) burckhardti* (O'Connell) [Fig. 3.21(11)]

3.3.9.12 **Rope**: Shell has a keel with cord-like ornamentation. Example: *Amaltheus margaritatus* Montfort [Fig. 3.21(12)]

3.3.9.13 **Roval**: Shell has no keel, groove(s) or crossing ribs, essentially smooth and oval. Example: *Psilophyllites hagenowi* (Dunker) [Fig. 3.21(13)].

3.3.10 Ornamentation

Ornamentation in cephalopods is represented by Ribs (Fig. 3.22) and is categorized on the basis of (a) Rib direction, (b) Rib distance and combination (the distance between ribs), (c) Rib type, and (d) Rib spreading. These are explained below and illustrated in Fig. 3.22.

3.3.10.1 **Rib direction** [Fig. 3.22(1–4)]

This categorization is based on the orientation of ribs with respect to the aperture.

3.3.10.1.1 **No ribs**: The whorls are smooth. Example: *Tmaegoceras latesulcatum* (Hauer) [Fig. 3.22(1)]

3.3.10.1.2 **Prorsiradiate**: The whorls have forward projecting ribs. Example: *Inyoites oweni* Hyatt and Smith [Fig. 3.22(2)]

3.3.10.1.3 **Rectiradiate**: The whorls have straight outward projecting ribs. Example: *Epideroceras roberti* Spath [Fig. 3.22(3)]

3.3.10.1.4 **Rusrsiradiate**: The whorls have backward projecting ribs. Example: *Popanoceras bowmani* (Böse) [Fig. 3.22(4)]

3.3.10.2 **Rib distance and combination** [Fig. 3.22(5–9)]

This categorization is based on the distance between of ribs and combinations thereof.

3.3.10.2.1 **No ribs**: The whorls are smooth. Example: *Saynella clypeiformis* (D'Orbigny) [Fig. 3.22(5)]

3.3.10.2.2 **Close bundled/Close Fascilutaed**: The ribs are bundled on the dorsal side and the distance between individual bundles is close. Example: *Lithacoceras ulmense* (Oppel) [Fig. 3.22(6)]

3.3.10.2.3 **Wide bundled/Wide Fascilutaed**: The ribs are bundled on the dorsal side and the distance between individual bundles

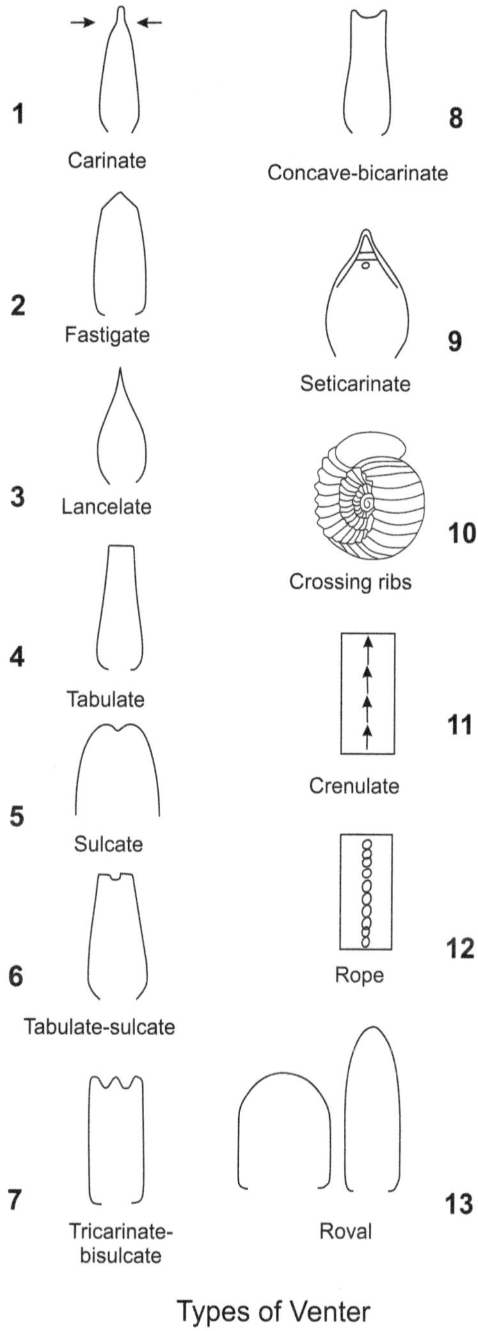

Fig. 3.21 Types of venter

is wide. Example: *Virgatosphinctes brollii* (Uhlig) [Fig. 3.22 (7)]

3.3.10.2.4 **Close**: The ribs are closely spaced. Example: *Blanfordiceras wallichi* (Gray) [Fig. 3.22(8)]

3.3.10.2.5 **Wide**: The ribs are widely spaced. Example: *Andiceras trigonostromum* Krantz [Fig. 3.22(9)]

3.3.10.3 **Rib type** [Fig. 3.22(10–19)]

This categorization is based on the type of ribs with respect to the aperture.

3.3.10.3.1 **No ribs**: The whorls are smooth. Example: *Haploceras rlimatusm* (Oppel) [Fig. 3.22(10)]

3.3.10.3.2 **Straight**: The ribs are straight with respect to the aperture. Example: *Tmetoceras scissum* (Benecke) [Fig. 3.22(11)]

3.3.10.3.3 **Concave**: The ribs are concave with respect to the aperture. Example: *Euhoploceras acanthodes* Buckman [Fig. 3.22 (12)]

3.3.10.3.4 **Proconcave**: The ribs are initially straight and the bend toward the aperture [Fig. 3.22(13)]

3.3.10.3.5 **Biconcave**: The ribs are double hollow toward the aperture. Example: *Neodimorphoceras texanum* (Smith) [Fig. 3.22 (14)]

3.3.10.3.6 **Convex**: The ribs are convex toward the aperture. Example: *Popanoceras scrobiculatum* (Gammellaro) [Fig. 3.22(15)]

3.3.10.3.7 **Biconvex**: The ribs are double convex toward the aperture [Fig. 3.22(16)]

3.3.10.3.8 **Falcoid**: The ribs form a clear wave with no or a weak wave through. Example: *Sublunuloceras lairense* (Waagen) [Fig. 3.22(17)]

3.3.10.3.9 **Falcate**: The ribs form a “reaping-hook” shape. Example: *Poecilomorphus cycloides* Buckman [Fig. 3.22(18)]

3.3.10.3.10 **Sinus/Sinuous**: The ribs are “S-shaped,” form a wave through. Example: *Campylites delmontanum* (Oppel) [Fig. 3.22(19)]

3.3.10.4 **Rib spreading** [Fig. 3.22(20–34)]

This categorization is based on the relative spread of ribs on the whorl.

3.3.10.4.1 **No ribs**: The whorl is smooth. Example: *Subpulchellia oechlerti* (Nicklès) [Fig. 3.22(20)]

3.3.10.4.2 **Single**: The ribs are single (i.e., Primary ribs). Example: *Bajocia farcyi* Brasil [Fig. 3.22(21)]

3.3.10.4.3 **Switched (Intercalated)**: Complete ribs with one or more incomplete ribs in between them (i.e., an Intercalatory). Example: *Hypacanthoplites plesiotypicus* (Fritel) [Fig. 3.22 (22)]

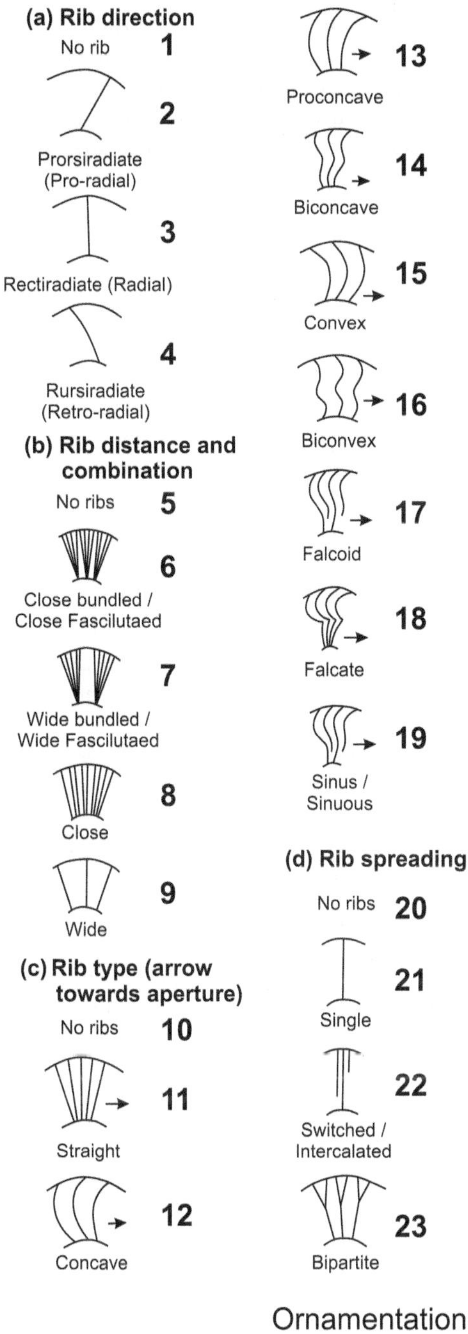

Fig. 3.22 Categorization of ribs (ornamentation) based on Rib direction (*1–4*); Rib distance (*5–9*); Rib type (*10–19*); and Rib spreading (*20–34*)

3.3.10.4.4 **Bipartite (branching into Primary and Secondary ribs)**: The ribs split into two ribs (two Secondaries). Example: *Subdichotomoceras lamplughi* Spath [Fig. 3.22(23)]

3.3.10.4.5 **Tripartite**: The ribs split into three ribs (three Secondaries). Example: *Kerberites kerbus* Buckman [Fig. 3.22(24)]

3.3.10.4.6 **Quadripartite**: The ribs split into four ribs (four Secondaries). *Campylotoxia campylotoxa campylotoxa* (Uhlig) [Fig. 3.22(25)]

3.3.10.4.7 **Multipartite**: The ribs splits into more than four ribs (more than four Secondaries). Example: *Zaraiskites zarajskensis* (Michalski) [Fig. 3.22(26)]

3.3.10.4.8 **Polyploke**: The primary rib splits into two secondaries, which split again, i.e., double splitting. Example: *Androgynoceras sparsicosta* (Trueman) [Fig. 3.22(27)]

3.3.10.4.9 **Fibulate/Looped**: The ribs are looped. Example: *Tramelliceras trachinotum* (Oppel) [Fig. 3.22(28)]

3.3.10.4.10 **Zigzag**: The ribs display a zigzag shape. Example: *Otohoplites raulinianus* (D'Orbigny) [Fig. 3.22(29)]

3.3.10.4.11 **Plicate**: The ribs are faint, folded, and radial. Example: *Ebrayiceras pseudoanceps* Buckman [Fig. 3.22(30)]

3.3.10.4.12 **Diverse plicate**: The ribs split multiple times in a certain direction. Example: *Juraphyllites mimatensis* (D'Orbigny) [Fig. 3.22(31)]

3.3.10.4.13 **Polygrate**: The ribs are straight with respect to the aperture. Example: *Orthosphinctes polygratus* (Reinecke) [Fig. 3.22 (20)]

3.3.10.4.14 **Parabolic**: It is the remainder of a former growth pause. Examples: Genera *Alligaticeras* and *Passendorferia* [Fig. 3.22(32)]

3.3.10.4.15 **Virgatome**: The ribs split into a variable number of ribs at the front side (toward the aperture) of the rib. Example: *Virgatites virgatus* (Buckman) [Fig. 3.22(33)]

3.3.10.2 **Modifications of Ornamentation**

The type of ornamentation varies ranging from faint (Lirae) to massive Spines. Various types are mentioned below and illustrated in Fig. 3.23.

3.3.10.2.1 **Lirae**: These are characterized by small transverse or longitudinal raised portions (parallel fine ridges or raised lines) on the shell's surface and are separated by the Striae [Fig. 3.23(1)]. Example: *Phyllopachyceras forbesianum* D'Orbigny

3.3.10.2.2 **Striae**: These are small longitudinal grooves. They are parallel, small to minute grooves or channels (either transverse or longitudinal) on the surface of the shell,

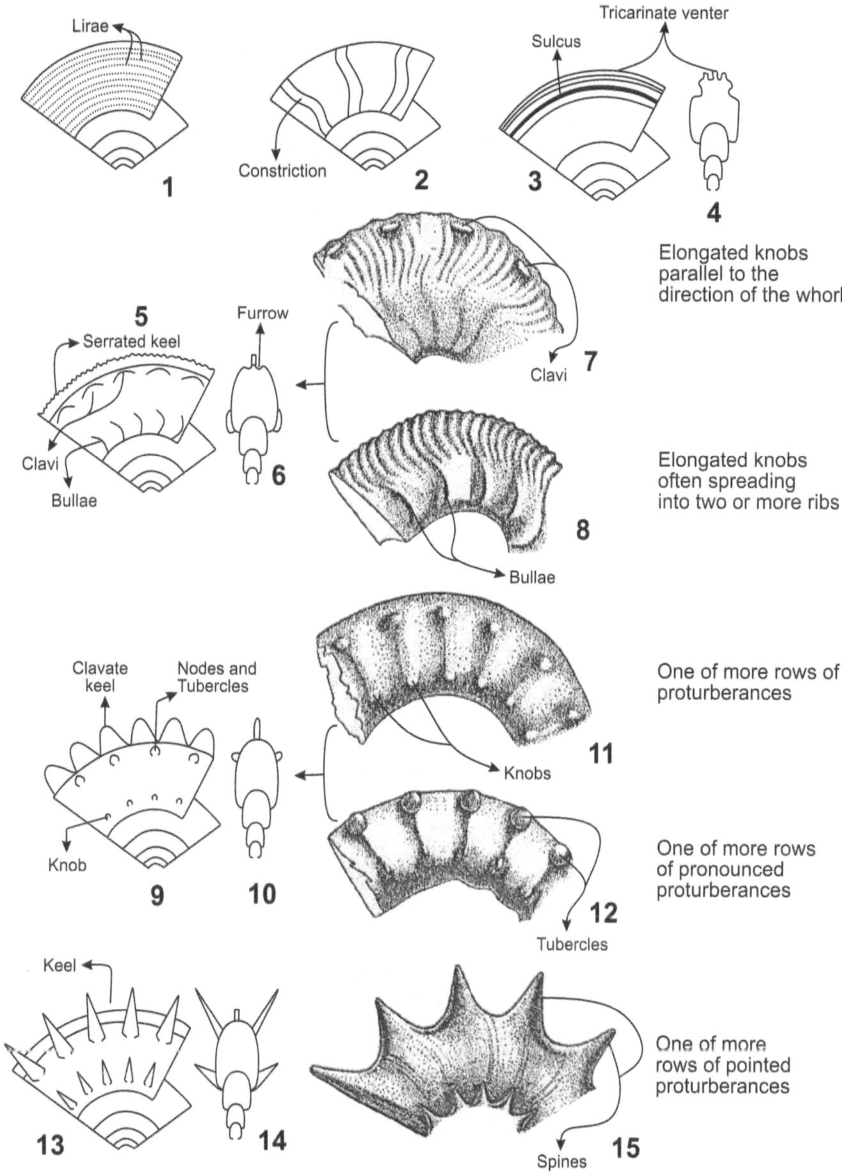

Fig. 3.23 Types of shell ornamentation

separated by Lirae. If strong enough, they are visible on the internal mold also and form ribs. Example: *Falciferella millbournei* Casey

3.3.10.2.3 **Tubercles**: One or more rows of pronounced protuberances. Example: *Collotia fraasi* (Oppel) [Fig. 3.22(12)]

3.3.10.2.4 **Nodes**: On an internal mold, the base of a spine, is a Node [Fig. 3.22(9–12)]. Some have considered them as large, blunt, or formless tubercle. Example: *Mosjsisoviczia ventanillensis* (Gabb)

3.3.10.2.5 **Knobs**: One or more rows of protuberances [Fig. 3.22(9–12)]. Example: *Pseudaspidoceras footeanum* (Stoliczka)

3.3.10.2.6 **Spines**: One or more rows of pointed protuberances [Fig. 3.22(13–15)]. Spines are initially hollow on the peristome but are sealed later as the shell grows. Example: *Epaspidoceras* (*E.*) *subdistractum* (Waagen)

3.3.10.2.7 **Bullae**: These are elongated knobs often spreading into two or more ribs. In other words, the Tubercles when elongated radially are called Bullae [Fig. 3.22(5–8)]. Example: *Reineckeia* (*R.*) *anceps* (Reinecke)

3.3.10.2.8 **Clavi**: These are longitudinally arranged (i.e., parallel to the direction of the whorl) elongated Tubercles [Fig. 3.22 (5)]. Example: *Tramelliceras trachinotum* (Oppel)

3.3.10.2.9 **Keel**: It is a raised longitudinal ridge on the venter [Fig. 3.22(9, 13)]. It could be Entire, Serrated, or Clavate

3.3.10.2.9.1 **Entire**: The keel is smooth. Example: *Oxytropidoceras* (*O.*) *roissyanum* (D'Orbigny)

3.3.10.2.9.2 **Serrated**: The keel is serrated. Example: *Amoebites* (*A.*) *kitchini* (Salfeld) [Fig. 3.22 (5)]

3.3.10.2.9.3 **Clavate**: The keel is clavate in outline. Example: *Eudiscoceras gabbi* Hyatt [Fig. 3.22(7)]

3.3.10.2.10 **Furrow or Groove**: Sometimes a Furrow or Groove [Fig. 3.22(3, 4)] can be found on each side of the keel. Example: *Arietoceltites arietitoides* (Diener)

3.3.10.2.11 **Sulcus**: In venteral or lateral positions of the shell, a large, deep, longitudinal groove exists, called the Sulcus [Fig. 3.22(3)]. Example: *Tmetoceras scissum* (Benecke)

3.3.10.2.12 **Constriction**: Constrictions are also present in some shells and range from total absence to nine (Figs. 3.22(2) and 3.10)

3.3.10.3 Position of ornamentation

The positions of ornamentation are varied and are often distinctive characters for a particular genus; a good example of this is the late Paleozoic Nautiloid genus *Cooperoceras* (large spines). Its living relatives include the squid, cuttlefish, octopus, and nautilus. Its spines were probably either for defence, for telling each other apart or possibly for attracting a mate. Various positions of ornamentation types are mentioned below and illustrated in Fig. 3.24.

3.3.10.3.1 **No ornamentation**: The shell is smooth [Fig. 3.24(1)]

3.3.10.3.2 **Ventral**: The ornamentation is on the external side of the whorl. Example: *Lytohoplites burckhardti* (Mayer) [Fig. 3.24(2)]

3.3.10.3.3 **Ventrolateral**: The ornamentation between the external side and the flank (lateral side) of the whorl. Example: *Hemihaploceras nobilis* (Neumayr) [Fig. 3.24(3)]

3.3.10.3.4 **Lateral**: The ornamentation is on the flank (lateral side) of the whorl. Example: *Himalayites treubi* Uhlig [Fig. 3.24(4)]

3.3.10.3.5 **Umbilical**: The ornamentation is on the internal side of the whorl. Example: *Spiticeras spitiense* (Blanford) [Fig. 3.24 (5)]

3.3.10.3.6 **Ventralumbilical**: The ornamentation is on the external and internal side of the whorl. Example: *Spathiceras antipodeum* (Etheridge) [Fig. 3.24(6)]

3.3.10.3.7 **Ventroumbilical**: The ornamentation is on the internal side and between the external side and the flank of the whorl. Example: *Pseudaspidoceras footeanum* (Stoliczka) [Fig. 3.24(7)]

3.3.10.3.8 **Combined**: Ornamentation present in all combinations. Example: *Texanites texanus* (Roemer) [Fig. 3.24(8)].

3.4 Size of a Cephalopod Shell

Among the molluscs, the cephalopods are large with an average size ranging between 6 and 70 cm (including tentacles). However, the *Architeuthis* (the giant modern day squid), can measure up to 16 m (including tentacles). The Ordovician straight-shelled endocerid nautiloid *Cameroceras* [Fig. 3.32(1–3)] was 10–11 m in length, and the Cretaceous ammonoid *Pachydiscus seppenradensis*, a planispirally coiled shell, measured 3 m, and is the largest known invertebrate, with a weight close to two tons. These giants were top predators, played the same ecological role of a top predator as those of the Devonian arthrodire placoderms, Mesozoic pliosaurs and Cenozoic toothed whales.

Fig. 3.24 Position of shell ornamentation

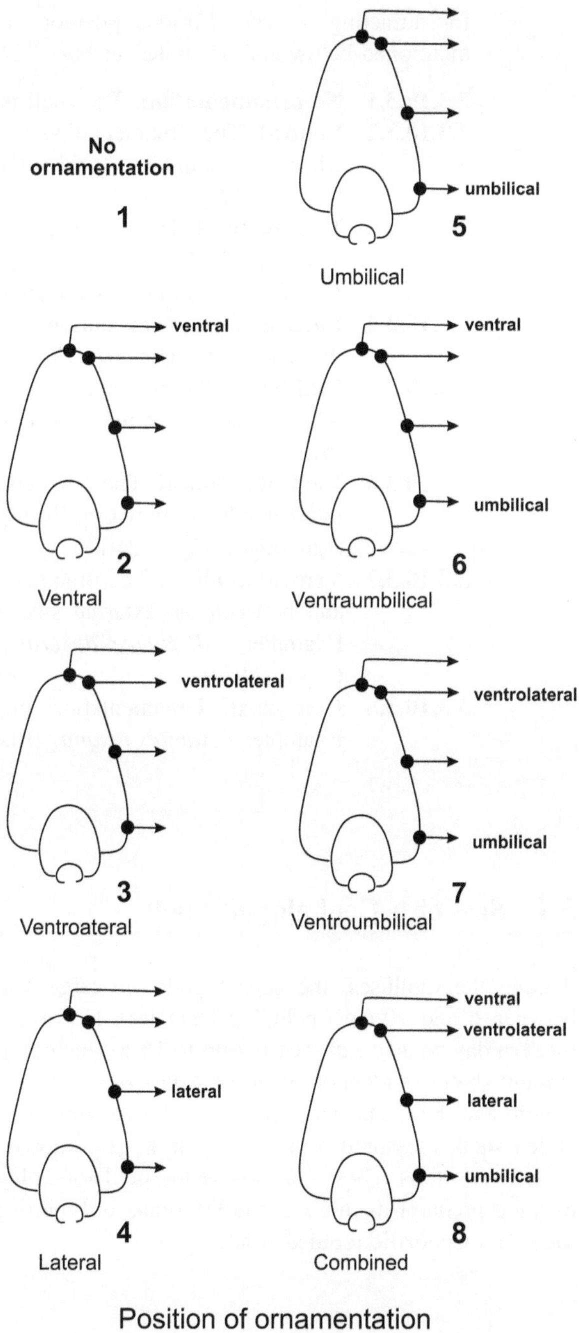

3.5 Classification

The extant cephalopods are categorized into three informal groups (after Pojeta 1987) (Fig. 3.25):

1. Those having external shells and a thin internal mantle, with as many as 94 tentacles, represented by the living genus *Nautilus* (six species).
2. Those having internal shells, a thick external mantle, and ten tentacles, represented by squids and cuttlefishes, with more than 450 living species.
3. Those having internal or no shells, a thick mantle, and eight tentacle, represented by octopuses and the paper Nautilus (*Argonaut*) with around 150 living species.

The classification of fossil cephalopods is after Arkell (1957) and Teichert (1964) (see also Moore 1957, 1964; Teichert and Moore 1964).

Class Cephalopoda

In this chapter only the major orders are discussed such as Endocerida (485–430 Ma), Actinocerida (480–312 Ma) and Bactritida (418.1–260.5 Ma) of Subclass Nautiloidea, Goniatitida, Ceratitida, and Ammonitida of Subclass Ammonoidea and Belemnitida of Subclass Coleoidea (see Table 3.8).

3.6 Geological History

Late Cambrian [$\sim$515 Ma; Fig. 3.25(1)] marks the first appearance of cephalopods with gently curved (horn shaped) shells like those of *Plectronoceras* [Fig. 3.1(1–3)]. During the Ordovician Nautiloids underwent a rapid evolutionary radiation, as the extinction of anomalocarids, a top predator, at the end of the Cambrian, provided new ecological niches for their diversification [Fig. 3.25(2)]. Tremendous diversity with varied shell forms (from long straight shells, to tightly coiled ones) is noted for the eight new Nautiloid orders that appeared during this time. This paraphyletic class Nautiloidea also showed great variety in the internal structure of the shell, primarily in the structure of the siphuncle. The early forms were slow movers, as compared to today's agile forms. They were also the ones that possessed large shells, straight shells that reached 3–5 or even 10 m in length. The Nautiloid dominated the Ordovician and Silurian seas, but by Late Devonian, gives way to the ever increasing presence of large predatory fishes. About this time the ammonoids began to take over from the nautiloids. Although rare during the early Devonian, but by the start of the Carboniferous, they proliferated. During this period of increased ammonite diversity, only two nautiloid orders persisted. Meanwhile, the Coleoidea made their first appearance in the Late Mississippian (Middle Carboniferous) but remained rare.

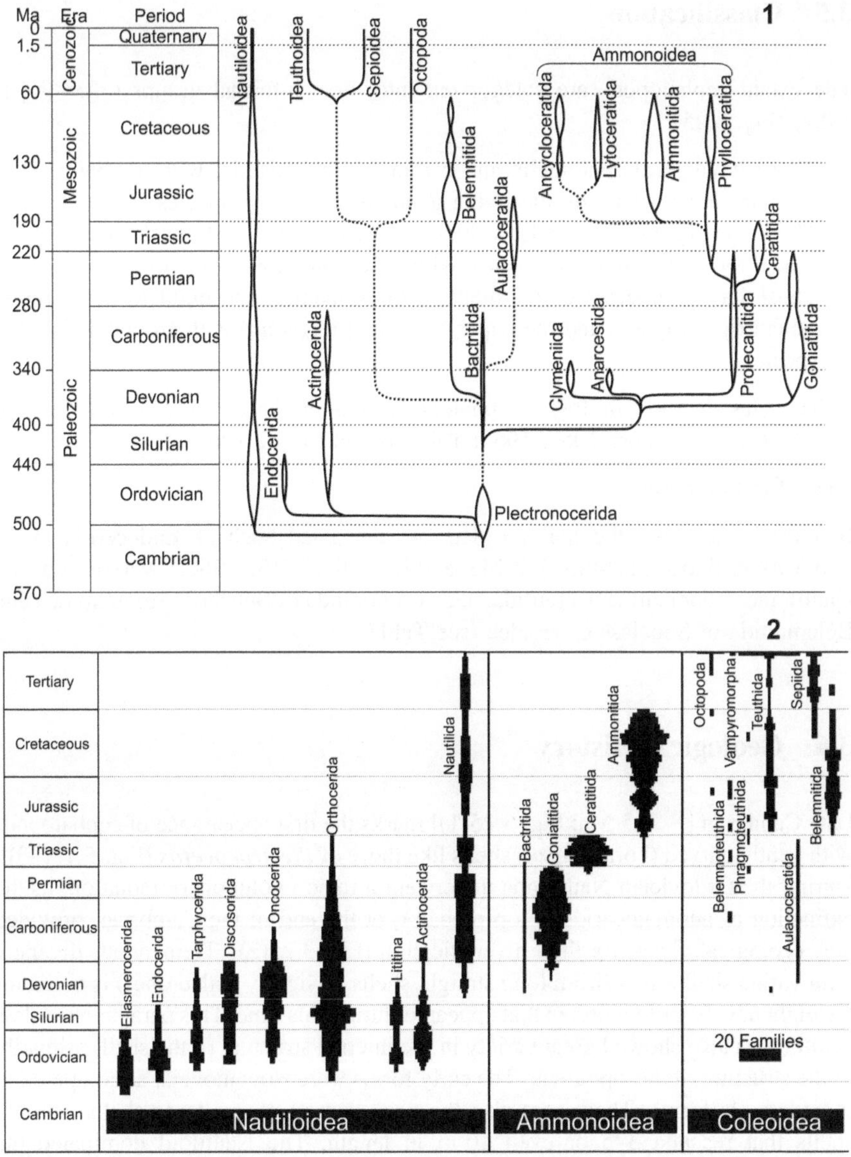

Fig. 3.25 *1* Cephalopod phylogeny. *2* Cephalopod diversity through time

Only one ammonoid family was able to survive the end-Permian extinction. However, by Mesozoic, they recovered quickly and strongly. The Ceratite lineage reached its acme (over 80 families) in the Triassic so much so that the period is called "The Age of *Ceratites*" (Table 3.8). The mass extinction at the end of the Triassic saw the final demise of the *Ceratites*, and also of the remaining

Table 3.8 Cephalopod classification

Subclass Nautiloidea
Order Plectronocerida (Cambrian)
Order Ellesmerocerida (500–470 Ma)
Order Endocerida (485–430 Ma)
Order Actinocerida (480–312 Ma)
Order Discosorida (482–392 Ma)
Order Pseudorthocerida (432–272 Ma)
Order Tarphycerida (485–386 Ma)
Order Oncocerida (478.5–324 Ma)
Order Nautilida (extant; 410.5–0 Ma)
Order Orthocerida (482.5–211.5 Ma)
Order Ascocerida (478–412 Ma)
Order Bactritida (418.1–260.5 Ma)
Subclass Ammonoidea (479–66 Ma)
Order Goniatitida (388.5–252 Ma)
Order Ceratitida (254–200 Ma)
Order Ammonitida (215–66 Ma)
Subclass Coleoidea (410.0 Ma-Rec)
Order Belemnoidea: Belemnites and kin
Genus *Jeletzkya*
Order Aulacocerida (265–183 Ma)
Order Phragmoteuthida (189.6–183 Ma)
Order Hematitida (339.4–318.1 Ma)
Order Belemnitida (339.4–66 Ma)
Genus *Belemnoteuthis* (189.6–183 Ma)

Only important groups (in bold) are discussed in this chapter

straight-shelled nautiloids (the *Pseudorthocerids*). Thereafter, new groups of ammonoids with more complex sutures ("*ammonites*" in the strict sense) took over. At the start of the Jurassic, like the ammonites, the squid-like belemnites, representing the Coleoidea, also underwent a massive evolutionary radiation. It is during the Jurassic that the first representatives of modern coleoid groups like octopus and squid, appeared. The ammonoids and belemnoids, the proto-modern-style coleoids formed a substantial part of Jurassic and Cretaceous nektonic marine ecosystem. Both the ammonoids and belemnoids proliferated until end Cretaceous. The end Cretaceous asteroid that caused the dinosaur extinction also killed of the ammonoids. However, a few belemnoids straggled on until the Eocene, but they were heavily outcompeted by the modern day Coleoidea (octopus, squid, cuttlefish, etc.), that remained an important and successful group. The once important and dominant Nautiloidea survived only through the six species of the pearly *Nautilus*.

In terms of evolutionary history, the Jurassic suborder Phylloceratina and its daughter suborder Lytoceratina, acted together as root stock for all the superfamilies from which suborder Ammonitina sprung. Each stock remained more or less unchanged, periodically budding off a subgroup. This pattern of evolution, with the main stock continuously branching sideways but remaining intact by itself is called

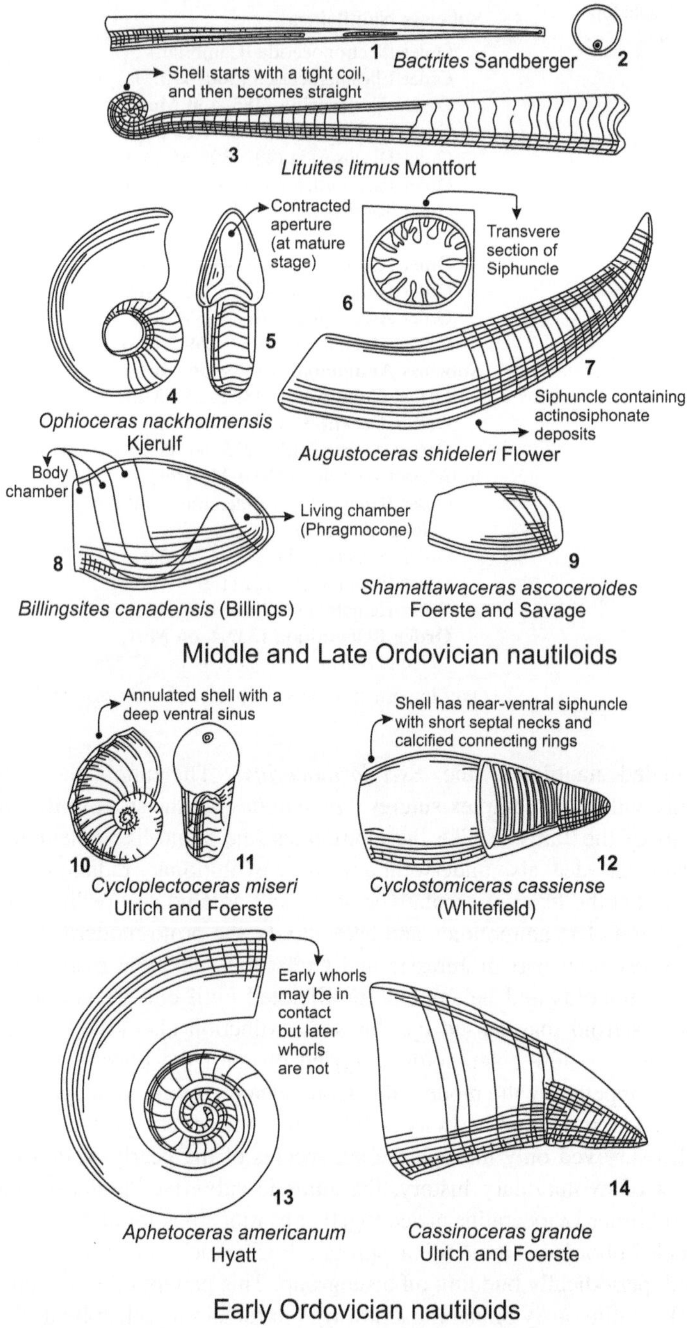

Fig. 3.26 Ordovician nautiloids and their major distinguishing characters

Iterative Evolution. Since Ammonitina is derived from two ancestral stocks, it is therefore a polyphyletic group.

3.7 Distribution Through Time

3.7.1 Subclass Nautiloidea

Salient features:

1. Age: Late Cambrian to Holocene
2. Number of genera: ~700 (Representative genera are illustrated in Figs. 3.26, 3.27, 3.28, 3.29, 3.30 and 3.31)

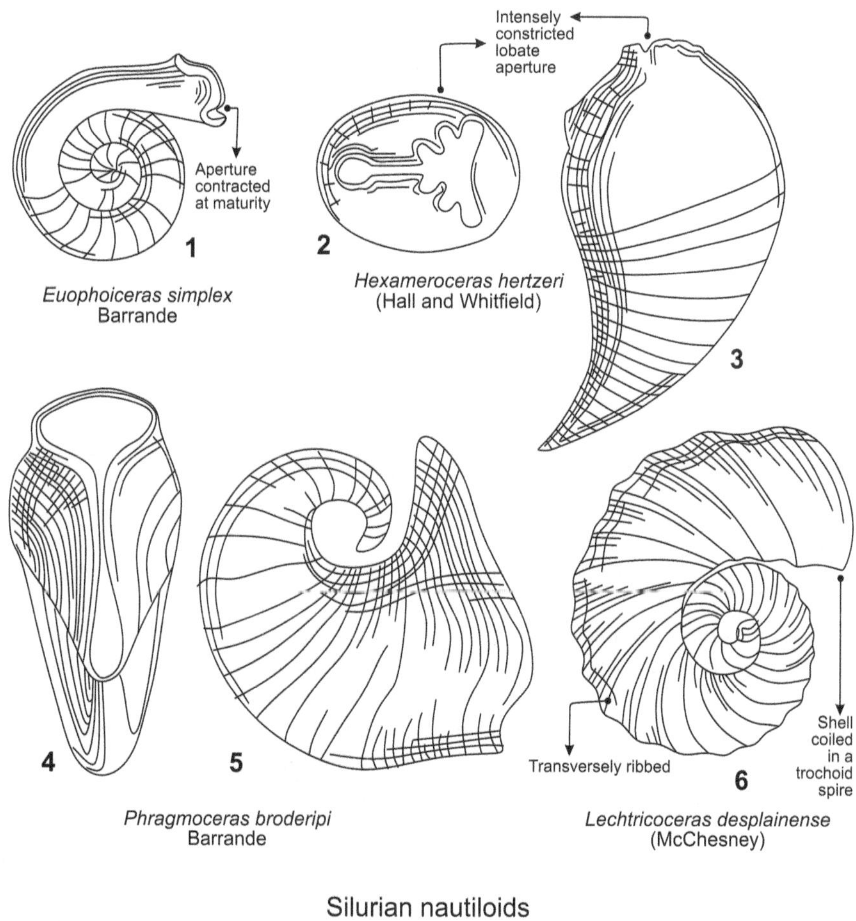

Fig. 3.27 Silurian nautiloids and their major distinguishing characters

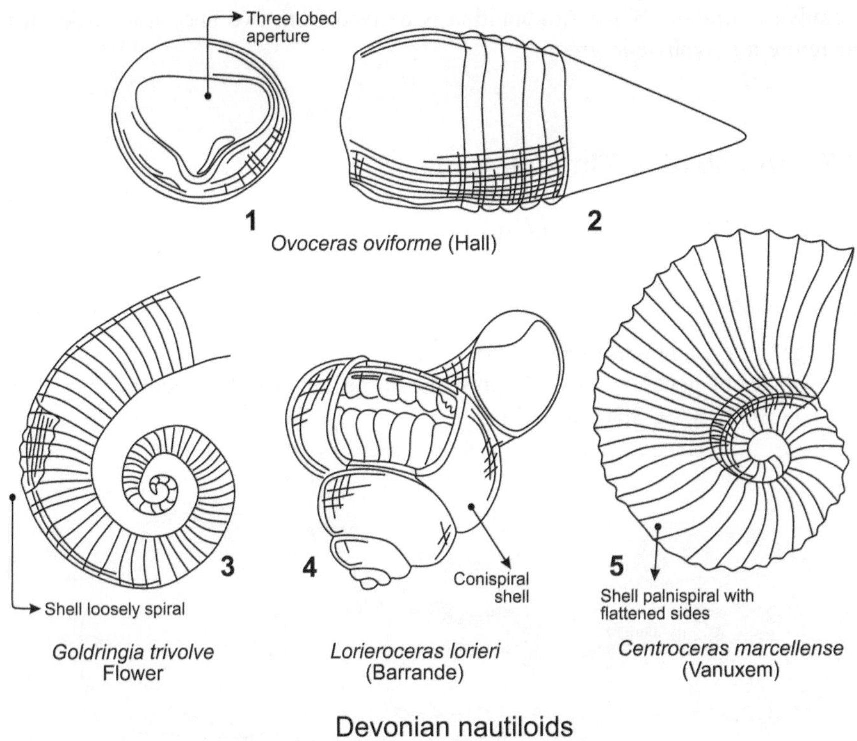

Fig. 3.28 Devonian nautiloids and their major distinguishing characters

3. Orthoconic to involute shell
4. Simple to slightly undulatory sutures
5. Retrochoanitic septal necks
6. Prominent straight connecting rings
7. Cameral deposits
8. Siphuncle central and relatively small
9. Coiled or straight in fossils; coiled in modern.

Nautiloids have orthoconic to tightly coiled planispiral shells. They started with a straight-shell (orthoconic; siphuncle running down the center of the septa) or slightly curved (Cyrtoconic) and evolved into a strongly coiled involute form as the present day *Nautilus*, somewhere during the Ordovician. Sutures are Orthoceratitic, generally simple and rarely with prominent lobes and saddles. Septal neck is Retrochoanitic. Siphuncle is not large in diameter, and is mostly marginal and ventral to subdorsal in position. The Ordovician and Silurian early nautiloids were grazers on algal mats. However, gradually they fed on arthropods, worms, and other molluscs, thereby adopting a more carnivorous habit. Nautiloids, the dominant cephalopods of the early and middle Paleozoic, declined in numbers and species since the Ordovician, barring few episodes of diversification and also extinction.

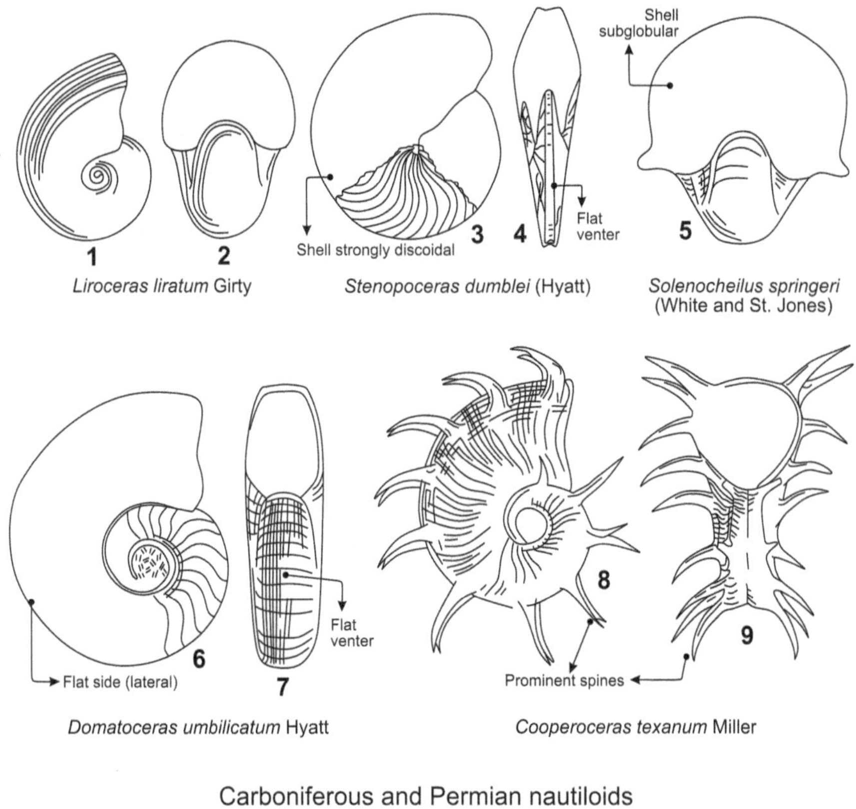

Fig. 3.29 Carboniferous and Permian nautiloids and their major distinguishing characters

One genus, *Nautilus* (a living fossil) lives today with six species restricted to the deep waters of the SW Pacific and in tropical waters form depths ranging from 5 to 550 m. *Nautilus* is a nocturnal carnivore, strong active swimmer and an excellent predator. Additionally, its ability to change position within the water column also made it successful. The squids and cuttlefishes, the ten-armed cephalopods, occur in all oceans from depths up to 3000 m, whereas the eight-armed living cephalopods (the octopuses) live at depths up to 5000 m. Both occur in all oceans. Their average size ranges between 5 cm and 10 m. They have 90 tentacles and migrate vertically, daily. They are also the only living cephalopod with an external shell.

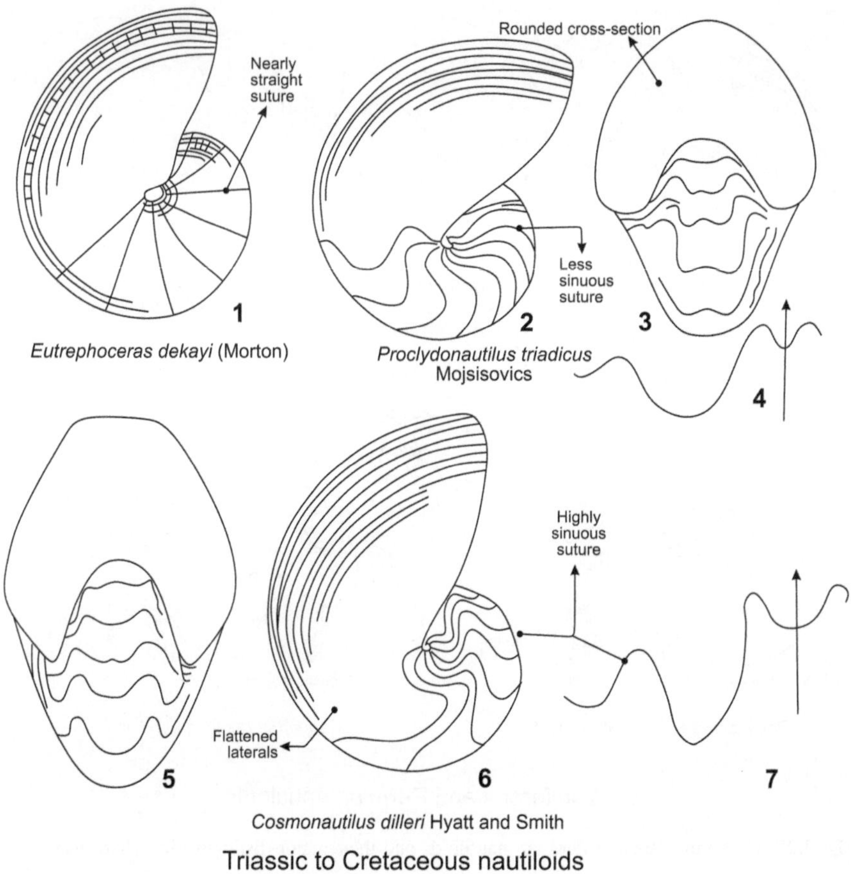

Fig. 3.30 Triassic to Cretaceous nautiloids and their major distinguishing characters

3.7.2 Order Endocerida (485–430 Ma)

Salient features

1. Age: Early Ordovician to Silurian
2. Number of genera: ~80 (Representative genera are illustrated in Fig. 3.32)
3. Straight to slightly curved
4. Simple to slightly undulatory sutures
5. Large subcentral siphuncle (up to 1/4 shell diameter)
6. Retrochoanitic septal necks
7. Prominent straight connecting ring
8. Endocones within siphuncle
9. No cameral deposits
10. Some very large forms such as Cameroceras (10 m!).

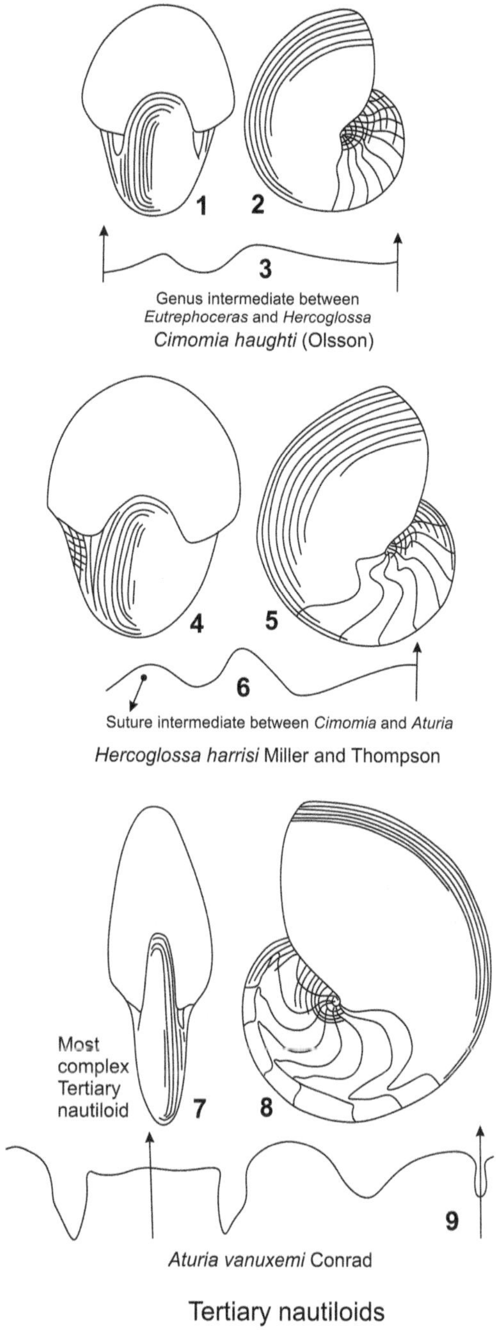

Fig. 3.31 Tertiary nautiloids and their major distinguishing characters

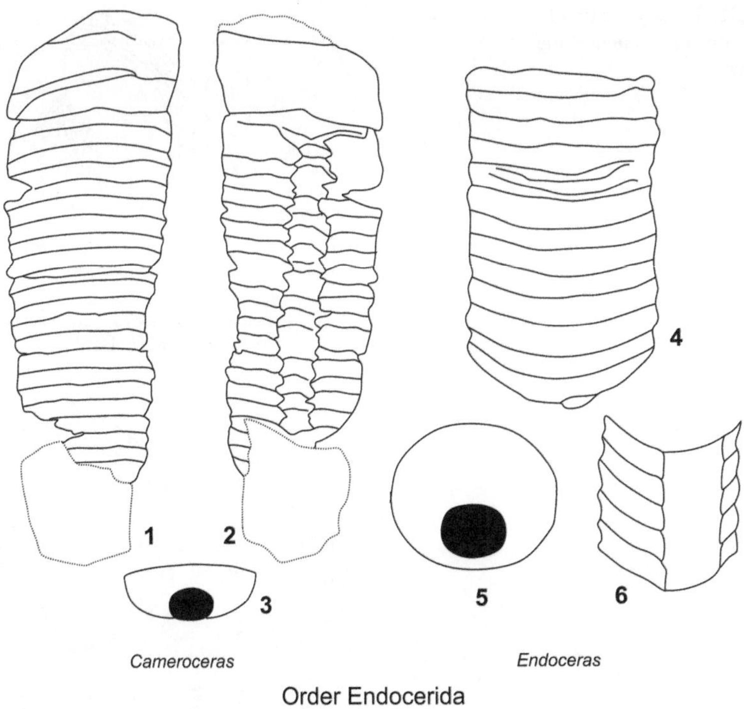

Fig. 3.32 Representative genera of Order Endocerida

These are medium to large orthoconic (rarely Cyrtoconic)-shelled cephalopods. Suture is orthoceratitic, usually simple, but less commonly having midventral lobe or saddle. Siphuncle is large; marginal to submarginal and ventral (Exogastric) in most species. Septal necks are Retrochoanitic. They were major predators in the Ordovician but died out in the beginning of Silurian. *Cameroceras* is the largest known Middle Ordovician Endoceratoid [Fig. 3.32(1–3)] that reached a length of 10 m (see also Teichert and Kummel 1960). *Endoceras* [Fig. 3.32(4–6)] attained lengths as much as 3.5 m (~ 13 ft) (see Flower 1955).

3.7.3 Order Actinocerida (480–312 Ma)

1. Age: Middle Ordovician to Late Mississippian
2. Number of genera: ~ 40 (Representative genera are illustrated in Fig. 3.33)
3. Straight to Cyrtoconic
4. Simple to slightly undulatory sutures
5. Siphuncle large with bulges between septa
6. Cameral deposits common.

Fig. 3.33 Representative genera of Order Actinocerida

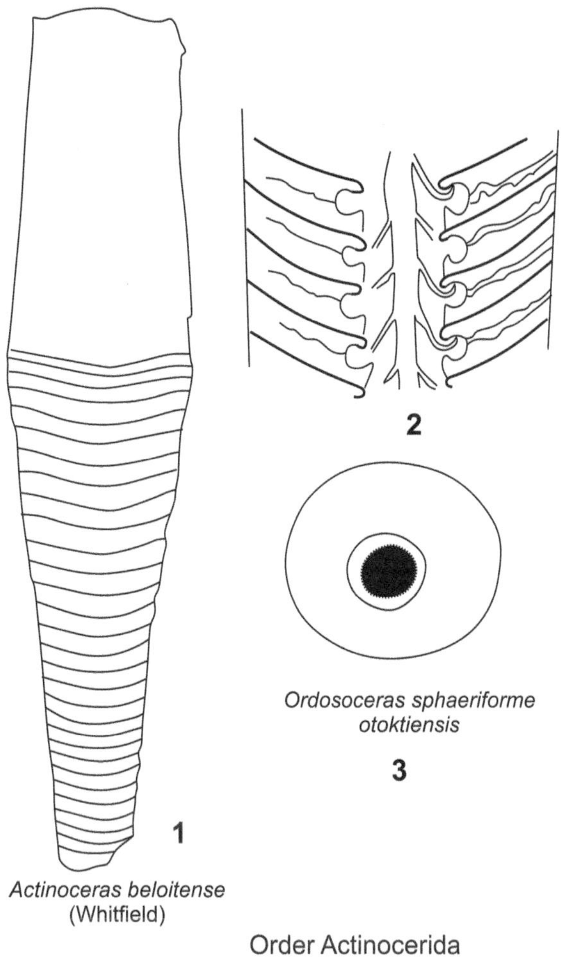

These were medium-sized to very large cephalopods having orthoconic shells with Orthoceratitic sutures. Septal necks are Retrochoanitic. Cameral deposits are present. Connecting rings are inflated and filled with siphuncular deposits. Siphuncle is large and exogastic.

3.7.4 Order Bactritida (418.1–260.5 Ma)

1. Age: Latest Silurian (or Ordovician) to Late Triassic
2. Number of genera: ~27 (Representative genera are illustrated in Fig. 3.34)
3. Seem to have come out of Nautiloidea
4. But have characteristics of later groups (esp. ammonoids)

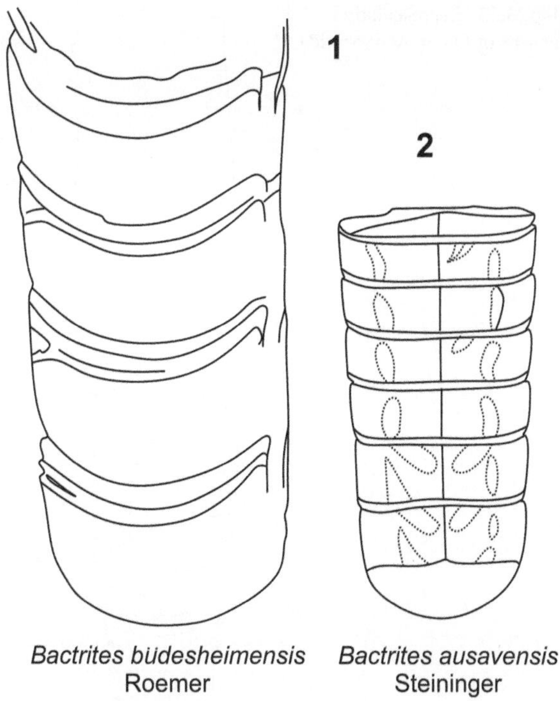

Fig. 3.34 Representative genera of Order Bactritida

5. Shell small, Orthoconic to Cyrtoconic
6. Bulbous protoconch
7. Siphuncle small and ventral, possibly prochoanitic
8. No secondary deposits.

These are small Orthoconic (or Cyrtoconic) forms with Exogastric external shells. Sutures are orthoceratitic. Septal necks are Retrochoanitic. Both siphuncular and cameral deposits are absent in most. The Bactritoides are a link between the nautiloids and the ammonoids. They gave rise to the ammonoids by evolving planispiral coiling, a homeomorphic feature, having been independently developed in various nautiloids as well as in ammonoids.

3.7.5 Subclass Ammonoidea (479–66 Ma)

1. Age: Early Devonian to Late Cretaceous
2. Number of genera: ~2000 (Representative genera are illustrated in Figs. 3.35, 3.36, 3.37, 3.38, 3.39, 3.40, 3.41, 3.42, 3.43, 3.44, 3.45, 3.46, 3.47 and 3.48)
3. Mostly planispiral
4. Biostratigraphically most useful (resolution to 1 Ma)
5. Siphuncle small, ventral, and prochoanitic

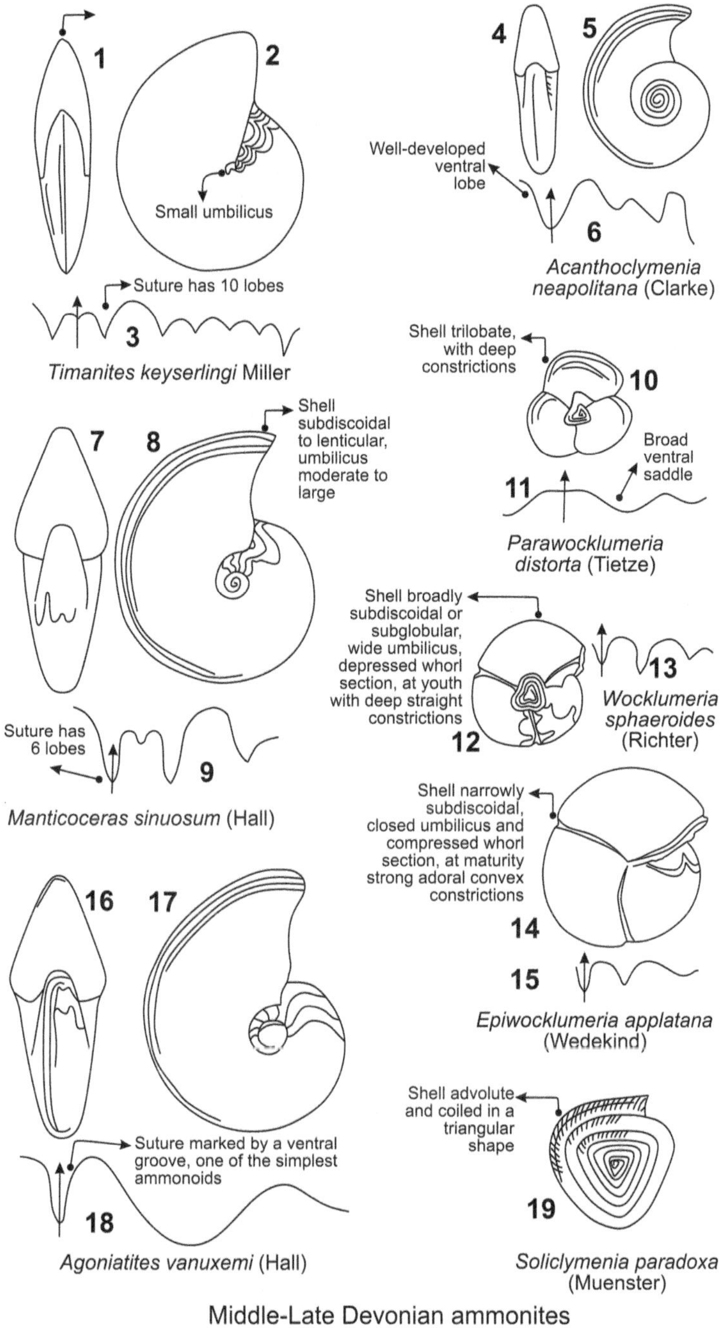

Fig. 3.35 Representative genera of Middle-Late Devonian ammonites and their major distinguishing characters

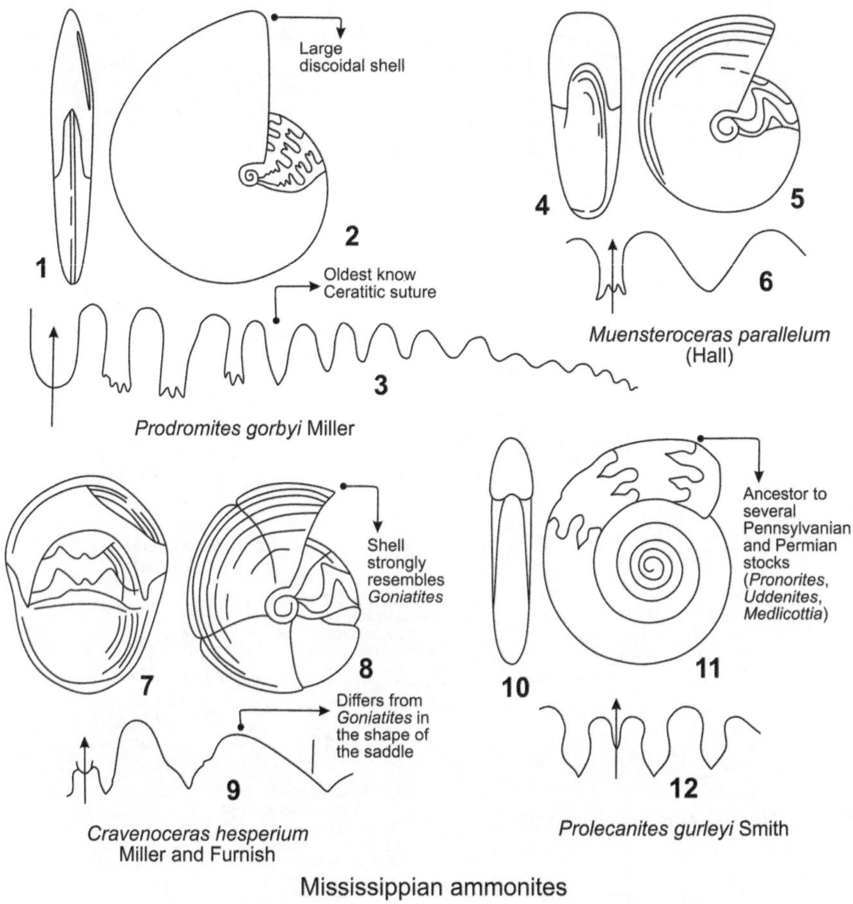

Fig. 3.36 Representative genera of Mississippian ammonites and their major distinguishing characters

6. Informally division based on sutural morphology
7. Goniatitic (in the Palaeozoic)
8. Ceratitic (in the Triassic)
9. Ammonitic (in the Jurassic-Cretaceous).

During the Devonian, these cephalopods arose from the nautiloids through an intermediate stock such as the straight-shelled bactritid nautiloids. These are Exogastric, and small to large in size, with planispirally coiled external shells

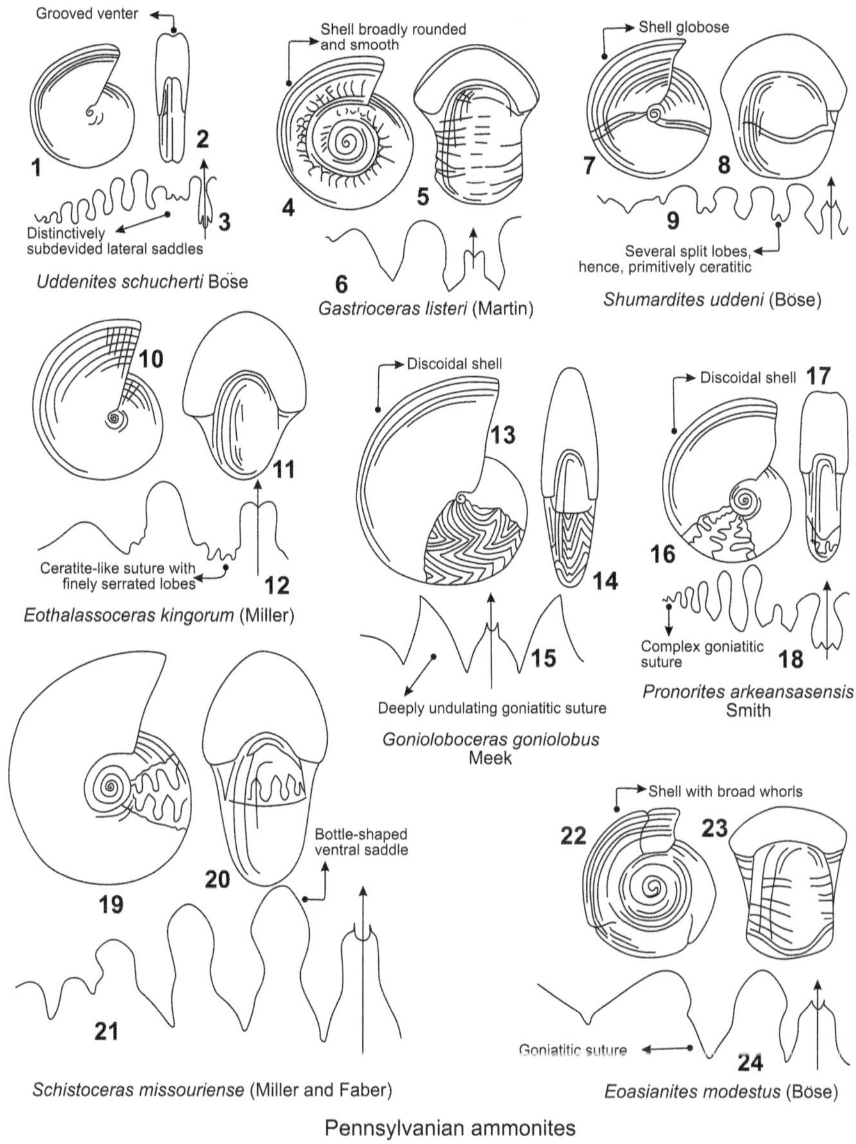

Fig. 3.37 Representative genera of Pennsylvanian ammonites and their major distinguishing characters

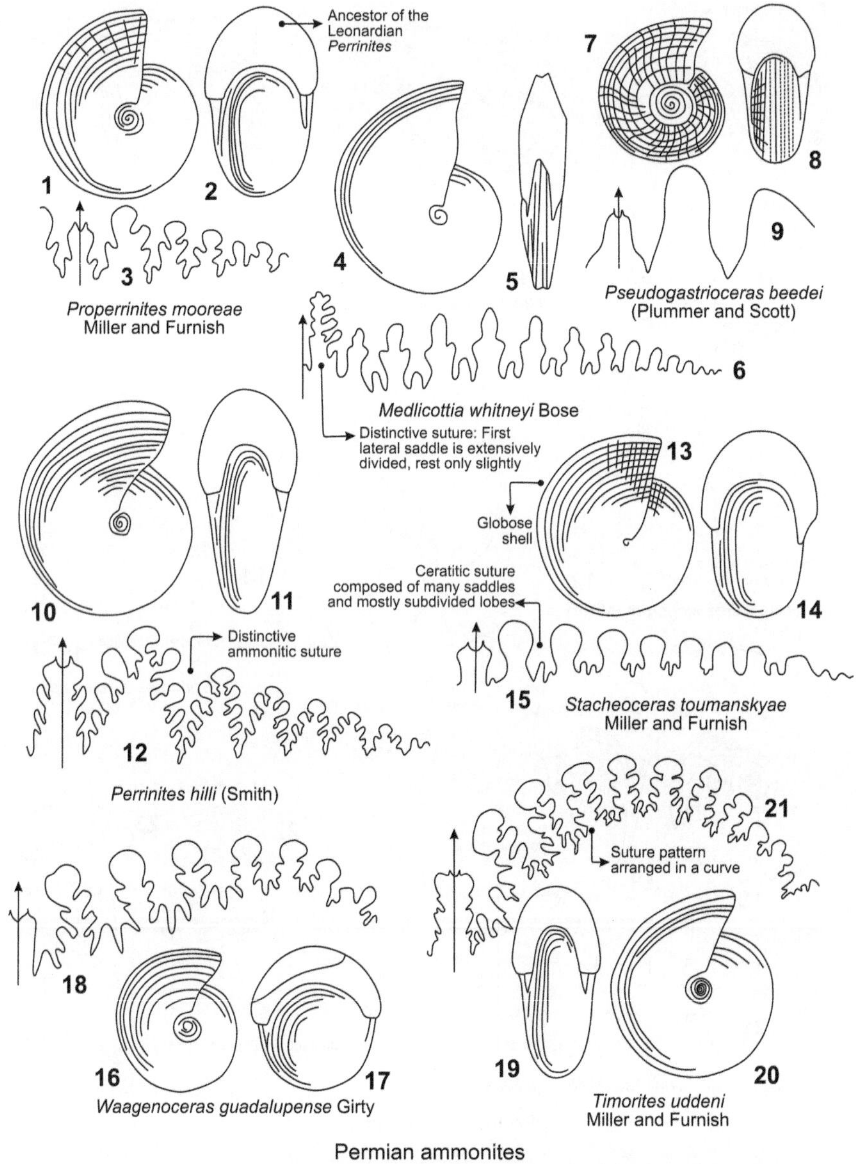

Fig. 3.38 Representative genera of Permian ammonites and their major distinguishing characters

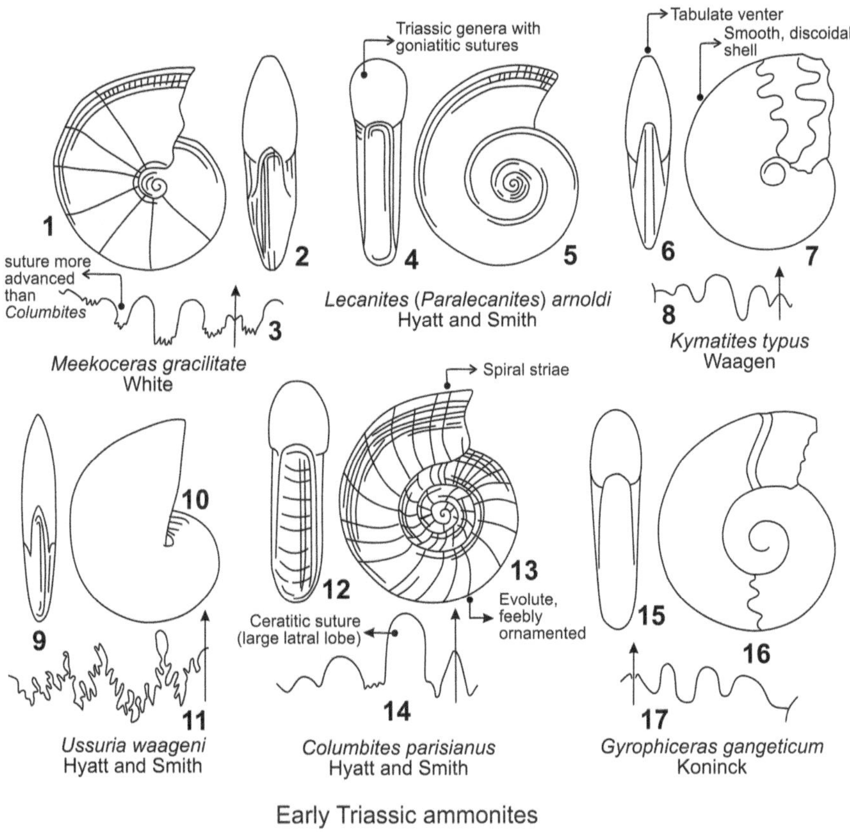

Fig. 3.39 Representative genera of Early Triassic ammonites and their major distinguishing characters

(mostly). Septal necks are most commonly prochoanitic in adult stages. Siphuncular deposits are rare and cameral deposits are absent. Siphuncle is small in diameter, usually marginal and ventral in mature stages; in most, ventral throughout ontogeny; in few, dorsal. Sutures are Agoniatitic, Goniatitic, Ceratitic, or Ammonitic. The ammonoid shells that are not planispiral are called Heteromorphs [Fig. 3.18(1, 3 and 5)]. These are recorded from Devonian sediments, but appear in significant numbers only in Late Triassic.

The shells of adult ammonoids range from 10 mm to 3 m in diameter. They show great variety in size, shape, style of coiling, thickness of shell external

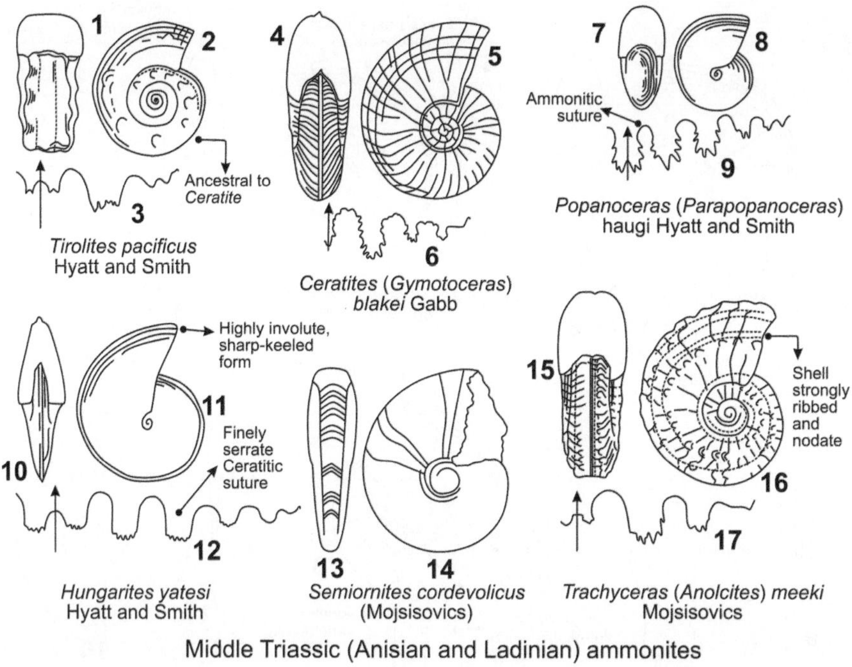

Fig. 3.40 Representative genera of Middle Triassic (Anisian and Ladinian) ammonites and their major distinguishing characters

ornament, and sutural complexity suggesting that ammonoids were well adapted to many modes of life. In Carboniferous, ammonoids with shark bite marks are recorded; in Jurassic, specimens are known with bony fish bite marks; and in Cretaceous, specimens are known with Mosasaur bite marks (Kase et al. 1998; Kauffman 2004). The Plesiosaurs may also have preyed on ammonoids.

Some ammonoids have a pair of calcareous chitinous plates called Aptychi (Aptychus: singular; Fig. 3.49). It is a heart shaped structure [Fig. 3.49(1, 2)] and is sometimes found in the final body chamber [Fig. 3.49(3–6)]; interpreted as jaws or a hood with which the creature protected itself once it withdrew into its shell. They may have closed the shell, as Operculum does so in Gastropods. Sometime they are massive [Fig. 3.49(6)]. However, their exact function is debated. In some Mesozoic rocks, they are more common than the ammonoid shells themselves.

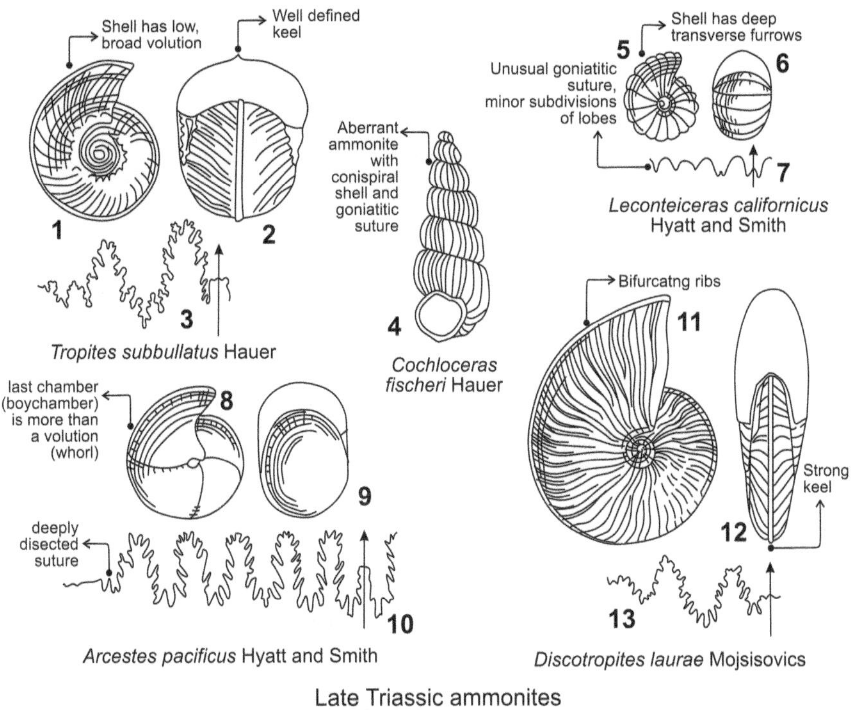

Fig. 3.41 Representative genera of Late Triassic ammonites and their major distinguishing characters

3.7.6 Subclass Coleoidea (410 Ma to Recent)

1. Age: Early Devonian to Holocene
2. Number of genera: ~250 (Representative genera are illustrated in Fig. 3.50)
3. Internal to no shell at all
4. 2 gills (instead of four as in *Nautilus*)
5. *Sepia/Spirula* (cuttlefish), squids, octopods
6. *Belemnites* (Late Carboniferous to Late Cretaceous).

Except *Nautilus*, all living cephalopods (including cuttlefish, octopus and squids) are Coleoids. They either have a highly reduced internal shell (as the Cuttle bone of the cuttlefish and the thin flexible rod as in squids) or none at all (as in

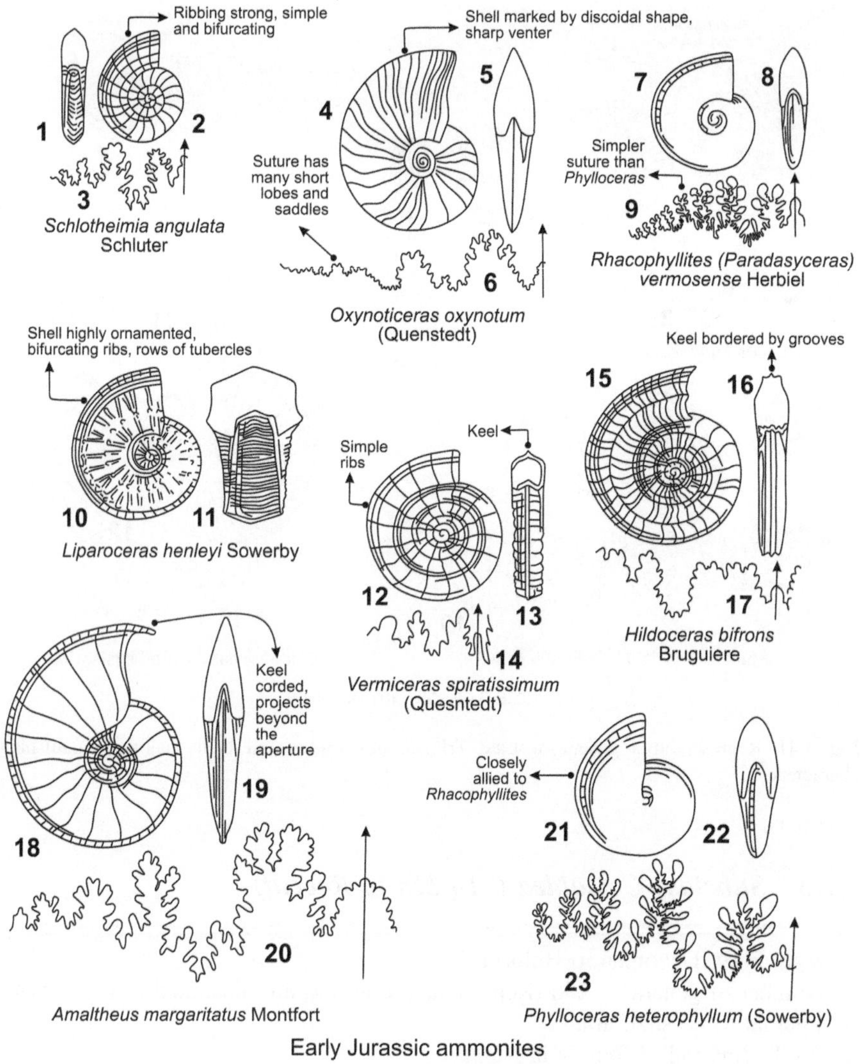

Fig. 3.42 Representative genera of Early Jurassic ammonites and their major distinguishing characters

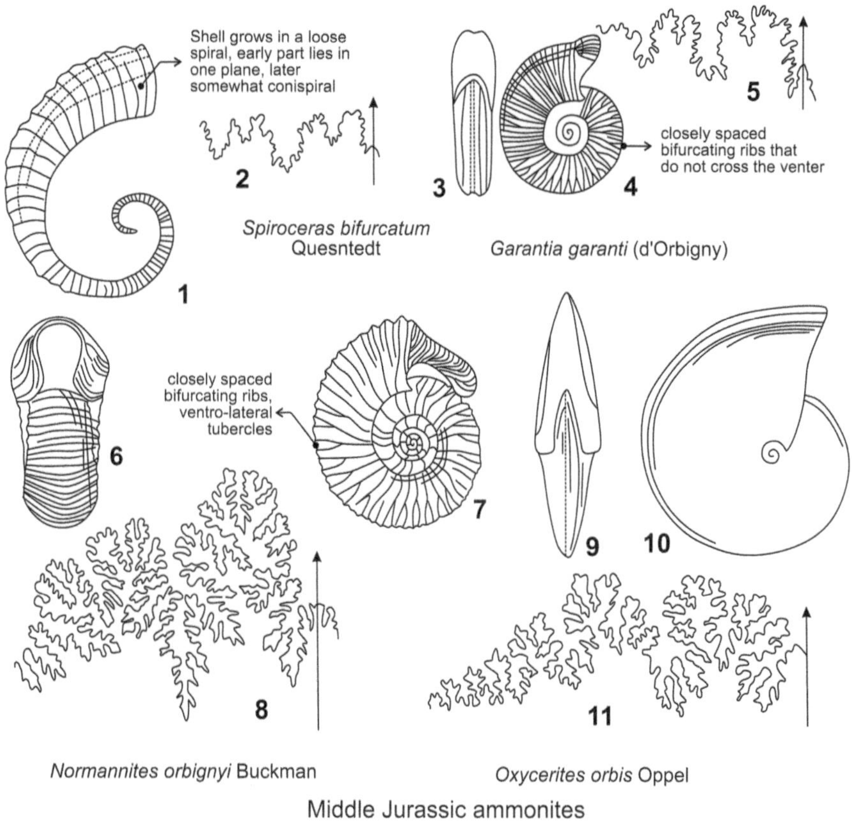

Fig. 3.43 Representative genera of Middle Jurassic ammonites and their major distinguishing characters

octopus). Hence, their bodies are not preserved well enough. Shell is orthoconic, cyrtoconic, or rarely coiled. Head has 8 or 10 tentacles. The most important group of coleoids is the order Belemnitida, which ranges from the Late Mississippian to Late Cretaceous and is particularly well represented in the Mesozoic rocks. Like the ammonoids, coleoids may have evolved from the bactritoids. The heavy internal shelled extinct belemnite, resembled squids with a cylindrical body, a head and a set of anterior arms (Fig. 3.51). Morphologically, they resemble the look of a 0.50

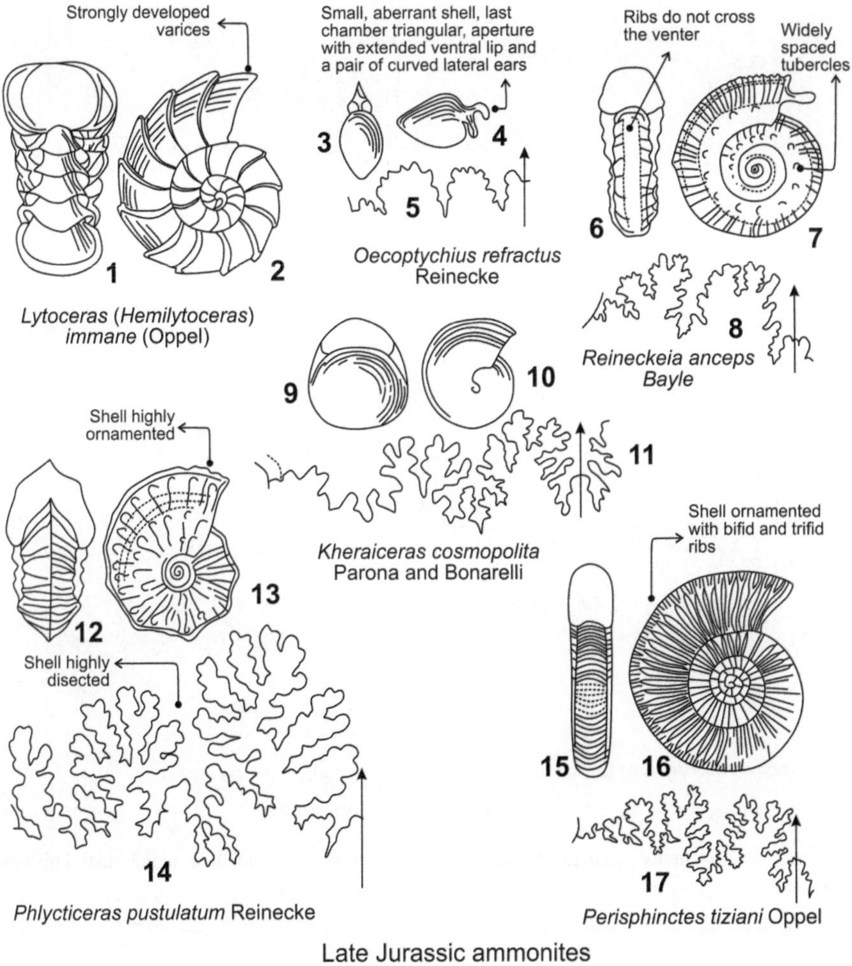

Fig. 3.44 Representative genera of Late Jurassic ammonites and their major distinguishing characters

caliber bullet. The belemnites possess a phragmocone, a chambered area, a rudimentary siphuncle, and a massive rostrum that acted as a counterweight (Fig. 3.51). Earliest belemnites have been recorded in Mississippian sediments, became

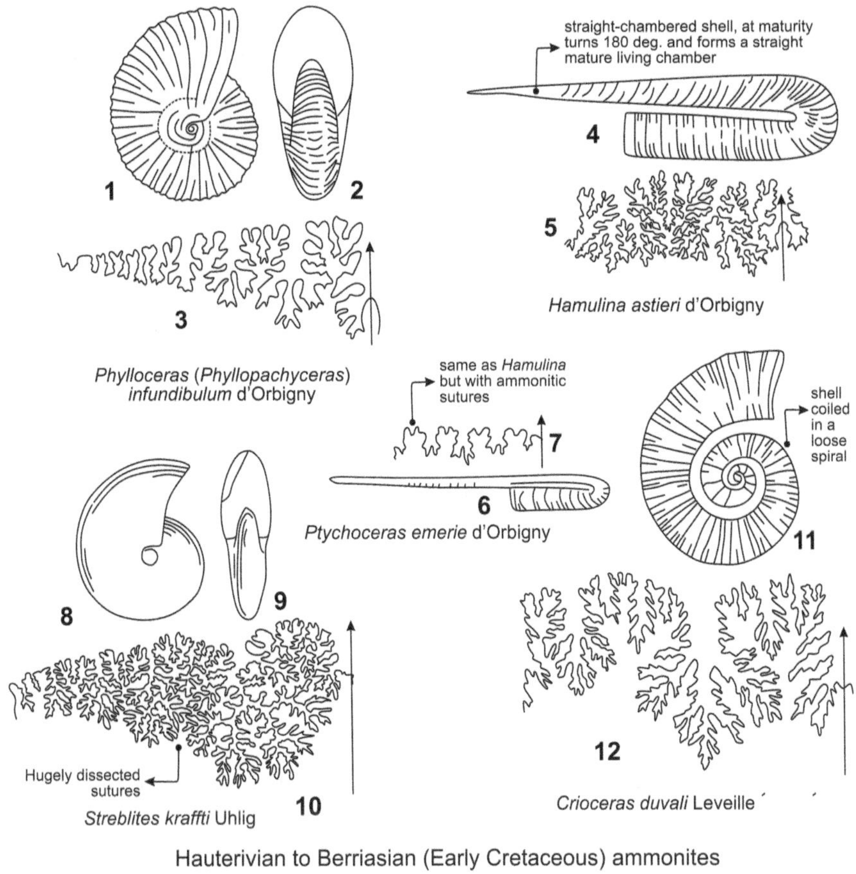

Fig. 3.45 Representative genera of Early Cretaceous (Hauterivian to Berriasian) ammonites and their major distinguishing characters

common in middle and late Mesozoic deposits and died out by the end of the Cretaceous. A doubtful example has been recorded from the Eocene.

Appendix 1 gives the list of illustrated specimens mentioning the chapter number, species name, age, and locality along with its figure number within the said chapter.

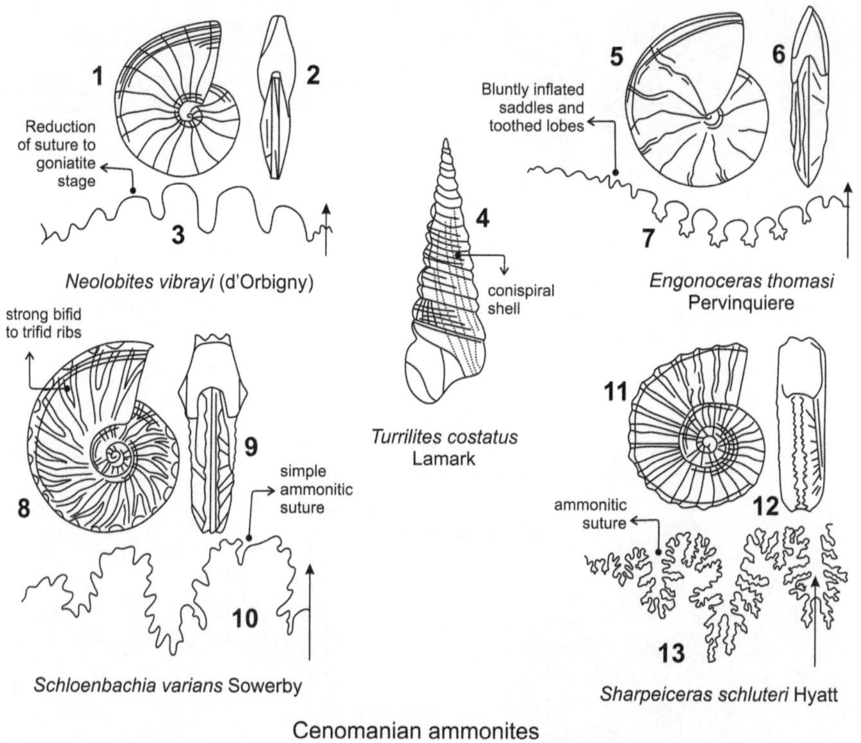

Fig. 3.46 Representative genera of Early Cretaceous (Aptian and Albian) ammonites and their major distinguishing characters

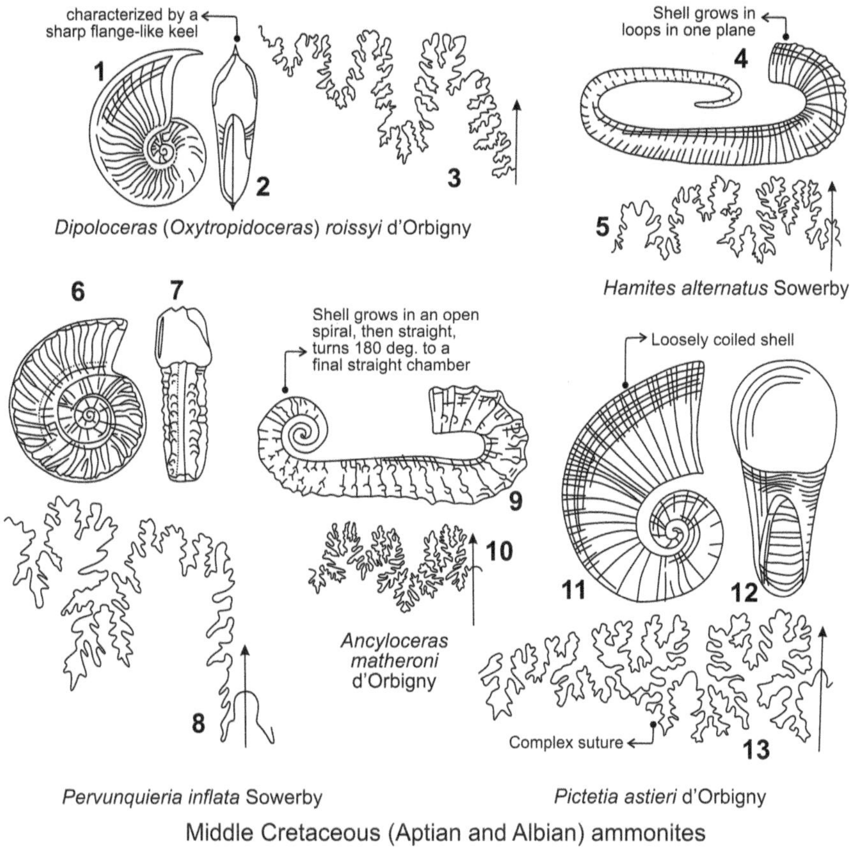

Fig. 3.47 Representative genera of Late Cretaceous (Cenomanian) ammonites and their major distinguishing characters

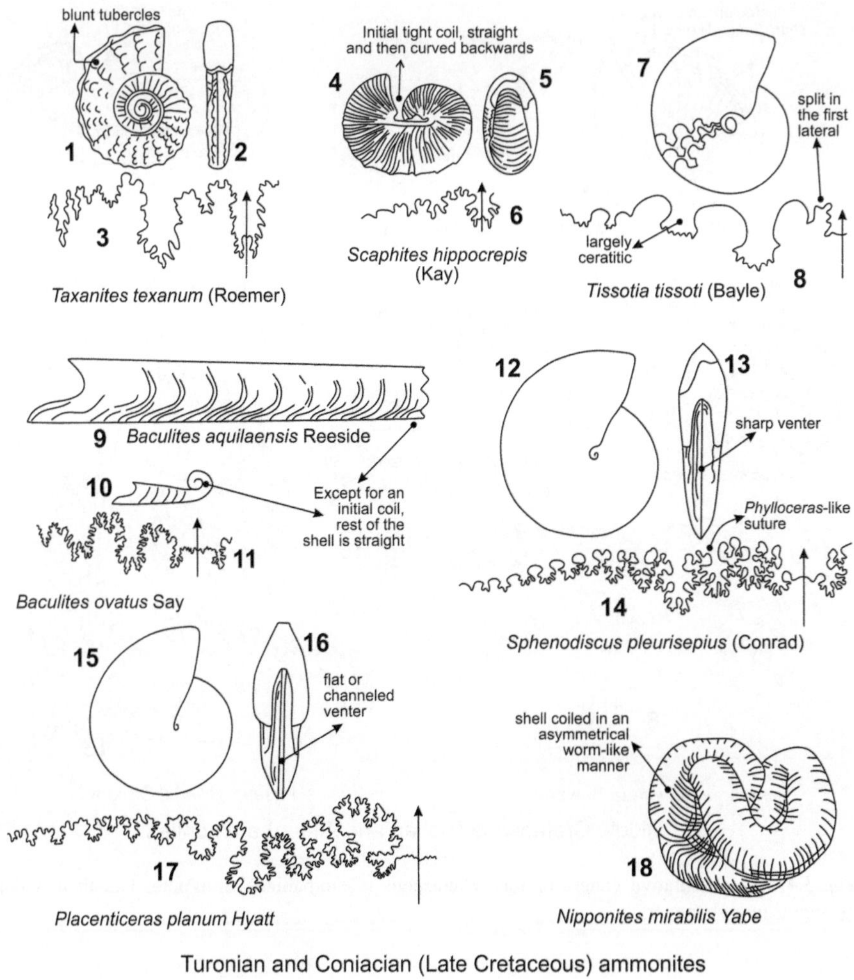

Fig. 3.48 Representative genera of Late Cretaceous (Turonian and Coniacian) ammonites and their major distinguishing characters

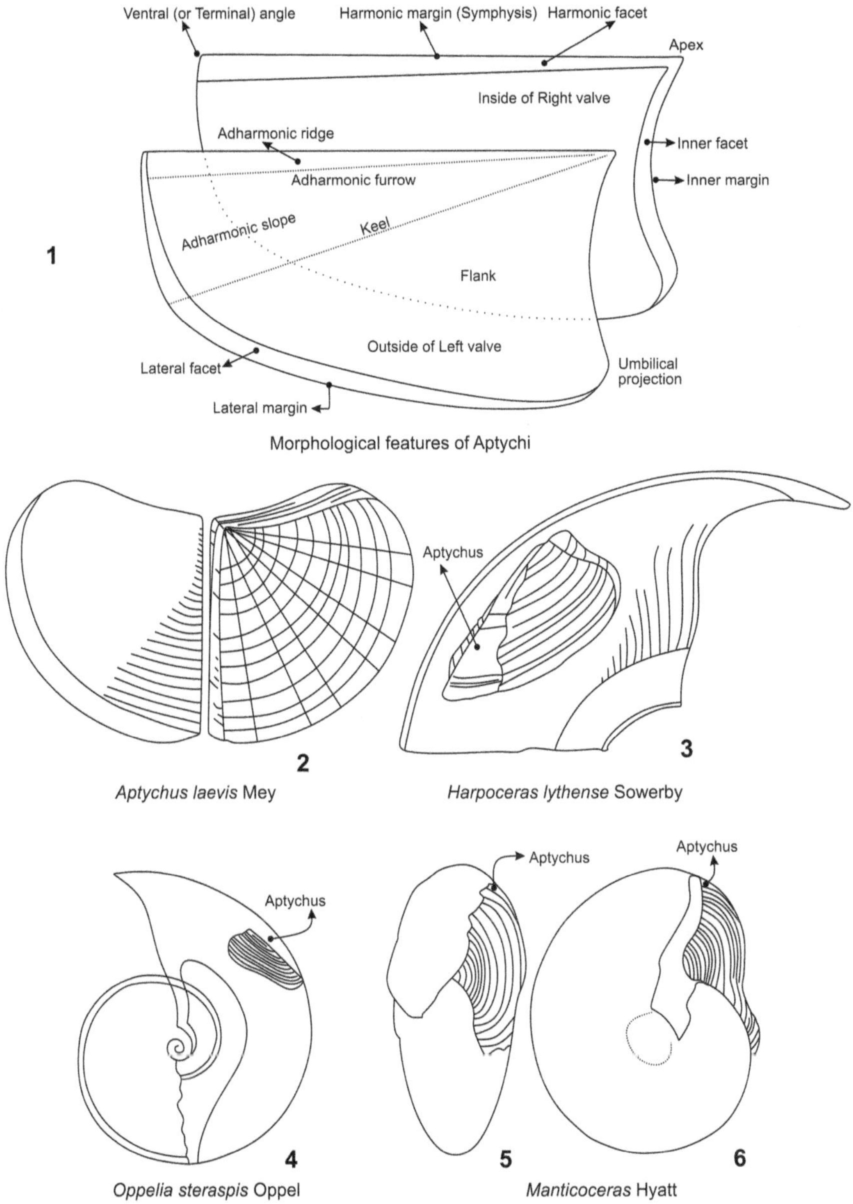

Fig. 3.49 Aptychus. *1* General morphology and descriptive terms (actual Aptychus), external surface (*left side*), and internal side (*right side*). *2–6* Aptychus preserved in body chamber

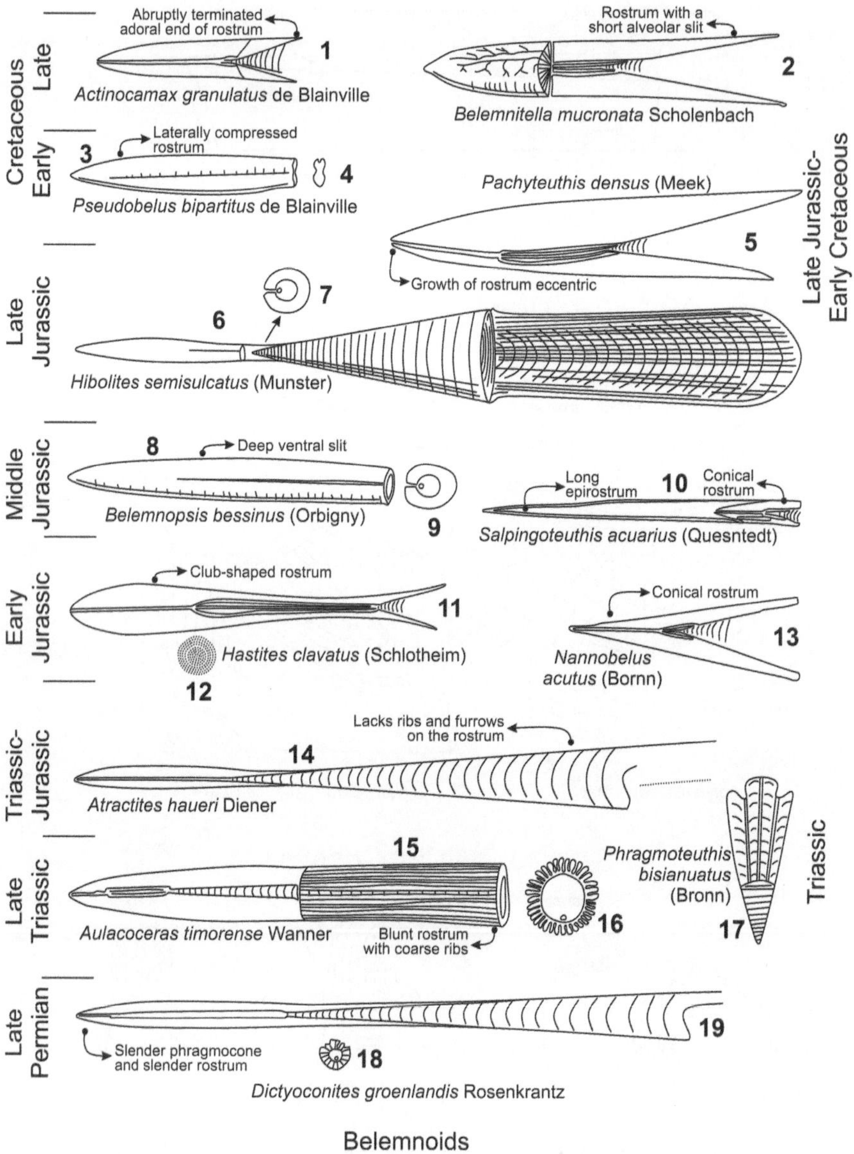

Fig. 3.50 Representative genera of Subclass Coleoidea and their major distinguishing characters, through time

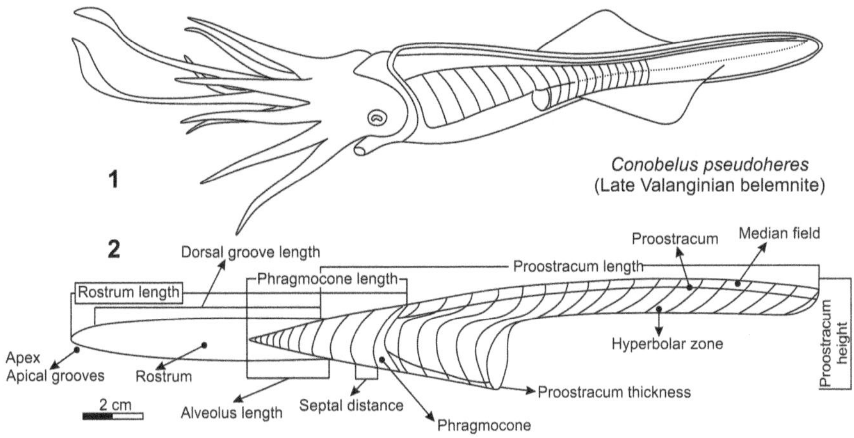

Fig. 3.51 Morphological features of a Belemnite. *1* Living example. *2* Fossil example

References

Arkell, W.J. 1957. Introduction to Mesozoic Ammonoidea. In: *Treatise on Invertebrate Paleontology, Part L, Mollusca 4, Cephalopoda-Ammonoidea*, ed. R.C. Moore, L80–L100. Geological Society of America and University of Kansas Press.

Callomon, J.H.C. 1963. Sexual dimorphism in Jurassic ammonites. Trans. Leicester Lit. *Philosophical Society* 51:21–56. Leicester.

Callomon, J.H.C. 1969. Dimorphism in Jurassic ammonites. In: *Sexual Dimorphism in Fossil Metazoa and Taxonomic Implications, Ser. A*, ed. G.E.G. Westermann, vol. 1, 111–125.

Clarke, M.R., and E.R. Trueman. 1985. Main features of cephalopod evolution. In *The Mollusca, 12. Paleontology and Neontology of Cephalopods*, ed. K.M. Wilbur, M.R. Clarke, and E.R. Trueman. New York: Academic Press.

Dzik, J. 1981. Origin of the cephalopoda. *Acta Palaeontologica Polonica* 26(2): 161–191.

Flower, R.H. 1954. *Cambrian Cephalopods*, vol. 40, 1–58. New Mexico: State Bureau of Mines and Mineral Resources New Mexico Institute of Mining and Technology, Campus Station Socorro.

Flower, R.H. 1955. Status of endoceroid classification. *Journal of Paleontology* 29(3): 327–370.

Kase, T., P.A. Johnston., A. Seilacher., and J.B. Boyce. 1998. Alleged mosasaur bite marks on Late Cretaceous ammonites are limpet (patellogastropod) home scars. *Geology* 26: 947–950.

Kauffman, E.G. 2004. Mosasaur predation on Upper Cretaceous nautiliods and ammonites from the United States Pacific coast. *Palios* 19: 96–100.

Lehmann, U. 1981. *The Ammonites: Their Life and Their World*, 246. Cambridge: Cambridge University Press.

Makowski, H. 1962. Recherches sur la dimorphisme sexuel chez les Ammonoides; Note preliminaire. Księga pam, ku czci Prof. J. Samsonowicza [Book in memory of Professor J. Samsonowicz], 31–42. Warszawa.

Makowski, H. 1963. Problem of sexual dimorphism in ammonites. *Palaeontologia Polonica* 12:1–492. Warszawa.

Moore, R.C. (ed.). 1957. *Treatise on Invertebrate Paleontology, Part L, Mollusca 4—Cephalopoda and Ammonoidea*, 490. Kansas: The Geological Society of America and University of Kansas Press.

Moore, R.C. (ed.). 1964. *Treatise on Invertebrate Paleontology, Part K, Mollusca 3*. Kansas: The Geological Society of America and University of Kansas Press, Lawrence. 519 p.

Pojeta, J. 1987. *Phylum Mollusca in Fossil Invertebrates*, eds. R.S. Boardman, A.H. Cheetham, and A.J. Rowell, 713. Blackwell Scientific Publishing.

Pojeta, J. Jr., and Gordon, M. Jr. 1987. Class Cephalopoda. In: *Fossil Invertebrates*, eds. R.S. Boardman, A.H. Cheetham, and A.J. Rowell, 329–358. Blackwell Scientific Publications.

Saunders, W.B., and E.A. Shapiro. 1986. Calculation and simulation of ammonoid hydrostatics. *Paleobiology* 12(1): 64–79.

Teichert, C. 1964. Morphology of hard parts. In: *Treatise on Invertebrate Paleontology, Part K, Mollusca 3, Cephalopoda, General Features, Endoceratoidea-Actinoceratoidea-Nautiloidea-Bactritoidea*, ed. R.C. Moore, K13–K59. Geological Society of America and University of Kansas Press.

Teichert, C., and B. Kummel. 1960. Size of endocerid cephalopods. *Breviora Museum of Comparative Zoology* 128: 1–7.

Teichert, C., and R.C. Moore. 1964. Classification and Stratigraphic Distribution. Treatise on Invertebrate Paleontology, Vol. K., Geological Society of America and University of Kansas Press.

Trueman, A.E. 1941. The ammonoid body-chamber with special reference to the buoyancy and mode of life of the living ammonoid. *Quarternary Journal of the Geological Society of London* 96: 339–383.

Westermann, G.E.G. 1996. Ammonoid life and habitat. In: *Ammonoid Paleobiology, Topics in Geobiology*, vol. 13, 607–707. New York: Plenum Press.

Chapter 4
Pelecypoda

4.1 Introduction

The name "*Bivalve*" was first coined by Linnaeus in 1758 (later known as "Carl von Linné"; see Bonnani 1681). "*Pelecypoda*" and "*Lamellibranchia*" are later and more commonly used terms; "*Pelecypoda*" is used in this book. Pelecypods evolved from the Early Cambrian (Tommotian) molluscan class Rostroconchia (Pojeta et al. 1972; though this hypothesis is debatable). Rostroconchia are a group of Palaeozoic bivalved molluscs that now belong to a distinct class of their own (Pojeta and Runnegar 1976).

The Cambrian genera, *Fordilla* and *Pojetaia*, are the earliest known pelecypod representatives (Elicki and Gürsu 2009). *Fordilla troyensis* Barrande (Fig. 4.1) and have been recorded from the Early Cambrian rocks (513–520 Ma) of North America, Greenland, Europe, Middle East, and Asia. *Pojetaia runnegari* has been recorded from Australia (Barrande 1881; Pojeta 1975; Elicki and Gürsu 2009). Both are small forms (few mm in length), and shallow burrowers. This is followed by a stratigraphical hiatus from Middle to Late Cambrian (521–485.4 Ma), before pelecypods reappeared in the Earliest Ordovician (Tremadoc: 485.4–477.7 Ma).

4.2 Basic Morphology

These bilaterally symmetrical, equivalved (two similar valves on either side of the commissure) aquatic molluscs (pelecypods; = the modern mussels, cockles, oysters, and scallops) are often elongated in the anteroposterior direction as left and right valves [Fig. 4.2(2)]. The shell, secreted by the mantle through marginal accretion, in most forms, is external. The presence of growth lines on the valve's surface [Fig. 4.2(1, 4, 5)] are evidence of shell growth by mantle section. The Ligament [Fig. 4.2(5)], an elastic structure, connects the valves, dorsally and opens and closes by hinging along the valve that passes through or close to it (Moore

S. Jain, *Fundamentals of Invertebrate Palaeontology*, Springer Geology,
DOI 10.1007/978-81-322-3658-0_4

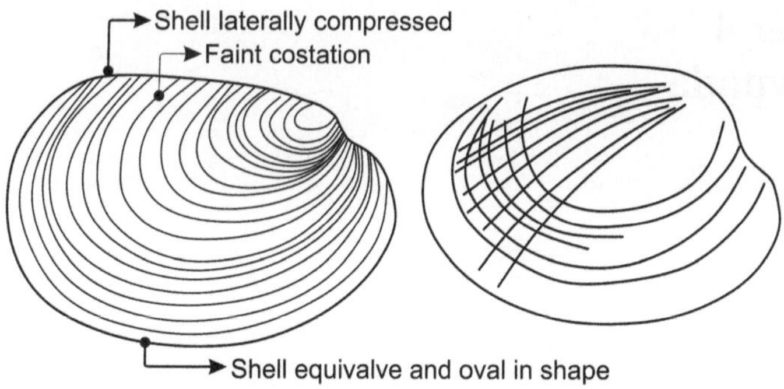

Fig. 4.1 *Fordilla troyensis* Barrande is the earliest known pelecypod representatives and comes from Early Cambrian rocks (513–520 Ma) of North America, Greenland, Europe, the Middle East and Asia. Genus *Fordilla* contains three species, *F. germanica*, *F. sibirica*, and the type species *F. troyensis*. *Pojetaia* has two species, *P. runnegari*, the type species, and *P. sarhroensis*. The genera *Buluniella*, *Jellia*, and *Oryzoconcha* are considered synonyms of *Pojetaia* (see also Elicki and Gürsu 2009)

1969a, b) (see also ligament area and hinge ligament; Fig. 4.2(3, 4), respectively; see Sect. 4.3 for details).

Within molluscs, pelecypods represent one end of the spectrum (lacking a head, radula, and anterior sense organs) whereas cephalopods, the other, with intelligence, agility, and cephalization (Pojeta 1987).

4.3 Terminology

The pelecypod terminology used in this chapter is explained briefly under the following eight subheads, and illustrated in Fig. 4.2:

4.3.1 General shell morphological
4.3.2 Shell form
4.3.3 Ornamentation
4.3.4 Umbo/Beak position
4.3.5 Shell structure
4.3.6 Dentition
4.3.7 Ligament
4.3.8 Muscle scars
4.3.9 Gills

Each subhead is detailed below:

4.3.1 General Shell Morphological Terms

The morphological features that are visible externally on the shell are enumerated here and illustrated in Fig. 4.2.

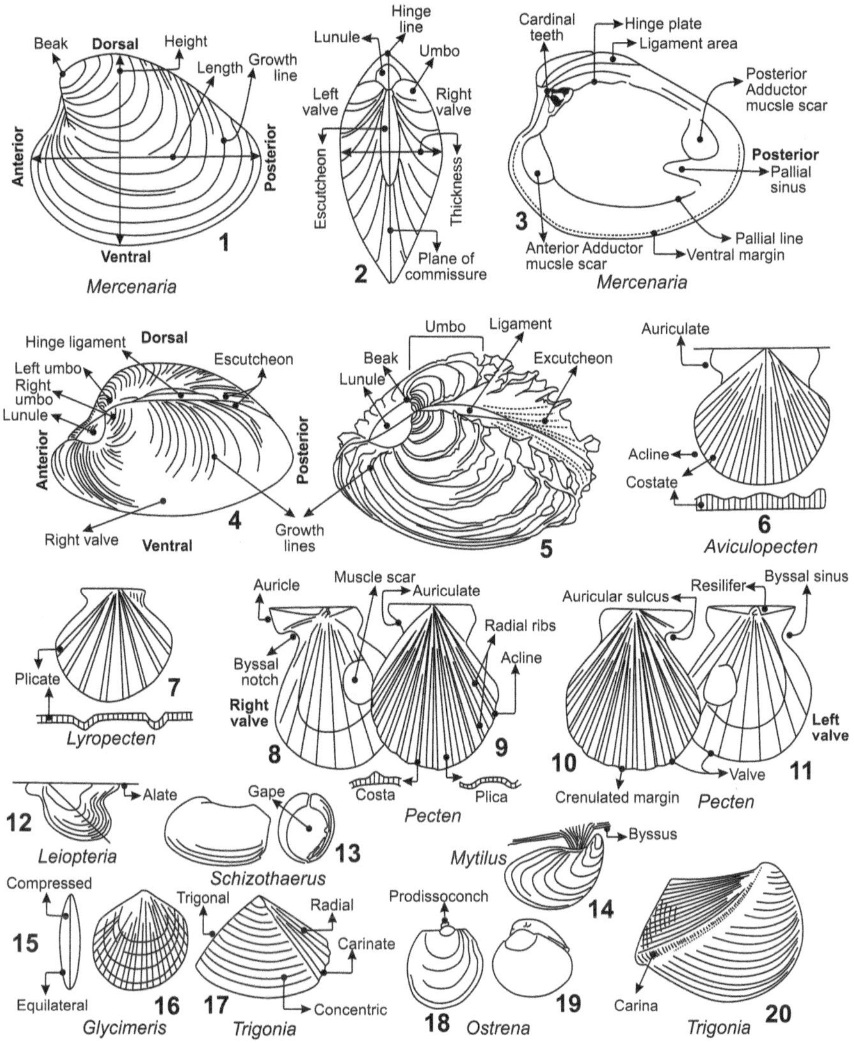

Fig. 4.2 Morphological features and terminology used to describe a bivalve

4.3.1.1 **Alate**: Shell characterized by possession of wings or auricles [Fig. 4.2 (12)].

4.3.1.2 **Anterior**: This is the part of shell that contains the mouth. Beaks of most bivalves point forward, i.e., are inclined anteriorly and when muscle scars are unequal, the anterior is the smaller one [Fig. 4.2(1, 4)].

4.3.1.3 **Auricle**: Forward or backward projection of shell along hinge line (i.e., wing-shaped or ear-like extensions of the valve, anteriorly and posteriorly) [Fig. 4.2(8)]. Best noted in Pectinidae and Teredinidae.

4.3.1.4 **Auricular sulcus**: Furrow or groove of shell exterior separating shell exterior form remainder of the valve (i.e., junction of auricle with the body of shell) [Fig. 4.2(10)].

4.3.1.5 **Auriculate**: Shell possessing auricles, equivalent to Alate [Fig. 4.2(9)].

4.3.1.6 **Beak**: The tip of the valve and is the oldest part of the shell. It is a more or less sharp pointed projection located along or above the hinge line [Fig. 4.2(1, 5)].

4.3.1.7 **Byssal notch**: Indentation of the anterior edge (between the anterior auricle of the right valve with Lunula) for protrusion of thread-like attachment called Byssus; most common in pectinoid shell on the right valve, which is lowermost, allowing protrusion of the small foot without opening valve widely [Fig. 4.2(8)]. Byssal notch is best noted in Pectinacea.

4.3.1.8 **Byssal sinus**: It is the indentation beneath the auricle of left valve of pectinoid shells on the anterior margin and corresponds to the byssal notch of right valve but is usually shallower [Fig. 4.2(11)]. Byssal sinus is best noted in Pectinacea.

4.3.1.9 **Byssus**: A muscular structure for burrowing and locomotion composed of a bundle of hair-like strands used for attachment [Fig. 4.2(14)].

4.3.1.10 **Commissure**: This is the line of junction of both valves and makes the Plane of commissure [Fig. 4.2(2)].

4.3.1.11 **Compressed**: Transversely flattened shell having small thickness [Fig. 4.2(15)].

4.3.1.12 **Dorsal**: Directed toward that part of shell which contains the hinge line [Fig. 4.2(1, 4)].

4.3.1.13 **Equilateral**: When the growth of the shell (generally symmetrical or almost so) is on either side of the beaks. In such equilateral shells, the beak is positioned close to the middle of the length of the shell [Fig. 4.2 (15)].

4.3.1.14 **Equivalve**: With two valves are of same size and shape, i.e., they are bilaterally symmetrical; they are termed left and right valves [Fig. 4.2 (6)].

4.3.1.15 **Escutcheon**: It is the posterior part of cardinal area, a depression (or a curved area), along the hinge and behind the beak marked by a change in sculpture or color [Fig. 4.2(4, 5)].

4.3.1.16 **Gape**: It is the space left between the valves (anterior or posterior) when the adductor muscles are fully contracted [Fig. 4.2(13)].

4.3.1.17 **Growth line**: These irregularly arranged concentric lines (parallel to the shell margin), represent growth stages [Fig. 4.2(4, 5)].

4.3.1.18 **Height**: The total distance, measured as a straight line, between the apical and basal extremities of a shell (i.e., perpendicular to the plane of commissure that just touches the most dorsal and ventral parts of the shell) [Fig. 4.2(1)].

4.3.1.19 **Inequilateral**: Anterior and posterior parts of valve unequal and lacking symmetry. A condition when the growth on either side of the beaks is asymmetrical. In such forms, the beak is closer to one end of the shell than the other [Fig. 4.2(1, 3)].

4.3.1.20 **Inequivalve**: The valves differ in size or shape or both, i.e., the valves are asymmetrical about the commissure.

4.3.1.21 **Left valve**: It is the valve lying on the left-hand side when the shell is placed with the anterior end pointing away from the observer and the commissure is vertical; the hinge being uppermost [Fig. 4.2(2)].

4.3.1.22 **Length**: Distance from anterior to posterior margin at farthest points or measured parallel to the hinge line [Fig. 4.2(1)].

4.3.1.23 **Lunule**: A depressed plane or curved area (commonly cordate in shape) along hinge line in front of the beak (in one or both valves), marked by a change in sculpture or color [Fig. 4.2(4, 5)].

4.3.1.24 **Plane of commissure**: A surface that approximately coincides with the margins of the valve [Fig. 4.2(2)].

4.3.1.25 **Posterior**: Direction or part of shell toward position of anus and siphonal opening; in most bivalves it is opposite to inclination of beak. The pallial sinus and ligament are always posterior in position [Fig. 4.2(1, 4)].

4.3.1.26 **Prodissoconch**: It is the earliest formed part of shell, secreted by the larva or embryo and generally preserved at tip of beak (in some adult shells) [Fig. 4.2(18)].

4.3.1.27 **Right valve**: Shell on the right side of the antero-posterior axis; it is generally lowermost in pectinoids and uppermost in oysters and pachyodonts. Alternatively, if the shell is considered to be in the hands of the observer, with the beaks and umbones uppermost, the right valve will always be in the right hand and the left valve in the left hand [Fig. 4.2(2)] when:

(a) the external ligament is placed between the beaks and the observer's body; or when,
(b) the opening of the pallial sinus is toward the observer's body; or when
(c) the center of the single adductor muscle scar in monomyarian forms is placed on the observer's side of the midline of the shell

To identify the right valve from the left, the anterior and posterior margins needs to be ascertained [Fig. 4.2(1)]. Then, if the shell is held with the hinge margin uppermost and the posterior and anterior in the same plane as the observer, then the right valve will be in the right hand of the observer and vice versa [see Fig. 4.2(2)].

4.3.1.28 **Thickness**: Maximum dimension of a bivalve shell measured normal to plane of commissure [Fig. 4.2(2)].

4.3.1.29 **Umbo**: Very strong convex part of valve that is adjacent to the beak [Fig. 4.2(4, 5)]. The term "umbo" is often used interchangeably with "beak", however, for most shells two distinct terms are required.

4.3.1.30 **Valve**: Part of calcareous shell lying on either side of hinge line [Fig. 4.2 (3)].

4.3.1.31 **Ventral**: Direction or part of shell lying opposite the hinge line; generally located lowermost [Fig. 4.2(1, 4)].

4.3.2 Shell Form

Broadly, 34 types of shell forms are illustrated (Fig. 4.3). However, these are by no means an end-all, as minor variations and varied combinations of these would exist. Descriptive terminology used in this book is elaborated and illustrated in Fig. 4.3 (1).

4.3.3 Shell Ornamentation

Shell ornamentation is a widely used character for species-level identification. Figure 4.4 provides a broad (certainly not all encompassing) pattern of ornamentation that is common to most pelecypods. However, like shell form (Fig. 4.3), minor variations and varied combinations of these would exist.

General descriptive terms explaining the pattern of shell ornamentation are enumerated first followed by a brief description and illustration of the types of shell ornamentation (Fig. 4.4).

4.3.3.1 General Terms

4.3.3.1.1 **Carina**: It is a prominent keel-like ridge [Fig. 4.2(20)].

4.3.3.1.2 **Carinate**: The shell surface is marked by a sharp-angled edge extending outward from the beak [Fig. 4.2(17)].

4.3.3.1.3 **Costae (Costa: sing.)**: The shell bears radial ribs formed by localized thickening. These thickenings are moderately broad and prominent elevation of surface of shell, directed radially or otherwise [Fig. 4.2(9)].

4.3.3.1.4 **Costellae**: These are narrow linear elevations on shell's surface [Fig. 4.2 (16)].

4.3.3.1.5 **Plica**: These are radially disposed ribs that are formed by fold or costa that involves the entire thickness of shell [Fig. 4.2(9)].

4.3.3.1.6 **Plicate**: The shell is radially folded to form ribs that also increase the shell's surface area [Fig. 4.2(7)].

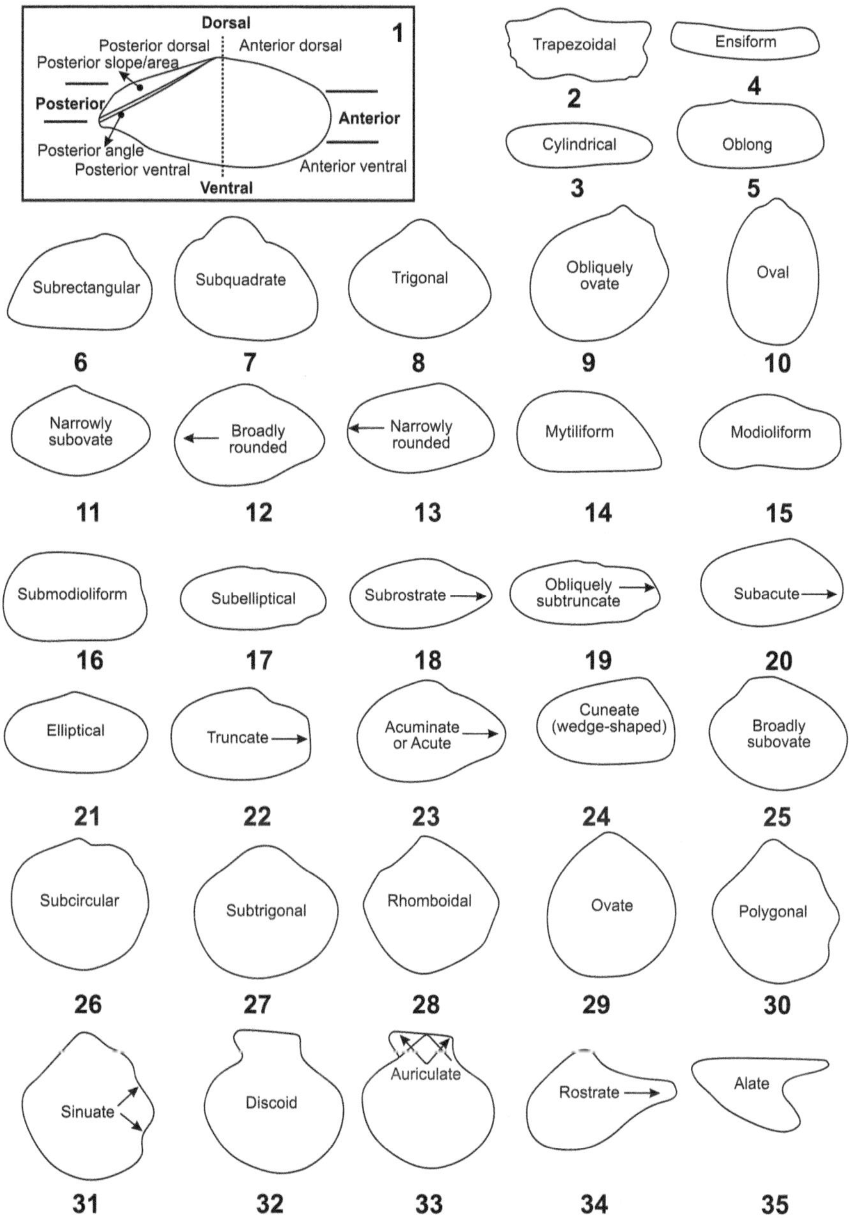

Fig. 4.3 Shell form terminology (1) and varied shell forms of a pelecypod (2–35)

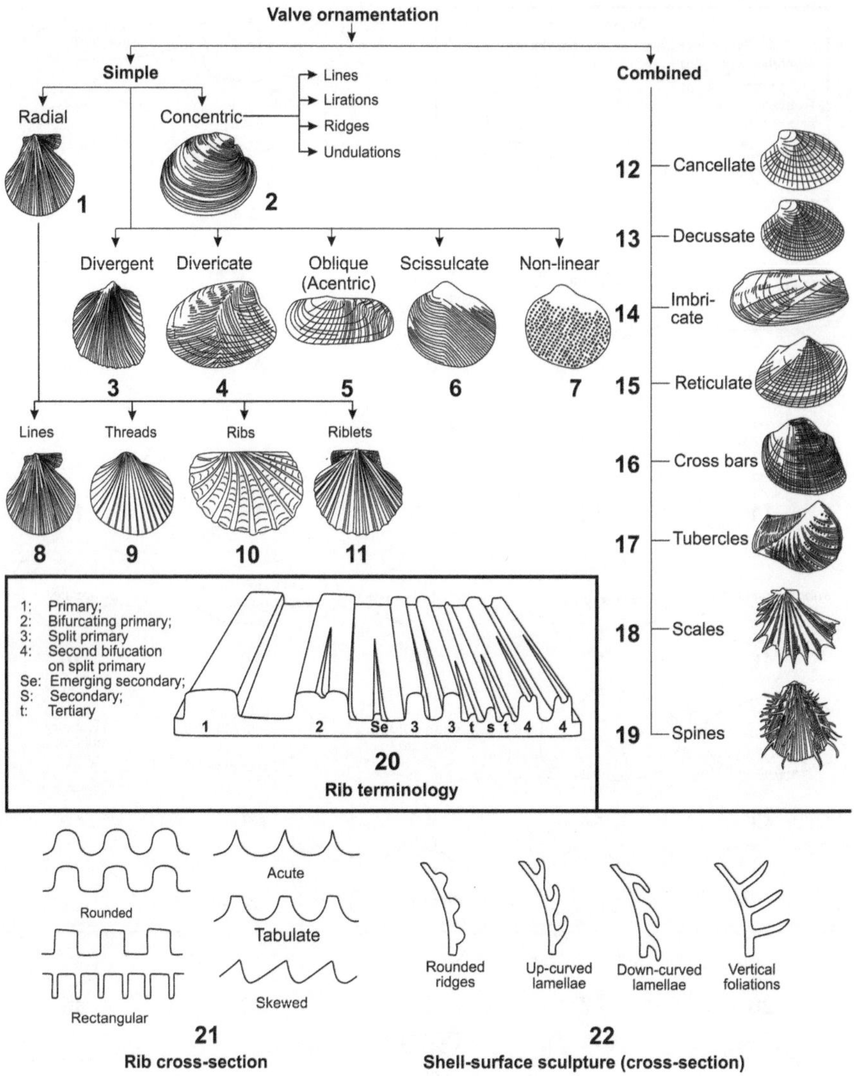

Fig. 4.4 Types of pelecypod ornamentation (1–19), Rib terminology (20), terminology to describe rib cross section (21), and types of external rib sculpture (22)

4.3.3.2 Types of Shell Ornamentation

Shell ornamentation broadly consists of two elements: Simple and Combined (Fig. 4.4).

4.3.3.2.1 Simple

4.3.3.2.1.1 Radial: A linear pattern originating from the beak/umbo and radiating toward the margins of the shell [Fig. 4.4(1)].

This is commonly represented by the direction of costa (a broad prominent elevation) or other elements of ornamentation. The strength of sculpture may vary from faint to strong thorough a series—Lines, Threads, Riblets, and Ribs [Fig. 4.4(8–11)]. Incised sculpture is usually very fine designated as Striations or Grooves.

4.3.3.2.1.2 **Co-marginal or Concentric**: An ornamentation pattern (ridges) that follows the margins of the shell [Fig. 4.4(2)], coinciding with growth lines. The strength of raised concentric pattern (ridges) ranges from faint to strong thorough a series termed Lines, Lirations, Ridges, and Undulations.

Other less common patterns [Figs. 4.4(3–7)] that do not conform to the above two are:

4.3.3.2.1.3 **Divergent**: The radial elements diverge from their normal position; secondary elements appear [Fig. 4.4(3)].

4.3.3.2.1.4 **Divaricate**: The divergent radial element are angled in opposite directions, the angulations in a line medially or post-medially [Fig. 4.4 (4)].

4.3.3.2.1.5 **Oblique or Acentric**: This is a linear feature and is angled across the shell. The pattern does not radiate from the beaks [Fig. 4.4(5)].

4.3.3.2.1.6 **Scissulate**: These are dense oblique striations often abruptly angled to concentric striations [Fig. 4.4(6)].

4.3.3.2.1.7 **Nonlinear**: The shell is wholly or partly textured, if the pattern is raised then it is granular or pustulose, but if not sunken then it is pitted [Fig. 4.4(7)].

4.3.3.2.2 **Combined patterns**: It is formed by a combination of both radial and co-marginal patterns. At times, the radial pattern dominates with the co-marginal forming a variety of structures on the riblets or ribs [Fig. 4.4 (12–19)].

4.3.3.2.2.1 **Cancellate**: These are regular rough rectangular blocks made by the intersection of radial and co-marginal elements [Fig. 4.4(12)].

4.3.3.2.2.2 **Decussate**: These are angular block pattern ornamentation that is finer than the Cancellate made by the intersection of radial and co-marginal elements [Fig. 4.4(13)].

4.3.3.2.2.3 **Imbricate**: These are interrupted series of co-marginal scales crossing radial elements [Fig. 4.4(14)].

4.3.3.2.2.4 **Reticulate**: In this, the shell surface is characterized by a fine network of intersecting raised threads [Fig. 4.4(15)].

4.3.3.2.2.5 **Cross bars**: These are closely spaced narrow ridges set at right angles on radial ribs [Fig. 4.4(16)].

4.3.3.2.2.6 **Tubercles**: These are rounded raised ornamentation set on radial ribs [Fig. 4.4(17)].

4.3.3.2.2.7 **Scales**: These erect or flush like tiles are borne by the radial ribs [Fig. 4.4(18)].

4.3.3.2.2.8 **Spines**: These are erect pointed structures arising from the ribs [Fig. 4.4(19)].

4.3.3.2.3 **Ribbing pattern**: In most pelecypods shells, the ornamentation (ribs) remains simple and consistent (in terms of the number of ribs), throughout shell growth [Fig. 4.4(20)]. Though, in some, the primary ribs divide or new secondary ribs appear between primaries; this process is repeated, at times, and gives rise to Tertiary ribs [see Fig. 4.4(20)].

4.3.3.2.4 **Cross sections and spacing**: The ribs have a variety of cross sections [see Fig. 4.4(21)], from Rounded, Rectangular, Acute, Tabulate to Skewed [see Fig. 4.4(21)]. Additionally, rib spacing also varies; from equal to, wider than or narrower than the ribs. Cross sections vary from broad low rounded ridges to down-curved or up-curved lamellae to vertical foliations [see Fig. 4.4(22)].

4.3.4 Umbo/Beak Position

4.3.4.1 **Acline**: Shell having neither forward nor backward obliquity, i.e., the midline of umbo is normal to the hinge line [Fig. 4.2(6)].

4.3.4.2 **Opisthocline**: Shell having backward obliquity, i.e., the approach along midline to beak is pointed backward [Fig. 4.5(1)].

4.3.4.3 **Opisthogyral (Opisthogyrate)**: The shell is curved; the beak points in the posterior direction (a term applied to umbos) [Fig. 4.5(2)].

4.3.4.4 **Prosocline**: The shell possess forward obliquity; the approach to beak along midline of shell is inclined forward [Fig. 4.5(3)].

4.3.4.5 **Prosogyral (Prosogyrate)**: The shell is curved so that the beaks point in the anterior direction, i.e., the beaks are directed forward [Fig. 4.5(4)].

4.3.4.6 **Rostrate**: The shell possesses prominent beaks [Fig. 4.5(5)].

4.3.5 Shell Structure

4.3.5.1 **Ostracum**: Calcareous structure that the pelecypod shell is composed of, except for the thin outer conchiolin layer (periostracum) (Fig. 4.6). Conchiolin is a proteinaceous layer.

4.3.5.2 **Periostracum**: This is the outer thin layer of conchiolin. It is a nonliving, horny layer overlying the external, calcareous, surface of the shell; sometimes called the Epidermis. In some forms, it also covers the calcareous Ostracum. Periostracum gives rise to shell sculpture and ornamentation.

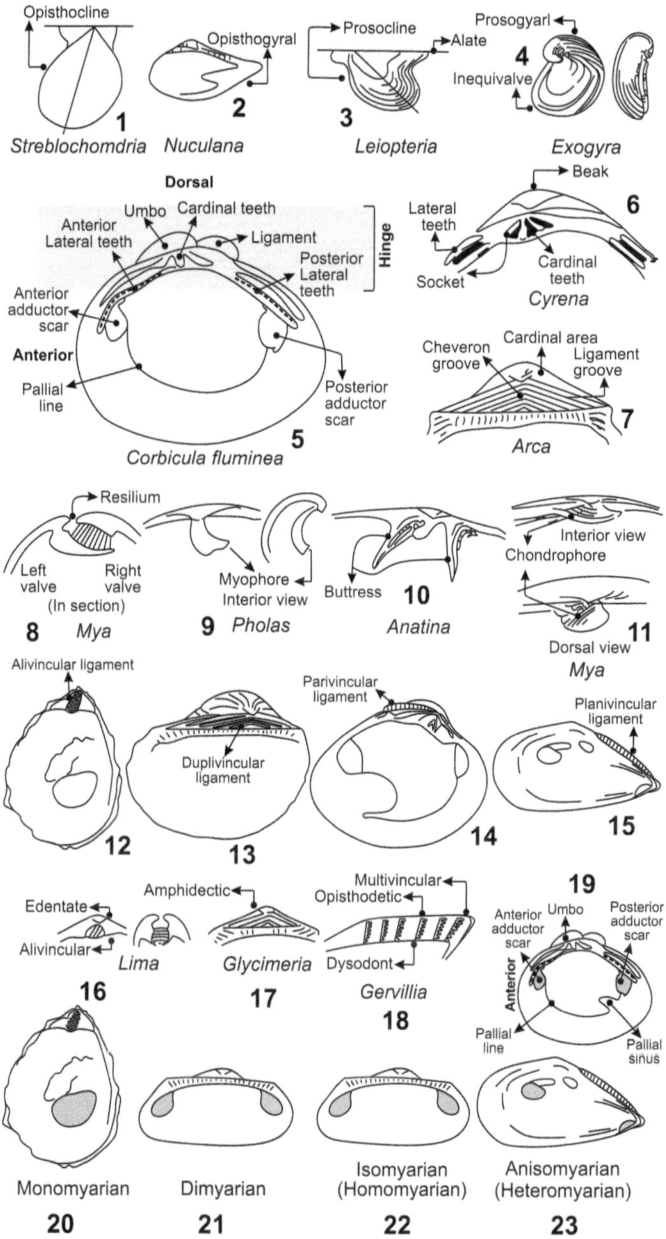

Fig. 4.5 Morphological features and terminology used to describe a pelecypod (continued from Fig. 4.2)

Fig. 4.6 Shell layers

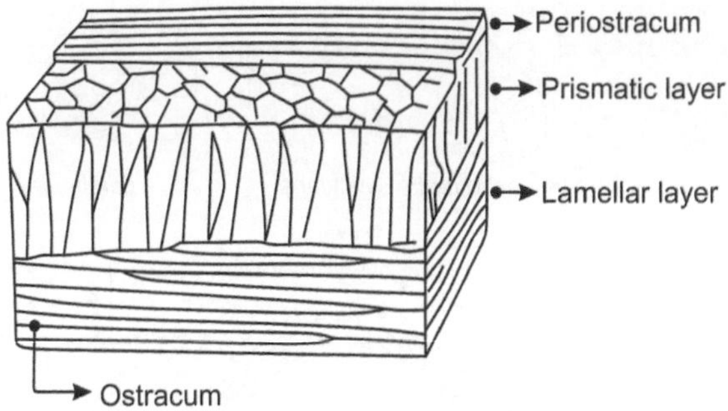

4.3.5.3 Prismatic layer: This is the outer part of the Ostracum and consists of closely spaced polygonal prisms of calcite.

4.3.5.4 Lamellar layer: It is the innermost part of the pelecypod shell, consisting of microscopically thin sheets of calcite or aragonite separated by layers of conchiolin.

4.3.6 *Dentition*

Dentition, considered collectively, is hinge teeth and sockets [Fig. 4.5(5)]. At the dorsal margin, the valves are united to one another for a shorter or longer distance, along a line, which is called the Hinge line [Fig. 4.2(2)]. The teeth, ligament, and all structures that support and attach these to the shell, make up the Hinge [Fig. 4.5(5)].

The hinge line is mostly curved, but it may be quite straight. The teeth perform three functions: enables alignment of the two valves when they close, interlocks them to prevent shearing, and above all, during burrowing, maintains contact between valves. Sockets are depressions on the opposite side of the valve where the teeth fit [Fig. 4.5(6)]. Two types of teeth are noted—Cardinal teeth, located below the beak (umbo), and Lateral teeth, located at a distance from the beak and located on either side of the Cardinal teeth [Fig. 4.5(5, 6)]. Sometimes there may be lateral teeth only; sometimes the cardinal teeth alone are present; in some cases (as in Arcadce) there is a row of similar and equal teeth or they could just be absent (as in Edentulous). However, when teeth are present, they do differ much in their form and arrangement (see also Sect. 4.3.6.2—Types of dentition).

4.3.6.1 General Terms

General descriptive terms explaining dentition and associated structures (see Fig. 4.5) are enumerated first followed by a brief description and illustration of the types of dentition in Fig. 4.7.

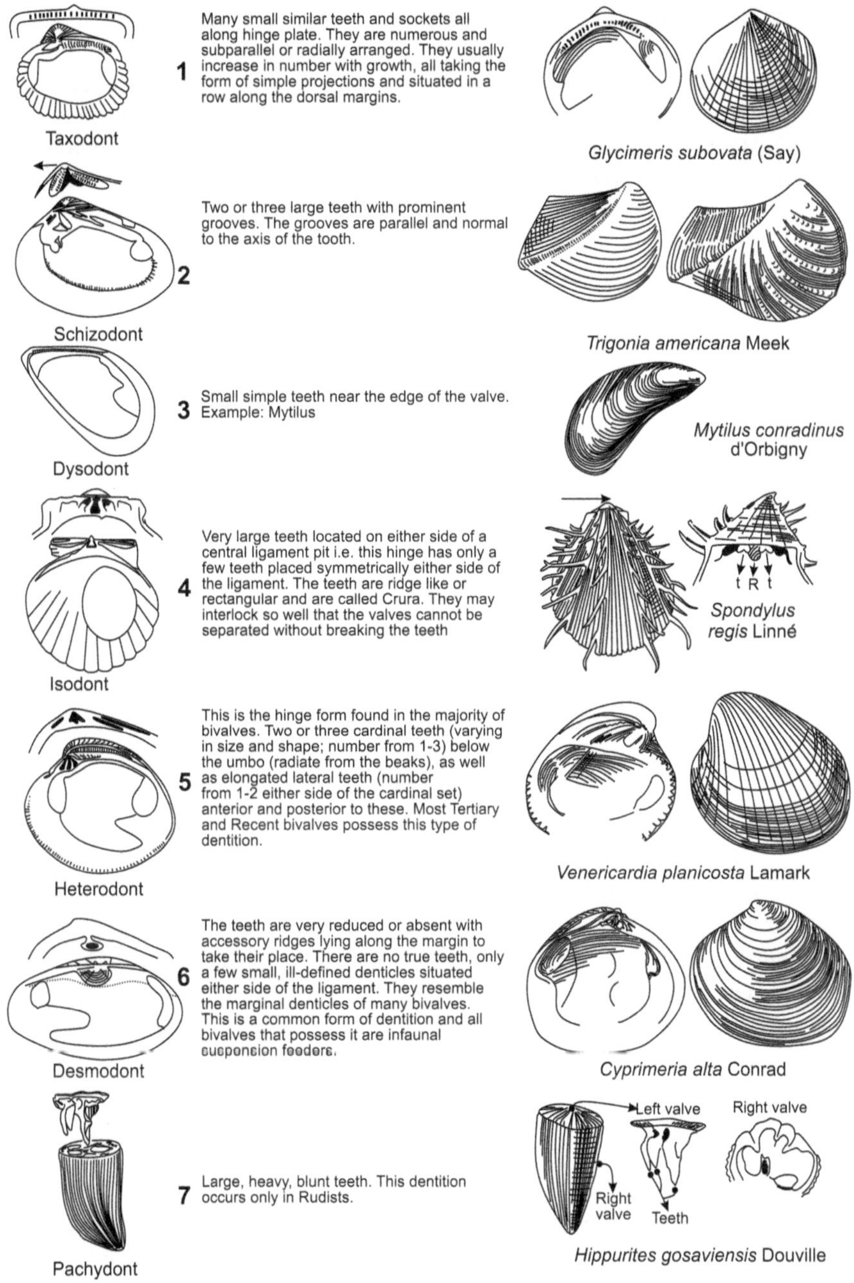

Fig. 4.7 Types of pelecypod dentition

4.3.6.1.1 **Buttress**: It is a ridge on the inner surface of a valve, which serves as support for the hinge plate or a Chondrophore [Fig. 4.5(10, 11)].

4.3.6.1.2 **Cardinal area**: A triangular area amidst beak and hinge margin and occupied partly or wholly by a ligament [Fig. 4.5(7)]. It is distinguished from the remainder of valve exterior by a sharply angulated border.

4.3.6.1.3 **Cardinal teeth**: These are vertical or oblique projections of the hinge line and are placed directly beneath or at close proximity to the beak. The teeth fit into sockets of the opposite valve [Fig. 4.5(5, 6)].

4.3.6.1.4 **Hinge line**: Along dorsal margin of the shell, it is the edge of valve that is in contact with the opposite valve [Fig. 4.5(5)].

4.3.6.1.5 **Hinge plate**: It is the internal shelly surface adjacent to hinge line along which the hinge teeth project [Fig. 4.2(3)].

4.3.6.1.6 **Hinge**: It is the dorsal region of the shell along which the valves meet and where they may be held together by interlocking teeth; these may sit directly on the hinge line or on an infolding of it called the Hinge plate. Hence, structures that are in the dorsal region of the shell, and those that enable opening and closing of valves, are collectively called Hinge [Fig. 4.5(5)].

4.3.6.1.7 **Lateral teeth**: These are placed in front or behind the cardinal teeth, and nearly parallel to the hinge line, these bony projections from hinge plate are the Lateral teeth [Fig. 4.5(5, 6)]. Those on the anterior side of the valve are called the Anterior lateral tooth and on the posterior, the Posterior lateral tooth [Fig. 4.5(5)].

4.3.6.1.8 **Socket**: This is a depression in the hinge plate where the hinge tooth of the opposite valve fits [Fig. 4.5(6)].

4.3.6.2 Types of Dentition (Fig. 4.7)

There are seven types of dentition namely Taxodont, Schizodont, Dysodont, Isodont, Heterodont, Desmodont and Pachydont, and these are illustrated and briefly described in Fig. 4.7. Pelecypods that lack teeth are called Edentate (or Edentulous or Anodont) [see Fig. 4.5(16)].

4.3.7 Ligament

Posterior to the umbones, is another structure passing between the valves called the Ligament [Fig. 4.5(5–7)]. It is usually composed of two parts—External ligament, and the Cartilage (Internal ligament) which enables the opening of the shell. The ligament is horny and elastic and its main function is to join valves along the hinge margin. It also serves to open the shell when the adductor muscles relax. The ligament is difficult to interpret, as it is a complex structure but is an important diagnostic shell character.

Below, general descriptive terms are enumerated first [see Fig. 4.5(5–11)] and are followed by a brief description and illustration of the types of ligament in Fig. 4.7.

4.3.7.1 General Terms

4.3.7.1.1 **Chevron groove**: It is a narrow depression on the cardinal area having an inverted V shape, marking ligament attachment. Examples: Some Arcacea and early Pectinacea [Fig. 4.5(7)].

4.3.7.1.2 **Chondrophore**: This is a relatively prominent internal spoon-shaped structure (which may or may not project). It holds the internal ligament [Fig. 4.5(11)].

4.3.7.1.3 **Ligament groove**: A narrow linear depression in the cardinal area or ligament area marking the attachment of ligament fibers [Fig. 4.5(7)].

4.3.7.1.4 **Ligament**: It is a horny and elastic structure (or sometimes, structures). It joins the two valves of the shell dorsally, and acts as spring causing them to open when the adductor muscles relax [Fig. 4.5(8)].

4.3.7.1.5 **Resilifer**: It is the portion of the hinge plate that bears the internal ligament (the Resilium). The Resilifer is generally a simple shallow pit; a process for the attachment of the internal ligament [Fig. 4.2(11)].

4.3.7.1.6 **Resilium**: It is part of the ligament below valve margins; an internal cartilage of the hinge best noted in Pectinidae [Fig. 4.5(8)].

4.3.7.2 Types of Ligament

There are four main types of ligament—Alivincular, Duplivincular, Parivincular, and Planivincular [see Fig. 4.5(12–15)].

4.3.7.2.1 **Alivincular**: It is the type of external ligament with greatest length transverse to the plane of commissure [see Fig. 4.5(12)]. It is located between the cardinal areas (where present) of respective valves. Lying between the beaks on a flat cardinal area, alivincular is a flattened, usually triangular, structure bounded by lamellar layers on either side. External alivincular ligaments are typical of *Lima* (see Fig. 3.44), *Limopsis* and *Ostrea.*

4.3.7.2.2 **Duplivincular**: It is a ligament with a lamellar component. The latter is repeated as a series of bands, each with its two edges inserted in narrow grooves within cardinal areas of respective valves, forming V-shaped chevrons [see Fig. 4.5(13)]. The ligament is composed partly of fibrous (compressional) tissue and partly of lamellar (tensional) tissue. Example: best noted in Arcidae, Glycymeridae and Noetiidae.

4.3.7.2.3 **Parivincular:** It is a curved structure positioned behind the beaks with its long axis parallel to the hinge line, and consisting mainly of lamellar (tensional) tissue [Fig. 4.5(14)]. Example: typical of Veneridae and Tellinidae.

4.3.7.2.4 **Planivincular**: It is a long slightly bent (arched) ligament extending posteriorly as a narrow band [Fig. 4.5(15)]. Example: typical of Mytiloidea.

Additionally, the ligaments may lie symmetrically both in front (anterior) and behind (posterior) the beak (Amphidetic; Fig. 4.5(17) or generally behind (posterior) to the beak (Opisthodetic; Fig. 4.5(18) and very frequently in front of the beaks (Prosodetic).

4.3.8 Muscle Scars

The adductors muscles leave a distinct muscular impressions or scars, in the interior of the shell, so that it is easy in any given specimen to determine where there was only one adductor, or whether two were present [Fig. 4.5(19)].

4.3.8.1 General Terms

4.3.8.1.1 **Adductor muscle scar**: Scars (impressions on inside of valve; normally one or two) left by muscles that close the valves as well noted in species of the family Pectinidae and Pholadidae. A third accessory adductor scar may be present where the pallial line is expanded as it bends back into the pallial sinus (=ventral adductor scar) [Fig. 4.2(3)].

4.3.8.1.2 **Myophone**: Plate or rodlike structure on inside of shell for attachment of muscle (usually adductor) [Fig. 4.5(9)].

4.3.8.1.3 **Pallial line**: The mantle lobes when attached to the inside of a shell leave a mark at their position of attachment; it is called a Pallial line. It is normally concentric and follows the ventral margin as it connects the anterior and posterior adductor scars. The Pallial line possesses an indentation called the Sinus [Fig. 4.5(19)].

4.3.8.1.4 **Pallial sinus**: It is the inward deflection of the pallial line in the posterior part of the shell, defining a space for the retraction of siphons [Fig. 4.5 (19)].

4.3.8.2 Types of Ligament

4.3.8.2.1 **Monomyarian**: Having only one adductor muscle, originally posterior but tending to be central in position [Fig. 4.5(20)].

4.3.8.2.2 **Dimyarian**: Valves having two adductor muscle scars in each valve, whether equal or unequal [Fig. 4.5(21)].

4.3.8.2.3 **Isomyarian (Homomyarian)**: Having two adductor muscles of approximately equal size [Fig. 4.5(22)].

4.3.8.2.4 **Anisomyarian (Heteromyarian)**: Adductor muscle scars conspicuously unequal. The anterior adductor muscle is much reduced or absent [Fig. 4.5(23)].

4.3.9 Gills

Gills are flat, thin layers of tissue, ciliated and attached to the two sides of the visceral mass and/or to the proximal part of the foot [Fig. 4.8(1)]. Their ciliated nature generates inhalant currents that brings oxygenated and food-laden water into the mantle cavity. Hence, they function both as a respiratory and a food gathering organ. Gills are rarely preserved in the fossil record but are an important component for taxonomic identification. Fossilized gills have been noted in the Late Jurassic *Trigonia* and are similar to the present day *Neotrigonia*!

Four basic gill types are recognized: Filibranch, Eulamellibranch, Protobranch, and Septibranch; these are illustrated with their major characteristics given in Fig. 4.8(2–5), respectively. The Filibranch and Eulamellibranch are the most common types (see also Clarkson 1993).

4.4 Classification

Major taxonomic characters used for pelecypod classification includes shell microstructure, dentition, hinge structure, and the type of gills (see Table 4.1). Lesser characters (and rarely used or for extant forms) include stomach anatomy and nature of labial palps (see also Clarkson 1993). Classification followed here follows Treatise on Invertebrate Paleontology Part N, Volume 1. Representative examples mentioning their major distinguishing characters are illustrated in Figs. 4.9, 4.10, 4.11, 4.12, 4.13, 4.14, 4.15, 4.16, 4.17, and 4.18.

4.5 Geological History

Pelecypods, since their inception, have lived in the marine environment at depths ranging from intertidal to abyssal (Pojeta 1987). During Cambrian and Ordovician times, few species were present, but thereafter, they became abundant and by

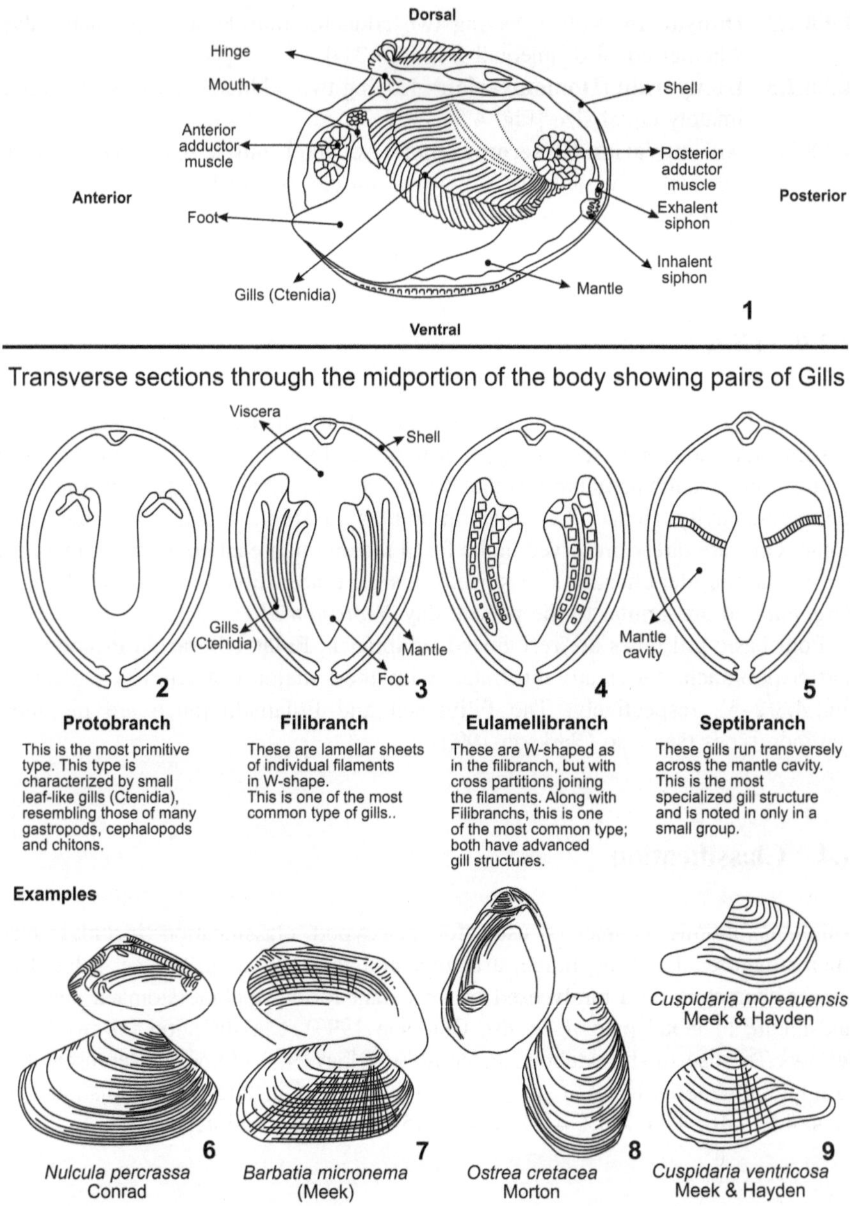

Fig. 4.8 Pelecypod gill terminology and types of gills (2–5)

Middle Ordovician all major groups were established. Post Early Ordovician, the pelecypods became the dominant benthic fauna, at scattered stratigraphic levels. During the Mesozoic and Cenozoic, they were common to dominant shallow water

Table 4.1 Bivalve classification (after Treatise on Invertebrate Paleontology, Part N, Volume 1)

Subclass	Order	Name	Age	Major characteristics							Examples	References
				Shell characteristics	Shell type	Shell composition	Muscle scars	Dentition	Gills	Ecological characteristics		
Subclass 1		Paleotaxodonta	Early Cambrian to Holocene	In most, shell anteriorly or posteriorly elongated	Equivalved and inequilateral	Wholly aragonitic of nacre and prisms	–	Taxodont	Living forms with small protobranch gills	Deposit feeders and infaunal	*Nucula*	Pojeta (1987)
Order		Nuculoida	Ordovician—Holocene	Shell with closed margins. Foot present, but byssus absent in adults	Equivalved	–	–	Taxodont	Protobranch gills	–	–	Moore (1969a, b)
Subclass 2		Cryptodonta	Ordovician—Holocene	Thin shelled, mainly Paleozoic forms. Active burrowers and infaunal	Generally equivalved	Wholly aragonitic of nacre and prisms	–	Dysodont to toothless	Protobranch gills	–	*Solemya, Cardiola*	Clarkson (1993)
Order 1		Soloemyoida	Devonian—Holocene		–	Aragonitic	–		Protobranch gills	–	–	Moore (1969a, b)
Order 2		Praecardiodia	Late Ordovician—Late Mississippian	Shell circular to oval to elongate, and thin	–	–	–	Taxodont	–	–	–	Moore (1969a, b)
Subclass 3		Pteriomorphia	Ordovician—Holocene	Adult byssus present in many. Foot commonly reduced, or absent in some (free-living or cemented). Mantle mostly unfused and lacking siphons	Many inequivalved, some equivalved; beak is close to the anterior end	Aragonitic, calcitic, or both. All include some aragonite within the shell (nacre, crossed lamellar microstructures)	Possesses a range of different hinge architectures and musculature (iso and heteromyarian)	Dentition variable	Gills fillibranch to eulamellibranch	Free-living habits	*Modiolus, Pinna, Ostrea, Gryphaea, Pecten, Spondylus, Isognomon, Imoceramu*	Pojeta (1987)
Order 1		Arcoida	Late Ordovician—Holocene	Shell circular to trapezoidal	Equivalved	–	–	Taxodont	Gills fillibranch in living forms	–	*Arca, Glycimeris, Limopsis*	Moore (1969a, b)
Order 2		Mytiloida	Devonian—Holocene	Shell with or without well-developed siphons. Weak anterior muscle scars	Equivalved and very inequilateral	–	–	Moreorless toothless hinge line	Gills filibranchiate or eulamellibranchiate	–	*Mytilus, Lithophaga, Modiolus*	Moore (1969a, b)
Order 3		Pterioida	Ordovician—Holocene	Adults fixed by byssus or cementation. One posterior adductor scar		–	–	–	Gills filibranchiate or eulamellibranchiate	–	Eurydesma	Moore (1969a, b)

(continued)

Table 4.1 (continued)

Subclass	Order	Name	Age	Major characteristics							Examples	References
				Shell characteristics	Shell type	Shell composition	Muscle scars	Dentition	Gills	Ecological characteristics		
Subclass 4		Paleoheterodonta	Middle Cambrian—Holocene	Dominantly Paleozoic	Equivalved with closed margins	Wholly aragonitic with nacre and prisms	–	Heterodont teeth; rarely becoming taxodont in a few genera	–	Burrowers. Includes some freshwater forms	*Modiolopsis, Unio, Trigonia*	Moore (1969a, b)
	Order 1	Modiomorphoida	Middle Cambrian—Late Permian	Generally unornamented. Extinct Paleozoic marine group		–	–		–	–	Calyptogena	Moore (1969a, b)
	Order 2	Unionoida	Devonian—Holocene	Shell flatly compressed to globular	Inequilateral and equivalved to subequivalved	–	–	Dentition, when present, consist of only one or two subumbonal teeth	–	–	–	Moore (1969a, b)
	Order 3	Trigoniodia	Middle Ordovician?, Devonian—Holocene	Shell surface smooth ornamented	Equivalved	–	–	Hinge teeth radiating from beak	–	–	–	Moore (1969a, b)
Subclass 5		Heterodonta	Middle Ordovician—Holocene	–	Equivalved and inequivalved forms	Wholly aragonitic with mainly crossed lamellar and complex crossed lamellar microstructures	–	Hinge structure may degenerate to a desmodont condition. Mostly large heterodont teeth	Gills eulamellibranchiate	Majority are shallow and deep burrowers, but the extinct rudists were either cementing or free-living	*Modiolopsis, Unio, Trigonia,* cockles, clams, shipworms, and giant clams	Moore (1969a, b), Clarkson (1993)
	Order 1	Veneroida	Middle Ordovician—Holocene	–	Commonly equivalved	–	–		–	Habit active or nestling, rarely sedentary burrowers or sessile	*Corbicula, Sphaerium* and *Dreissena*	Moore (1969a, b)

(continued)

Table 4.1 (continued)

Subclass	Order	Name	Age	Major characteristics							Examples	References
				Shell characteristics	Shell type	Shell composition	Muscle scars	Dentition	Gills	Ecological characteristics		
Order 2		Myoida	Carboniferous —Holocene	Hinge with one cardinal tooth on each valve			–		–		Saltwater clams (geoducks, and shipworms)	Moore (1969a, b)
Order 3		Hippuritoida	Middle Silurian - Late Cretaceous (Maastrichtian)	Rudists and pachydonts, thick-shelled, abberant heterodonts	Primitive forms equivalved	–	–	Few thick and amorphous hinge teeth - Pachydont dentition	–	Most attached and strongly inequivalved. Sizes up to 2 m long	*Diceras*, *Hippurites*, *Radiolites*	Moore (1969a, b), Pojeta (1987)
Subclass 6		Anomalodesmata	Early Ordovician?, Middle Ordovician—Holocene	–	equivalved to subequivlaved	Wholly aragonitic (largely prisms and nacre)	Mainly Isomyarian	Desmodont dentition. Largely toothless	Gills eulamellibranchiate	Burrowing and cementing, byssate	*Cuspidaria*, *Poromya*	Moore (1969a, b), Pojeta (1987)
Order		Pholadomyoida	Early Ordovician?, Middle Ordovician—Holocene	Ordinal characters same as those of subclass Anomalodesmata		–					*Thracia*	Moore (1969a, b)

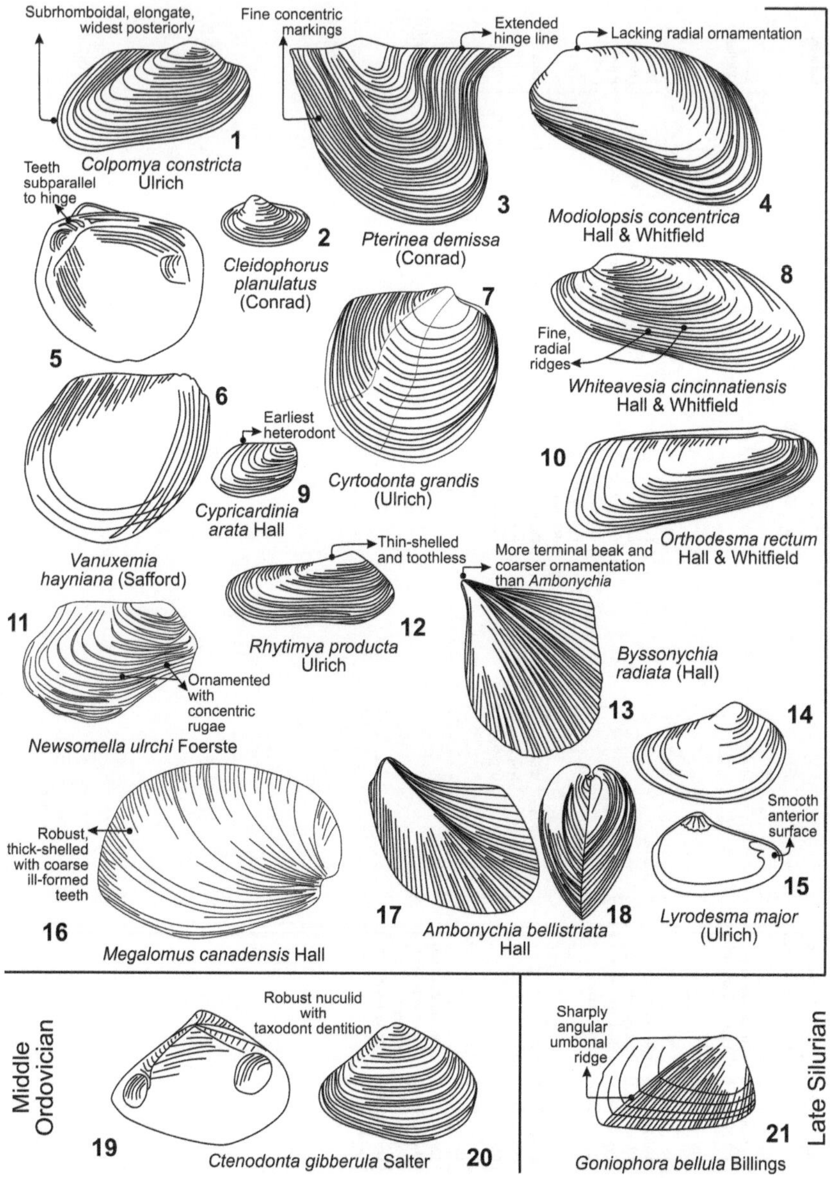

Fig. 4.9 Examples of select Ordovician and Silurian pelecypods and their major distinguishing characters

benthos (Pojeta 1987; Clarkson 1993). Most modern pelecypods belong to one of two subclasses: the largely epifaunal Pteriomorphia and the infaunal Heterodonta (see also Table 4.1) (Fig. 4.19).

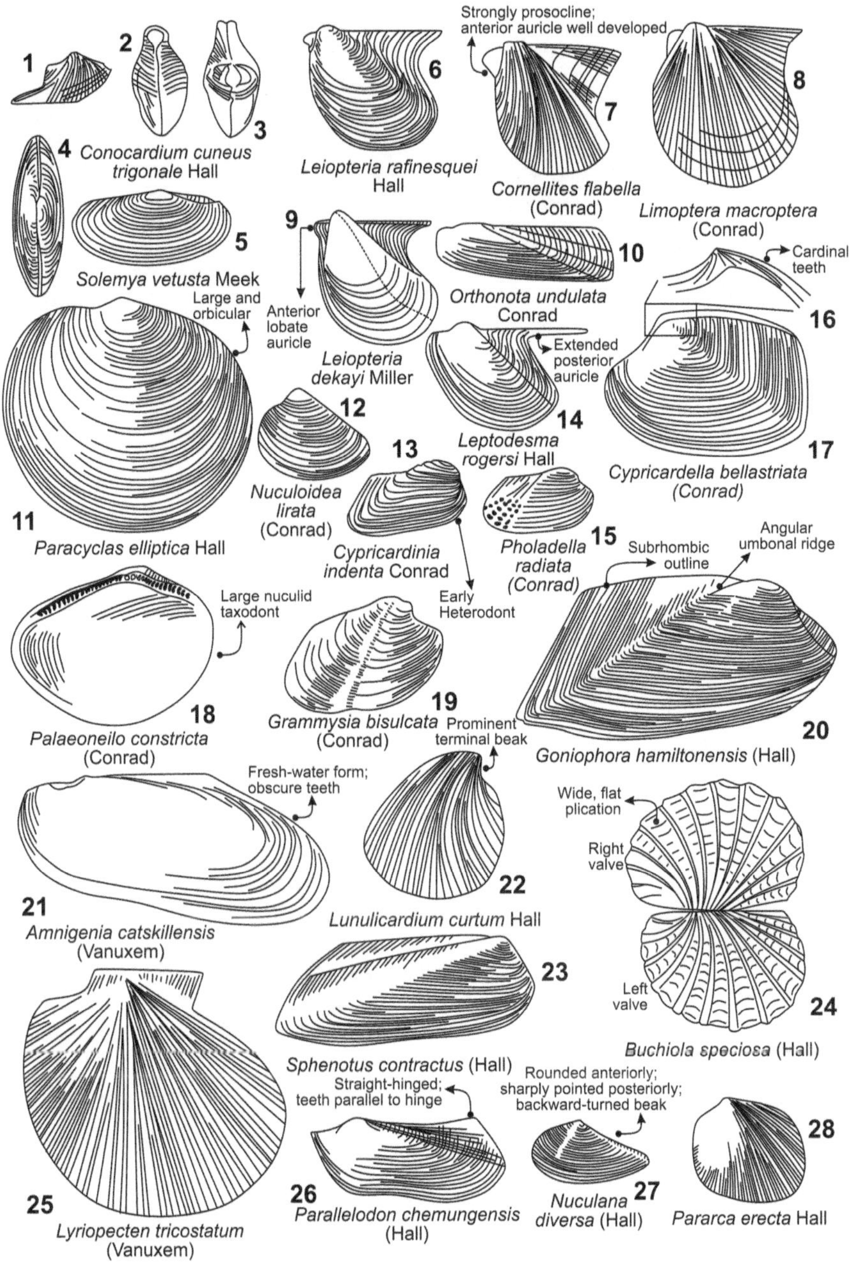

Fig. 4.10 Examples of select Devonian pelecypods and their major distinguishing characters

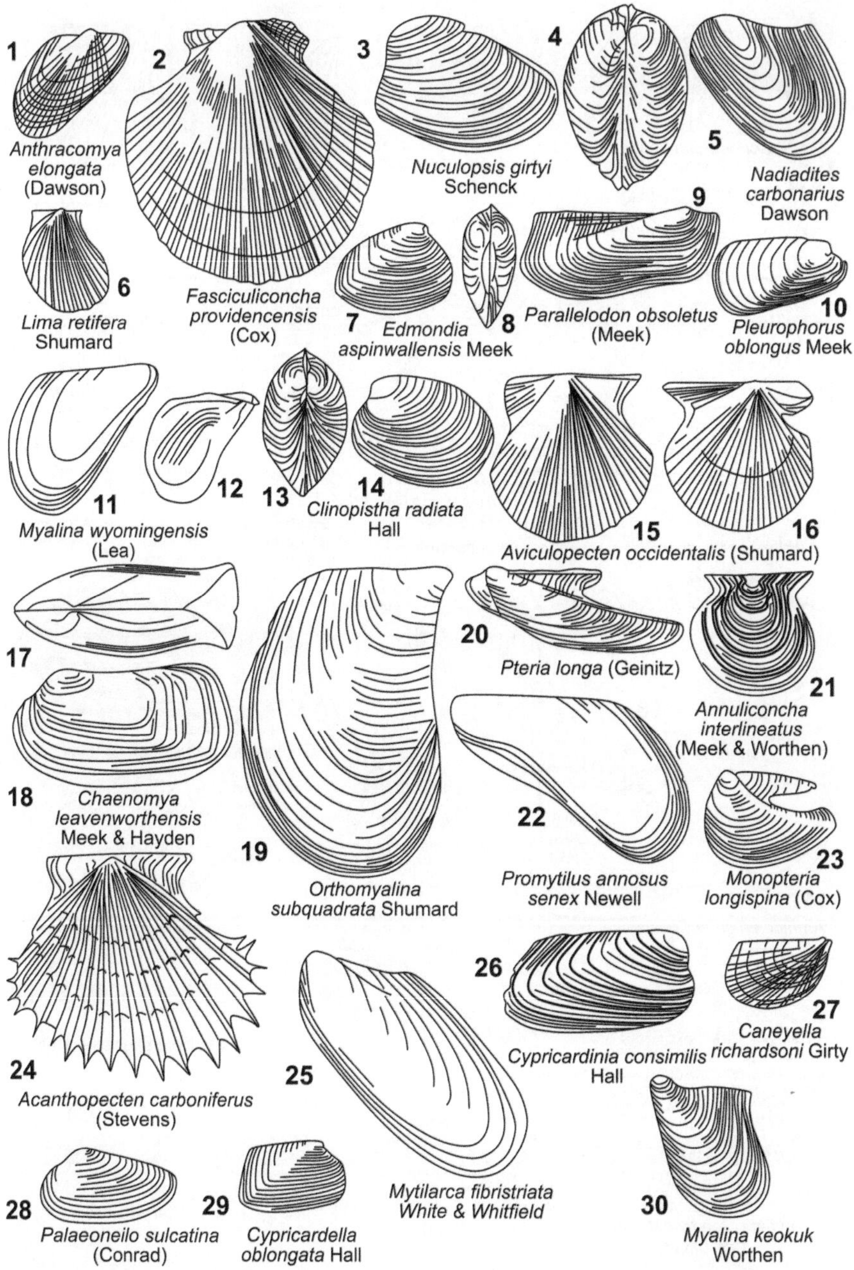

Fig. 4.11 Examples of select Early Mississippian to Late Pennsylvanian pelecypods and their major distinguishing characters

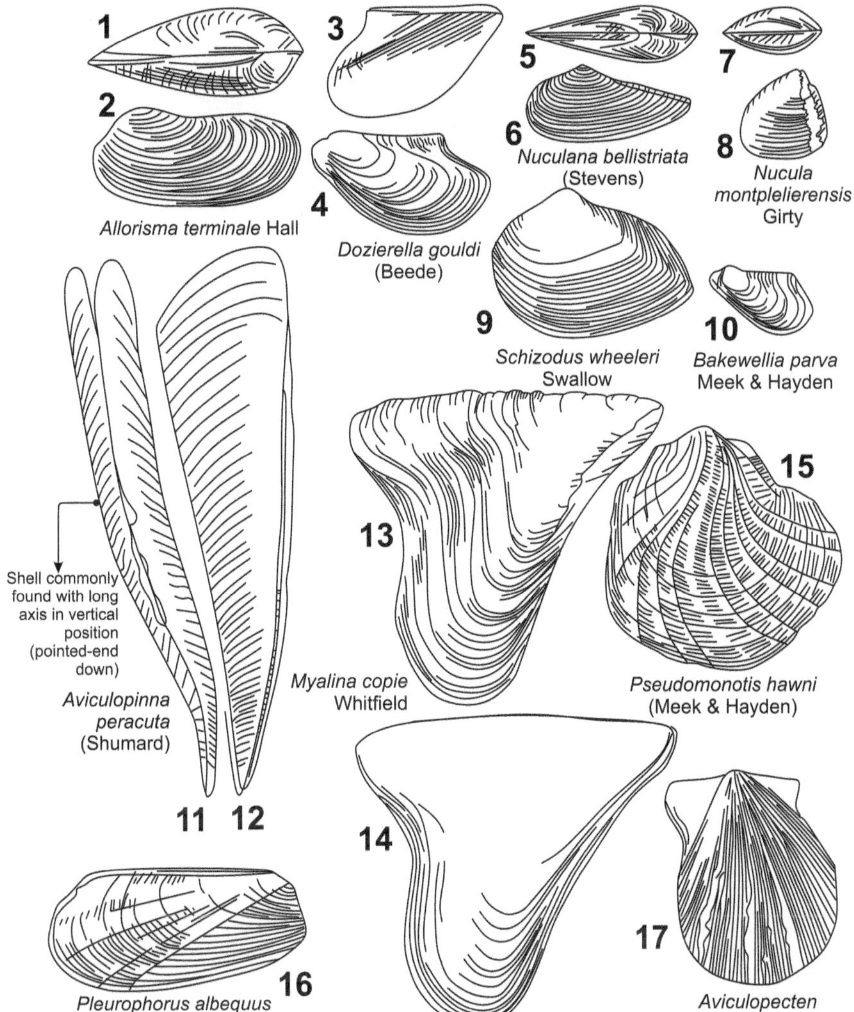

Fig. 4.12 Examples of select Early Permian pelecypods and their major distinguishing characters

Appendix gives the list of illustrated specimens mentioning the chapter number, species name, age, and locality along with its figure number within the said chapter.

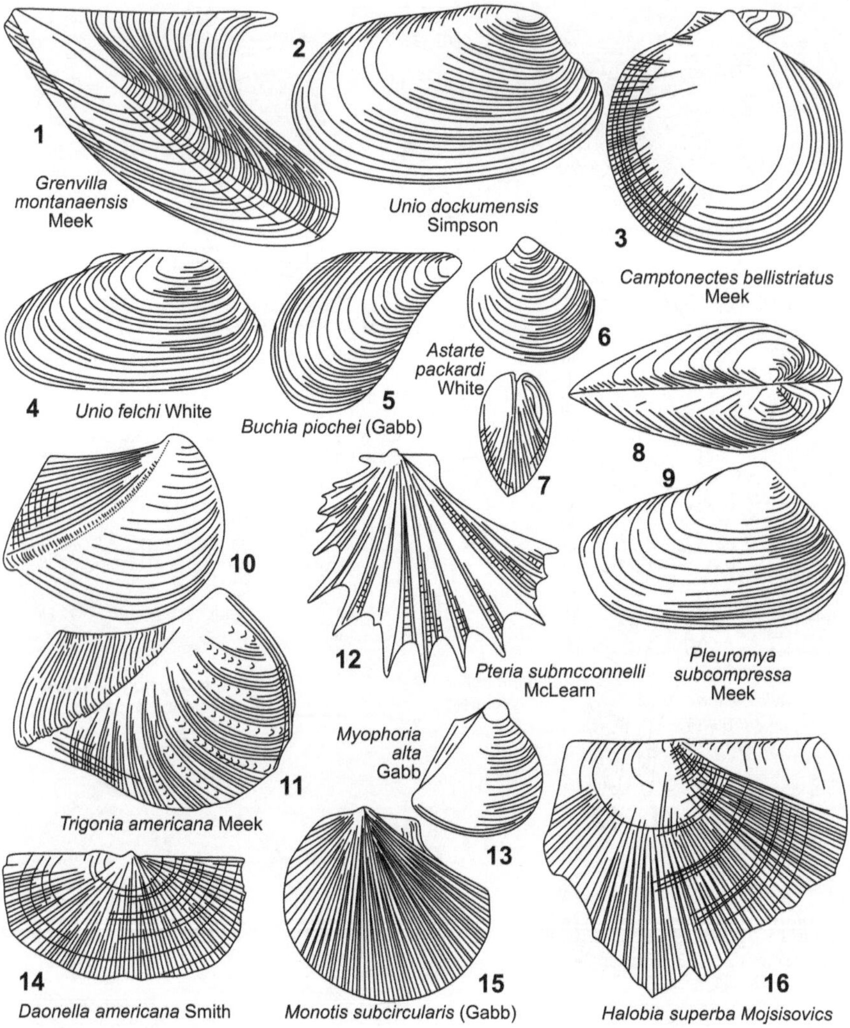

Fig. 4.13 Examples of select Middle Triassic to Late Jurassic pelecypods and their major distinguishing characters

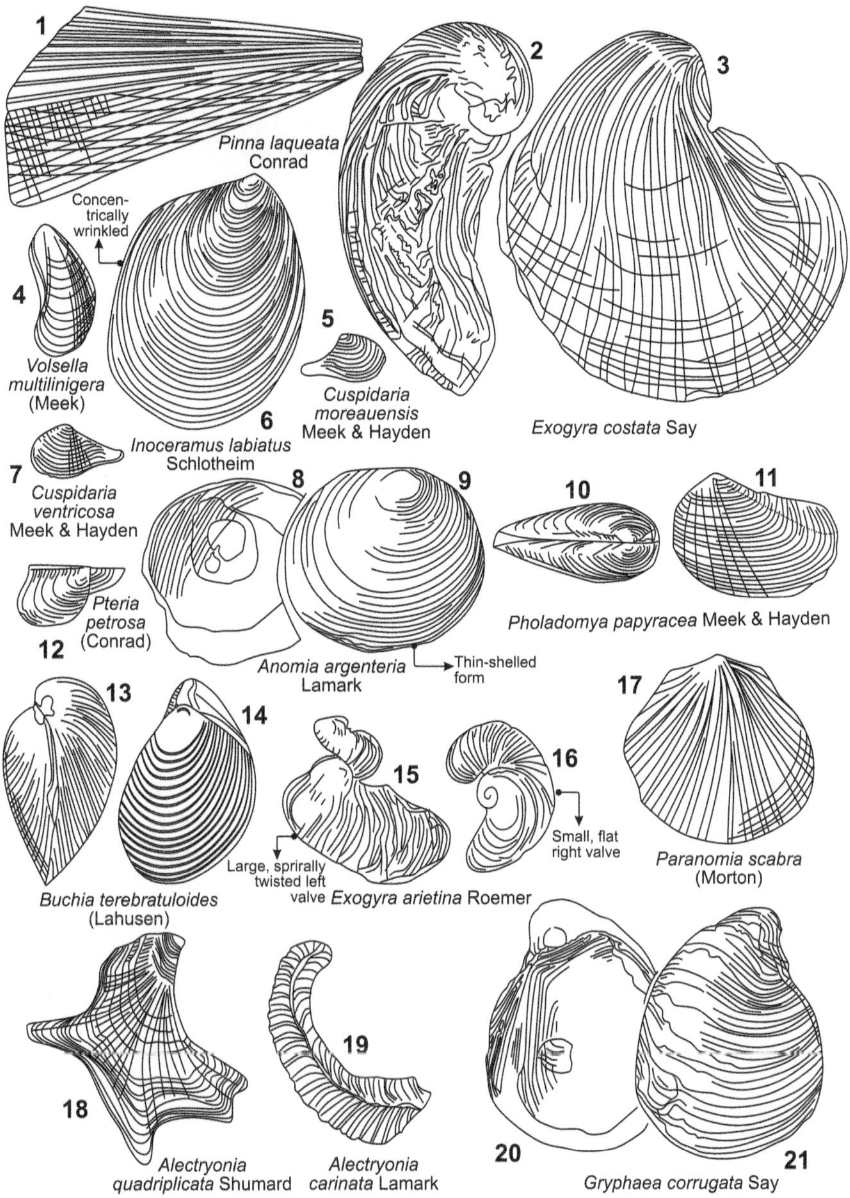

Fig. 4.14 Examples of select Cretaceous pelecypods and their major distinguishing characters

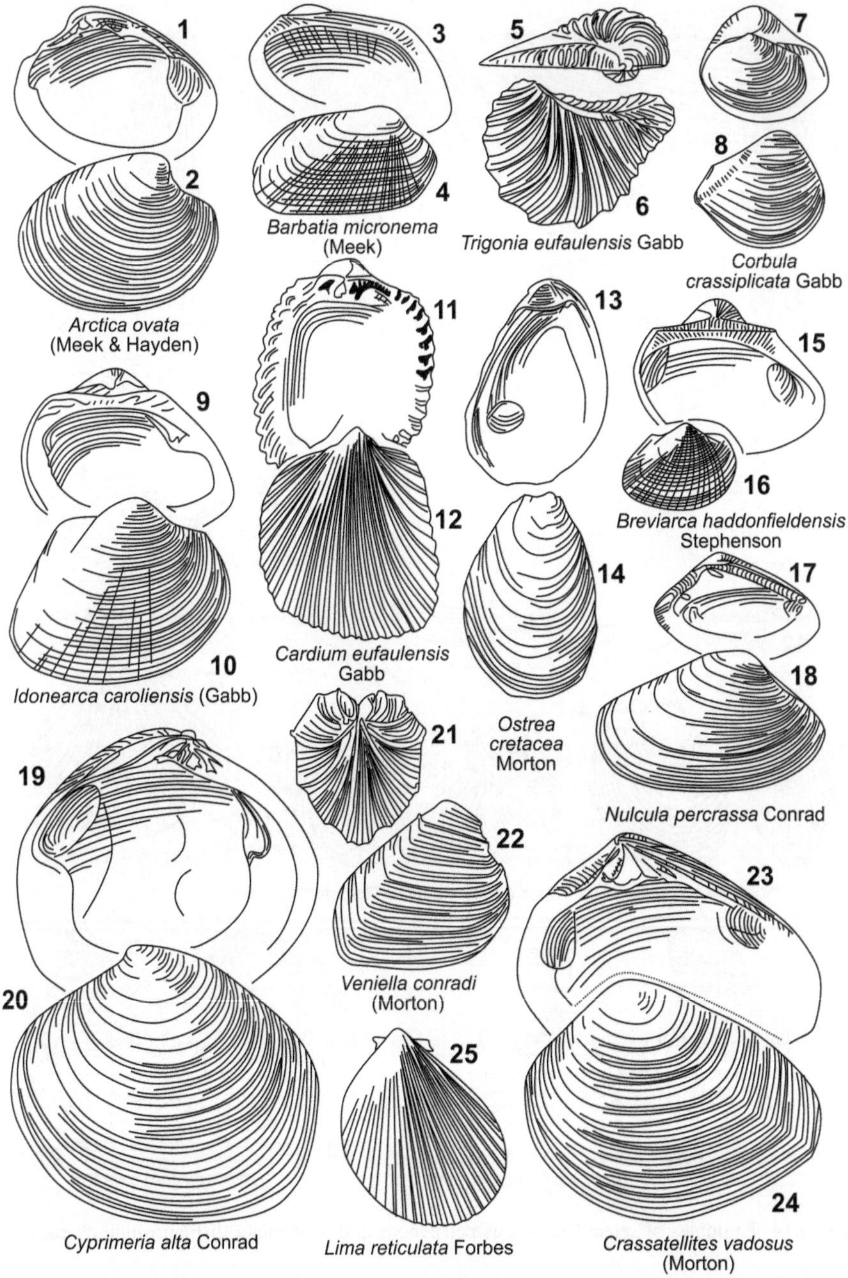

Fig. 4.15 Examples of select Late Cretaceous pelecypods and their major distinguishing characters

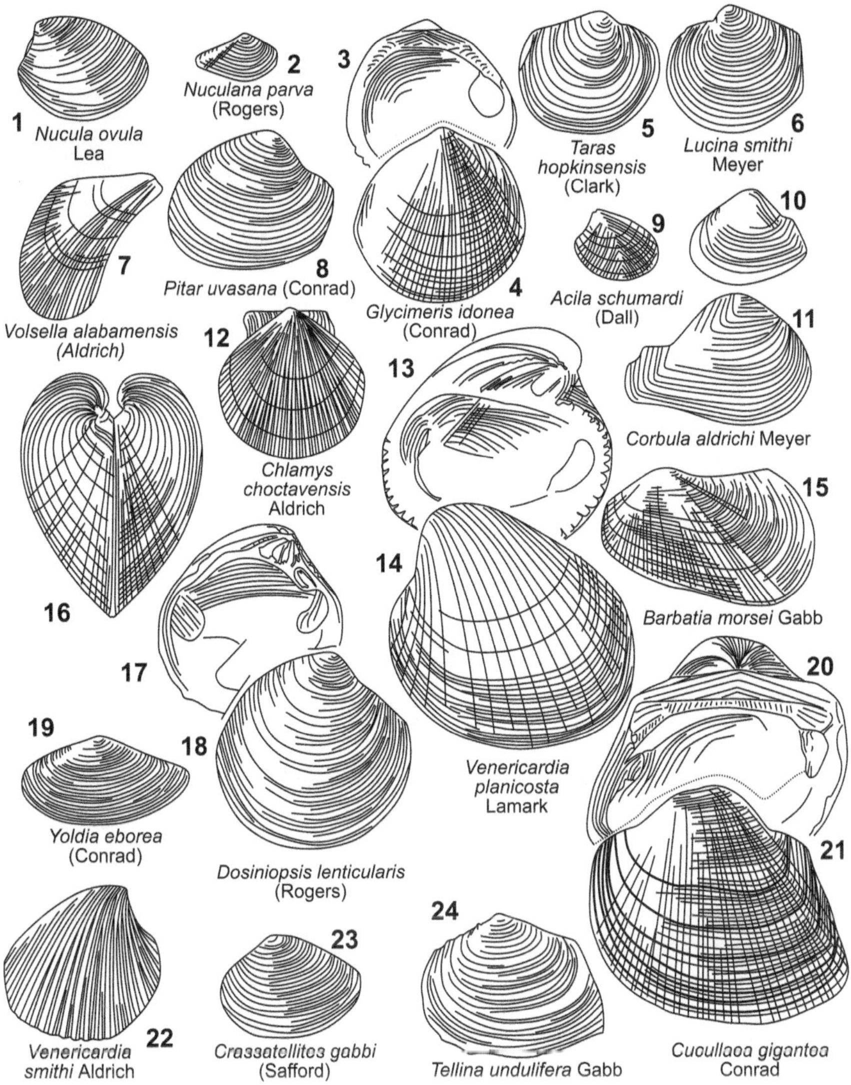

Fig. 4.16 Examples of select Paleogene pelecypods and their major distinguishing characters

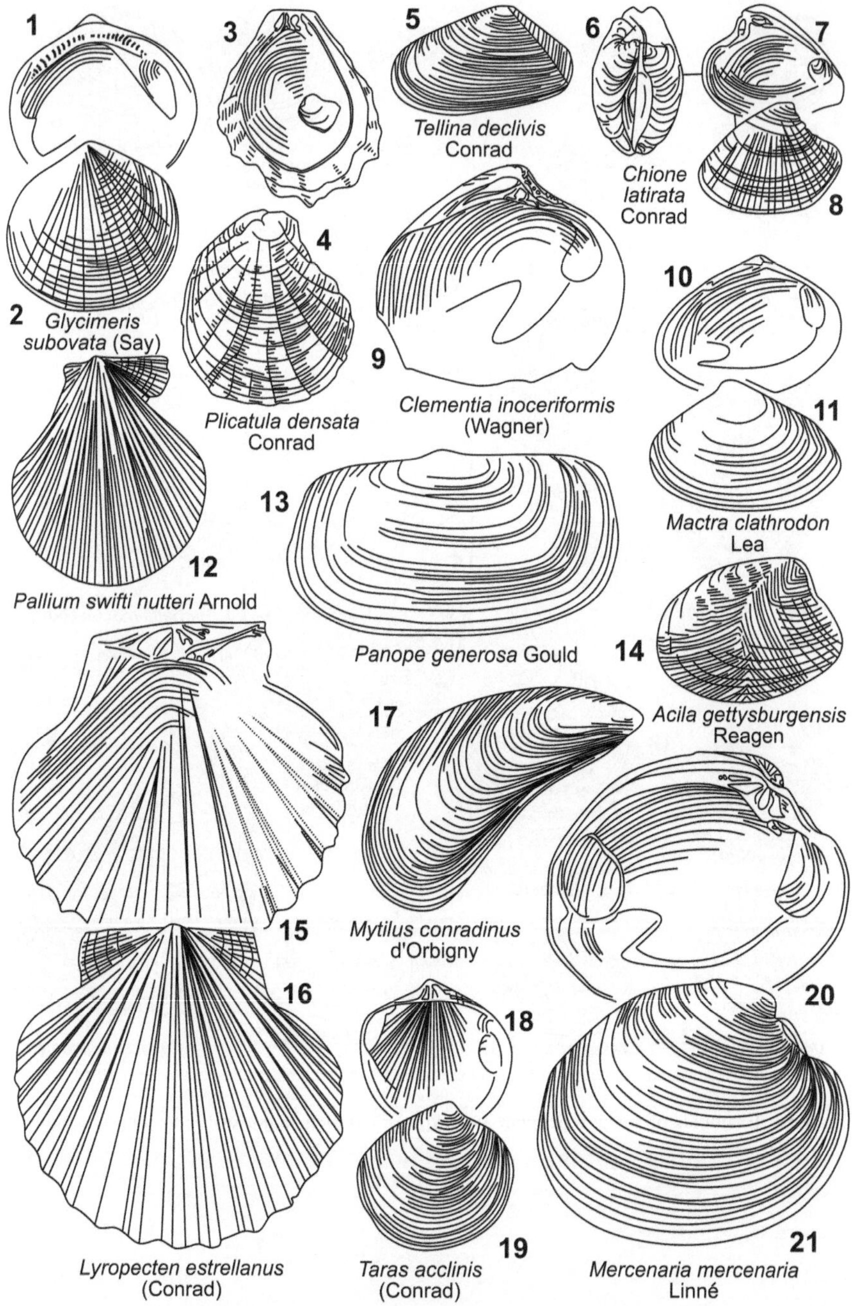

Fig. 4.17 Examples of select Miocene pelecypods and their major distinguishing characters

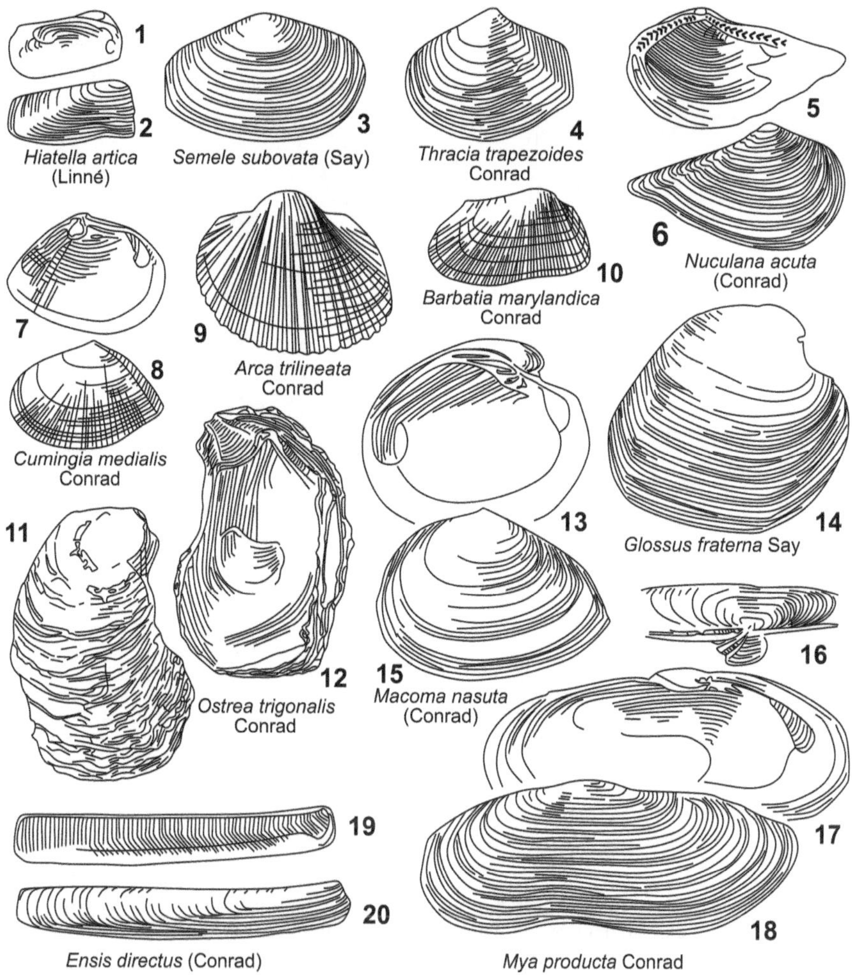

Fig. 4.18 Examples of select Miocene pelecypods and their major distinguishing characters (continued from Fig. 4.17)

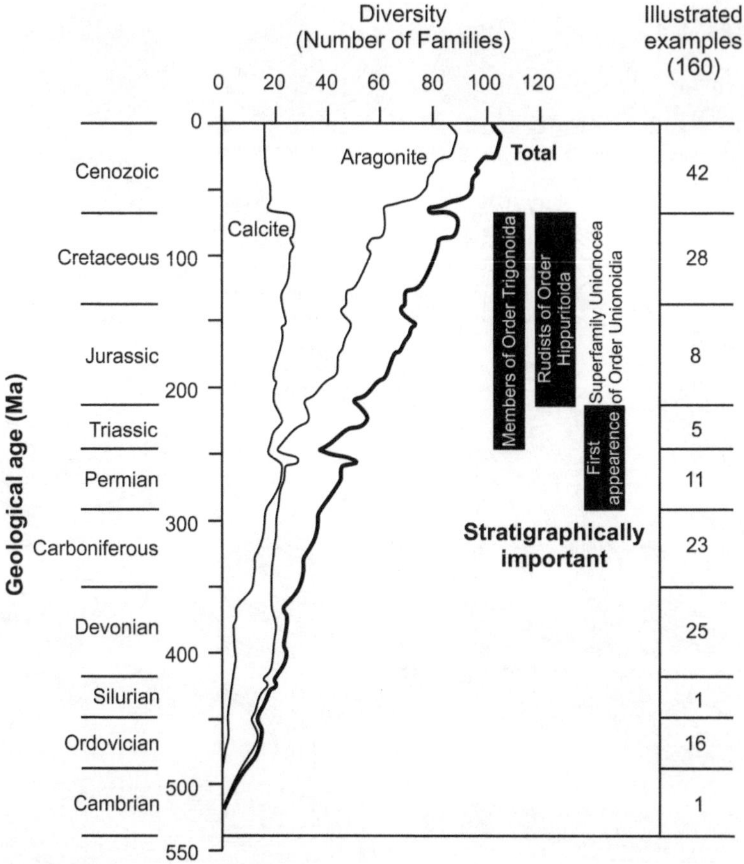

Fig. 4.19 Pelecypod diversity, mineralogy (aragonite, calcite; after Skelton and Benton 1993; Amler et al. 2000) and stratigraphically important taxa. Only marine families, no freshwater ones

References

Amler, M., R. Fischer, and N. Rogalla. 2000. Systematische Einheiten. In *Muscheln, Haeckel-Bücherei*, ed. H.K. Erben, G. Hillmer and H. Ristedt, vol. 5, 70–133.

Barrande, J. 1881. *Système Silurien de Centre de la Bohême. Volume VI: Acéphales.* Système Silurien de Centre de la Bohême, 1–342.

Bonnani, 1681. *Ricreatione dell'occhio.*

Clarkson, E.N.K. 1993. Invertebrate palaeontology and evolution. London: Chapman and Hall. 434 p.

Elicki, O., and S. Gürsu. 2009. First record of *Pojetaia runnegari* Jell, 1980 and *Fordilla* Barrande, 1881 from the Middle East (Taurus Mountains, Turkey) and critical review of Cambrian bivalves. *Paläontologische Zeitschrift* 83(2): 267–291.

Moore, R.C., ed. 1969a. Treatise on invertebrate paleontology. Part 1. *Mollusca 6, Bivalvia*, 1–489. Geological Society of America and University of Kansa Press.

Moore, R.C., ed. 1969b. Treatise on invertebrate paleontology. Part 2. *Mollusca 6, Bivalvia*, 490–952. Geological Society of America and University of Kansa Press.

Pojeta, J. 1975. *Fordilla troyensis* Barrande and early pelecypod phylogeny. *Bulletins of American Paleontology* 67: 363–384.

Pojeta, J., Jr. 1987. Class Pelecypoda. In: Boardman, R.S., Cheetham, A.H., and Rowell, A. J. (eds.), *Fossil Invertebrates*, Blackwell Scientific, Cambridge, Mass., pp. 385–435.

Pojeta, J., Jr., and B. Runnegar. 1976. The paleontology of rostroconch mollusks and the early history of the phylum Mollusca. USGS Numbered Series, Professional Paper 968, 88 pp.

Pojeta, J., B. Runnegar, N.J. Morris, and N.D. Newell. 1972. Rostroconchia: A new class of bivalved mollusks. *Science* 177(4045): 264–267.

Skelton, P.W., and M.J. Benton. 1993. Mollusca: Rostroconchia, Scaphopoda and Bivalvia. In *The fossil record*, ed. M.J. Benton, vol. 2, 237–263.

Chapter 5
Trilobites

5.1 Introduction

Trilobites, marine arthropods, belong to phylum Arthropoda, the most diverse phylum that includes all insects, millipedes, centipedes, scorpions, and crustaceans, such as crabs and lobsters.

The first trilobites appeared in Early Cambrian sediments (early part of Series 2, $\sim$521 Ma; Fig. 5.1) in Africa (Morocco), Europe (Siberia, Spain), and North America (Laurentia) (Hollingsworth 2008). The earliest forms include *Profallotaspis jakutensis* Repina from SE Siberia (in the Atdabanian Stage = Stage 3 of Series 2; see Fig. 5.1; Table 5.1), *Fritzaspis* sp. from the Esmeralda Basin of western Nevada and eastern California (USA), *Hupetina antiqua* (Marruecos) and *Eofallotaspis tioutensis* Sdzuy from the Anti-Atlas Mountains of Morocco and *Serrania gordaensis* Linani from Spain (Hughes 2007; Hollingsworth 2008; Liñán et al. 2008; Clarkson et al. 2006). Thus, the Olenellids are the earliest occurring trilobites that include the members of order Redlichiida (Fig. 5.2; see also Hughes 2007), suborder Olenellina, and particularly that of Fallotaspididae (see also Harrington et al. 1959; Fortey and Owens 1997; Brezinski 1999; Chatterton and Speyer 1997; Fortey 2000, 2001; Hollingsworth 2008; Gon 2014). Trilobite's long geological record (of $\sim$270 Ma; see Figs. 5.1 and 5.2), until their end-Permian extinction (251 Ma), is largely due to their mineralized exoskeleton (of calcite and phosphate) which provided the required preservational edge. Albeit this long geological history, the trilobites also witnessed a continuous decline throughout the Late Palaeozoic; from their high of 63 families in Late Cambrian to a low of 2 in Late Permian (Fig. 5.1). Orders Redlichiida (Fallotaspidoidea) and Ptychopariida (Ellipsocephaloidea) were the first Early Cambrian orders of trilobites [Fig. 5.5(2)]. Since then, trilobites proliferated, diversified, and gave rise to 10 orders, $\sim$150 families, $\sim$5000 genera, and $\sim$20,000 species (Mikulic et al. 2007; Lieberman and Karim 2010; Gon 2014).

S. Jain, *Fundamentals of Invertebrate Palaeontology*, Springer Geology,
DOI 10.1007/978-81-322-3658-0_5

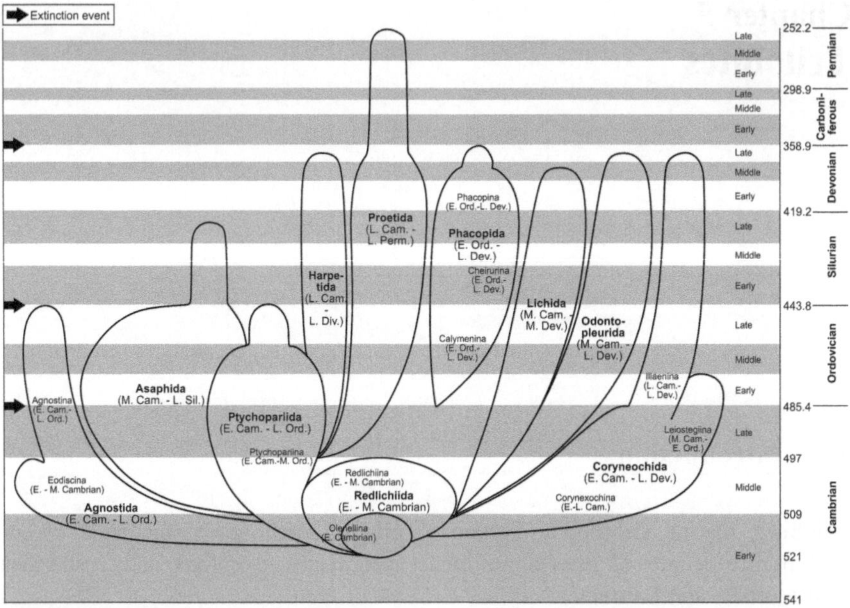

Fig. 5.1 Summary of the evolutionary history of the major trilobite clades plotted against stratigraphic time (see also Hughes 2007). The *y*-axis is in log scale to permit detailed illustration of Cambrian and Ordovician diversifications. No singular character (such as facial sutures, numbers of thoracic segments, etc.) define the mode of classification. Instead, several characters (such as facial sutures, glabellar shape and pattern of lobation, eyes, thoracic features and numbers of thoracic segments, pygidial shape, size and segmentation, spinosity, hypostomal conditions and shared ontogeny) play a role in defining trilobite orders. With permission, figure modified after Gon (2014). All numerical ages have been recalibrated to the latest Gradstein et al. (2012) timescale. *Black arrows* mark extinction events where trilobites suffered (see also Fig. 5.2)

Trilobites were mainly bottom dwellers, and developed marked provincialism in their faunas (see Sect. 5.7) that also made them useful paleobiogeographic indicators, especially in the Cambrian and Early Ordovician times. For rocks of this time period, they are also very useful for biostratigraphy, more so for the Cambrian (especially agnostoid trilobites, particularly in Late Cambrian). In fact, most of the Cambrian stratigraphy is based on the use of trilobite marker fossils (see Tables 5.1 and 5.2) (see also Geyer 1998; Babcock et al. 2005; Peng et al. 2012).

5.2 Shell

The exoskeleton of trilobites is longitudinally divided into three parts—a Cephalon (cephalic shield or head-shield) with a pair of compound eyes, a jointed Thorax (composed of 2–40 segments) and a Pygidium (tail) made up of fused segments (Fig. 5.3). Typically, trilobites measure between 2–8 cm (20–800 mm) in length;

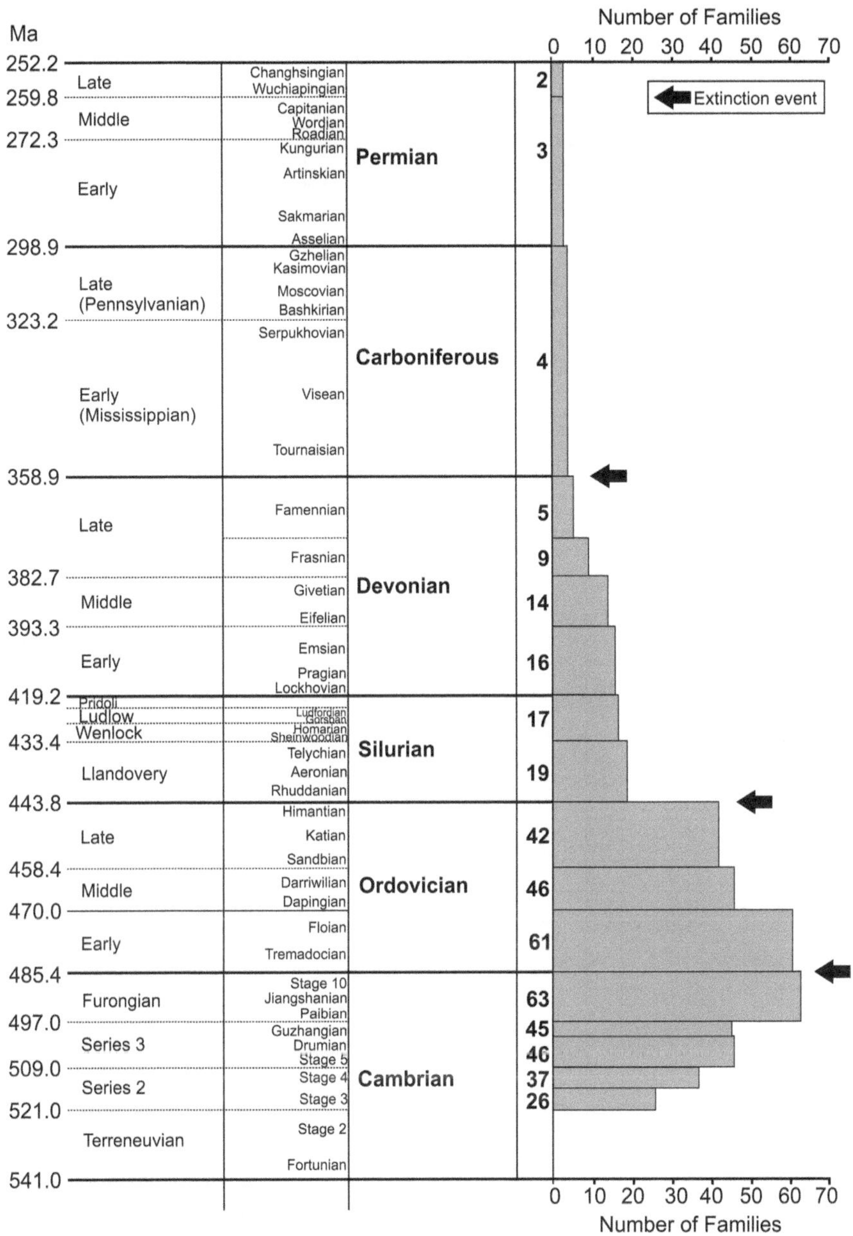

Fig. 5.2 The stratigraphic distribution and diversity of Trilobite families. Only to two families (Proetidae and Brachymetopidae, both in the order Proetida) remained before their end-Permian extinction. With permission, figure modified after Gon (2014). All numerical ages have been recalibrated to the latest Gradstein et al. (2012) timescale. *Black arrows* mark extinction events where trilobites suffered

Table 5.1 Trilobite-based regional biostratigraphic zonal schemes of the Cambrian, 505–483 Ma (modified after Peng et al. 2012; see also Gradstein et al. 2012)

Cambrian Time Scale

Ma	Epoch/Age (Stage)	South China	Serbia	Australia	Laurentia
		Trilobite Zonation			
485	Ordovician		*Eo[atokephalus nyaicus*	Conodonts	*Symphysurina bulbosa*
	485.4	*Hysterolenus asiaticus*	*Loparella loparica-Plethopeltides magnus*	Conodonts	*Symphysurina brevispicata*
	Furongian – Stage 10	*Leiostegium constrictum-Shenjiawania brevis*	*Dolgeuloma abunda - D. dolganensis*	Conodonts	*Missisquoia*
		Mictosaukia striata-Fatocephalus	*Lotagnostus americanus* / *Ketyna ketiensis-Monosulcatina laeve*	*Mictosaukia perplexa*	*Eurekia apopsis*
		Leioagn. cf. bexelli Archaeul. taoyuanens	*Lotagnostus americanus* / *Ketyna ketiensis-Monosulcatina laeve*	*Neoagnostus quasibilobus-Shergoldia nomas*	*Saukiella serotina*
	489.5	*Lotagnostus americanus*	*Lotagnostus americanus* / *Ketyna ketiensis-Monosulcatina laeve*	*Sinosaukia impages*	*Saukiella junia / Saukiella pyrene-Rasettia magna*
490	Jiangshanian	*Probinacunaspis nasalis-Peichiashania hunanensis*	*Parabolinites rectus* / *Ketyna ketiensis-Monosulcatina laeve*	*Rhhaptagnostus clarki-maximus-Rh. papilio* / *Rhhaptagnostus bifax-Neoagn. denticulatus*	*Saukiella junia / Saukiella pyrene-Rasettia magna*
		Eolotagnostus decoratus - Kaolishaniella	*Ketyna ketiensis-Monosulcatina laeve*	*Rh. clarki prolatus - Caz. sectatrix*	*Saukiella junia / Saukiella pyrene-Rasettia magna*
		Eolotagnostus decoratus - Kaolishaniella	*Plicatolina perlata*	*Rh. c. patulus-C. squamosa - H. lilyensis* / *Peichiashania tertia-Peichiashania quarta*	*Ellipsocephaloides - Idahoia*
		Rhaptagnostus ciliensis / Onchonotellus cf. kuruktagensis	*Maladioidella abdita*	*Peichaish. secunda-Prochuangia glabella* / *Wentsuia iota - Rhaptagnostus apis*	*Taenicephalus*
	494	*Agnostotes orientalis*	*Agnostotes orientalis*	*Irvingella tropica*	*Elvinia*
495	Paibian	*Tomagnostus orientalis - Corynexochus plumula*	*Erixanium sentum*	*Stigmatoa diloma* / *Erixanium sentum*	*Dunderbergia*
		Agnostus inexpectans - Proceratopyge protracta	*Stigmatoa destruncta*	*Proceratopyge cryptica*	*Aphelaspis*
	497	*Glyptagnostus reticulatus*	*Glyptagnostus reticulatus*	*Glyptagnostus reticulatus*	*Aphelaspis*
	Series 3 – Guzhangian	*Glyptagnostus stolidotus*	*Glyptagnostus stolidotus*	*Glyptagnostus stolidotus*	*Crepicephalus*
		Linguagnostis reconditus	*Clavagnostus spinosus*	*Achmarhachis quasivespa*	*Crepicephalus*
		Proagnostus bulbus	*Proagnostus bulbus*	*Erediaspis eretes*	*Proagn. bulbus* / *Cedaria*
500	500.5	*Lejopyge laevigata*	*Lejopyge laevigata*	*Damesella torosa - Ferenepea janitrix* / *Lejopyge laevigata*	*Lejopyge laevigata* / *Bolapidella*
	Drumian	*Lejopyge armata*	*Anomocarioides limbataeformis*	*Goniagnostus nathorsti*	*Ptychagnostus punctuosus* / *Bolapidella*
		Goniagnostus nathorsti	*Anomocarioides limbataeformis*	*Goniagnostus nathorsti*	*Ptychagnostus punctuosus* / *Bolapidella*
		Ptychagnostus punctuosus	*Anopolenus henrici-Corynexochus perforatus*	*Doryagnostus deltoides* / *Ptychagnostus punctuosus*	*Ptychagnostus punctuosus* / *Bolapidella*
		Ptychagnostus atavus	*Anopolenus henrici-Corynexochus perforatus*	*Euragnostus optimus*	*Ptychagnostus atavus* / *Bolapidella*
505	504.5	*Ptychagnostus atavus*	*Tomagnostus fissus*	*Ptychagnostus atavus*	*Ptychagnostus atavus* / *Bolapidella*

Table 5.2 Trilobite-based regional biostratigraphic zonal schemes of the Cambrian, 521–503 Ma (modified after Peng et al. 2012; see also Gradstein et al. 2012)

Cambrian Time Scale

Ma	Epoch/Age (Stage)		Trilobite Zonation: South China	Serbia	Australia	Laurentia	
	Series 3	504.5 Drumian	*Ptychaagnostus atavus*	*Tomagnostus fissus*	*Ptychagnostus atavus*	*P. atavus*	
505	Series 3	Stage 5	*Ptychaagnostus gibbus* *Peronopsis taijiangensis* *Oryctocephalus indicus*	*Ptychaagnostus gibbus* *Kounamkites*	*Ptychaagnostus gibbus* *Pentagnostus shergoldi* *Ptychagnostus praecurrens* *Pentagnostus anabarensis* *Xystridura negrina / Redlichia forresti*	*Ptychagnostus gibbus* *Ptychagnostus praecurrens* *Oryctocephalus indicus*	*Ehmaniella* *Glossopleura* *Albertella* *Plagiura - Poliella*
		509					
510	Series 2	Stage 4	*Ovatoryctocara granulata - Bathynotus holopygus* *Protoryctocephalus wuxuensis* *Arthricocephalus taijiangensis* *Arthricocephalus chauveaui*	*Ovatoryctocara granulata - Schistocephalus antiquus* *Anabaraspis splendens* *Lermontovia grandis* *Bergeroniellus ketemensis* *Bergeroniellus ornata* *Bergeroniellus asiaticus*	*Xystridura negrina / Redlichia forresti* *Pararaia janeae*	*Amecephalus arrojosensis* *Eokochaspis nodosa* *Olenellus*	
		514					
515	Series 2	Stage 3	*Arthricocephalus jiangkouensis* *Szechaunolenus - Paokannia* *Ushaspis* *Sinodiscus - Hupeidiscus* *Tsunyidiscus niutitangensis*	*Bergeroniellus gurarii* *Bergeroniellus micmacciformis - Erbiella* *Judomia* *Pagetiellus anabarus* *Fallotaspis* *Profallotaspis jakutensis*	*Pararaia janeae* *Pararaia bunyerooensis* *Pararaia tatei* *Abadiella huoi*	*Olenellus* *Nevadella* *Fallotaspis* *Fritzaspis*	
520		521	?	?		?	

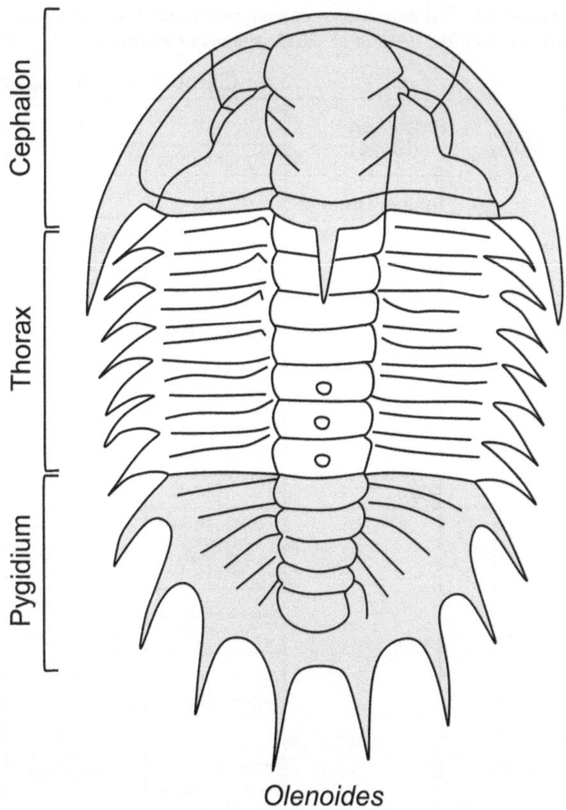

Fig. 5.3 The Trilobites are characterized by the longitudinal division of their exoskeleton into three parts—a Cephalon (cephalic shield or head-shield) with a pair of compound eyes, a jointed Thorax and a Pygidium (tail)

Acanthopleurella, when fully grown, is just over 1 mm in length, whereas, the largest known trilobite, *Isotelus*, is 72 cm long (720 mm). Some of the largest forms are illustrated in Fig. 5.4 (see Reimann 1942; Rabano 1989; Whittington 1992; Kaesler 1997; Rudkin et al. 2003).

The trilobite head is covered by the unsegmented, semicircular or triangular Cephalon (Fig. 5.3). The cephalon is characterized by a median Glabella [Fig. 5.5 (1–5)] and two lateral portions, called Cheeks [Fixed and Free; Fig. 5.5(4)]. The Axial furrow separates the glabella from the cheeks [that are more or less triangular in shape and less convex than the glabella; Fig. 5.5(4)], by means of a furrow on either side [Fig. 5.5(1)]. The posterior angles of the cheeks [the Genal angle; Fig. 5.5(7)] may be rounded (as in *Calymene*) but is often pointed or produced into spines [called the Genal spines; as in *Trinucleus*; Figs. 5.5(1 and 4)]. A suture [Facial suture; Fig. 5.5(4)] divides each cheek into two halves. The Fixed cheek which is immovable is the inner part between the facial suture and the glabella, whereas the Free cheek is movable on the Fixed cheek [Fig. 5.5(4)]. Cephalon has eyes on its upper surface, one on each free cheek near the facial suture, and near the middle of the cheek [Fig. 5.5(4)]; the eyes are compound, each consisting of a number of lenses. In a few genera, the eyes are absent such as in *Agnostus* and

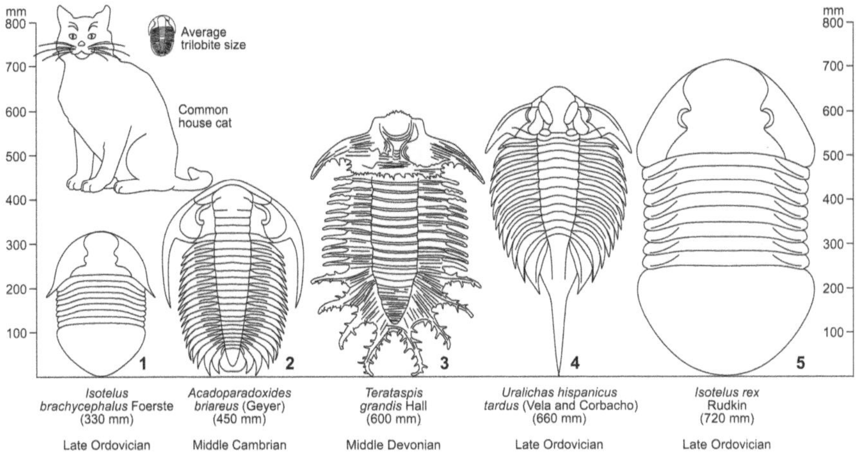

Fig. 5.4 Comparative scales depicting the largest recorded trilobites. For reference a household common cat is shown. Trilobite gigantism is well noted in Orders of Asaphida, Redlichiida, and Lichida. The Late Ordovician *Isotelus rex*, as asaphid, from Churchill, Manitoba (Canada) is the largest of them all (see Rudkin et al. 2003)

Ampyx. The Cephalon continues on the lower surface of the head forming a marginal rim. Attached to the rim and situated just in front of the mouth, is an oval-shaped plate called the Hypostome [Fig. 5.5(5)]. A more detailed cephalic nomenclature is given in Fig. 5.5 and corresponding terms are briefly described under Terminology (Sect. 5.3).

The Thorax [Figs. 5.3 and 5.6(1 and 7)] is formed of a series of segments, which varies in number, from 2 to 40. The segments are movable upon one another, and in some cases sufficiently spaced to enable the animal to roll itself like a wood-louse. Two Axial furrows [Fig. 5.6(1)], divide each segment into a median and two lateral parts. The more convex median (central) part forms the Axis [Fig. 5.6(3)]; the lateral parts are called Pleurae [singular: Pleura; Fig. 5.6(1)]. The pleura is curved downwards and backwards, slightly away from the median axis [Fig. 5.6(1)]; this point of curvature is called the Fulcrum [Fig. 5.6(1)]. Generally the anterior part of each pleura overlaps the succeeding one and the rounded fulcrum forms an articulating facet [Fig. 5.6(2)]. The terminal end of each pleura is modified; either rounded or produced into a spine, called a Pleural spine [Fig. 5.6(1)]. A more detailed thorax nomenclature is illustrated in Fig. 5.6(1–7) and corresponding terms are briefly described below under Terminology (Sect. 5.3).

The triangular or semicircular Pygidium [Fig. 5.6(8 and 9)] is made up by a variable number of segments that are immovable and fused, hence different from those of the thorax. The region of segmentation in the pygidium is marked by a groove called the Interpleural groove [Fig. 5.6(8)].

Segments in both thorax and pygidium possess an axis (a central part) and lateral portions; the axis may reach to the posterior extremity or only part of the way

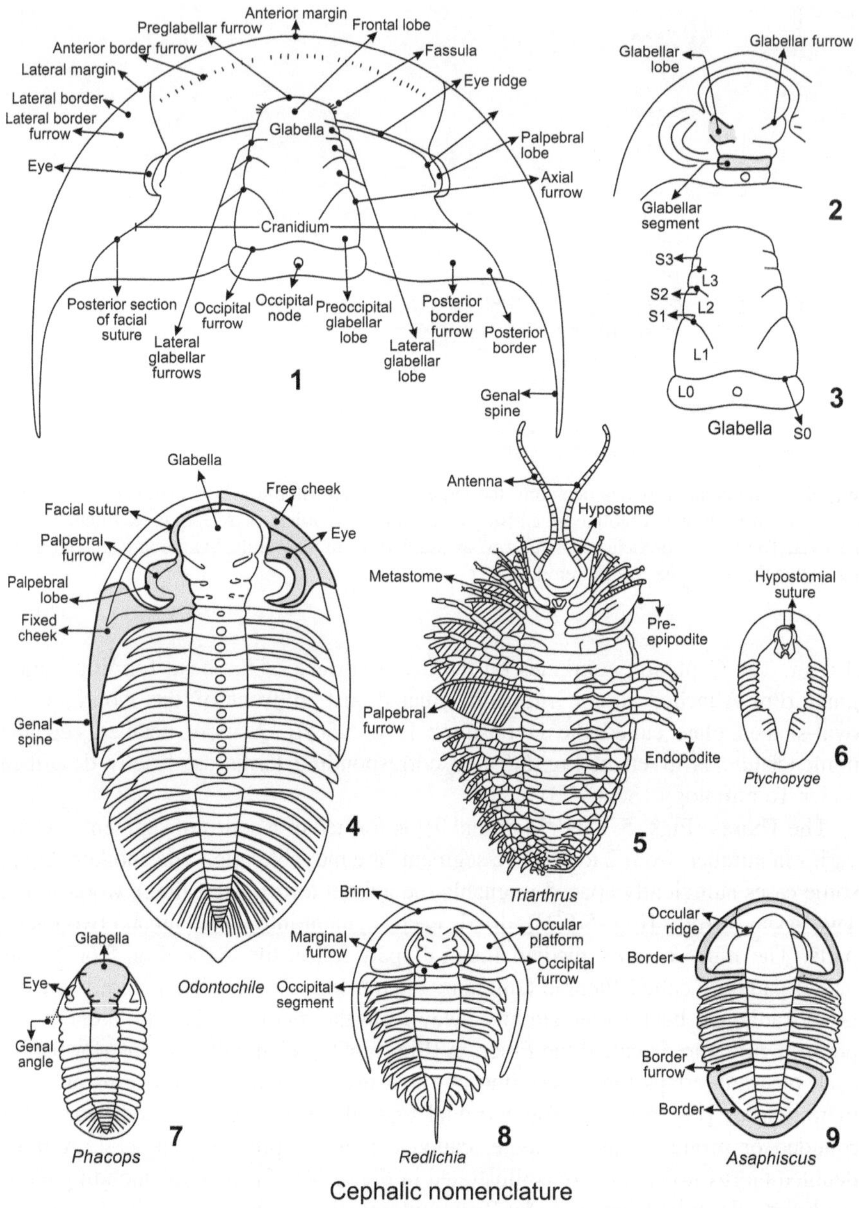

Fig. 5.5 Cephalic nomenclature

[Fig. 5.6(8 and 9)]. The pygidial margin may be even or entire [Fig. 5.6(8 and 9)], or it may be modified into a posterior spine [Fig. 5.6(8)] or a Telson [Fig. 5.6(1)] or have lateral spines [Fig. 5.6(8)].

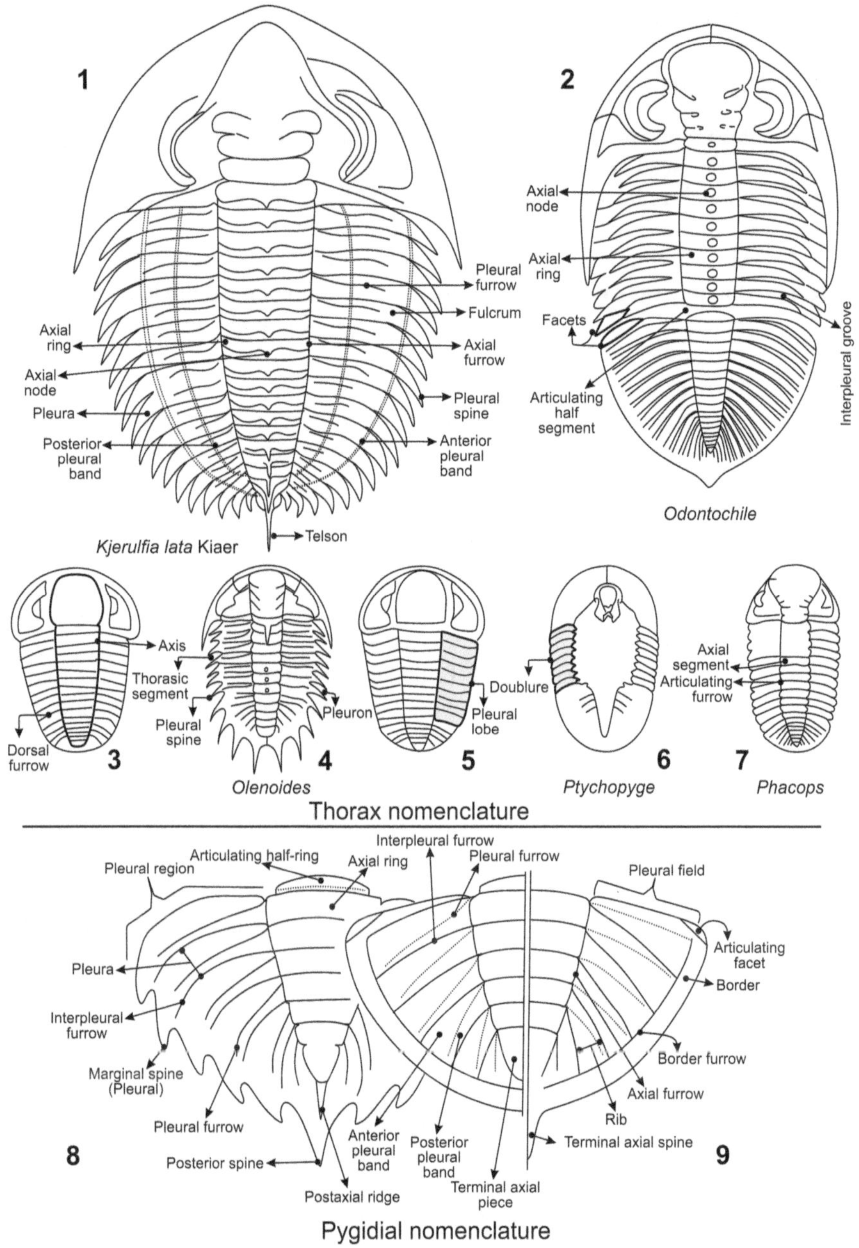

Fig. 5.6 Thorasic and Pygidial nomenclature

A more detailed pygidial nomenclature is illustrated in Fig. 5.6(8 and 9) and corresponding terms are briefly described under Terminology (Sect. 5.3).

5.3 Terminology

5.3.1 Cephalon

5.3.1.1 **Antenna**: Many-segmented sensory appendage attached to the front part of the head [Fig. 5.5(5)]

5.3.1.2 **Anterior border furrow**: Furrow defining the adaxially border of cephalon [Fig. 5.5(1)]

5.3.1.3 **Anterior margin**: The anterior portion of the cephalon [Fig. 5.5(1)]

5.3.1.4 **Border furrow**: It is a furrow that marks the adaxially border of cephalon, pygidium, and hypostome (syn., marginal furrow) [Figs. 5.5(9) and 5.6(9)]

5.3.1.5 **Border**: Outer raised peripheral parts (dorsal in case of cephalon and pygidium and ventral in case of hypostome) [Figs. 5.5(9) and 5.6(9)], and is generally bounded by a border furrow (also called as Marginal rim, or Marginal limb and rarely Rim)

5.3.1.6 **Brim**: Part of cranidium bounded anteriorly by marginal furrow and posteriorly by front of glabella and ocular ridges, or lines running from front of eyes to glabella [Fig. 5.5(8)]

5.3.1.7 **Cephalon (pl., cephala)**: This is the part of carapace in front of thorax (Fig. 5.3)

5.3.1.8 **Cranidium (pl., cranidia)**: Bounded laterally by facial sutures, it is the central part of the cephalon; those having marginal sutures, it includes the dorsal part of the cephalon [Fig. 5.5(1)]

5.3.1.9 **Eye**: Visual area containing one or many lenses, located on either side of the glabella; generally curved in plan and sloping steeply outwards; among proparians and opisthoparians borne by inner margins of free cheek [Fig. 5.5(4)]. Three types of eyes are noted: Holocroal, Schizocroal, and Abathocoral (Fig. 5.7). Their distribution in time is illustrated in Fig. 5.8 (see also Clarkson 1979, Clarkson et al. 2006) and their evolutionary trends are given in Fig. 5.9(1) (see also Feist and Clarkson 1989; Thomas 2005; Clarkson et al. 2006)

5.3.1.10 **Facial suture**: This is the line of junction between cranidium and free cheeks; it may be wholly marginal [Protoparian; see Fig. 5.10(1)], partly marginal [Hypoparian; see Fig. 5.10(2)] and partly dorsal [Proparian; see Fig. 5.10(3)], or wholly dorsal [Opisthoparian; see Fig. 5.10(4)]. The facial suture separates free cheek from fixed cheek and may even be present when the eye is absent. The position of the facial sutures determines the relative size of the fixed and free cheeks. After the death of the animal, or after molting, the cephalic shield frequently falls into pieces, dividing along these suture lines

5.3.1.11 **Fassula**: Ciliated growth at the margins of the preglabellar furrow [Fig. 5.5(1)]

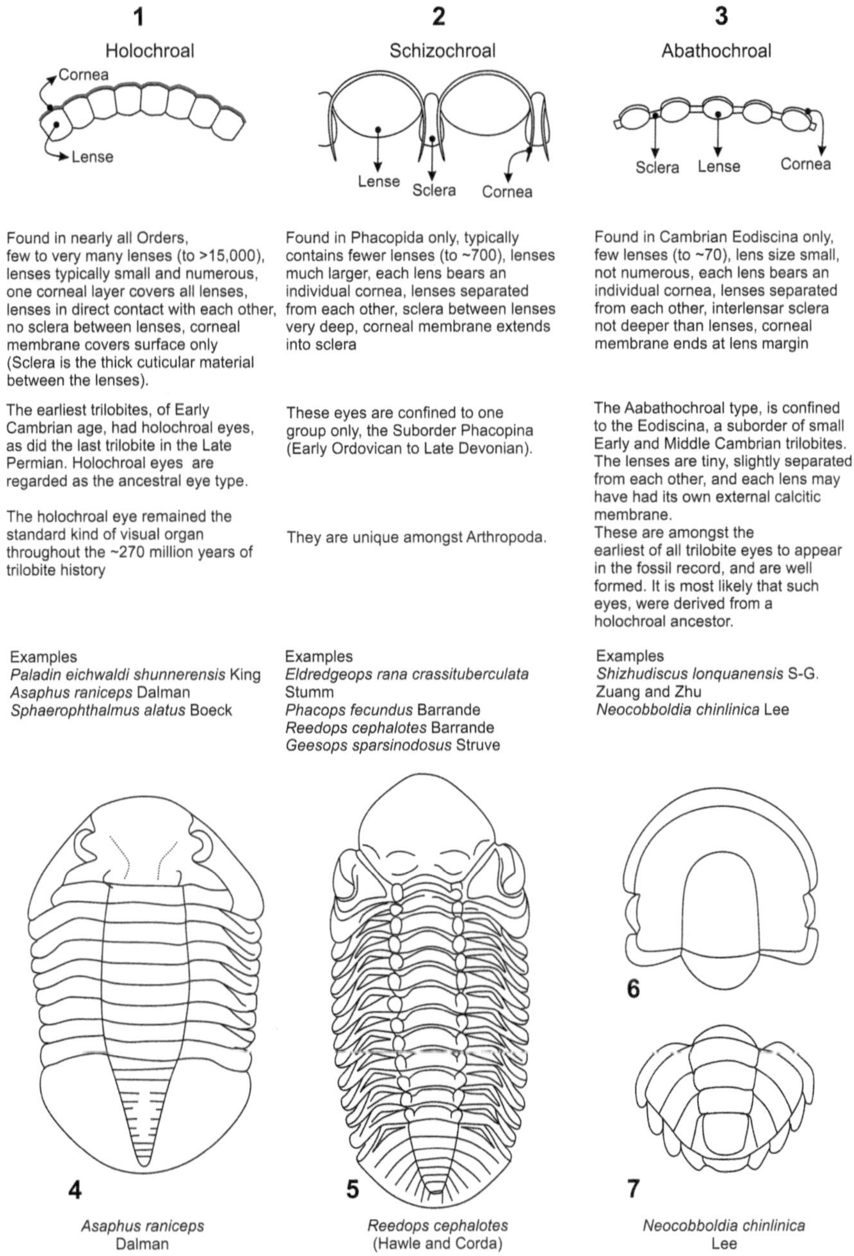

Fig. 5.7 Types of Trilobite eyes

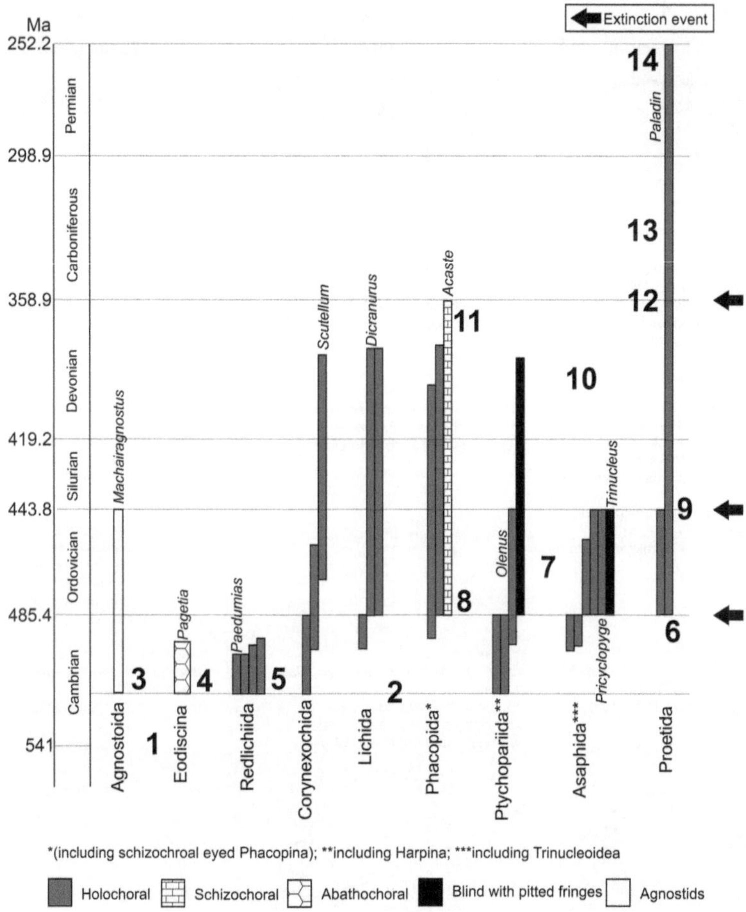

Fig. 5.8 The geological history of trilobites, showing various eye-types and major historical events. Single vertical bars represent suborders, grouped bundles represent orders (modified from Clarkson et al. 2006). Major events numbered 1–13 are: *1* Base of Cambrian System; *2* First appearance of trilobites; *3* Origin of the agnostine system (median eye and reduced ventral compound eyes); *4* Cryptic origin of eodiscid abathochroal eyes; *5* The earliest holochroal eyes in Redlichiida; *6* Extinction event at the end of the Cambrian; *7* Acme of trilobites and proliferation, amongst others, of blind, pitted-fringe taxa—Harpetidae (Ha) and Trinucleoidea (Tr); *8* Origin of schizochroal eyes by paedomorphosis; *9* End-Ordovician major extinction event; *10* Gradual decline and final extinction of many taxa; *11* Loss of eyes in many proetids and phacopids; *12* Late Devonian major extinction event; *13* Proetida continue to the latest Permian; *14* Final extinction of trilobites. All numerical ages have been recalibrated to the latest Gradstein et al. (2012) timescale

5.3.1.12 Fixed cheek: Part of cranidium on either side of glabella; the two fixed cheeks comprise all of cranidium exclusive of glabella and may be confluent in front of it [Fig. 5.5(4)]. The fixed cheeks are the lateral extension of the glabella, to which they are firmly joined, forming the central portion of the cephalon. They may occupy more than two-thirds

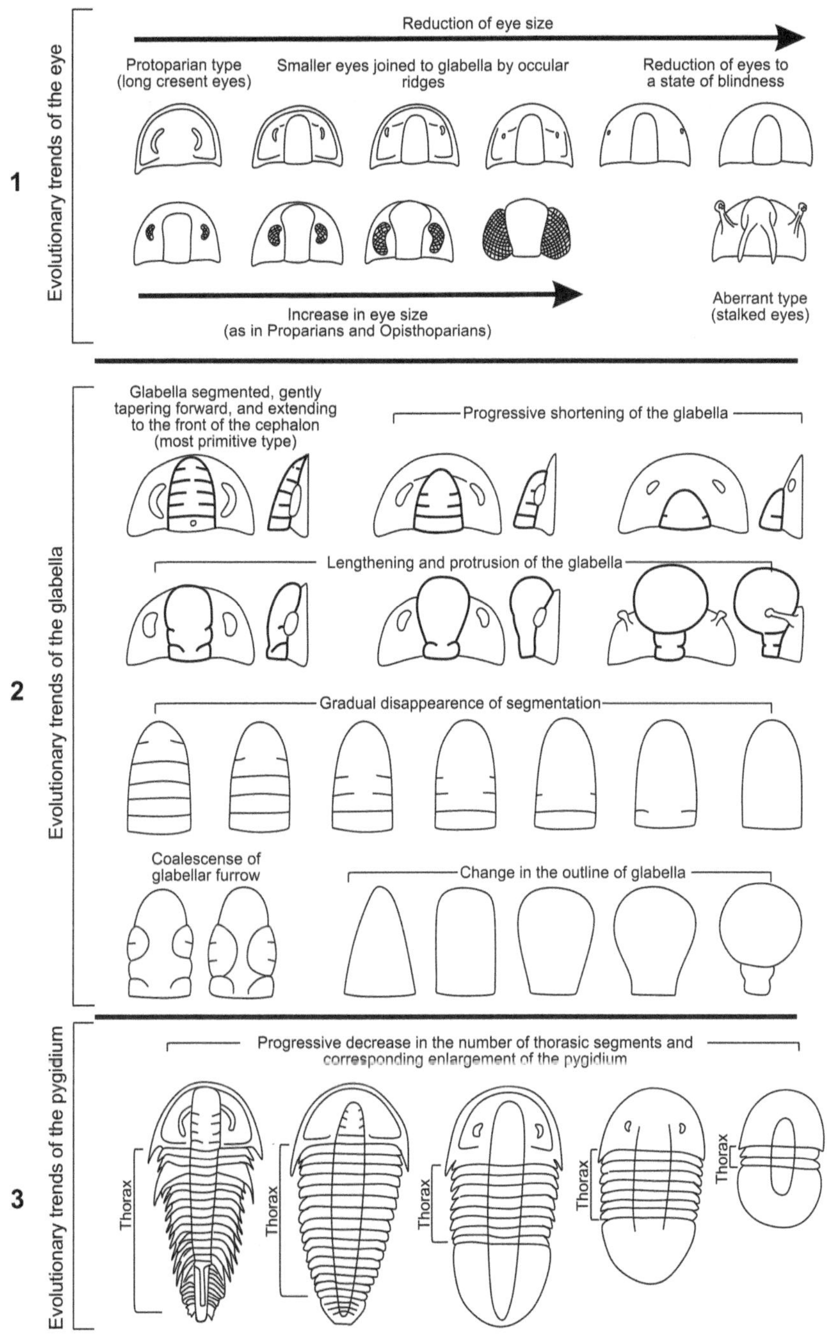

Fig. 5.9 Evolutionary trends of the Trilobite eye (*1*), glabella (*2*) and pygidium (*3*)

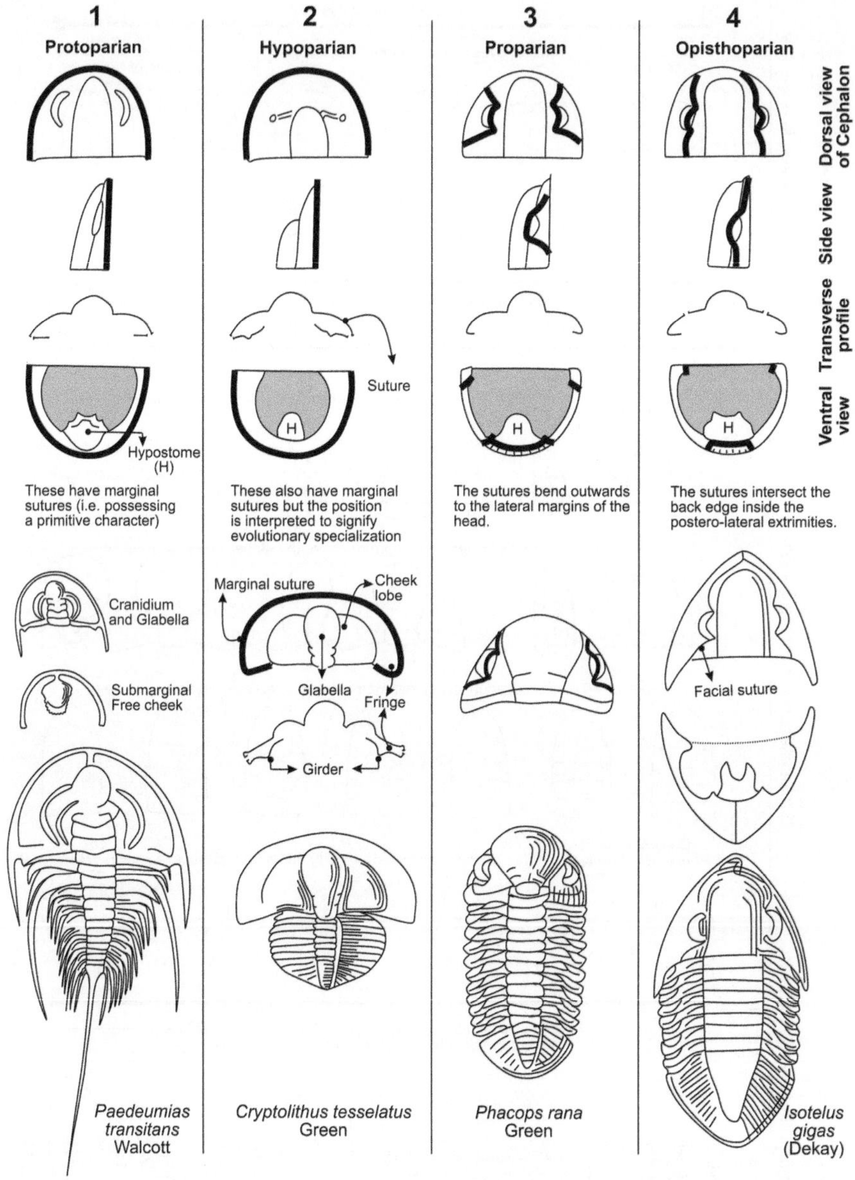

Fig. 5.10 Types of facial sutures and their corresponding examples

of the cephalon, as in *Conocoryphe*, or become greatly reduced, as in *Asaphus*, *Lichas*, or *Proetus*

5.3.1.13 **Free cheek**: Part of cephalon separated by facial suture from cranidium, including part of all of doublure on one side of cephalon; the two free cheeks may be confluent in front of glabella divided by a suture or separated by an accessory plate (Rostrum). In other words, it is the portion of cephalon abaxial to the facial suture [Fig. 5.5(4)]. The free cheeks carry the compound eyes, and are separated from the cranidium by a suture. They may form (a) a continuous ventral plate, as in *Harpes*, *Agnotus*, and *Trinucleus*; they may include (b) a greater or lesser portion of the dorsal surface, being either entirely separated by the cranidium or (c) meeting, and (d) sometimes coalescing in front. They are widely separated in *Ptychoparia*, in juxtaposition in *Asaphus*, and continuous in *Dalmanites*

5.3.1.14 **Frontal lobe**: It is entire portion of the glabella that lies in front of the anterior lateral furrows, and is often somewhat enlarged laterally [Fig. 5.5(1)]

5.3.1.15 **Genal angle**: These are the posterior lateral angles of the cephalon [Fig. 5.5(7)]. They may be rounded, as in *Illaenus*, angular, as in *Bronteus*, or spiniform, as in *Trinucleus* and *Dalmanites*. They belong either to the fixed cheeks, as in *Dalmanites*, or to the free cheeks, as in *Illaenus*, *Bronteus*, and *Proetus*

5.3.1.16 **Genal spine**: Backward extension of the posterolateral corner of cephalon in the form of a spine [Fig. 5.5(1 and 4)]. It is a hollow, posteriorly directed, pointed projection and the axial part of cranidium bounded by the dorsal furrow at its front and sides [Fig. 5.5(1 and 4)]

5.3.1.17 **Glabella**: Bounded by axial and preglabellar furrows, it is the axial part of the cephalon, [Figs. 5.5(1–4), 5.6 and 5.7]. The glabella may constitute nearly the whole of the cephalon, as in *Deiphon* or *Aeglina*, or it may be narrow, as in *Harpes*, and *Eurycare*. In some cases it does not extend over half the length of the cephalon, as in *Harpes* and *Arethusina*, but it may extend to the frontal border, as in *Placoparia* or *Calymmene*, or even beyond, as in *Phacops*, *Ampyx*, and *Conolichas*. The evolutionary trends of glabella are given in Fig. 5.9(2)

5.3.1.18 **Glabellar furrow**: Straight or curved groove extending inward from side of glabella [Fig. 5.5(2)]

5.3.1.19 **Glabellar lobe**: Part of glabella bounded in front or behind, or on both of these sides by short glabellar furrow [Fig. 5.5(2)]

5.3.1.20 **Glabellar segment**: Part of glabella bounded in front and behind by long glabellar furrows [Fig. 5.5(2)]

5.3.1.21 **Hypostome (Labrum)**: Plate or underside of cephalon in front of mouth [Fig. 5.5(5)]

5.3.1.22 **Hypostomial suture**: Line at anterior edge of hypostome where it joins doublure or rostrum [Fig. 5.5(6)]

5.3.1.23 Lateral glabellar furrow: These narrow and pairs of bilaterally symmetrical grooves on external surface extend partway across glabella, from (or near) the axial furrow (Fig. 5.1). They are formed by fold in the exoskeleton. Furrows show great variation in terms of length, depth, and direction. Additionally, they may be short, limited to pits in or close to axial furrow, or isolated from the axial furrow. When in three pairs, they are called anterior, median, and basal (syn., posterior glabellar furrow, preoccipital glabellar furrow). Their numbering is from back forward—occipital as S0, preoccipital as S1, and continued forward as S2, S3, etc. [Fig. 5.5(3)]

5.3.1.24 Lateral glabellar lobe: It is the inflated (generally) portion of the glabella and separated by successive lateral glabellar furrows [Fig. 5.5 (1)]. Numbering is from posterior end forward: L1, L2, etc. [Fig. 5.5(3)]. When in three pairs, anterior, median, and basal lobes are used to designate them [Fig. 5.5(1)]

5.3.1.25 Marginal furrow: Groove or abrupt inflection of surface along inner edge of border of cephalon or pygidium [Fig. 5.5(8)]

5.3.1.26 Marginal spine: Sharp projection at edge of Pygidium [Fig. 5.6(8)]

5.3.1.27 Metastome: Small plate behind mouth [Fig. 5.5(5)]

5.3.1.28 Occipital furrow: A furrow (transverse in nature) that separates the occipital ring from the remainder of the glabella [Fig. 5.5(1) and 5.8]. It is the transverse groove in front of the hindmost glabellar segment

5.3.1.29 Occipital node: Tubercle on mid-portion of occipital segment, in some trilobites observed to have structure of a simple eye [Fig. 5.5(1)]

5.3.1.30 Occipital ring: Axial region of most posterior segment of cephalon, bounded at sides by axial furrows, at front by occipital furrow, and at back by posterior margin. Considered part of glabella in all trilobites [Fig. 5.5(1)]

5.3.1.31 Occipital segment: Hindmost part of glabella bounded in front by a complete transverse groove (occipital furrow) [Fig. 5.5(8)]

5.3.1.32 Ocular platform: Part of fixed cheek behind brim and extending laterally outwards from eye [Fig. 5.5(8)]

5.3.1.33 Ocular ridge: Narrow elevation extending from front edge of each eye to glabella; lacking in many genera [Fig. 5.5(9)]

5.3.1.34 Palpebral furrow: Groove or abrupt inflection of surface along inner edge of palpebral lobe [Fig. 5.5(4)]

5.3.1.35 Palpebral lobe: It is the raised portion (a protruding subsemicircular flange) of fixed cheek along the inner edge of the visual area of eye. It is distally bounded by the palpebral suture [Figs. 5.5(1) and 5.4]

5.3.1.36 Preglabellar furrow (= Border furrow): It is the portion of the axial furrow outlining front of glabella [Fig. 5.5(1)]

5.3.2 Thorax

5.3.2.1 **Articulating facet**: Sharply down bent areas along outer front edges of pleura and pygidium; in articular movement providing for impingement on adjacent parts of skeleton [Fig. 5.6(2 and 9)]

5.3.2.2 **Articulating furrow**: Transverse grove between axial segment of thorax [Fig. 5.6(7)]

5.3.2.3 **Articulating half segment**: Arched anterior extension of axial segments of thorax [Fig. 5.6(4)]

5.3.2.4 **Axial furrow**: It defines the axial region of cephalon, thorax, and pygidium. Formed by the fold in the exoskeleton, axial furrow is a groove on external surface [Fig. 5.6(3 and 9)]

5.3.2.5 **Axial lobe**: It is the central region of the dorsal exoskeleton, bordered by axial furrow including the preglabellar furrow [Fig. 5.5(1)]

5.3.2.6 **Axial node**: Centrally located tubercle on an axial segment [Fig. 5.6(2)]

5.3.2.7 **Axial ring**: Central portion of the thoracic of pygidium segment, bounded laterally by an axial furrow [Fig. 5.6(2)]

5.3.2.8 **Axial segment**: Transverse division of axis of thorax or pygidium [Fig. 5.6(7)]

5.3.2.9 **Axis**: Longitudinal central part of cephalon, thorax, and pygidium bounded by dorsal furrow [Fig. 5.6(3)]

5.3.2.10 **Dorsal furrow**: Groove bounding axis; it is located along sides and front of glabella, sides of axial lobe of thorax, and sides and rear of axial lobe of pygidium [Fig. 5.6(3)]

5.3.2.11 **Doublure**: Reflexed continuation of dorsal exoskeleton along ventral margins of cephalon, pleura, and pygidium [Fig. 5.6(6)]

5.3.2.12 **Endopodite**: Inner branch (walking leg) of biramous paired appendages attached to each post-antennal segment [Fig. 5.5(5)]

5.3.2.13 **Interpleural furrow**: These is a transverse groove indicating the boundary of fused pleurae. The furrow extends from axial furrow across the pleural region of the pygidium, (syn., interpleural groove, rib furrow) [Fig. 5.6(9)]

5.3.2.14 **Interpleural groove**: Transverse furrow between adjoining pleura of thoracic region or crossing a pleural lobe of pygidium [Fig. 5.5(9)]

5.3.2.15 **Pleura (pl., pleurae)**: Lateral portion of a thoracic segment [Fig. 5.6(4 and 8)]

5.3.2.16 **Pleural furrow**: Groove extending outward and generally backward from inner front edge of each pleuron; interpreted as a trace of primary segmentation, formed by fold in exoskeleton [Fig. 5.6(9)]

5.3.2.17 **Pleural lobe**: Lateral portion of thorax or pygidium and abaxial (away from the axis) to the axial furrow [Fig. 5.6(5)]

5.3.2.18 **Pleural spine**: Sharp-pointed extremity of a pleuron [Fig. 5.6(4)]

5.3.2.19 **Pre-epodotide**: Outer and upper branch of paired biramous post-antennal appendages on ventral side [Fig. 5.5(5)]

5.3.2.20 **Thoracic segment**: Transverse division of thorax, consisting of an axial and two pleural portions [Fig. 5.6(4)]

5.3.2.21 **Thorax**: Post-cephalic part of body [Fig. 5.6(1–7)] composed of individually movable segments (articulation of successive somites (somites are the transverse divisions of the arthropod body); region between cephalon and pygidium. The thoracic segments are movable upon one another, in some cases enough so that the animal can roll itself up like a wood-louse (see Enrolment). Each segment is divided into a median (central) and two lateral parts by means of two furrows. The median part is more convex and forms the Axis; the lateral parts are called Pleurae (see above). The number of thoracic segments varies exceedingly among different genera. The smallest number (2) occurs in *Agnostus*. The largest number so far observed (29) is found in some species of *Harpes*. A variation is to be noted even among the species of a single genus; hence this character is not of general application for purposes of classification. For example, there are species of *Ampyx* and *Aeglina* with 5–6 thoracic segments, *Phillipsia* with 9–15, *Cheirurus* with 10–12, *Cyphaspis* with 10–17, and *Paradoxide* 16–20. In general, there seems to be a sort of mutual relationship between the number of thoracic segments and the size of the pygidium. When the latter is large, the thoracic segments are usually few; but if small, the number of segments is large. The evolutionary trends of thorax are given in Fig. 5.9(2).

5.3.3 *Pygidium*

5.3.3.1 **Pygidial segment**: Transverse division of pygidium representing one of fused body segments composing it; homologous to thoracic segment [Fig. 5.6(8 and 9)]

5.3.3.2 **Pygidium (pl., pygidia)**: Posterior part of trilobite carapace (the exoskeleton) generally formed by fusion of several body segments (fused somites) [Fig. 5.6(8 and 9)]. The outline of the pygidium is most frequently semicircular, parabolic, or elliptical; more rarely it is triangular or trapezoidal. The evolutionary pattern of pygidium is given in Fig. 5.9(3). Based on size, four categories of pygidium are noted: these are Micropygous (pygidium < cephalon), Subisopygous (pygidium ≠ cephalon), Isopygous (pygidium = cephalon) and Macropygous (pygidium > cephalon) (Fig. 5.11)

5.3.3.3 **Telson**: Prominent backwardly directed spike borne by an axial segment at or near rear extremity; may or may not be equivalent to true telson of other arthropods [Fig. 5.6(1)].

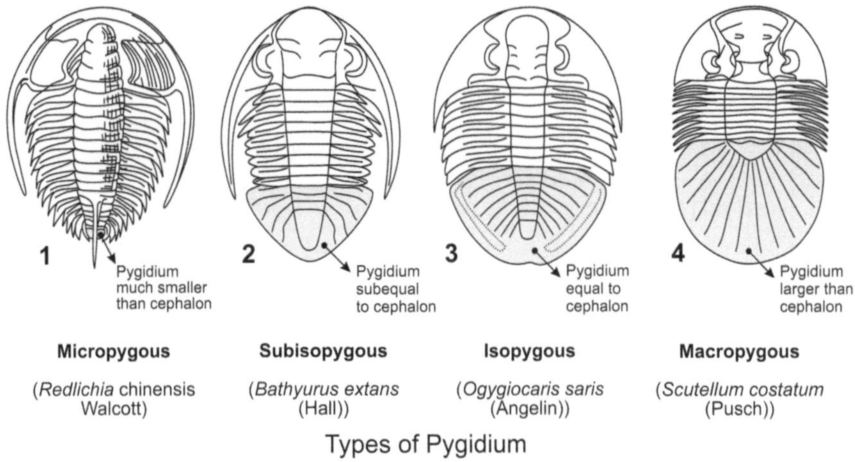

Fig. 5.11 Types of pygidium

5.4 Growth Stages

The Trilobite development begins when each molt adds articulated segments (the Anamorphic mode) and then switches to another mode in which no more segments are added (the Epimorphic mode) (Fig. 5.12). This half anamorphic, half epimorphic mode is called Hemianamorphic mode. The addition of articulated segments for each molt is best noted during the meraspid period, whereas the holaspid period includes the epimorphic growth during which the unarticulated segments might be added to the fused pygidium, but the total number of articulated units remain stable (Fig. 5.12). For demarcating the developmental patterns in trilobites (and enabling their higher grouping), the understanding of the timing of segment additions versus articulations is fundamental (see also Chatterton and Speyer 1990; Levi-Setti 1993; Hughes 2005; Hughes et al. 2006). Based on this distinction, three developmental stages are recognized—Protaspid, Meraspid, and Holaspid (Fig. 5.12). These are briefly described below

5.4.1 **Protaspid**: The Protaspis is a larva with a planktonic habit. It is made by an unarticulated exoskeletal shield, and is a fraction of a millimeter in length. As molting progresses, cranidium develops and is followed by the addition of a protopygidium with fused thoracic segments. Later, at the posterior end of the pygidium, successive segments are added. They grow forward, and are released into the thorax (Meraspid stages) until the full complement of segments is achieved (and is now represented by the Holaspid stage (see Fig. 5.12)

5.4.2 **Meraspid**: Two or more articulated segments characterize this stage (Fig. 5.12). Each molt adds one or two, but rarely greater numbers of articulated thoracic segments to the body. Several molts occur, until the

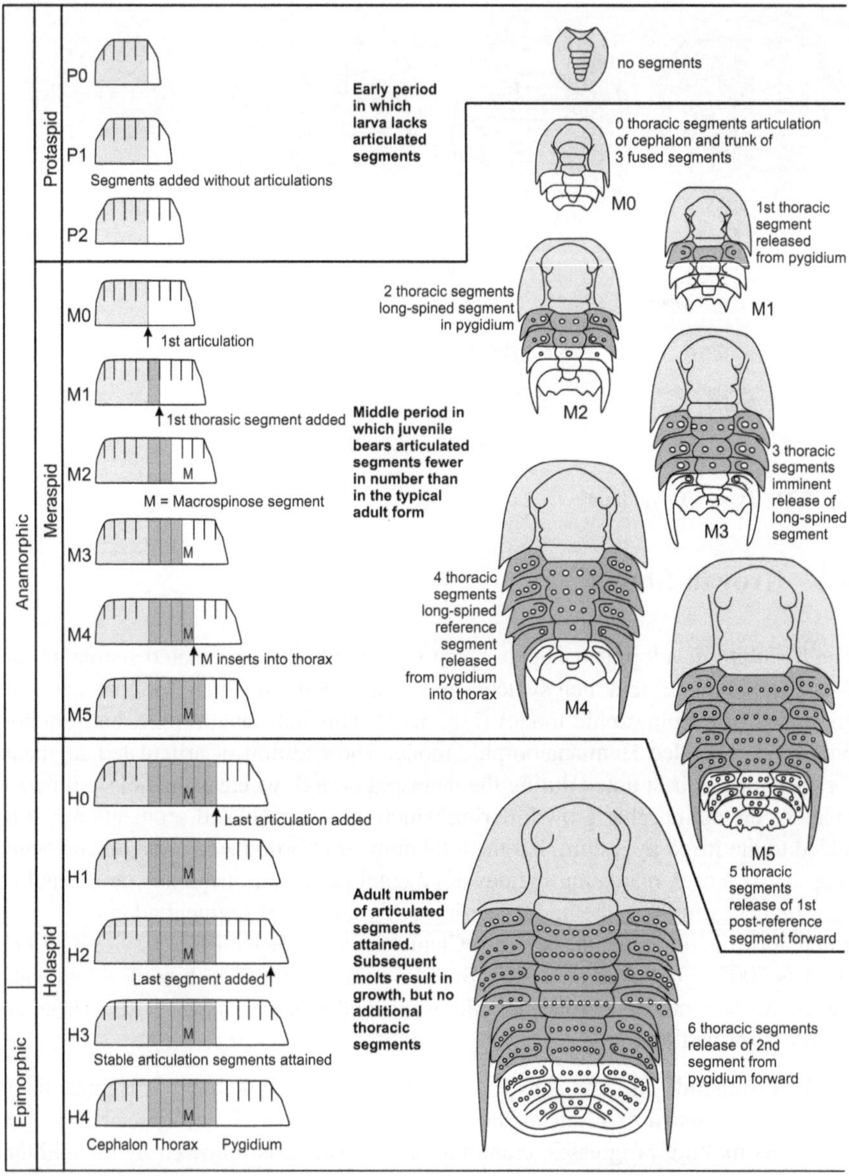

Fig. 5.12 Growth stages of a typical Trilobite (see Sect. 5.4. for explanation). With permission, figure modified after Gon (2014)

number of thoracic segments added to the meraspis achieves the number typical of the adult form of the species. Thus, by now, the general pattern of the body morphology (as evidenced by its shape and ornamentation) resembles the adult of the species

The first complete articulation between cephalon and trunk is noted as Degree 0 (M0; see Fig. 5.12; the trunk is shown as a white background); the Thoracopygon, a combined thorax and pygidium is formed. By M1, the first thoracic segment appears (shown as dark gray in Fig. 5.12). By M2, the long-spined segment is still not obvious, albeit a hint of it is noted in the expression of a pair of larger spines in meraspis degree 1. It moves forward in the pygidium (shown as white background in Fig. 5.12) until it is released into the thorax in meraspid degree 4 (see Fig. 5.12). The addition of two more segments in subsequent molts follows. Each segment is added from behind the reference segment (i.e., released from the pygidium into the thorax) until the last complete articulation yields six thoracic segments, and the holaspid stage is attained (Fig. 5.12)

5.4.3 Holaspid: At this stage, with each molt, no articulated segments are added and is the last stage of trilobite development (Fig. 5.12). Now, there is no major form change, but only increase in size. At this stage, the fused pygidium may have some increase in the number of segments, but no further articulated segments are added to the thorax. Hence, maximum size increase in the life cycle of a trilobite occurs during this stage, only (Fig. 5.12).

5.5 Enrolment

The bodies of most Trilobites are capable of being rolled up completely like many of the Isopods (a crustacean order that includes woodlice, sea slaters and their relatives). In the enrolled condition the margin of the pygidium is closely applied to the doublure of the cephalon, thus entirely concealing the ventral side of the body [Figs. 5.13(6–8) and 5.17(11–14)]. The thoracic segments overlap, and have an overlapping motion upon one another. The pleura also imbricate, and their fulcra are provided with facets upon which the fulcra of adjacent segments impinge. Thus, when the animal is enrolled, the ends of the pleura protect the ventral surface, along the sides. But, only few forms had this ability. In them, the organism is usually found extended, and the facets on the fulcra are either rudimentary or absent. In Isopygous condition (Fig. 5.11), the cephala and pygidia are of comparable size and thus, they enroll spherically as in Agnostida, Proetida, Asaphida, Illaenina, and some Phacopida (see Fortey 2001).

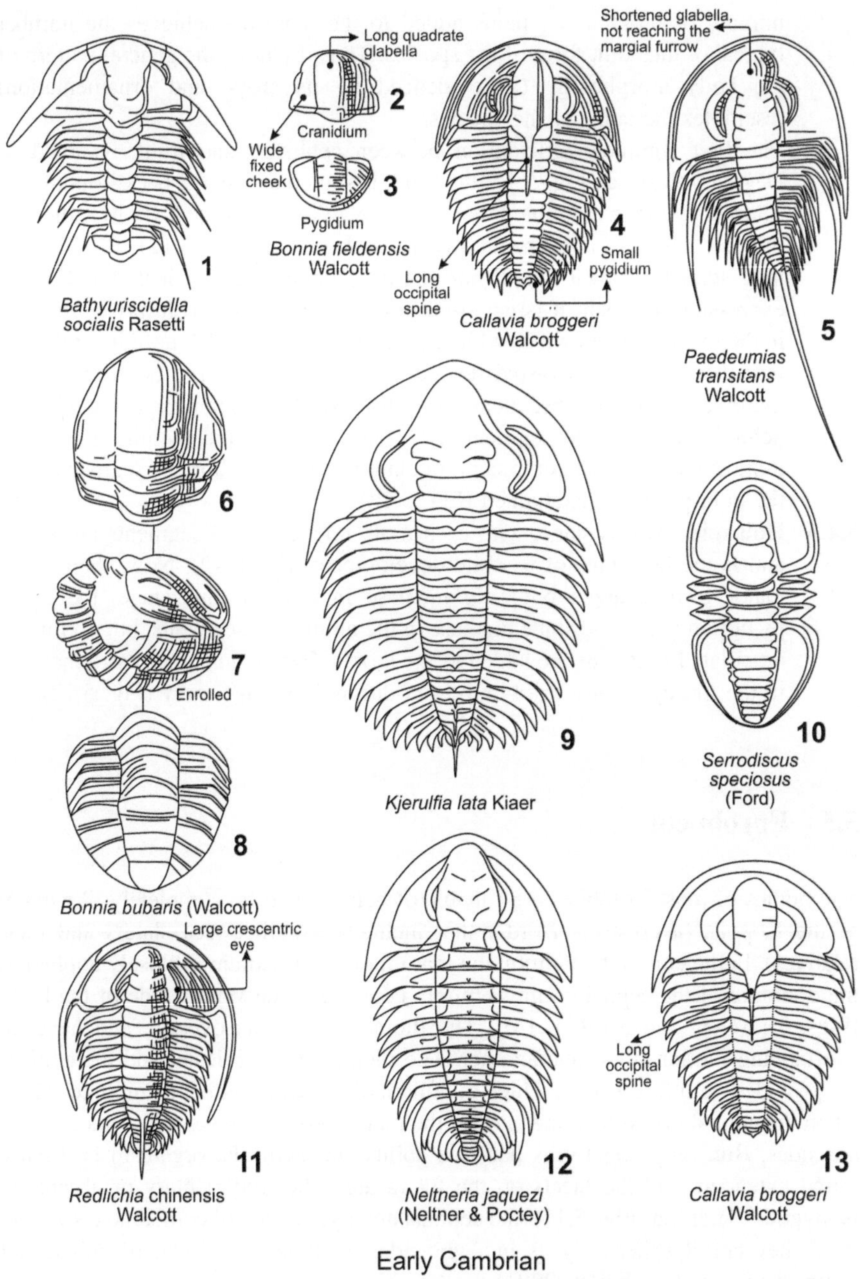

Fig. 5.13 Selected Early Cambrian trilobites and their major distinguishing characters

5.6 Classification

In higher level classification, "no single character (e.g., facial sutures) dominates. Instead, such characters as facial sutures, glabellar shape and pattern of lobation, eyes, thoracic features and numbers of thoracic segments, pygidial shape, size and segmentation, and spinosity all play a role in helping define the orders. In addition, hypostomal conditions and shared ontogeny *also* play an important role in defining the orders of trilobites" (see Hughes 2007; Gon 2014) (see also Fig. 5.1; Tables 5.3 and 5.4). Table 5.3 lay out the classification of Trilobites and major characters of the Orders and Table 5.4 gives their age range (modified after Gon 2014).

5.7 Geological History and Distribution

Trilobite diversity have been hardest by mass extinctions (Lieberman and Karim 2010) with major losses during end Ordovician (Melott et al. 2004) and Late Devonian (McGhee 1996). After this, they failed to recover fully (Brezinski 1999) and the group was eventually wiped out during the largest mass extinction of all time, at the end of the Permian (Fortey and Owens 1997).

5.7.1 Cambrian: Trilobite's early history is still a mystery. The earliest form appeared in shallow marine settings, and assigned to a number of separate clades, suggesting that trilobites had a prior history, probably as unmineralized or weakly mineralized forms. The earliest trilobites were recorded around 521 Ma in Morocco (Africa), Siberia, Spain, and Laurentia (North America) (see Lieberman 1998, 2002; Hollingsworth 2008; Liñán et al. 2008; Brasier 2009). These include *Profallotaspis jakutensis* from Siberia; *Fritzaspis* sp. from the Esmeralda Basin of western Nevada and eastern California; *Hupetina antiqua* from Morocco and *Serrania gordaensis* (opisthoparian bigotinid; an endemic olenelloids) from Spain. The earliest of the clade, the Olenellina went on to radiate particularly in the Early Cambrian of USA. During the same time, Redlichiida dominated in south China and extended their spread to Morocco, where they were associated with the endemic olenelloids. This Early Cambrian provinciality of shallow-water assemblages continued well into the Middle Cambrian, but as trilobites colonized deeper waters, the Arthricocephalus, certain Agnostida, and Centropleura crossed from one continental margin to another; now trilobites had become cosmopolitan.

The Late Cambrian saw an expansion of ptychopariid deposit feeders, with endemic faunas in Laurentia, north and south China, Australia, Kazakhstan, and NW Siberia, while the dysaerobic environment of the 'olenid sea' around Baltica provided conditions that allowed the Olenidae to flourish. The widely distributed pelagic taxa, *Glyptagnostus* and *Irvingella*, reached

Table 5.3 Classification of Trilobites with major characters of the Orders (modified from Gon 2014)

Class	Order	Major characters			Suborders	Sperfamilies
		Cephalon	Thorax	Pygidium		
Trilobita	Agnostida	Cephalic shield with deeply parabolic outline, sutures proparianor lacking; glabella fusiform, most species eyeless; hypostome natant, sometimes specialized with ribbon – like wings; rostral plate lacking (or uncalcified)	2 segments in Agnostina or 3 in some Eodiscina, axis typically broad, short fulcrate pleurae	Strongly is opygous; pygidial margin typically closely matching cephalic margin	Agnostina	*Agnostoidea*
					Eodiscina	*Eodiscoidea*
	Redlichiida	Large and semicircular; glabella typically long, well-segmented; genal spines typically present, strong; eyes typically large, crescentic; hypostome typically conterminant, very wide rostral plate	With numerous segments (up to 60+), pleasure usually with spinose tips	Typically tiny (micropygous), one or very few segments	Olenellina	*Olenelloidea*
						Fallotaspidoidea
					Redlichiina	*Emuelloidea*
						Redlichioidea
						Paradoxidoidea
	Corynexochida	With opisthoparian sutures; glabella elongate; hypostome conterminant or (in derived forms) impendent; eyes typically large, in some gently arcute	Typically with 7–8 segments (but range for order is 2–12, rarely to 18 in primitive general), pleural tips often spinose	Typically large (isopygous or subisopygous), of variable form, some spinose	Corynexochina	*Corynexochoidea*
					Illaenina	*Illaenoidea*
					Leiostegiina	*Leiostegioidea*
	Lichida	Opisthoparian sutures; glabella broad, large, extending to anterior border; eyes typically present, holochroal, usually not large; conterminant hypostome	Variable, 8-13 segments, usually spine-tipped, sometimes with distinctive spines(Odontopleuroidea)	Typically isopygous to macropygous, often with 3 pairs of furrowed pleurae, typically ending in spinose tips	Lichina	*Lichoidea*
	Odontopleurida	Opisthoparian sutures; glabella extending to anterior border; simple holochroal eyes, usually not large; conterminant hypostome, not particularly large	Variable, 8–13 segments, usually spine-tipped, sometimes developed with distinctive spines patterns (Odontopleuroidea)	Macropygous to subisopygous, with multiple pairs of spines, even in primitive forms (Damesselloidea)	Odontopleurina	*Odontopleuroidea*
						Damesselloidea
	Phacopida	Typically proparian (Phacopina and Cheirurina) or gonatoparian (Calymenina), 4 or fewer pairs of glabellar furrows (these sometimes fused); eyes, when present, schizochroal (Phacopina) or holochroal (Cheirurina and Calymenina); with rostral plates (Calymenina and Cheirurina) or without (some Phacopina); hypostome conterminant (all suborders) to impendent (some Phacopina)	8–19 segments, sometimes distinctly furrowed, axis sometimes broad (e.g., Homalonotidae)	Typically macropygous (most Calymenina and Phacopina), but variable (e.g., subisopygous in Dalminitoidea and Acastoidea), may be lobed or spiny (e.g., Cheirurina, some Dalminitoidea, Acastoidea), or smooth-margined, with round or subtriangular outline (Calymenina, Phacopoidea)	Phacopina	*Phacopoidea*
						Dalmanitoidea
					Calymenina	*Acastoidea*
					Cheirurina	Calymenoidea
						Cheiruroidea

(continued)

Table 5.3 (continued)

Class	Order	Major characters			Suborders	Sperfamilies
		Cephalon	Thorax	Pygidium		
	Ptychopariida	Typically with opisthoparian facial sutures, with gently forward-tapering simple glabella bearing a broad, rounded front, usually with 3 pairs of rather narrow parallel glabellar furrows; natant hypostome	typically large with 8+ thoracic segments	Quite variable, but typically with a small pygidium bearing a border (Cambrian) or a larger pygidium with or without border (post-Cambrian)	Ptychopaiina	*Ellipsocephaloidea*
						Ptychoparioidea
					Olenina	Olenoidea
	Asaphida	Often equal/subequal to pygidium (Asaphoidea), but some not so (Trinucleioidea); eyes usually large (some forms secondarily blind); cephalic doublure often wide; dorsal anterior facial sutures often curve adaxially to meet in front of the glabella; sutures opisthoparian; hypostome conterminent or impedent, with only primitive forms (Anomocaroidea) natant	Typically 5–12 segments, but 2–3 in a few Trinucleioidea, 13+ i some Anomocaroidea, up to 30 in an Alsataspidid(Trincleioidea)	Typically large(subisopygous to macropygous), with a wide doublure	Asaphina	*Anomocaroidea*
						Asaphoidea
						Dikelokephaloidea
						Remopleuridoidea
						Cyclopygoidea
						Trincleioidea
	Proetida	Opisthoparian sutures; glabella large, vaulted, well-defined, typically 4 pairs of glabellar furrows; eyes, usually present, holochroal, often large, convex; rostral plate narrow and backward tapering; long hypostome, most species natant, but some secondarily conterminant (late Proetidae & Bathyuridae); typically with genal spines	8–22(typically 10) segments, tips variable, blunt to long-spined	Micropygous to subisopygous, often spineless, and usually with 4-10+ distinct pleural furrows	Proetina	*Aulacopleuroidea*
						Proetoidea
						Bathyuroidea
	Harpetida	Semicircular to ovate; facial sutures marginal; glabella convex, narrowing forwards, with 1 to 3 pairs of furrows; occipital ring convex; eyes commonly reduced to prominent tuberles, centrally located on genae; external surface of cephalon may be tuberculose or granulose	With 12 or (frequently) more segments, pleurae flattened, with broad axial furrows	Subtriangular, elongate to short	Harpetina	*Harpetioidea*

Table 5.4 Classification of Trilobites with their age range (modified from Gon 2014)

Class	Order	Age	Suborder	Age	Sperfamilies	Age
Trilobita	Agnostida	Early Cambarian-Late Ordovician	Agnostina	Early Cambarian-Late Ordovician	Agnostoidea	Early Cambarian-Late Ordovician
					Condylopygoidea	Early Cambarian-Middle Cambrian
			Eodiscina	Early Cambarian-Middle Cambrian	Eodiscoidea	Early Cambarian-Middle Cambrian
	Redlichiida	Early Cambarian-Middle Cambrian	Olenellina	Early Cambarian	Olenelloidea	Early Cambarian
					Fallotaspidoided	Early Cambarian
			Redlichiina	Early Cambarian-Middle Cambrian	Emuelloidea	Early Cambarian
					Redlichioidea	Early Cambarian-Middle Cambrian
					Paradoxidoidea	Early Cambarian-Middle Cambrian
	Corynexochida	Early Cambarian-Late Devonian (Frasnian)	Corynexochina	Early Cambarian-LateCambarian	Corynexochoidea	Early Cambarian-LateCambarian
			Illaenina	Early Cambarian-Late Devonian (Frasnian)	Illaenoidea	Early Cambarian-Late Devonian (Frasnian)
			Leiostegiina	Middle Cambrian-Early Ordovician	Leiostegioidea	Middle Cambrian-Early Ordovician
	Lichida	Middle Cambrian-Middle Devonian	Lichina	Middle Cambrian-Middle Devonian	Lichoidea	Middle Cambrian-Middle Devonian
	Odontopleurida	Middle Cambrian-Middle Devonian (Frasnian)	Odontopleurina	Middle Cambarian-Late Devonian (Frasnian)	Odontopleuroidea	Early Ordovician?-Late Devonian (Frasnian)
					Dameselloidea	Middle Cambrian-LateCambrian?
	Phacopida	Early Ordovician- Late Devonian	Phacopina	Early Ordovician- Late Devonian	Phacopoidea	Early Ordovician- Late Devonian

(continued)

Table 5.4 (continued)

Class	Order	Age	Suborder	Age	Sperfamilies	Age
		(Fammenian)		(Fammenian)		(Fammenian)
					Dalmanitoidea	Early Ordovician-Early Devonian
			Calymenina	Early Ordovician- Middle Devonian	Acastoidea	Early Ordovician- Middle Devonian
			Cheirurina	Early Ordovician- Middle Devonian	Calymenoidea	Early Ordovician- Middle Devonian
					Cheiruroidea	Early Ordovician- Middle Devonian
	Ptychopariida	Early Cambrian Late Ordovician	Ptychopariina	Early Cambrian Middle Ordovician	Ellipsocephaloidea	Early Cambrian Middle Cambrian
					Ptychoparioidea	Early Cambrian -Middle Ordovician
			Olenina	Late Cambrian LateOrdovician	Olenoidea	Late Cambrian Late Ordovician
	Asaphida	Middle Cambrian-Late Silurian	Asaphina	Middle Cambrian-Late Silurian	Anomocaroidea	Middle Cambrian-Late Cambrian
					Asaphoidea	Early Ordovician-Late Ordovician
					Dikelokephaloidea	Late Cambrian
					Remopleuridoidea	Late Cambrian?-Late Ordovician
					Cyclopygoidea	Early Ordovician-Late Ordovician
					Trinucleioidea	Late Cambrian-Late Silurian
	Proetida	Late Cambrian-Late Permian	Proetina	Late Cambrian-Late Permian	Aulacopleuroidea	Early Ordovician-Late Permian
					Proetoidea	Early Ordovician-Late Permian
					Bathyuroidea	Late Cambrian-Middle Ordovician?
	Harpetida	Late Cambrian-Late Devonian (Frasnian)	Harpetina	Late Cambrian-Late Devonian (Frasnian	Harpetioidea	Late Cambrian-Late Devonian (Frasnian)

all these sites. Some important Cambrian forms are illustrated in Figs. 5.13, 5.14, 5.15 and 5.16

5.7.2 Ordovician: The trilobites reached their acme during the Ordovician. By now many Cambrian groups had disappeared, and many new ones appeared, characterized by Order Asaphida, including remopleuridids, trinucleids, and

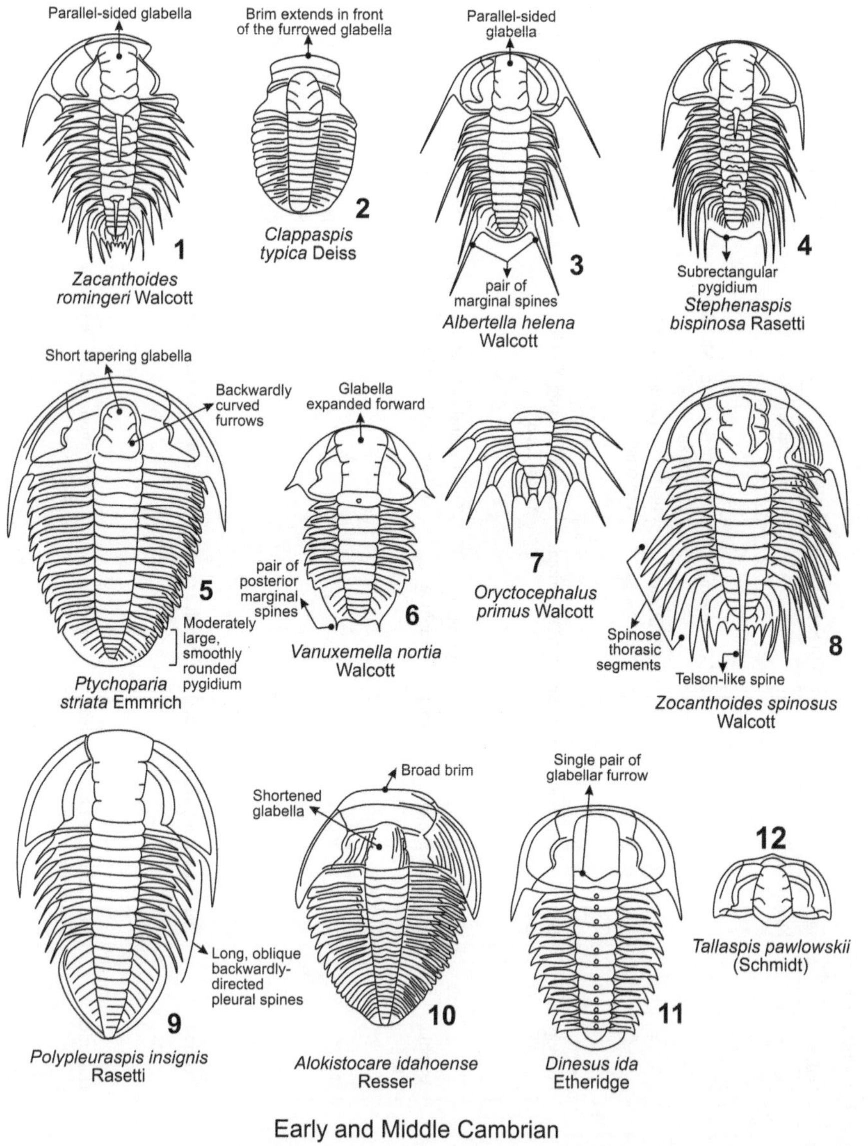

Fig. 5.14 Selected Early and Middle Cambrian trilobites and their major distinguishing characters

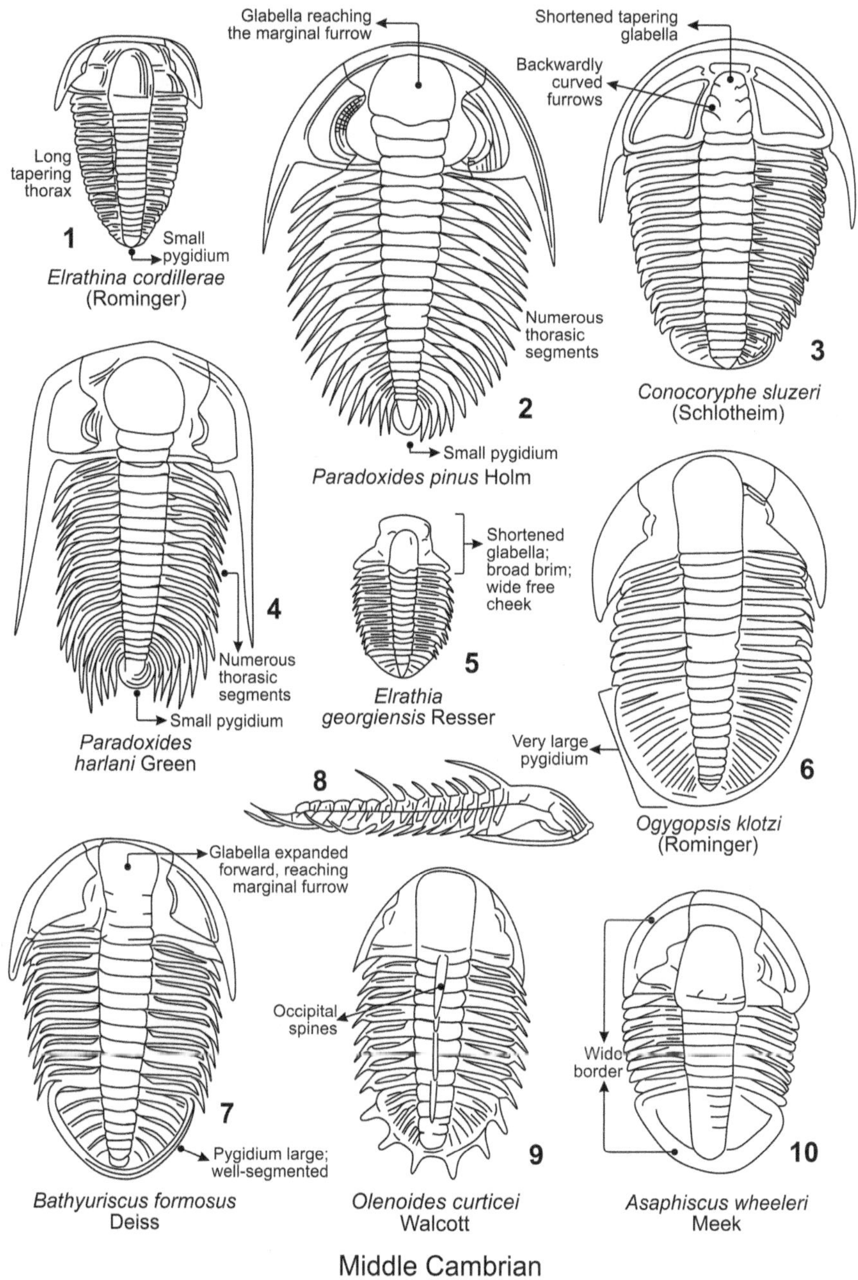

Fig. 5.15 Selected Middle Cambrian trilobites and their major distinguishing characters

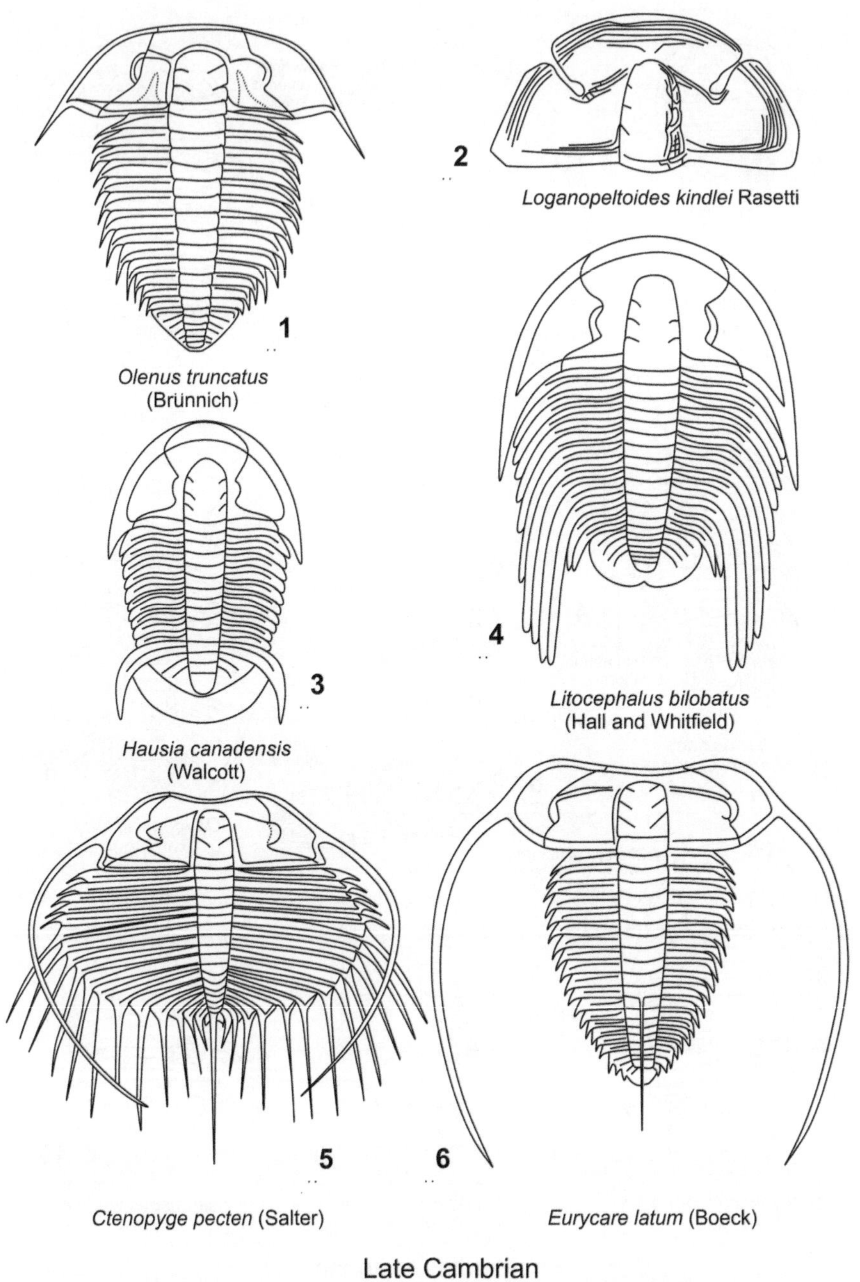

Fig. 5.16 Selected Late Cambrian trilobites and their major distinguishing characters

cyclopygids. Provincialism was somewhat maintained in shallow platformal areas with the bathyurids in Laurentia and Siberia, dikelokephalinids in eastern Gondwana (south China and Australia), and calymenoideans in the west Gondwana. But by mid- to late Ordovician, blurring of this provinciality occurred. Few Ordovician forms are illustrated in Fig. 5.17

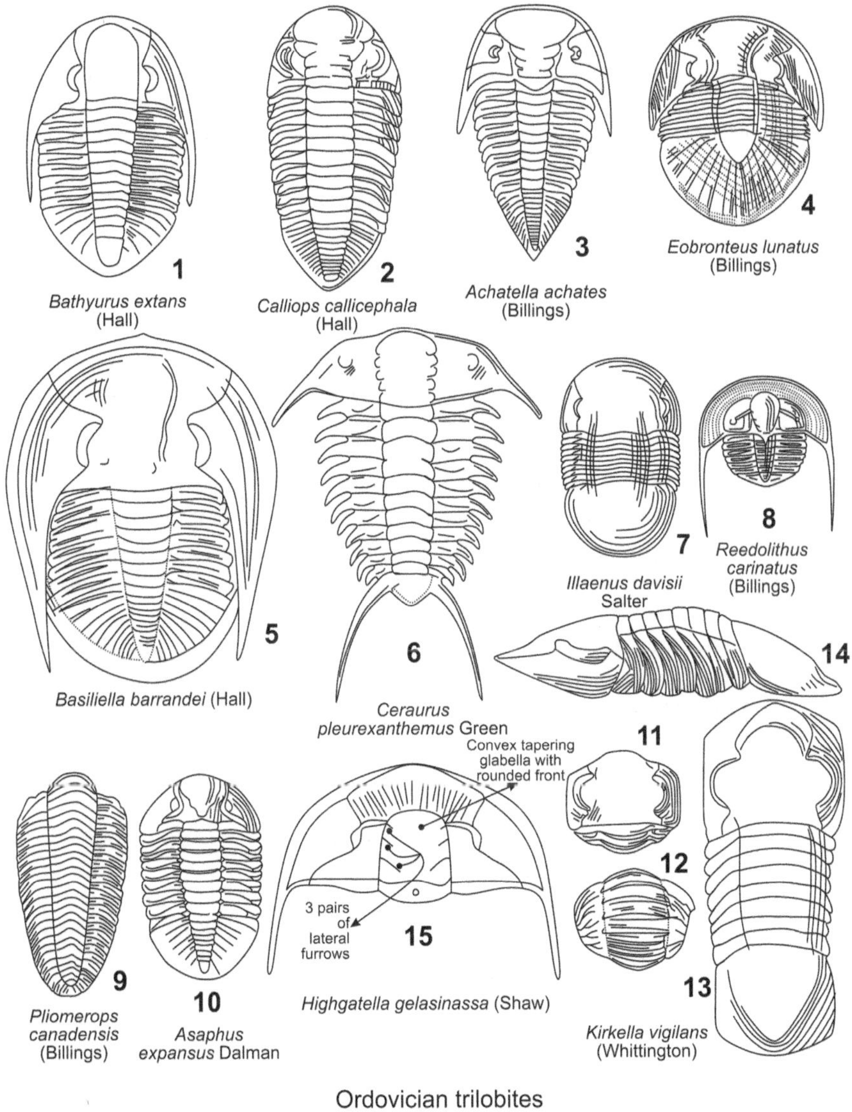

Fig. 5.17 Selected Early to Late Ordovician trilobites and their major distinguishing characters

5.7.3 Silurian: The phacopida, proetida, styginids, lichids, and odontopleurids (~74 % of trilobite fauna) survived the end-Ordovician extinction, and persisted well into the Late Devonian. Many of the genera were now cosmopolitan. In the cold waters of the southern hemisphere, a new province, the Malvinokaffric Province, arose. Few Silurian forms are illustrated in Fig. 5.18

5.7.4 Devonian: The Calmoniids (a clade of phacopids), homalonotids, aulacopleurids, and odontopleurids, part of the Silurian fauna of the Malvinokaffric Province dominated the Devonian seas

The Silurian fauna of the temperate and tropical latitudes, gradually diversified in warmer, more varied conditions of Early Devonian. The following extinction events due to eustatic changes led to the demise of cheirurids, calymenids, and lichids in the Givetian Stage; odontopleurids, harpetids, and styginids in the Frasnian; and the last phacopids at the end of the Famennian (see Fig. 5.2); only proetida survived to the Carboniferous. Few Devonian forms are illustrated in Figs. 5.18 and 5.19.

5.7.5 Carboniferous: Post-extinction, during the Mississippian (Early Carboniferous), the proetid trilobites (represented by just 4 families), diversified to occupy niches in the inner-shelf, carbonate mound, and outer-shelf settings. And post the mid-Carboniferous crisis, the Pennsylvanian (Late Carboniferous) and Permian trilobites now lived in shallow shelf settings, only. Few Carboniferous forms are illustrated in Fig. 5.20

5.7.6 Permian: The already diminished proetida, now represented by only three families, survived in the Permian. The mid-Permian saw the diversification of the phillipsiids. But by end Permian marine regression, which limited their habitats, wiped the trilobites out, just before the end of the Palaeozoic era. Few Permian forms are illustrated in Fig. 5.20.

Appendix 1 gives the list of illustrated specimens mentioning the chapter number, species name, age and locality along with its figure number within the said chapter.

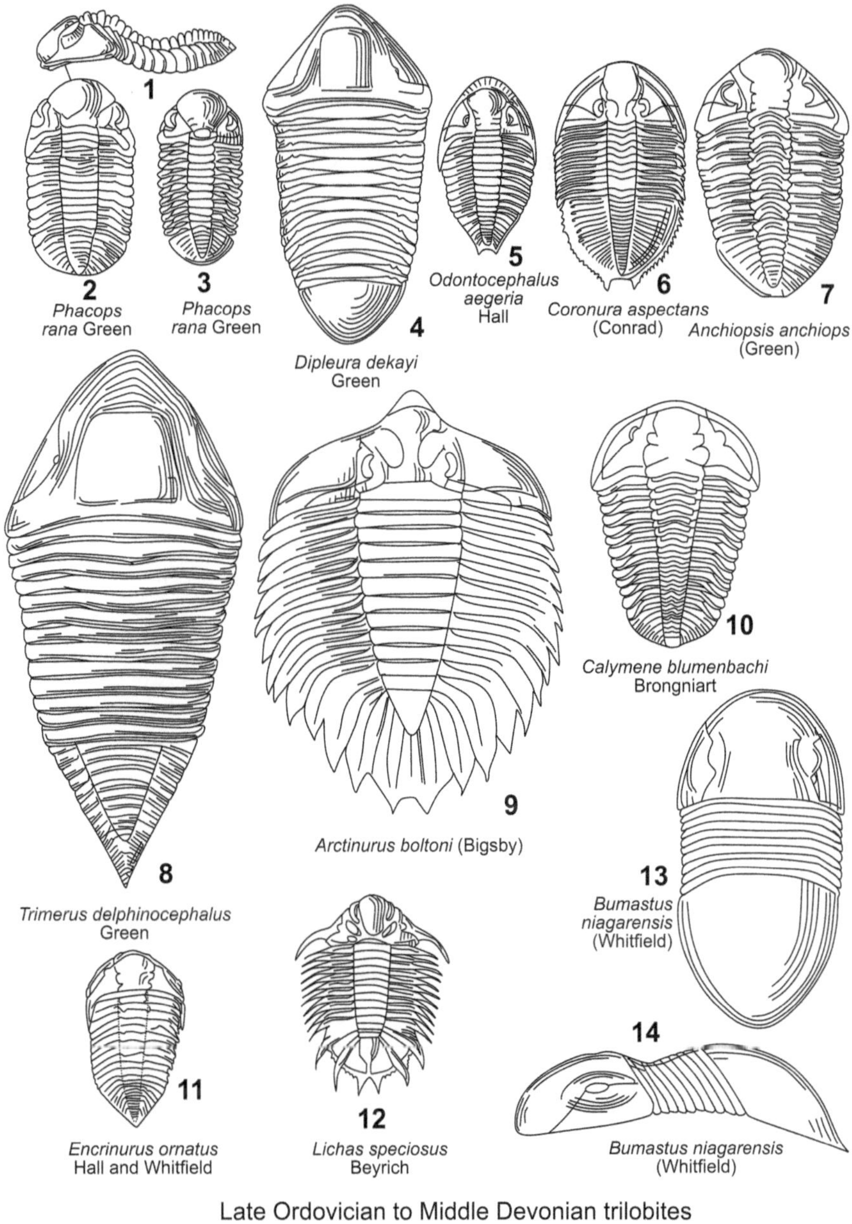

Fig. 5.18 Selected Late Ordovician to Middle Devonian trilobites and their major distinguishing characters

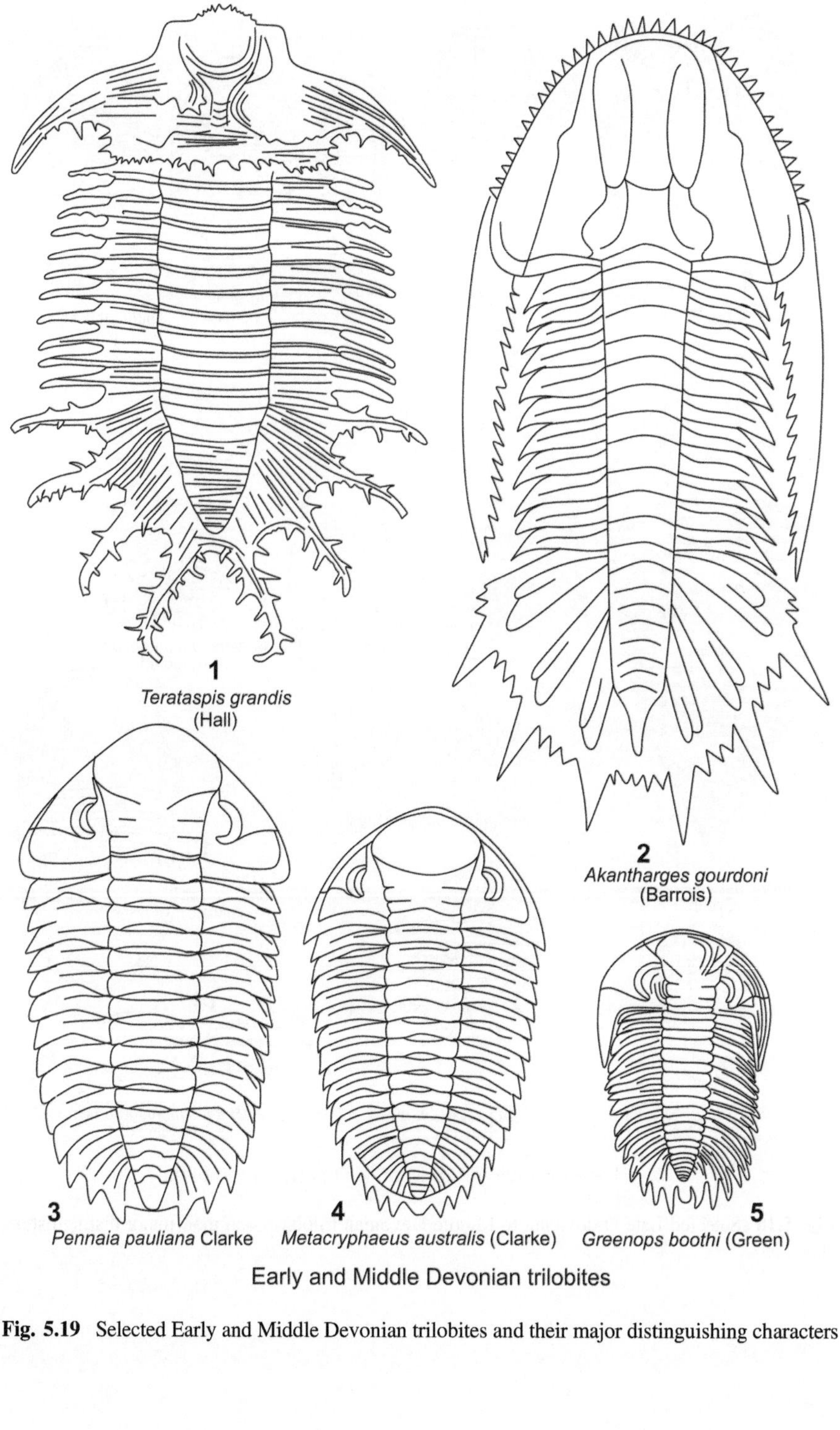

Fig. 5.19 Selected Early and Middle Devonian trilobites and their major distinguishing characters

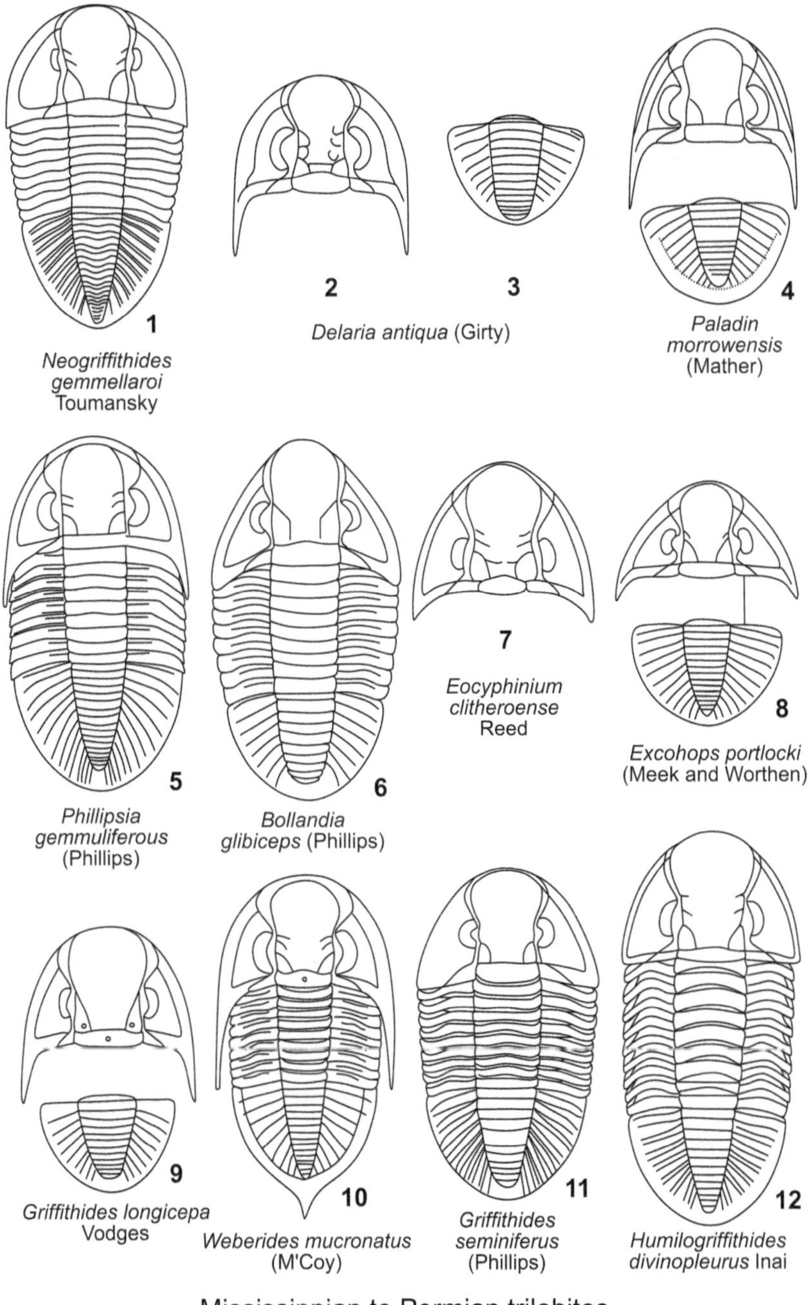

Fig. 5.20 Selected Mississippian to Early Permian trilobites and their major distinguishing characters

References

Babcock, L.E., S. Peng, G. Geyer, and J.H. Shergold. 2005. Changing perspectives on Cambrian chronostratigraphy and progress toward subdivision of the Cambrian system. *Geoscience Journal* 9(2): 101–106.

Brasier, M. 2009. Darwin's lost world - The hidden history of animal life. Oxford University Press. Brezinski, D. K. 1999. The rise and fall of late Paleozoic trilobites of the United States. *Journal of Paleontology* 73: 164–175.

Brezinski, D.K. 1999. The rise and fall of late Paleozoic trilobites of the United States. *Journal of Paleontology* 73: 164–175.

Chatterton, B.D.E., and S.E. Speyer. 1990. Applications of the study of trilobite ontogeny. *Short Courses in Paleontology, Paleontological Society* 3: 116–136.

Chatterton, B.D.E., and S.E. Speyer. 1997. Ontogeny. In *Treatise on Invertebrate Paleontology, Part O, Arthropoda 1, Trilobita*, ed. Kaesler, R.L., 173–247, revised. Volume 1: introduction, order agnostida, order redlichiida, 554. The Geological Society of America, Inc. and The University of Kansas. Boulder, Colorado and Lawrence, Kansas.

Clarkson, E.N.K. 1979. The visual systems of trilobites. *Palaeontology* 22: 1–22.

Clarkson, E., R. Levi-Setti, and G. Horváth. 2006. The eyes of trilobites: the oldest preserved visual system. *Arthropod Structure and Development* 35: 247–259.

Feist, R., and E.N.K. Clarkson. 1989. Environmentally controlled phyletic evolution, blindness and extinction in Late Devonian tropidocoryphine trilobites. *Lethaia* 22: 359–373.

Fortey, R.A. 2000. *Trilobite! eyewitness to evolution*. London: HarperCollins.

Fortey, R.A. 2001. Trilobite systematics: the last 75 years. *Journal of Paleontology* 75(6): 1141–1151.

Fortey, R.A., and R.M. Owens. 1997. Evolutionary history. In *Treatise on invertebrate paleontology, O (arthropoda 1, trilobita, revised)*, ed. R.L. Kaesler, 249–287. Lawrence, Kansas: Geological Society of America and University of Kansas Press.

Geyer, G. 1998. Intercontinental, trilobite-based correlation of the Moroccan early middle Cambrian. *Canadian Journal of Earth Sciences* 35(4): 374–401.

Gon, S.M., III. 2014. *A guide to the orders of trilobites*. http://www.trilobites.info/.

Gradstein, F.M., J.G. Ogg, M. Schmitz, and G. Ogg. 2012. The Geologic Time Scale 2012. doi:10.1016/B978-0-444-59425-9.00026-3.

Harrington, H.J., G. Henningsmoen, B.F. Howell, V. Jaanusson, Ch. Lochman-Balk, R.C. Moore, Ch. Poulsen, F. Rasetti, E. Richter, R. Richter, H. Schmidt, K. Sdzuy, W. Struve, L. Stprmer, C.J. Stubblefield, R. Tripp, J.M. Weller, and H.B. Whittington. 1959. Systematic descriptions. In *Treatise on invertebrate paleontology, Part 0, arthropoda 1*, ed. R.C. Moore, 170–539. Lawrance, Kansas: Geological Society of America and University of Kansas Press.

Hollingsworth, J.S. 2008. The first trilobites in Laurentia and elsewhere. In *Advances in trilobite research. Cuadernos del Museo Geominero* 9, ed. I. Rábano, R. Gozalo, and D. García-Bellido. Instituto Geológico y Minero de España, Madrid.

Hughes, N.C. 2007. The evolution of trilobite body patterning. *Annual Reviews in Earth and Planetary Sciences* 35: 401–434.

Hughes, N.C. 2005. Trilobite construction: building a bridge across the micro and macroevolutionary divide. In *Evolving form and function: fossils and development*, ed. D.E.G. Briggs, 138–158. Peabody Museum of Natural History: Yale University, New Haven, Conn.

Hughes, N.C., A. Minelli, and J. Fusco. 2006. The ontogeny of trilobite segmentation: a comparative approach. *Paleobiology* 32(4): 602–627.

Kaesler, R.L. (ed.). 1997. *Treatise on invertebrate paleontology, Pt. O, arthropoda 1, trilobita, revised*, vol. 1, 530. Lawrence: Geological Society of America and University of Kansas Press.

Levi-Setti, R. 1993. *Trilobites*, 2nd ed. University of Chicago Press.

Lieberman, B.S. 1998. Cladistic analysis of the early Cambrian olenelloid trilobites. *Journal of Paleontology* 72: 59–78.

Lieberman, B.S. 2002. Phylogenetic analysis of some basal early Cambrian Trilobites, the biogeographic origins of the Eutrilobites, and the timing of the Cambrian radiation. *Journal of Paleontology* 76(4): 692–708.

Lieberman, B.S., and T.S. Karim. 2010. Tracing the trilobite tree from the root to the tips: a model marriage of fossils and phylogeny. *Arthropod Structure and Development* 39: 111–123.

Liñán, E., R. Gozalo, M.E. Dies Álvarez, J.A. Gámez Vintaned, and S. Zamora. 2008. Nuevos trilobites del Ovetiense inferior (Cámbrico Inferior bajo) de Sierra Morena (España), Ameghiniana, 45(1).

McGhee, G.R. 1996. *The late Devonian mass extinction: the Frasnian/Famennian crisis*, 303. New York: Columbia University Press.

Melott, A.L., et al. 2004. Did a gamma-ray burst initiate the late Ordovician mass extinction? *International Journal of Astrobiology* 3: 55–61.

Mikulic, D.G., E. Landing, and J. Kluessendorf. 2007. *Fabulous fossils: 300 years of worldwide research on Trilobites*. USA: New York State Museum.

Peng, S., L.E. Babcock, and R.A. Cooper. 2012. The Cambrian period. In: *The Geologic Time Scale 2012*. ed. F.M. Gradstein, J.G. Ogg, M. Schmitz, and G. Ogg, 437–488. Elsevier.

Rabano, I. 1989. El genero *Uralichas* Delgado, 1892 (Trilobite, Lichida) en al Ordovicico de la Peninsula Iberica. *Boletin Geologico y Minero* 100(1): 21–47.

Reimann, I.G. 1942. A new restoration of *Terataspis*. *Bulletin of the Buffalo Society of Natural Sciences* 17: 39–51.

Rudkin, D.M., G.A. Young, R.J. Elias, and E.P. Dobrzanski. 2003. The world's biggest trilobite - *Isotelus rex* new species from the Upper Ordovician of Northern Manitoba. *Canada Journal of Paleontology* 77(1): 99–112.

Thomas, A.T. 2005. Developmental palaeobiology of trilobite eyes and its evolutionary significance. *Earth Science Reviews* 71: 77–93.

Whittington, H.B. 1992. *Fossils illustrated 2*. Trilobites: Woodbridge, Boydell Press.

Chapter 6
Echinoids

6.1 Introduction

Of the five classes of echinoderm (Fig. 6.1), echinoids are the most diverse and well represented, from shallow waters to abyssal depths. The echinoderm account for 90 % of the biomass even at the depth of ocean trenches; echinoderm are bottom-dwellers except for a few swimming sea cucumbers. They are radially symmetrical in adults, and range from a few mm to more than a meter in diameter and live with their mouth downward-facing. The echinoderm group includes animals such as sea urchins and sand dollars (Echinoids), sea lilies and feather stars (Crinoidea), sea stars (starfish; Asteroidea), brittle stars (Ophiuroidea), sea cucumbers (Holothuroidea), sea daisies (Concentricycloidea) and the extinct classes of Cystoidea and Blastoidea (blastoids) (Fig. 6.1). The echinoderm fossil species number ~13,000 and 3500 genera and the living number ~65,000 species grouped in 20 classes.

The Echinoids, one of the groups dealt in this chapter (Fig. 6.1), are exclusively marine benthic macroinvertebrates. These "spiny skin" forms have an endoskeleton made of hard calcium-rich plates (of single calcite crystal) just beneath their thin skin. Paleontologically, echinoids are, by far the most significant group. They range in size from just a few mm in diameter to over 350 mm (both size spectrum is noted in living sea cucumbers and brittle stars). They also occur in varied shapes (globular, heart-shaped, cylindrical, hemispherical, or even flattened discoidal). However, irrespective of their overall shape, the skeleton (test) is always constructed along the same standardized plan, as outlined later in the chapter.

S. Jain, *Fundamentals of Invertebrate Palaeontology*, Springer Geology,
DOI 10.1007/978-81-322-3658-0_6

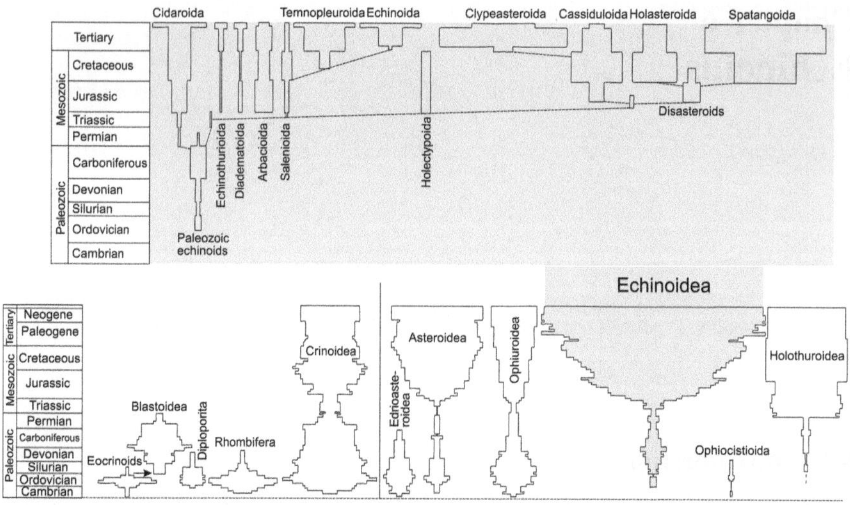

Stratigraphic ranges of the major groups of echinoderm. The widths of the bars are proportional to the taxic (family level) diversities. The cladogram shows their inferred phylogenetic relationships.

Fig. 6.1 The echinoderm group includes sea urchins and sand dollars (Echinoids), sea lilies and feather stars (Crinoidea), sea stars (starfish; Asteroidea), brittle stars (Ophiuroidea), sea cucumbers (Holothuroidea), sea daisies (Concentricycloidea) and the extinct classes of Cystoidea and Blastoidea (blastoids). In this chapter, however, only Echinoids are dealt (shaded)

6.2 General Morphology

In Echinoids, a rigid and robust calcitic test (skeleton) built by a mosaic of plates is firmly bound together, thus, making the skeleton is architecturally complex and mesodermal. Their excellent preservational history (good fossil record) is a testament of the rigid skeleton and hence, the echinoids are extensively studied for their phylogeny. Contextually, they are also a valuable palaeobiological tool.

The extant echinoid species are equally divided between regular forms (radial, fivefold symmetry; Figs. 6.2 and 6.3) whose anus opens in the aboral plated surface (i.e., the anus is located opposite to the mouth), and irregular forms (bilateral symmetry) whose anus is displaced away from the aboral plates into the posterior interambulacral zone (Fig. 6.4). The regular forms live epifaunally, whereas the irregular ones live predominantly infaunally. Table 6.1 lists major differences between Regular and Irregular echinoids.

In Echinoderms, instead of using “dorsal” and “ventral” (as in Brachiopods: Chap. 8), the position relative to the mouth is used; the side where the mouth lies is Oral, and the side opposite is Aboral. In regular echinoids the mouth is on the underside and the anus is at the top [see Fig. 6.2(1, 2)]. When viewed from the upper surface, the outline of the type of echinoid test just described is a circle, with the apical system at its center, and the double columns of plates forming the radii [Fig. 6.2(2, 9)]. The ambulacra are said to be Radial in position, since each overlies

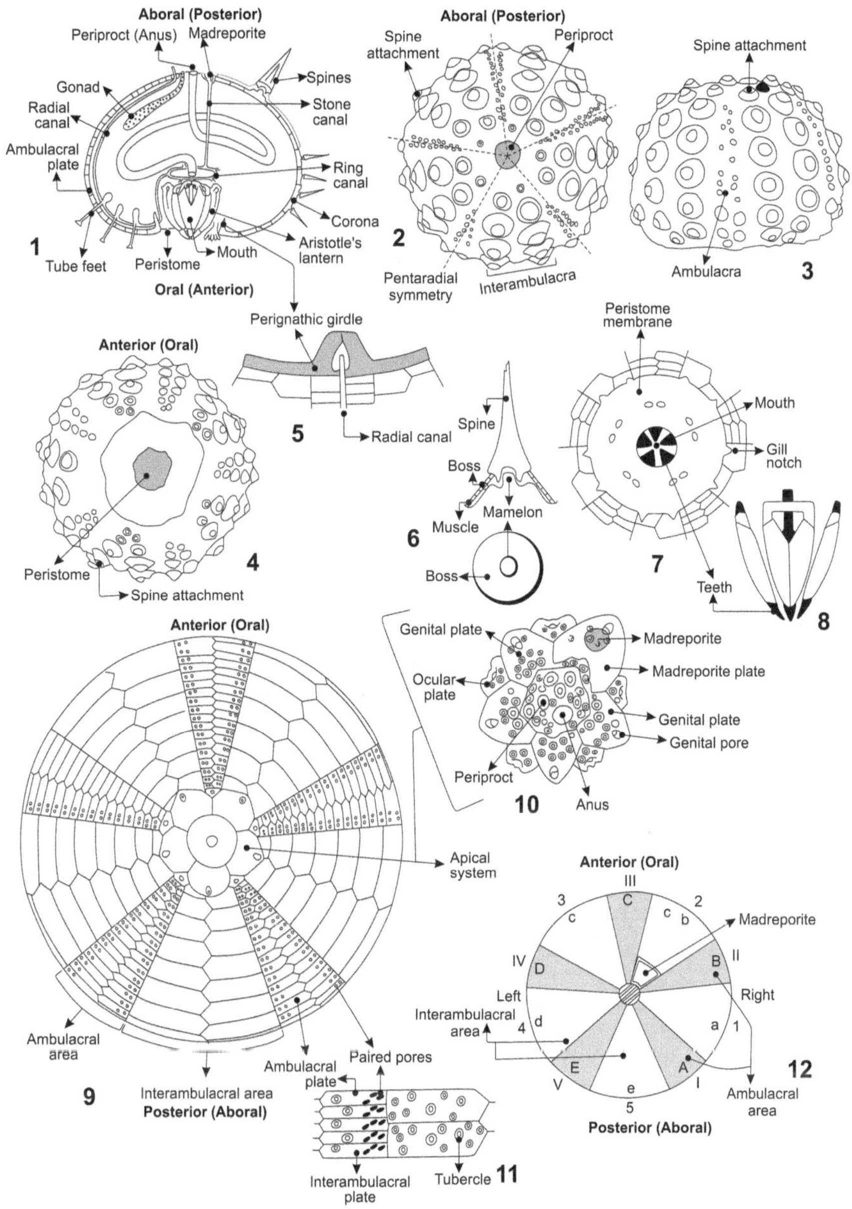

Fig. 6.2 Morphology of a regular echinoid. *1* Internal anatomy in cross section; *2–4* Regular echinoid—for Echinoderm, instead of using "dorsal" and "ventral" (as in Brachiopods), the position relative to the mouth is used; the side where the mouth lies is the Oral side, and the side opposite is called Aboral. Test shows pentaradial symmetry with ambulacra, inter ambulacra and spine attachments; *5* Perigrathic girdle (see Fig. 6.1), this serves to anchor muscles that raise and lower the complex jaw system (= the Aristotle's lantern; here Figs. 6.7 and 6.8); *6* Detail of a spine [see also Fig. 6.3(4)]; *9* Upper surface of a Regular echinoid (with no tubercles and spines) showing the Apical disc (*10*) of *Echinus esculentus* Linnaeus and paired pores (*11*); *12* Orientation of echinoids (Aboral side with shaded ambulacral areas). Echinoid orientation is based on the position of the Madreporite (M). Numerals or letters are used to designate various ambulacral and inter ambulacral areas; the areas may also be designated by reference to their position—anterior, lateral, right, left or posterior

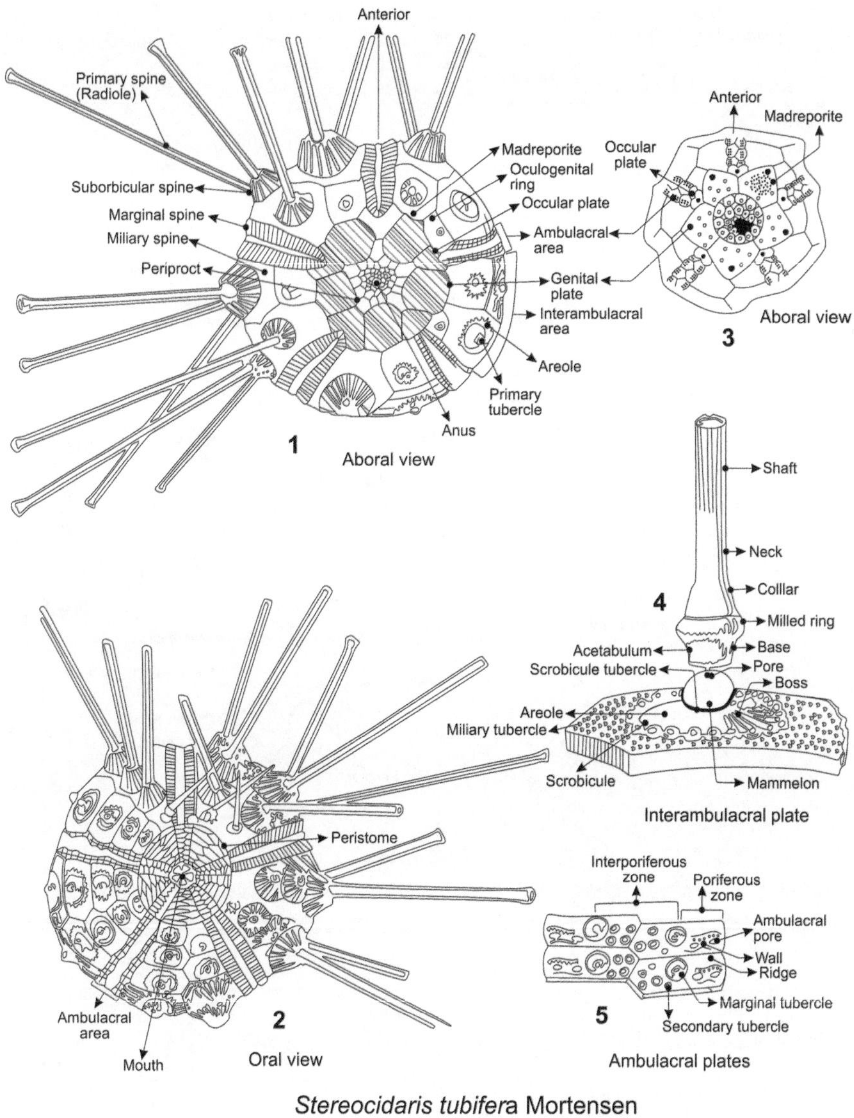

Fig. 6.3 Nomenclature of a regular echinoid *Sterocidaris tubifera* Mortensen

a radial canal of the water vascular system and the interambulacra are said to be Inter-Radial [Fig. 6.2(1, 9)]. The only asymmetric feature in the radial symmetry of this test is the position of the Madreporite [Fig. 6.2(1, 11)], and this is used to define the Anterior–Posterior orientation [Fig. 6.2(12)]. The test is conventionally aligned [Fig. 6.2(12)] with the madreporite on the right, toward the anterior end. The lower surface on which the mouth lies is the Oral surface; and the upper surface is the Aboral surface [Fig. 6.2(1)].

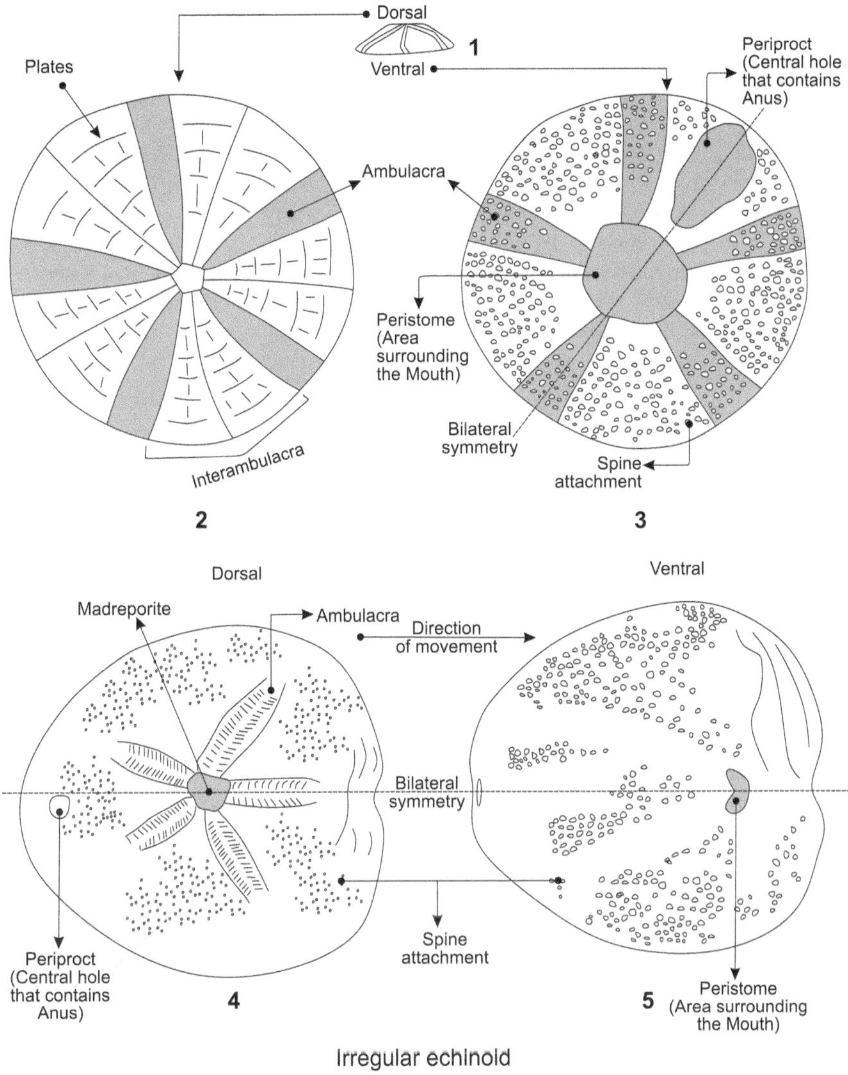

Fig. 6.4 Morphology of an irregular echinoid

The pattern of symmetry shown by the test is a convenient basis for separating the echinoids two groups—Regular forms in which the coronal plates show radial symmetry [Fig. 6.2(9)], and Irregular forms, in which the five rays are arranged in bilateral symmetry (Fig. 6.4). In regular form, the anus (surrounded by the Periproct) lies within the apical system [Fig. 6.2(10)], and the mouth is at the center of the oral surface and contains jaws (= the Aristotle's lantern) [Fig. 6.2(7–8)].

Table 6.1 Major differences between Regular and Irregular echinoids

Dissimilarities		
Parameters	Regular	Irregular
Test shape	Circular test	Distinctively elongate in adult stage
Mode life	Benthonic, epifaunal, seabead, high energy	Benthonic, infaunal, burial, low energy
Position of Anus	Dorsally (central location—directly above the mouth)	Posterior
Feeding	Aristotles lantern, grazes on seabed	Particle feeds, extracts from water
Symmetry	Fivefold (pentameral)	Bilateral
Ambulacrum	Runs from center of aboral to oral	Moddified over test, fused together
Tube feet function	Respiration, attachment and movement	Respiration, digging and maintaining a burrow
Fasicles	Absent	Two present to direct water to mouth and direct waste away
Tubercles	Small number of large tubercles	Uniform, fine, and dense tuberculation
Similarities		
Both live in a marine environment		
Both have 5 ambulacrum		
Tube feet used for respiration		
Both have anus and mouth		

In regular forms the anus lies outside the apical system in the posterior ambulacrum, and the mouth may lie either in the center of the oral surface and have jaws or toward the anterior margin and lack jaws.

The test of a typical echinoid is hemispherical in shape [Fig. 6.2(1–4)]. It consists of many interlocking plates, arranged in 10 double rows, which radiate from the apex of the upper surface to the mouth in the center of the lower surface [Fig. 6.2(2–4)]. Five of these columns carry tube feet and are known as Ambulacra plates [Fig. 6.2(3, 9)]. The other five, with no tube feet, are Interambulacra plates [Fig. 6.2(3, 9)]. Both ambulacra (ambs) and interambulacra (interambs) and are composed of two rows of plates meeting centrally at a zigzag suture [Fig. 6.2(9)].

The Ambulacral plates bear pore pairs near their outer edges, each pair, when the organism is alive, leads to a single tube foot [Fig. 6.2(11)]. Both ambs and interambs widen to the ambitus and narrow again on the oral surface (under side), where they terminate at the Peristome, i.e., around the mouth [Fig. 6.2(4)]. Peristome is a large hole where the mouth and the complex jaw apparatus known as Aristotle's lantern sit [Figs. 6.2(7–8) and 6.5(1–3)]. The other opening on the opposite side of the peristome (i.e., on the aboral side) is the Periproct (around the anus) [Fig. 6.2 (10)]. Both peristome and periproct make up the Corona (= Crown). The equatorial region of the corona is called the Ambitus and is the area of greatest width.

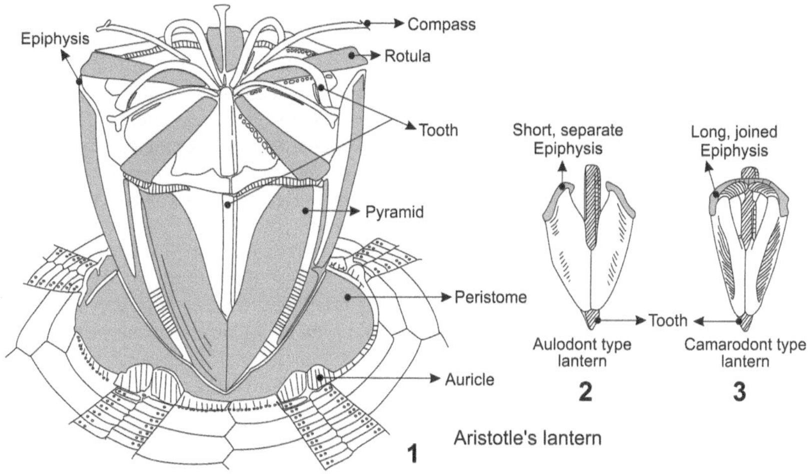

Aristotle's lantern

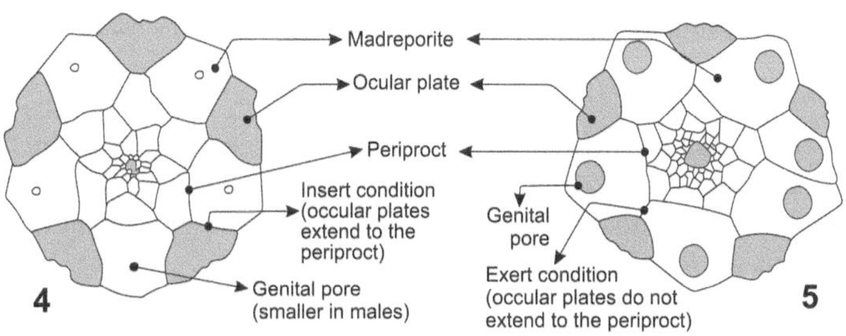

Apical system

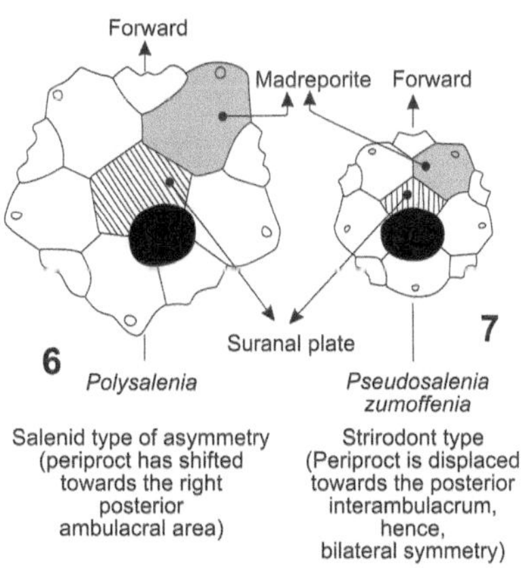

Excentricity of periproct

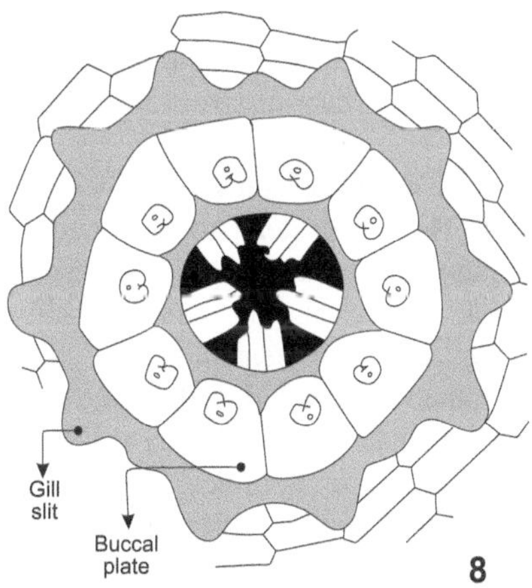

Gill slits and Buccal plate
(*Plesiodiadema*)

◄ **Fig. 6.5** Aristotle's lantern and Apical system. *1–3* Aristotle's lantern. This is the jaw of the echinoid and consists of no less than 40 individual bones (ossicles) operated by more than 60 individual muscles arranged in seven sets. In Cidaroids (as illustrated in Fig. 6.1) the projections around the peristome are interambulacral apophyses rather than ambulacral auricles, and adjacent apiphyses of the lantern are not joined; *2* Aulodont lantern; *3* Camarodont lantern. The Aulodont lantern has short separate epiphyses, whereas the Camarodont lantern has long epiphyses that join each other. *4–5* Apical system. The oculogenital ring surrounding the periproct consists of ambulacral ocular plates and interambulacral genital plates. The genital plates contain the Genital pores. The left anterior genital plate is also the Madreporite. The Apical systems are grouped as Insert, if the ocular plates extend to the periproct, and Exert, if they do not. Size of the genital pores marks as Male (see Fig. 6.4) and Female (see Fig. 6.5). *6–7* Excentricity of the periproct. *6* Saleniid type of symmetry in which the periproct has shifted toward the right posterior ambulacral area; the entire oculogenital ring has assumed a bilateral symmetry along the plane passing through the left anterior interambulacrum and the right posterior ambulacrum. *7* A Stirodont in which the periproct is displaced toward the posterior interambulacrum, hence, the plane to bilateral symmetry coincides with the conventionally defined anterior–posterior axis established by location of the madreporite (*shaded gray*). The strippled area (large central plate) is the Suranal plate. The arrow points forward. *8* Gill slits and Buccal plates. Two features common to all regular echinoids—embayments (gill slits) in the peristomial margin which make room for gills and a circle of large buccal plates on the peristome. The *shaded area* is the peristomial membrane

Both ambs and interambs that make up the corona are arranged in 10 double rows, terminate aborally at a genital plate [Fig. 6.2(10)]. The five large double rows of interambulacra plates make the Interambulacra area and five small double rows of ambulacra plates form the Ambulacra area.

The Interambulacral plates are imperforate and generally larger, and often bear knobs or tubercles [Fig. 6.2(2)]. These are the articulation bases for the many spines [Figs. 6.2(6) and 6.6(1–6)] which bristle from the test surface in life [Fig. 6.2(1)]. On both sides of the ocular plate, the Interambulacral plates arise [Fig. 6.2(9, 10)]. The tubercles that cover the outside surface of both ambulacral and interambulacral plates, greatly vary in size; these are also the sites for the articulation points for spines.

Spines are rarely preserved in place, but occur commonly in bioclastic limestones. The spines tend to be longest at the equatorial region of the test (= the Ambitus). On a smaller scale, the test surface is protected by pedicellariae [Fig. 6.6 (7–15)], minute pincers sometimes invested with poison glands. In detail, each Interabulacral plate bears a shallow pit called the Areole that bears a conical Boss, surmounted by a globular Mamelon [Figs. 6.2(6) and 6.3(4)]. Both the boss and the mamelon make up the Primary tubercle [Figs. 6.2(6) and 6.3(4)] which carries Primary spines or Radioles. The Mamelon, because of a dimple on its surface, is considered Perforate. The areole is surrounded by a ring of small Scrobicular tubercles [Fig. 6.3(4)]. These along with smaller tubercles that lack areole, are classed as Secondary tubercles and the microscopic ones are called Miliary tubercles [Fig. 6.3(1, 4)]. The scrobicular tubercles carry the scrobicular spines and the miliary tubercles carry tiny spines called Pedicellariae [Fig. 6.6(7–15)]. The

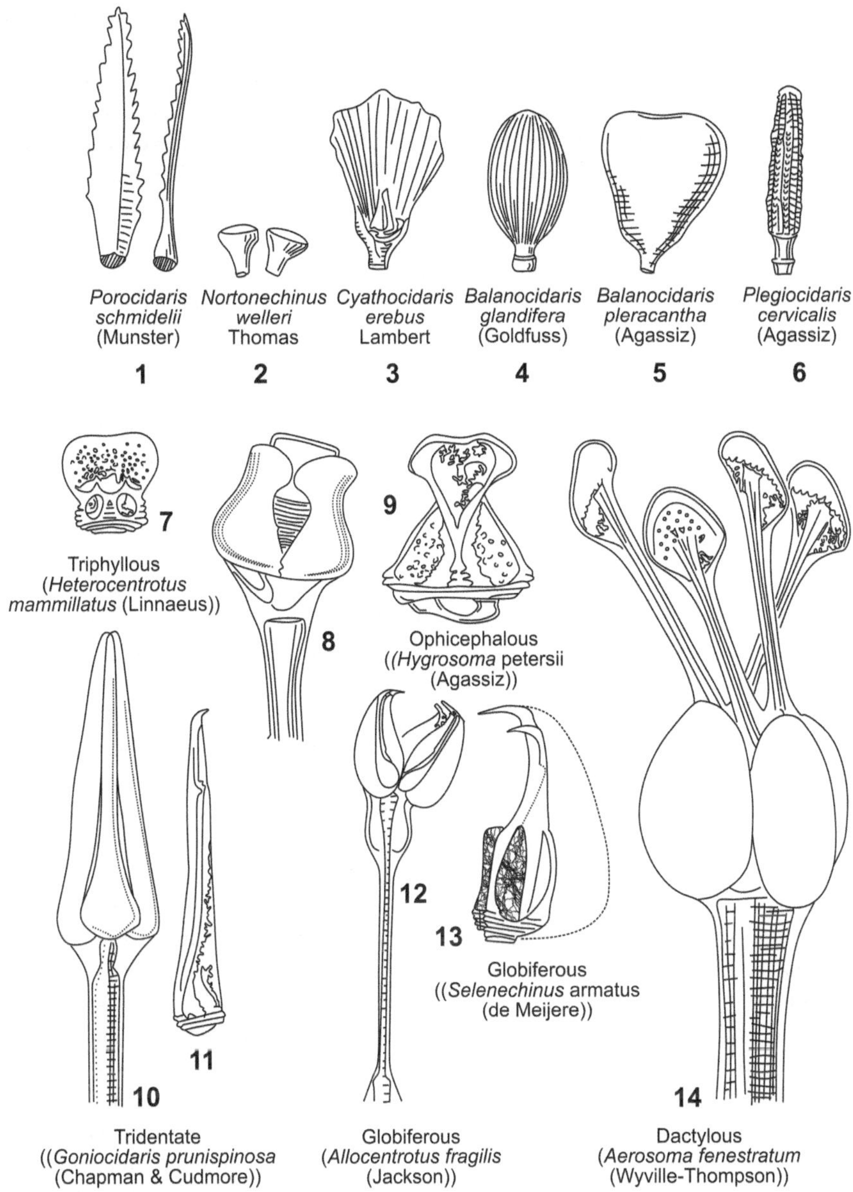

Fig. 6.6 Spines and pedicellariae. *1–6* Spines; *7–14* Pedicellariae of modern echinoids. The Pedicellariae are groups of two, three, or more minute snapping jaws (valves), mounted at the tip of small spines scattered over the surface of the test

primary spines or radioles are complex; their base is concave forming a socket called the Acetabulum which articulates with the Mamelon of the primary tubercle [Fig. 6.3(4)]. Above the Acetabulum, the spine expands into a Milled Ring to which the muscles are attached [Fig. 6.3(4)]. Above this, the Collar of the spine tapers into a smooth cylindrical Neck which is then succeeded by a rough longitudinally striated Shaft (which makes up most of the spine) [Fig. 6.3(4)]. Each Ambulacral plate bears a prominent Marginal tubercle on the edge between poriferous and inteperiferous zones [Fig. 6.3(1)]. The Secondary tubercles are, however, distributed in the inteperiferous zone. The marginal tubercles carry the flat Marginal spines (similar to the scrobicular spines of the Interambulacral plate) [Fig. 6.3(1)].

At the aboral (upper) extremity of the shell, each ambulacral area ends by a single Ocular plate and each interambulacral area by a Genital plate; these 10 plates form the Oculogenital Ring that surrounds the Periproct [Figs. 6.2(9–10) and 6.3 (3)]. Each of the ocular plates bears a small Ocular pore. Each genital plate is perforated by a Genital pore [Fig. 6.2(10)]; females have larger pores, enabling differentiation of sexes. One of the genital plates, the Madreporite is riddled with microscopic pores and acts as a sieve for the intake of water [Fig. 6.3(3)].

Situated at the top is the Apical System [Fig. 6.5] consisting of about 10 small plates that are connected with specialized functions, and one of which is the Madreporite [Fig. 6.5(4, 5)]. The central part of the apical system consists of a membrane, the Periproct, which surrounds the anus. In the center of the lower surface of the test, is a similar membrane, the Peristome which surrounds the mouth [Fig. 6.2(7)]. Most of the surface of the test is covered by spines which are attached to Tubercles.

The Aristotle's lantern is a highly complex structure with 50 skeletal elements and 60 muscles [Fig. 6.5(1–3)]. It is pentaradially symmetrical with the five teeth, braced in a hemipyramid [Fig. 6.5(1)]. Like a grab, the entire structure can open and close, and move in and out of the test. As the echinoids possess a rigid test with a fixed internal volume, a large and active lantern that moves in and out poses severe space constraints. To compensate for changes in internal volume there are 10 expandable soft tissue sacs around the edge of the peristome that connect directly to the interior and that accommodate any displaced body cavity fluids. For this, the echinoid body plan has come up with 10 expandable soft tissue sacs around the edge of the peristome. These connect to the interior directly and accommodate any displaced body cavity fluids. For each sac, there is a small notch in the exterior within the peristome called the Buccal notch. The latter are always absent in such echinoids that either lack a lantern or have an entirely internal lantern. The lantern is moved by primary muscles that are attached to the interior of the test around the rim of the peristome. Such sites of muscle attachment are modified into enlarged skeletal flanges; those developed from the adoral ambulacral plates are called Auricles while others from interambulacral plates are called Apophyses. All muscle attachments of the Aristotle's lantern form the Perignathic girdle [Fig. 6.2(1, 5)]. The latter serves to anchor muscles that raise and lower the complex jaw system (= the Aristotle's lantern).

In Regular echinoids, the apical system contains 10 plates, arranged in one or two rings around the periproct [Figs. 6.2(9, 11), 6.5(4, 5)]. Five of these, the Ocular plates, are situated radially; and alternating with them are five inter-radial Genital plates [Figs. 6.2(9, 11), 6.5(4, 5)]. The ocular plates each bear a pore through which passes the terminal tentacle of the radial water vessel. The genital plates are the larger and each has a pore through which eggs or sperms are discharged. One genital plate (the right anterior) is also the Madreporite and is finely perforated [Figs. 6.2(9, 11), 6.5(4, 5)]. In irregular echinoids [Fig. 6.4], the apical system does not enclose the periproct and it is small and compact; it may contain less than five genital plates, but always has five oculars [Fig. 6.2(9)].

The Ambulacral plates are small and each is pierced by one pair of pores [Figs. 6.2(11) and 6.5(5)], A Pore Pair [Fig. 6.2(11)], except in the gill-bearing regular echinoids are housed in the Compound ambulacral plates [Fig. 6.7]. The latter consist of two or more plates fused together and possess two or more pairs of pores. The pores are round and close-set, except in irregular echinoids that have respiratory tube feet; in these, one or both pores of each pair may be elongated [Fig. 6.4(5)]. Interambulacral plates are large and have no pores. Their surface is covered by many tubercles and granules to which, in life, movable spines are attached by muscles [Fig. 6.4(3)]. The spines are rarely preserved, situ, in fossil echinoids. The Tubercle consists of a round knob, the Mamelon, protruding from a shallow mound, called the Boss [Fig. 6.3(4)]. Tubercles occur on both ambulacral and interambulacral plates, but they are more numerous and larger on the latter. In

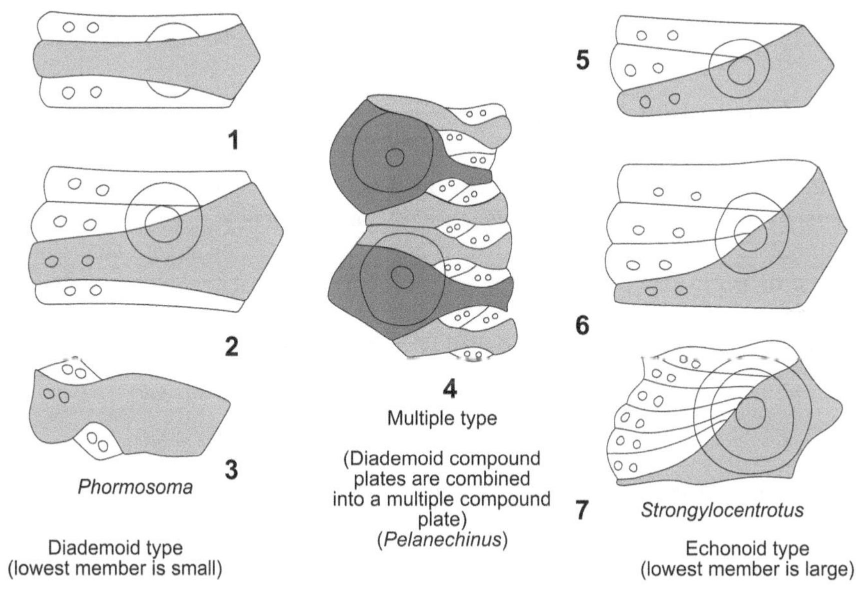

Fig. 6.7 Compound plates. Compound ambulacral plates are formed by the fusion of simple plates in which one (*gray*) remains larger than the others. Most Mesozoic and Cenozoic regular echinoids possess compound ambulacral plates. Two basic types of compound plates are noted: Diademoid and the Echinoid. Multiple type is a special condition where the diademoid compound plates are combined into a multiple compound plate

regular echinoids they vary in size from large primary tubercles to small granules, but they are rather small and close-set in irregular echinoids.

The Mesozoic and Cenozoic regular echinoids have compound ambulacral plates, instead of the normal monotonously added new rows of similar plates. Two basic styles are recognized: Diademoid and Echinoid (Fig. 6.7). In the diademoid, the middle one of the three component elements (or one below the middle in combination of 4 or 5) is largest, whereas in the Echinoid type, the lowest member is largest. The third style is the Multiple where the diademoid plates are combined into a multiple compound plate as in the Jurassic echinothurid *Pelanechinus* (Fig. 6.7).

The Spines and Pedicellariae are appendages which are attach to the test (Fig. 6.6). The Spines [Fig. 6.6(1–6)], based on their function (more so in irregular echinoids where maximum functional differentiation is noted), come in varied shapes and sizes and are generally long [Fig. 6.3(1, 2)]; they can be hollow or solid and smooth or ornamented, externally. A central ligament binds the spine to its articulation ball, and the associated tubercle is then, perforate; the surrounding platform may be smooth or crenulated. Pedicellariae which evolved from clusters of spines, resemble tulips in shape [Fig. 6.6(8–15)]. They are microscopic stalked, jawed appendages that are common in all echinoids from the Silurian onwards. They are used to deter small ectoparasites and occur in many different forms making them very useful species-level identification. However, they are rarely preserved as fossils.

6.3 Orientation of Echinoids

From the aboral view, the ambulacra are labeled with capital letters or roman numerals, in a counterclockwise sequence, the anterior ambulacrum being C or III [Fig. 6.2(12)]. The Interambulacral areas are numbered by lower case letters or by Arabic numerals or may simply be designated by listing the adjoining ambulara; thus, interambulacral “a” or “1” may likewise be listed as interambulacrum A-b or I-II. It must be noted that when the shell is viewed from the oral side, the directions are reversed, and the sequence of numbering is clockwise.

6.4 Irregular Echinoids

The morphological features of irregular echinoids have been taken up earlier. However, below each group is briefly dealt as they differ considerably from others. The morphological features of the following 4 (Holectypoids, Cassiduloida, Clypeasteroids and Spatangoids) are briefly described and illustrated (Figs. 6.8, 6.9, 6.10 and 6.11).

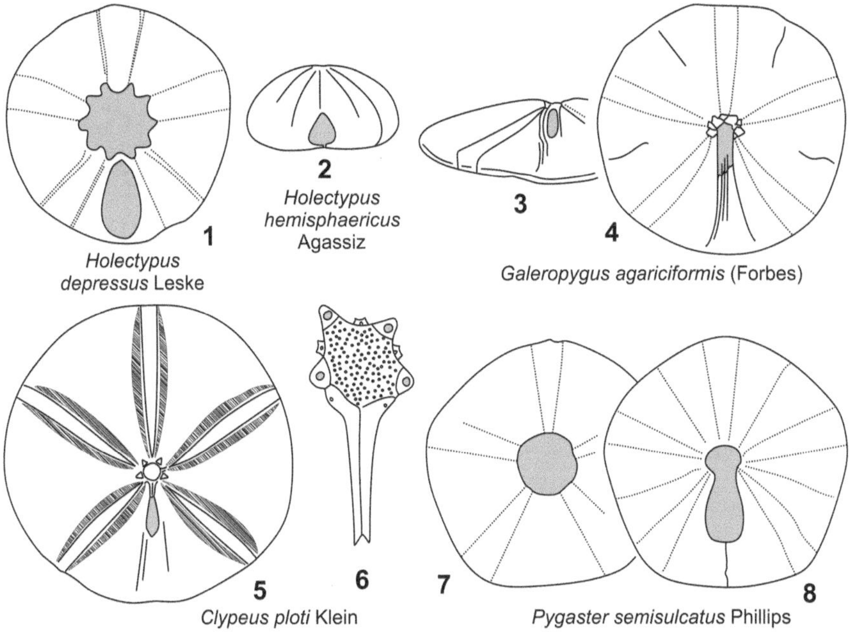

Fig. 6.8 Jurassic irregular echinoids. They first appeared in the Jurassic and are comparatively primitive in body plan. *1–2* Jurassic Holectypoid in which the peristome is extended into gill slits; the periproct is located at the margin (*2*) but in others (*1*), it is on the oral side and extends close to the mouth. *3–4* Jurassic Cassiduloid whose periproct is surrounded on three sides by oculogenital rings. *5–6* Jurassic Cassiduloid. This is an advanced irregular echinoid with well-developed petals and elongate posterior genital plates which border on the periproct. *7–8* Jurassic Holectypoid in which the periproct is a keyhole-shaped structure

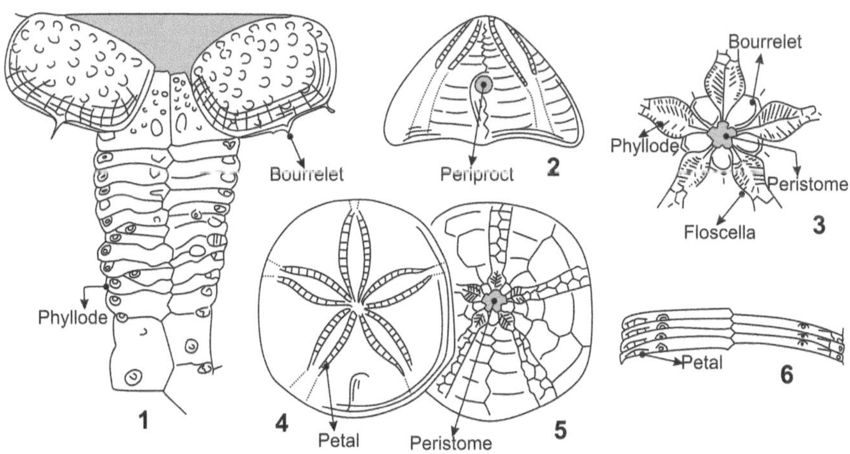

Fig. 6.9 Cassidulinoid structure. *1* Portion of an ambulacral area of *Cassidulus eugeniae* (Agassiz); *2* Rear view of *Cassiduls subconicus* Clarke showing the periproct; *3* Peristomial region of *Cassiduls subconicus* Clarke; *4–5 Cassiduls subconicus* Clarke *4* Aboral view; *5* Oral view; *6* Portions of a petal of *Cassiduls subconicus* Clarke as in *4*

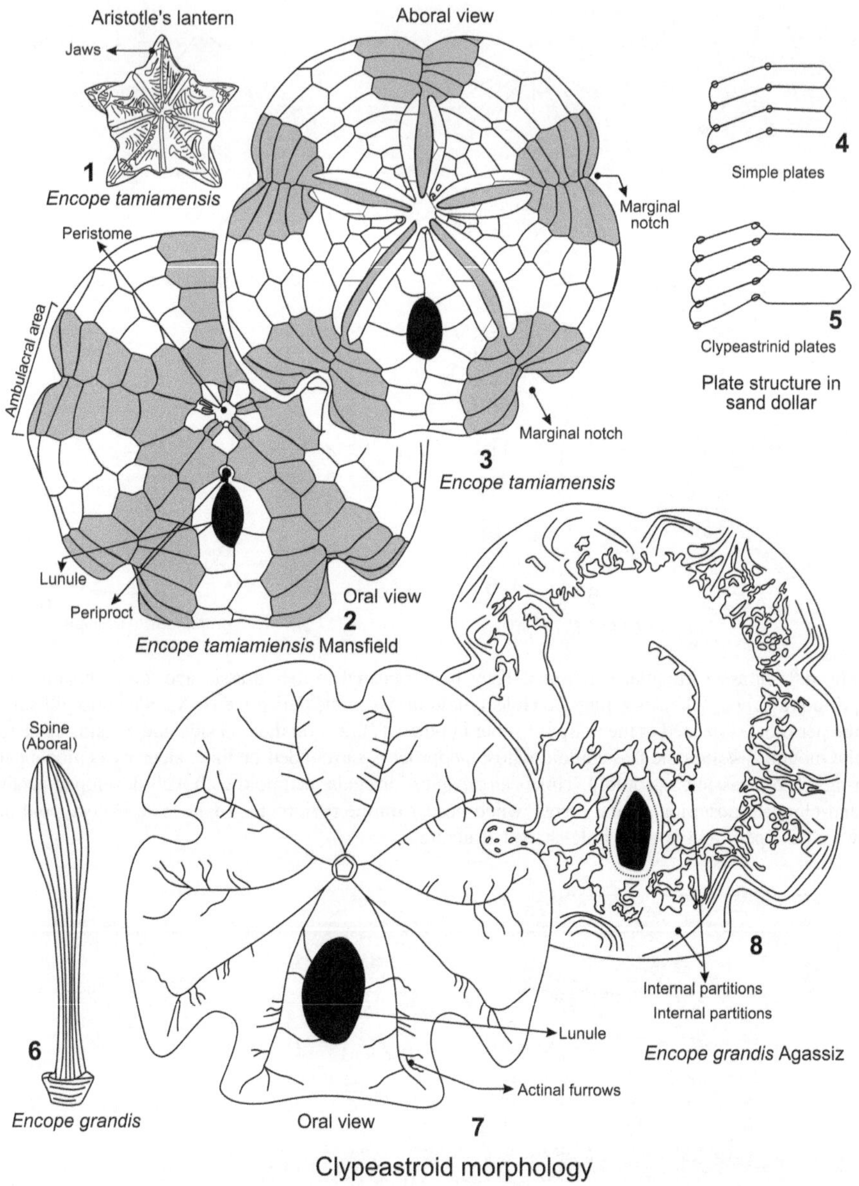

Fig. 6.10 Calypeasteroid morphology. *1* Aristotle's lantern; *2* Oral view; *3* Aboral view; *4–5* General plate structure among Calypeasteroids; *6* Aboral spine; *7* Oral view showing grooves; *8* Part of test showing internal structures

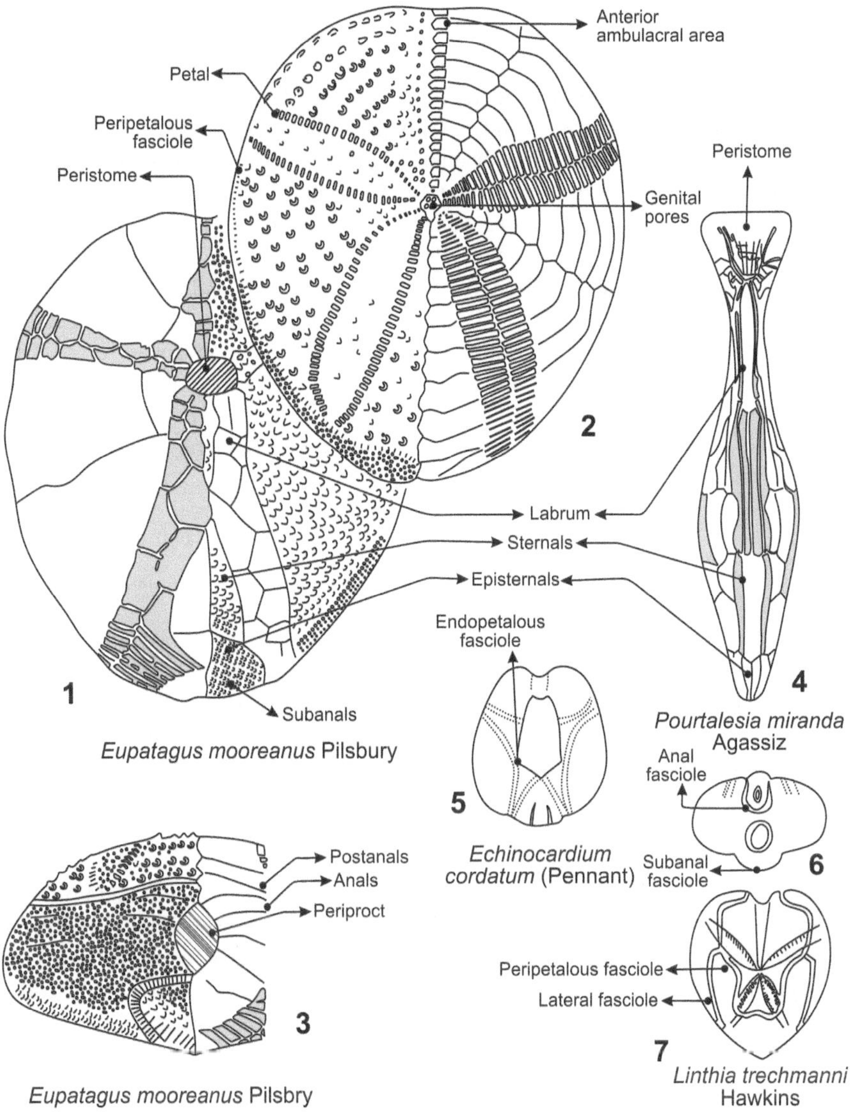

Fig. 6.11 Spatangoids morphology. *1–3 Eupatagus mooreanus* Pilsbury; *1* Oral view; *2* Aboral view; *3* rear view. *4 Pourtalesia miranda* Agassiz, a modern deep sea dweller, oral view, the ambulacral plates are shaded; *5–6 Echinocardium cordatum* (Pennant); *5* Oral view; *6* Rear view; *7 Linthia trechmanni* Hawkins, aboral view

6.4.1 Holectypoids

These are the most primitive (and least specialized) forms and are the closest to regular echinoids (Duncan 1889). Globally, they dominated the Jurassic–Cretaceous landscape, from Pliensbachian (Early Jurassic) to Maastrichtian (Late Cretaceous), and are best exemplified by Jurassic forms such as *Galeropygus agariciformis* (Forbes), *Holectypus depressus* Leske, *Holectypus hemisphaericus* Agassiz, *Clypeus ploti* Klein and *Pygaster semisulcatus* Phillips (Fig. 6.8). Holectypoids are small echinoids with a flattened oral side and a hemispherical to conical aboral surface. Except for the posterior displacement of the periproct, the test retains radial symmetry, becoming elliptical only in very advanced forms. They possess a functional jaw apparatus and a large central peristome. The periproct opens orally in *Holectypina*, but aborally in *Pygasterina*. The next group, the Cassiduloids, differ in having a much smaller peristome and have pore pairs crowded adorally to form phyllodes.

6.4.2 Cassiduloida

The Recent *Cassidulus* [*C. eugeniae* (Agassiz)] best exemplifies its structure (Fig. 6.9). It is characterized by the presence of flower-like Floscelle [Fig. 6.9(3)] centered around the peristome and is composed of depressed leaf-like areas called Phyllodes within the ambulacral segments separated by the bulging interambulacral Bourrelets [Fig. 6.9(1)]. The radiating arrangement of the five resulting ellipses on the aboral surface is aptly called Petals [Fig. 6.9(4)]. The most primitive ones do not possess petals whereas the advanced Cassidulinids, do. The jaws in Cassidulinids are present only in the early ontogeny stage. The Cassiduloida are considered paraphyletic. They range from Early Jurassic (Toarcian) to Recent. Cladistic analysis suggests that cassiduloids are a grade taxon comprising a small number of clades leading to the Clypeasteroida (see also Mooi 1990; Smith 2001). The majority of Jurassic cassiduloids belong to the common stem group of living cassiduloids plus clypeasteroids.

6.4.3 Clypeasteroids

The Clypeasteroids (Fig. 6.10) are characterized by possessing multiple microscopic pores to each ambulacral plate, usually arranged in broad bands. The primitive Clypeasteroids (small, and microscopic; such as the *Fibularia subglobosa*) resemble the most advanced Holectypoids in possessing the Aristotle's lantern (well-developed) and petals (*Fibularia* shows the beginning of the same). The advanced Clypeasteroids are mostly flat (like *Encope*; *Encope tamiamiensis* Mansfield; Fig. 6.10) with strongly petaloid ambulacra on the aboral side. The Actinal furrows (branching radial grooves) lead to the peristome which is small and surrounded by a floscelle. The test is covered by short spines. In some flat-disc or shield-shaped testes marginal indentations (Notches) are also noted for unknown functions. Clypeasteroid's sheer abundance, global distribution and short stratigraphic occurrence, especially within the Cenozoic (Late Paleocene to Recent) make them valuable index fossils.

6.4.4 Spatangoids

These heart-shaped sea urchins are highly specialized for a burrowing mode of life. *Eupatagus* (*E. mooreanus* Pilsbury; Fig. 6.11), a characteristic spatangoid, has a bun-shaped test, and is indented by the depressed Anterior ambulacrum. The peristome is a transversely elongated opening and the periproct is placed vertically at the end of the test. However, the most distinctive character of *Eupatagus* are the petals (4 in number); the anterior ambulacrum is non-petaloid. The posterior interambulacrum is highly modified; the single plate next to the mouth is called the Labrum and in some advanced spatangoids (as in *Eupatagus*), it is completely separated from the remainder of the posterior interambulacral area. This is succeeded by paired Sternal plates, followed by a pair of Episternal plates which are barely in contact with the succeeding pair of interambulacral plates called Subanals. The beading on the test is interrupted by two narrow depressed bands bearing miliary granules. One of these bands (Peripetalous fasciole) forms a ring around the petals; the other (Subanal fasciole) includes the subanal plates and parts of the adjacent ambulacral plates beneath the periproct. In some other spatangoids, the Anal fasciole extend part way around the periproct; Endopetalous fascioles lie within the petals. The Lateral fascioles lead from the peripetalous fasciole to the rear, along the flank of the test.

6.5 The Echinoderm-Backbone Connection

The Echinoderms are closely related to the vertebrate group, Chordates. This resemblance is based on Echinoderm's early cell division, embryonic development, and larvae. Additionally, the chemical similarities associated with muscle activity and the chemistry of oxygen-carrying pigments in the blood of echinoderms is also similar to that of chordates.

6.6 Terminology

6.6.1 **Abactinal (interchangeably used as Aboral, Posterior)**: body area opposite to the mouth [Fig. 6.2(1)]

6.6.2 **Aboral (interchangeably used as Abactinal, Posterior)**: Direction away from the mouth; the part of the body opposite the mouth [Fig. 6.2(1)]

6.6.3 **Adapical**: This is the highest part of an echinoids test [top part of Fig. 6.12 (2)]

6.6.4 **Ambulacral area**: Five narrow bands extending from periproct to peristome, composed of a double row of ambulacral plates [Fig. 6.2(9)]

6.6.5 **Ambulacral plates**: Skeletal elements which bear pores for passage of tube feet [Fig. 6.2(9, 11)]

6.6.6 **Ambulacral pores**: Perforation in each ambulacral plate, serving as passage way for a tube foot [Fig. 6.2(9, 11)]

6.6.7 **Ambulacrum (pl. ambulacra)**: An area of the body that carries tube feet; in echinoderms there are generally 5 ambulacra [Fig. 6.2(9)]

6.6.8 **Anal fasciole**: Groove encircling periproct [Fig. 6.11(6)]

6.6.9 **Anals**: Plates surrounding periproct [Fig. 6.11(3)]

6.6.10 **Anterior**: Direction or position of interambulacral area at left of madreporite [Fig. 6.2(12)]

6.6.11 **Anus**: Outlet of digestive tract, located at center of periproct, opposite mouth [Fig. 6.2(1)]

6.6.12 **Apical system**: It is a ring of specialized skeletal plates that includes genital and ocular plates. In echinoids, it is usually located on the highest point of the test [Figs. 6.2(10), 6.3(3) and 6.5(4)]

6.6.13 **Appendage**: A tube foot [Fig. 6.2(1)], spine (Fig. 6.6), pedicellaria (Fig. 6.6), or arm of an adult, or a projection from the larval body

6.6.14 **Areole**: Broad, shallow pit in each interambulacral plate, bearing a large tubercle which serves as base for a primary spine [Fig. 6.2(1)]

6.6.15 **Auricle**: Internal projection for attachment of lantern muscles, located in each ambulacral area at edge of peristome [Fig. 6.5(1)]

6.6.16 **Base**: Portion of a spine below milled ring [Fig. 6.3(4)]

6.6.17 **Bilateral Symmetry**: A pattern of symmetry in which the left side of the body is a mirror image of the right based upon an anterior–posterior axis [Fig. 6.4(3)]

6.6.18 **Boss**: Cone which support mamelon of a tubercle [Fig. 6.2(6) and 6.3(4)]

6.6.19 **Buccal**: Lying within the mouth [Fig. 6.5(8)]

6.6.20 **Clypeastrinids plates**: Structure characteristic of petals of Clypeastroids, in which alternate plates are shortened and thereby restricted to outer poriferous zones of ambulacral area [Fig. 6.2(10)]

6.6.21 **Collar**: Smooth, tapering portion of a spine located above the milled ring [Fig. 6.3(4)]

6.6.22 **Compass**: Slender arched radial rod in ambulacral position, at top of lantern [Fig. 6.5(1)]

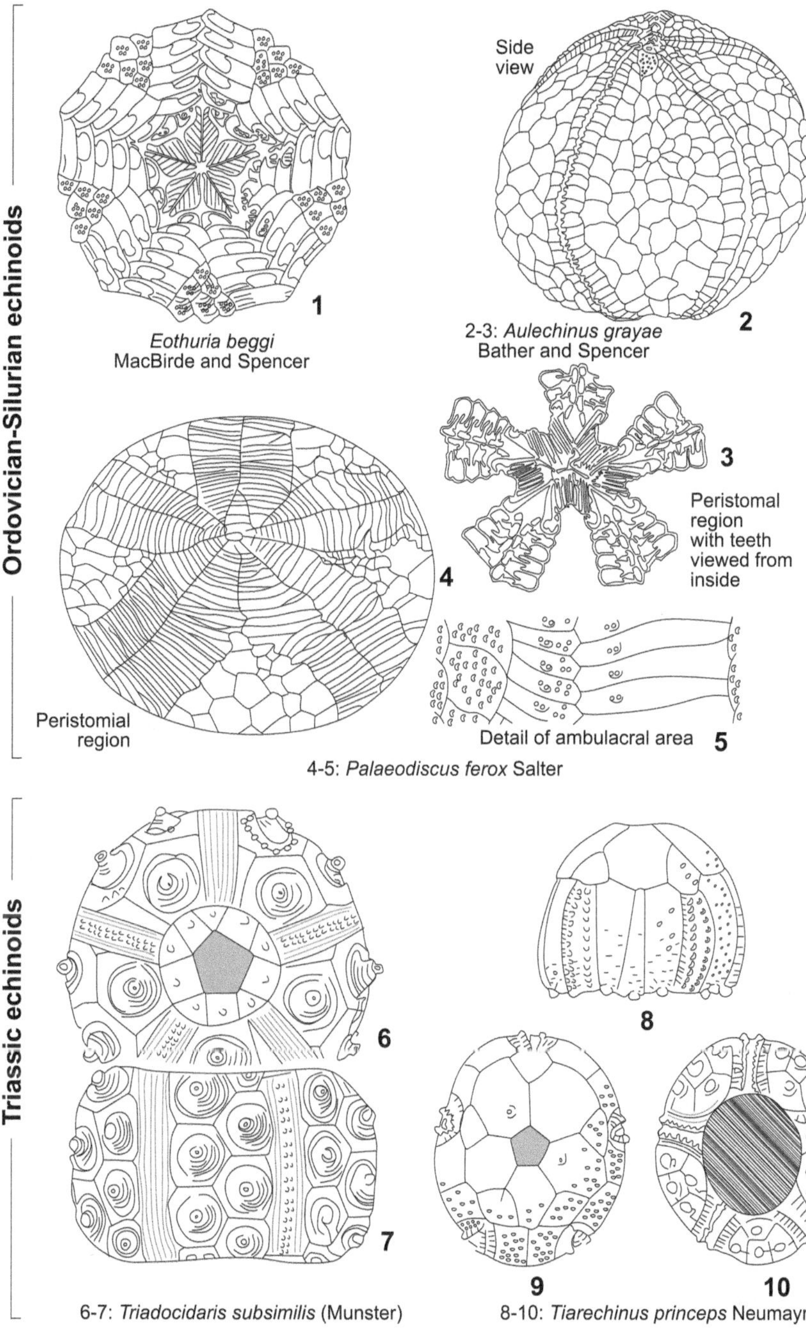

Fig. 6.12 Representative Ordovician-Silurian and Triassic echinoids. The flexible tests of most Silurian–Ordovician echinoids have ambulacral areas composed of two plate rows, interambulacral areas of irregularly overlapping plates. The Triassic echinoids are mostly very small and are rare

6.6.23 **Dactylous**: Having a variable number of stalked, spoon-shaped valves [Fig. 6.6(15)]

6.6.24 **Dorsal**: This term is variously applied. In asteroids, ophiuroids and echinoids, it refers to the surface opposite to the mouth (i.e., the uppermost surface). In holothuroids, with mouth and anus are at opposite ends, here, the uppermost surface is considered dorsal. In crinoids, the surface opposite the mouth is dorsal even though it is functionally the ventral (lower) side [Fig. 6.2(1)]

6.6.25 **Endopetalous fasciole**: Elliptical groove located within margins of petals [Fig. 6.11(5)]

6.6.26 **Epiphysis**: Rugged crossed member composed of two fused ossicles; occurs in each interambulacrum at top of lantern [Fig. 6.5(1)]

6.6.27 **Episternals**: Pair of interambulacral plates lying between sternals and subanals [Fig. 6.11(1, 4)]

6.6.28 **Fasciole**: Groove on test lacking larger tubercles and spines; minute spines located within these grooves are thickly covered with cilia, which move streams of mucus or currents of water for removal of foreign matter from the surface of the animal. These are narrow bands of small, specialized spines in many irregular echinoids; visible on the denuded test as bands of densely packed, tiny tubercles (Fig. 6.11)

6.6.29 **Genital Plate**: Large plate in each interambulaeral area, bordering periproct, and bearing and genital pore. In ophiuroids, a bar-like ossicle connecting the radial shield to the arm and supporting the radial edge of the bursal slit [Fig. 6.2(10)]

6.6.30 **Globiferous pedicellaria**: An echinoid pedicellaria (three-valved) equipped with venom sacs [Fig. 6.6(14)]

6.6.31 **Granules**: Minute and nearly equidimensional, small structures fixed to the surface of scales or plates [Fig. 6.6]

6.6.32 **Heart Urchin**: A burrowing echinoid (usually in the order Spatangoida) that is more or less heart-shaped (Fig. 6.11)

6.6.33 **Interambulacral areas**: Five broad bands trending from peristome to periproct, each composed of a double row of large interambulacral pores [Fig. 6.2(9)]

6.6.34 **Interambulacral plate**: Large, imperforate skeletal element composing larger part of interambulacral areas; each plate carries a primary spine [Figs. 6.2(9, 11)]

6.6.35 **Interambulacrum**: Lying between two ambulacra, it is the oral or aboral sector of the body [Fig. 6.2(9)]

6.6.36 **Internal partitions**: Pillars and anastomosing walls, which buttress interiors of Clypeastroids [Fig. 6.2(10)]

6.6.37 **Interporiferous zone**: Mid-region of each ambulacral area, between pore-bearing margins of ambulacral areas [Fig. 6.3(5)]

6.6.38 **Irregular echinoid**: These are heart-shaped or disc-shaped echinoid with very short spines (Fig. 6.4). They possess some degree of bilateral symmetry as the anus is not at the center of the dorsal surface

6.6.39 **Jaw**: It is a moveable triangular structure that extends into the mouth [Figs. 6.2(7, 8), 6.5(1, 3)]

6.6.40 **Lateral fasciole**: Groove alongside of test [Fig. 6.11(7)]

6.6.41 **Lunule**: "Keyhole-like" perforations in test of many flat sand dollars (as in the five or six-holed sand dollars) [Fig. 6.10(7)]

6.6.42 **Madreporite**: A plate (right anterior genital plate) with numerous perforations that is connected to the water vascular ring by a so-called stone canal. In some it is internal (holothuroids), whereas in others, it opens to the exterior on the dorsal surface of the body (asteroids and echinoids). In ophiuroids it opens near the mouth, on the ventral surface [Fig. 6.2(10, 12)]

6.6.43 **Mamelon**: Spheroidal summit of a tube feet [Figs. 6.2(6) and 6.3(4)]

6.6.44 **Marginal notches**: Embayments in margins of sand dollars [Fig. 6.10(3)]

6.6.45 **Marginal spines**: Flat, blade-like spines, occurring singly on ambulacral plates, protecting tube feet [Fig. 6.3(1)]

6.6.46 **Marginal tubercle**: Small elevation near an ambulacral pore, for attachment of marginal spine [Fig. 6.3(1)]

6.6.47 **Miliary spine**: Tiny spines scattered over surface of interambulacral and oculogenital plates [Fig. 6.3(1)]

6.6.48 **Milled ring**: Flange near base of a spine, serving for attachment of muscles that move spine [Fig. 6.3(4)]

6.6.49 **Mouth**: Opening located in center of underside [Fig. 6.2(1)]

6.6.50 **Neck**: Smooth cylindrical portion of a primary spine laying between collar and shaft [Fig. 6.3(4)]

6.6.51 **Ocular plate**: A plate in the apical system of echinoids, adjoining the periproct in ambulacral (radial) position (there are five in umber). Along with the genital plates, the ocular plates form the Oculogenital ring [Fig. 6.2(10) and 6.3(3)]

6.6.52 **Oculogenital ring**: Circlet of 10 plates (2 ambulacral oculars, and 5 interambulacral genitals, including one madreporite), which surround periproct [Fig. 6.2(9, 10)]

6.6.53 **Oral**: Part of the body that is on the same surface as the mouth [Fig. 6.2(1)]

6.6.54 **Pedicellariae**: Elements used for defense and grooming, these are small stalked or unstalked pincer-like organs [Fig. 6.6(8–15)]

6.6.55 **Perforate Tubercle**: A primary or secondary tubercle that has an apical perforation for the insertion of a ligament [Fig. 6.2(11)]

6.6.56 **Peripetalous fasciole**: Groove which encircles petals [Fig. 6.11(7)]

6.6.57 **Peripetalous**: A Fasciole, placed at the distal ends of the anterior and posterior petals (ambulacra I, II, IV and V) and that which crosses the ambulacrum III [Fig. 6.11(2)]

6.6.58 Periproct: Area surrounding anus and enclosed by oculogenital ring; covered with leathery skin in which small plates are embedded loosely [Fig. 6.2(1, 9, 10)]

6.6.59 Peristome: In echinoids, the area of the test which carries the mouth (the membranous area around mouth). The surrounding peristomial membrane is commonly plated (= leathery skin studded with small plates) [Fig. 6.2(1)]

6.6.60 Petal: Aboral portion of some or all of ambulacral areas, on which pore pairs are confluent in closely spaced, elongate slits [Fig. 6.11(2)]

6.6.61 Plates: Tubular structures with a characteristic shape and a fixed position [Fig. 6.6]

6.6.62 Pore (in a primary tubercle): Pit for attachment of a ligament which fastens spines to the tubercle [Fig. 6.2(9 and 11)]

6.6.63 Pore Pair: These are ambulacral pores divided by a skeletal wall, through which a tube foot passes [Fig. 6.2(11)]

6.6.64 Poriferous zone: Pore-bearing outer edges of ambulacral areas, as contrasted with non-perforate strip which extends down middle of each ambulacral plate [Fig. 6.3(5)]

6.6.65 Postanals: Interambulacral plates which lie above the periproct [Fig. 6.11 (3)]

6.6.66 Primary spines: Placed on the interambulacral plates, these are large movable projections that occur singly [Fig. 6.3(1)]

6.6.67 Primary tubercles: Prominent rounded elevations which bear primary spines [Fig. 6.3(1)]

6.6.68 Proximal: Toward the center of the body

6.6.69 Radial symmetry: A pattern where identical segments are arranged around a central axis; the echinoderms mostly possess a five-part (pentamerous) radial symmetry [Fig. 6.2(9)]

6.6.70 Regular echinoid: A more or less spherical echinoid characterized by anus situated at center of the aboral surface, and with long spines [Figs. 6.6(2, 3)]

6.6.71 Scales: These are flat, thin structures that are overlapping, tesselate, or haphazardly arranged [Fig. 6.6]

6.6.72 Scrobicular spines: Flat spines arranged in a ring around scrobicule of each areole, protecting muscles which move large primary spines [Fig. 6.3 (1)]

6.6.73 Scrobicular tubercles: Small elevations for attachment of scrobicular spines encircling areoles

6.6.74 Scrobicular: It is the area surrounding the base of a spine, in echinoids [Figs. 6.3(1, 4)]

6.6.75 Scrobicule: Depressed marginal area of an aureole, or any depressed ring around base of a tubercle, serving for attachment of muscles which move spines [Fig. 6.3(4)]

6.6.76 Secondary spines: These smaller spines are carried on secondary tubercles within the ambulacra and interambulacra, in echinoids [Fig. 6.3(5)]

6.6.77 **Secondary tubercles**: Tubercles which carry small (secondary) spines [Fig. 6.3(5)]

6.6.78 **Shaft**: Main part of a spine [Fig. 6.3(4)]

6.6.79 **Simple plates**: Ambulacral plate structure in which adjacent plates are equally developed, contrasting with more specialized clypeastrinid-type [Fig. 6.3(5)]

6.6.80 **Spinules**: Small, usually sharp-pointed, spines [Fig. 6.6].

6.6.81 **Spines**: These are long, slender, and attenuated moveable, articulating structures [Fig. 6.6]

6.6.82 **Spinelets (small spines)**: These are enlarged, elongate cylindrical, or angular granules, fixed to the surface of scales or plates (Fig. 6.6)

6.6.83 **Spinules**: They have various number of pointed apical projections (bifid, trifid, multifid), fixed to the surface of scales or plates (Fig. 6.6)

6.6.84 **Sternals**: Pair of plates or single plates in posterior interambulacrum, lying between labrum and episternals [Fig. 6.11(1, 4)]

6.6.85 **Stone canal**: Usually reinforced with ossicles, it is a tube, starting from the madreporite to the water vascular ring canal [Fig. 6.2(1)]

6.6.86 **Stumps**: Small structures, fixed to the surface of scales or plates, are relatively larger than granules; they are usually prickly (see also Fig. 6.6)

6.6.87 **Subanal fasciole**: Elliptical groove located below periproct [Fig. 6.11(6)]

6.6.88 **Subanals**: Pair of interambulacral plates located between episternals and annals [Fig. 6.11(1)]

6.6.89 **Teeth**: In echinoids, these are the five hard, sharp, and moveable ossicles incorporated in the Aristotle's Lantern [Figs. 6.2(7, 8) and 6.5(1–3)]

6.6.90 **Test**: The "shell" of an echinoid, made up of many small skeletal plates. A "naked" test is one from which soft tissue, and projecting structures such as spines, have been removed. This process occurs naturally after the death of a sea urchin. To identify some urchins, it is necessary to clean a portion of the test with bleach, to see the underlying plates [Fig. 6.2(1–4)]

6.6.91 **Tooth**: Rod located in each pyramid, having an uncalcified arched upper end but hard; nearly straight lower end; only the chisel-edged tip protrudes from lower end of pyramid [Fig. 6.2(7–8)]

6.6.92 **Tridentate**: Having three long, slender valves which generally have sharp, finely serrate edges [Fig. 6.6(11, 12)]

6.6.93 **Triphyllous**: Having three leaves or paddle-shaped valves (Fig. 6.8)

6.6.94 **Tube Feet**: These are fluid-filled, finger-like extensions of the water vascular system that protrude through openings in the skeleton or between skeletal elements [Fig. 6.2(1)]

6.6.95 **Tubercle**: A smooth, rounded and massive prominence on the skeleton. They can also be referred as outgrowths of plates, rather than to articulated elements (see also Fig. 6.6). In echinoids and some asteroids, a spine articulates with a tubercle [Fig. 6.2(11)]

6.6.96 **Tuberculated**: Carrying numerous tubercles. Various skeletal elements are also noted. These are: Plates, Scales, and Spines. Small structures fixed to the surface of scales or plates include: Granules, Spinelets, Spinules, Stumps and Tubercles

6.6.97 **Ventral**: In echinoderms, this term is variously applied. In asteroids, echinoids, and ophiuroids, it denoted the surface of the body that carries the mouth and in contact with the substrate. In holothuroids, where the mouth and anus are at opposite ends of a cylindrical body, the ventral surface is lowermost—and also in contact with the substrate. In crinoids, the ventral surface carries the mouth, and is functionally the uppermost surface [see Fig. 6.2(1–4)].

6.7 Classification

Phylum Echinodermata is divided into 5 major groups (Subphylum) that contain 20 Classes, and about half of which are known only from the Paleozoic (see Table 6.2). The current classification is based on multiple characters such as general

Table 6.2 Phylum Echinodermata is divided into 5 major groups (Subphylum)

Subphylum	Class	Genera	Species	Age range
Asterozoa	**Asteroidea**	430	1500	Early Ordovician–Recent
	Ophiuroidea	325	2000	Early Ordovician–Recent
Echinozoa	**Echinoidea**	765	940 (living)	Late Ordovician–Recent
	Holothuroidea	200	1150 (living)	Middle Cambrian–Recent
	Edrioasteroidea	35	–	Early Cambrian–Pennsylvanian
	Helicoplacoidea	4	–	Early Cambrian
	Cyclocystoidea	–	–	Middle Cambrian–Devonian
	Edrioblastoidea	1	–	Ordovician
Crinozoa	**Crinoidea**	~1000	625 (living)	Cambrian–Recent
Blastozoa	**Blastoidea**	95	–	Middle Ordovician–Late Permian
	Rhombifera	60	–	Early Ordovician–Late Devonian
	Diploporita	42	–	Early Ordovician- Early Devonian
	Eocrinoidea	32	–	Early Cambrian–Late Silurian
	Parablastoidea	–	–	Early–Middle Ordovician
	Paracrinoidea	~15	–	Early Ordovician- Early Silurian
Homalozoa	Stylophora	50	–	Middle Cambrian–Pennsylvanian
	Homoiostelea	–	–	Middle Cambrian–Silurian
	Homostelea	–	–	Middle Cambrian
	Ctenocystoidea	–	–	earliest Middle Cambrian

Those in bold are briefly discussed in the text

morphology, ossicle structure, arrangement of the water vascular system, and embryology. Major classes are briefly discussed below.

6.7.1 *Class Asteroidea (Sea Stars; Starfishes)*

The star-shaped echinoderms, the Asteroidea (=Sea stars or Starfishes), is the largest, most speciose ($\sim$1600 extant species), diverse and the common class within Phylum Echinodermata (see Duncan 1889; Beaver et al. 1978) and prime predators within many marine ecosystem. Their size varies from a centimeter to a meter across. They thrive in the intertidal zone, although, they also occur at depths as great as 10,000 m. The carnivorous asteroids are very sluggish movers and crawl by the concerted actions of their podia. The Asteroids generally have hollow arms and into this the coelomic cavity extends. The radial canals are located on the skeleton's exterior. Their skeleton is rarely robust and consists of a series of small elements embedded in a collagenous membrane. Hence, after death, they disarticulate quickly, leaving a poor fossil record. The asteroids first appeared in Early Ordovician but were never common or abundant in the fossil record, since then. They experienced major faunal transitions concurrently with two large extinction events; one in Late Devonian and the other in Late Permian.

6.7.2 *Class Ophiuroidea (the Brittle Stars)*

The Ophiuroids are a large group (over 1600 species) that include brittle stars (Ophiurida) and basket stars (Euryalida). The Ophiuroids are remarkable in that they are able to release their arms at ossicle sutures to escape predation, hence, they are also called brittle stars. This process of releasing a limb is called Autotomy; the lost limb is eventually regenerated. The Ophiuroids are somewhat different from the Asteroids (see Table 6.3), the sea stars, from which they are often mistaken, and

Table 6.3 Comparative account of Ophiuroids and Asteroids

	Ophiuroids	Asteroids
Similarity	Both have a stellate body plan	
	Both have five or more arms radiating from a small robust circular disc that contains the viscera	
	Both have a skeleton that is fragile and readily disintegrates upon death, hence, both they have left a relatively sparse fossil record	
Difference	The ophiuroid arms are solid, being supported by a series of internal disc-like ossicles termed vertebrae	The asteroids have hollow arms

hence, often lumped with them in higher taxa, Stelleroidea or Asterozoa. Unlike asteroids, the ophiuroids are active crawlers with thin whip-like five arms that wriggle like snakes, hence, their name (*ophi* = snake, in Greek); they are sometimes also called Serpent stars. The Ophiuroids are recorded from the Early Ordovician and achieved global distribution in the Paleozoic. They survived the Permian–Triassic extinction and increased in numbers during the Mesozoic–Cenozoic (see also Jagt 2000a, b).

6.7.3 *Classes Holothuroidea and Concentricycloidea (Sea Cucumbers and Sea Daisies)*

These soft-bodied sea cucumbers (~200 genera) are without arms (see also Gilliland 1993). They are the most recently discovered class of echinoderms; sea daisies are tiny, primitive echinoderms that live at great depths. Sea cucumbers resemble cucumbers. They possess Sclerites, microscopic hard parts that occur in various shapes resembling hooks, wheels and anchors. Although, the sea cucumber possess pentaradial symmetry, the anus is opposite to the mouth on an elongated oral–aboral axis. The Holothurians are the most diverse of the five extant classes of echinoderm, with over 2000 extant species. However, like the asteroids and the ophiuroids, they also have left a very poor fossil record and after death, their skeleton is reduced to thousands of microscopic spicules. The only fossilized part are the 10 ossicles that surrounds the mouth and provide an anchorage for the oral tentacles; forming the Circumoral ring. The earliest holothurian body fossil is recorded from Late Silurian, however, spicules attributable to holothurians are known from Ordovician onwards.

6.7.4 *Edrioasteroids*

These discoidal, clavate, or subglobular echinoderms are an extinct group of sessile stemgroup eleutherozoans (see also Bell 1976; Guensburg and Sprinkle 1994). They are characterized by possessing five ambulacra radiating from a central mouth. The Stromatocystitids, the earliest forms were completely plated and not attached to the substrate. They represent the basal eleutherozoans, ancestral to all later forms. The Stromatocystitids first appeared in Early Cambrian, while isorophids (a derived clade and specialized as hard ground colonizers) appeared a little later, in Late Cambrian, and survived through to Late Carboniferous.

6.7.5 *Class Crinoidea (Sea Lilies and Feather Stars; Crinoids)*

The Crinoids are unusual looking animals as they look more like plants than animals. Hence their name "sea lilies". Superficially, the stem or column of a crinoid resembles the stalk of a flower, the calyx or head resembles the sepals of a flower, and the arms resemble the petals of a flower. The Crinoids (with over 1000 known genera), are the only echinoderms that live attached to the sea bottom for most of their lives. Two groups of living crinoids are noted—those with columns, the living stalked crinoids, and those without, the Comatulids. Typically, a crinoid has a long stem with "roots" or a holdfast (an attachment device) at the bottom, and a cup-shaped thecum at the top. Some crinoids have become mobile by losing stem; living crinoids are swimmers, and not attached. Recent crinoids have both a wide vertical distribution (from bathyal depths to tropical reef and shallow cave habitats) and spatial (from polar to tropical latitudes). The Paleozoic forms were very common on shallow carbonate platforms. Except in the Cambrian, crinoids were common fossils in the Paleozoic.

6.7.6 *Class Blastoidea (Blastoids)*

Blastoids are the best known stalked echinoderm of the Paleozoic and widely distributed in Early and Middle Paleozoic rocks with 95 genera, known, so far. The Extinct Blastoids (such as *Pentremites*) are characterized by an armless bud-like calyx on a stem. The primitive stemmed blastoids have pentaradial symmetry with a very regular distribution of thecal plates; there are 13 thecal plates arranged in three circlets. The theca is the size of a rosebud, and a commonly preserved. Although they show pentaradiality, however, diverse body forms, symmetries, and ambulacral architectures are noted in this group. The hydrospire, a respiratory device, is a distinctive structure in all blastoids hat hangs into the body cavity beneath each ambulacrum. The blastoids resemble crinoids in appearance and lifestyle. The blastoids first appeared in Middle Ordovician, peaked in abundance by Mississippian (Early Carboniferous), with a general decline in the Paleozoic and the final extinction in the Permian.

6.7.7 Eocrinoids

These stalked arm-bearing echinoderms are a paraphyletic assemblage of basal pelmatozoans with irregularly plated theca. The eocrinoids include the ancestors of all other blastozoan groups and probably of the crinoids also. The eocrinoids were the likely ancestors to Diploporita, Rhombifera, Coronoidea, Blastoidea, Parablastoidea and Paracrinoidea. The Diploporites are characterized by larger and regular plating with no ambulacral flooring plates; the brachioles arise directly from the thecal plates. The thecal plates are pierced by numerous pairs of pores (diplopores) that had a respiratory role. The Rhombiferans, arising directly from around the mouth, had stout arms. The most regular thecal plating is noted in the pentameral symmetry bearing Blastoids that possess three basals, five radials and five lancet plates. The well-developed ambulacra form an essential part of the theca, giving rise to a dense fan of brachioles. The oldest blastozoans are eocrinoids of Early Cambrian age. Diploporites and rhombiferans appeared at beginning of the Ordovician, while the blastoids in the Silurian. The Eocrinoids finally went extinct by end Permian.

6.7.8 Class Paracrinoidea (Allied to Crinoids)

The Paracrinoid plates show no definite arrangement but commonly contain pores that are covered (as noted in deeply weathered specimens). In *Camarocystites*, the best know genus, the plates of the calyx are concave. The two arms bear side branchlets which are relatively thick. The Paracrinoid are recorded in Ordovician and Silurian sediments. They are abundant in mid-Ordovician strata. Some workers believe that Paracrinoid are highly modified blastozoans.

6.7.9 Carpoids

The Carpoids (Stylophorans; Class Ctenocystoidea; see also Robison and Sprinkle 1969) are unusual forms of the Cambrian strata, and common in the fossil record from 500 to 300 million years ago. Three competing hypotheses are proposed for the group's evolutionary origins:

1. Carpoids are very primitive echinoderms with a mobile stalk or single arm filled with muscle
2. Stylophorans are advanced echinoderms related to crinoids having an ambulacrum with a tube feet and an oral tegmen with pharynx, and
3. Carpoids are neither of the above two, but primitive chordates that retained a calcite exoskeleton from an older common ancestor of echinoderms and chordates, with the stalk possessing muscle, notochord and brain.

Recent studies, based on Middle Cambrian ceratocystid stylophoran from Morocco supports the first hypothesis (for details see Clausen and Smith 2005).

6.8 Geological History

The early Paleozoic is marked by a large number of echinoderm classes with low diversity, followed by moderate improvement in mid-Paleozoic and by Late Paleozoic, few classes had become common [such as crinoids and blastoids; see Fig. 6.1] (see also Kier 1965, 1966; Sprinkle 1982; Smith 1984; Littlewood et al. 1997; Waters and Maples 1997; Smith et al. 2004). Select fossil examples are illustrated in Figs. 6.8, 6.9, 6.10, 6.11, 6.12, 6.13, 6.14, 6.15, 6.16.

After the great Permo-Triassic extinction, the echinoids [sea urchins (regular echinoids), heart urchins (spatangoids), sea biscuits and sand dollars (clypeasteroids)] became the most prolific class of echinoderms as their hard test fossilized exceedingly well (Fig. 6.1) (see also Sprinkle 1980; Clarkson 1993). Echinoids are, in fact, the third most skeletonized phylum after arthropods and molluscs within the marine realm. Echinoderms are dominantly calcitic, and so tend to fossilize well, but after their death, they often disintegrate. Echinoderm plates and spines are important sediment-forming materials as a result of this, and are especially abundant in certain limestones.

The echinoderms, known since the Cambrian, evolved from bilaterally symmetrical ancestor(s) as they have a bilateral larvae. The characteristic radial symmetry developed later, in the adult body. The Holothurians are the closest living relative of the echinoids and their ancestry presumably lies amongst the "Asterozoan" taxa of the Early Ordovician. However, the early interval in echinoid's history is still not fully known.

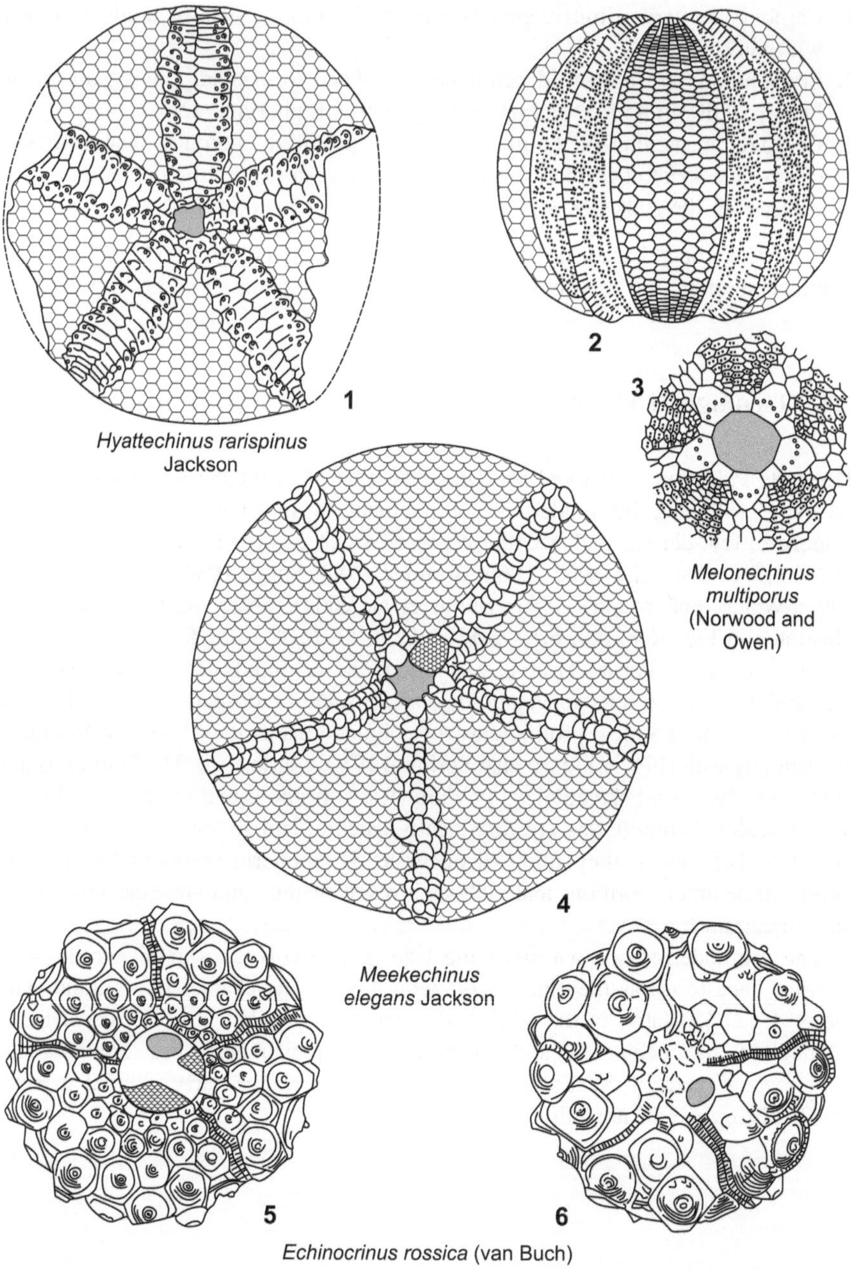

Fig. 6.13 Representative Early Mississippian echinoids. The tests are flexible. Their radial vessels lay on the inner surface of the skeleton, but were generally bordered by skeletal ridges. The mouth contained an Aristotle's lantern

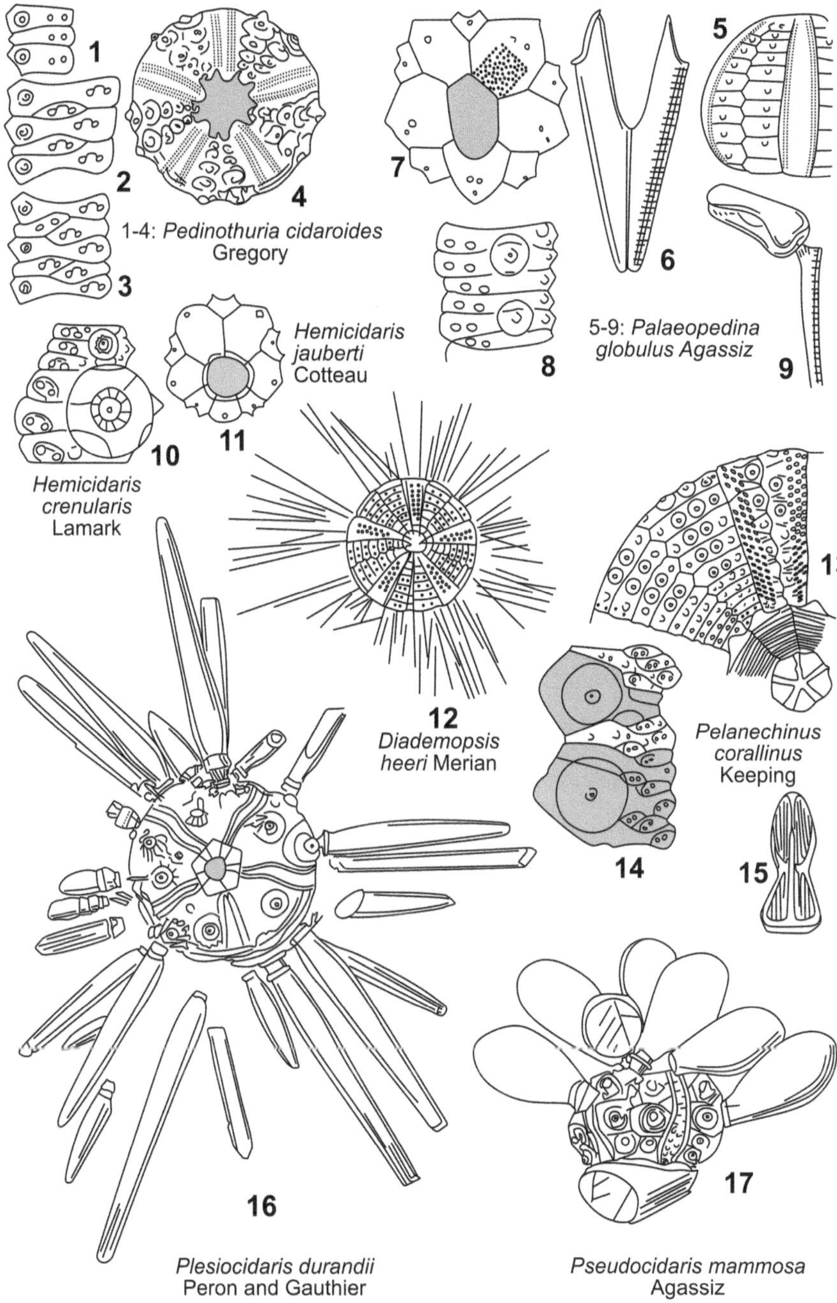

Fig. 6.14 Representative Jurassic regular echinoids. *1–4 Pedinothuria cidaroides* Gregory, *1* ambulacral plates from aboral side, *2* from ambitus, *3–4* oral side; *5–9 Palaeopedina globulus* Agassiz, *5* side view, *6* Aristotle's lantern, *7* apical system, *8* oral ambulacral plates, *9* pedicellaria; *10 Hemicidaris crenularis* Lamark; *11 Hemicidaris jauberti* Cotteau; *12 Diademopsis heeri* Merian; *13–15 Pelanechinus corallinus* Keeping, *13* oral view, *14* supercompound plate, *15* pedicellaria; *16 Plesiocidaris durandii* Peron and Gauthier; *17 Pseudocidaris mammosa* Agassiz

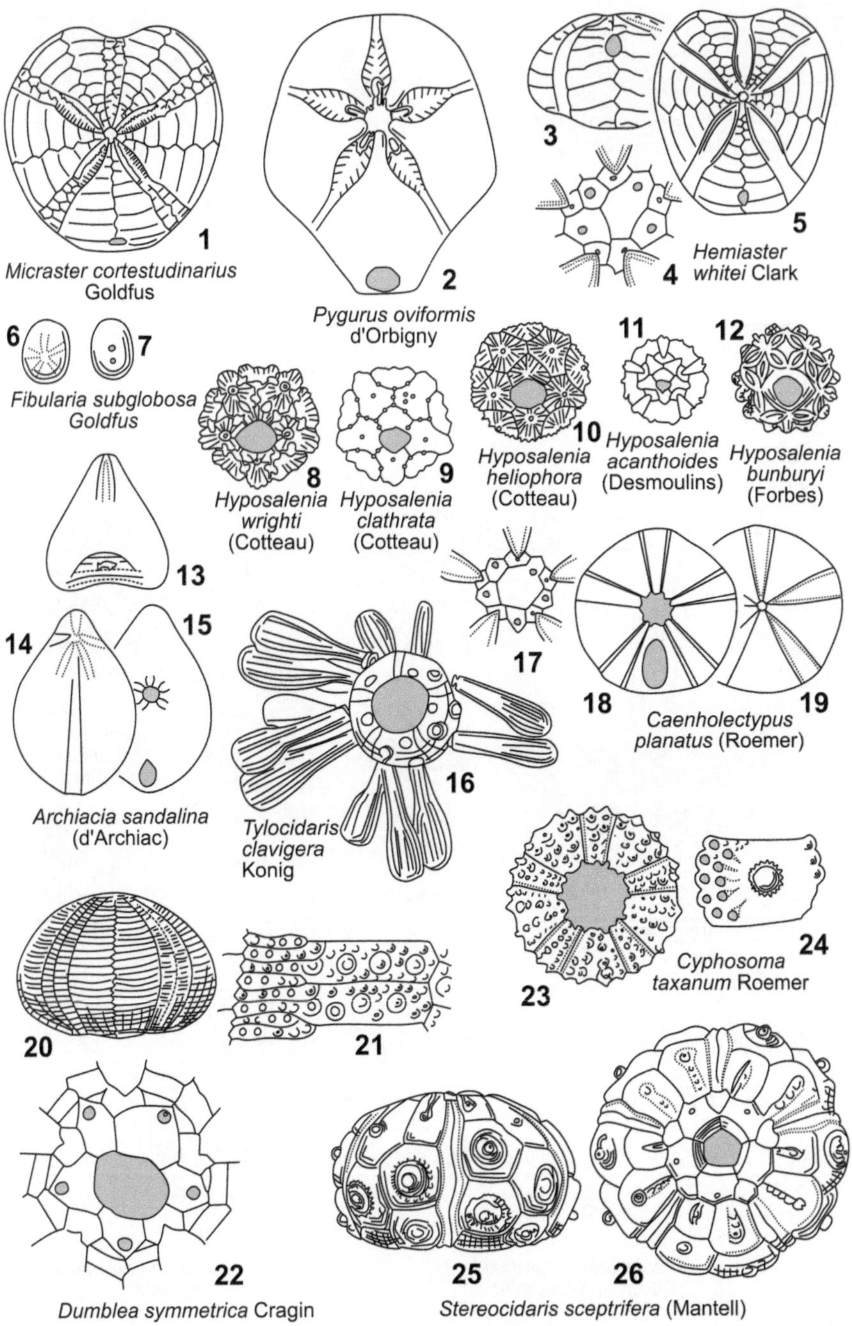
1
Micraster cortestudinarius Goldfus
2
Pygurus oviformis d'Orbigny
3
4
5
Hemiaster whitei Clark
6
7
Fibularia subglobosa Goldfus
8
Hyposalenia wrighti (Cotteau)
9
Hyposalenia clathrata (Cotteau)
10
Hyposalenia heliophora (Cotteau)
11
Hyposalenia acanthoides (Desmoulins)
12
Hyposalenia bunburyi (Forbes)
13
14
15
Archiacia sandalina (d'Archiac)
16
Tylocidaris clavigera Konig
17
18
19
Caenholectypus planatus (Roemer)
20
21
22
Dumblea symmetrica Cragin
23
24
Cyphosoma taxanum Roemer
25
26
Stereocidaris sceptrifera (Mantell)

◀ **Fig. 6.15** Representative Cretaceous echinoids. *1 Micraster cortestudinarius* Goldfus, aboral side; *2 Pygurus oviformis* d'Orbigny, oral side; *3–5 Hemiaster whitei* Clark, *3* rear view, *4* apical system, *5* aboral side; *6–7 Fibularia subglobosa* Goldfus, *6* aboral side, *7* oral side; *8 Hyposalenia wrighti* (Cotteau); *9 Hyposalenia clathrata* (Cotteau); *10 Hyposalenia heliophora* (Cotteau); *11 Hyposalenia acanthoides* (Desmoulins); *12 Hyposalenia bunburyi* (Forbes); *13–15 Archiacia sandalina* (d'Archiac), *13* frontal view, *14* aboral, *15* oral; *16 Tylocidaris clavigera* Köing; *17–19 Caenholectypus planatus* (Roemer), *17* apical system, *18* oral, *19* aboral; *20–22 Dumblea symmetrica* Cragin, *20* side view, *21* ambulacral plates, *22* apical system; *23–24 Cyphosoma taxanum* Roemer, *23* oral view, *24* enlarged oral ambulacral plate; *25–26 Stereocidaris sceptrifera* (Mantell), *25* side view, *26* aboral view

The echinoids, during the Early Palaeozoic, were a minor constituent of the marine benthic community. But, by Devonian, as they developed enlarged adoral tube feet and became specialized deposit feeders, they thrived through the Permian, before going extinct. By Carboniferous, the archaeocidarids, made their appearance. They possessed a single large tubercle on each interambulacral plate with long highly muscular spines; they were the first active predators. The Palaeozoic *Miocidaris*, is the only echinoid that has the test architecture of post-Palaeozoic forms, and which evolved directly from the Archaeocidaris by the reduction of plating columns in each interambulacral zone.

All echinoids in the Paleozoic were regular forms; the irregular ones underwent a spectacular radiation in the Mesozoic and were much more common and with a much better fossil record. The Irregular echinoids, on the other hand, appeared only in the Early Jurassic but evolved quickly as deposit feeders; they possessed a lantern like that of the regular echinoids. By the Middle Jurassic, the cassiduloids and spatangoids evolved; the sand dollars arose in the Early Tertiary from cassiduloid ancestors. The modern heart urchins, appeared in the Early Cretaceous and have diversified constantly, since then. The end-Cretaceous extinction did affect the echinoids, but selectively; the deposit feeders were the hardest hit. The regular echinoids thrived throughout the Mesozoic; but only became a major group (as also today), only by the Late Cretaceous.

Appendix 1 gives the list of illustrated specimens mentioning the chapter number, species name, age and locality along with its figure number within the said chapter.

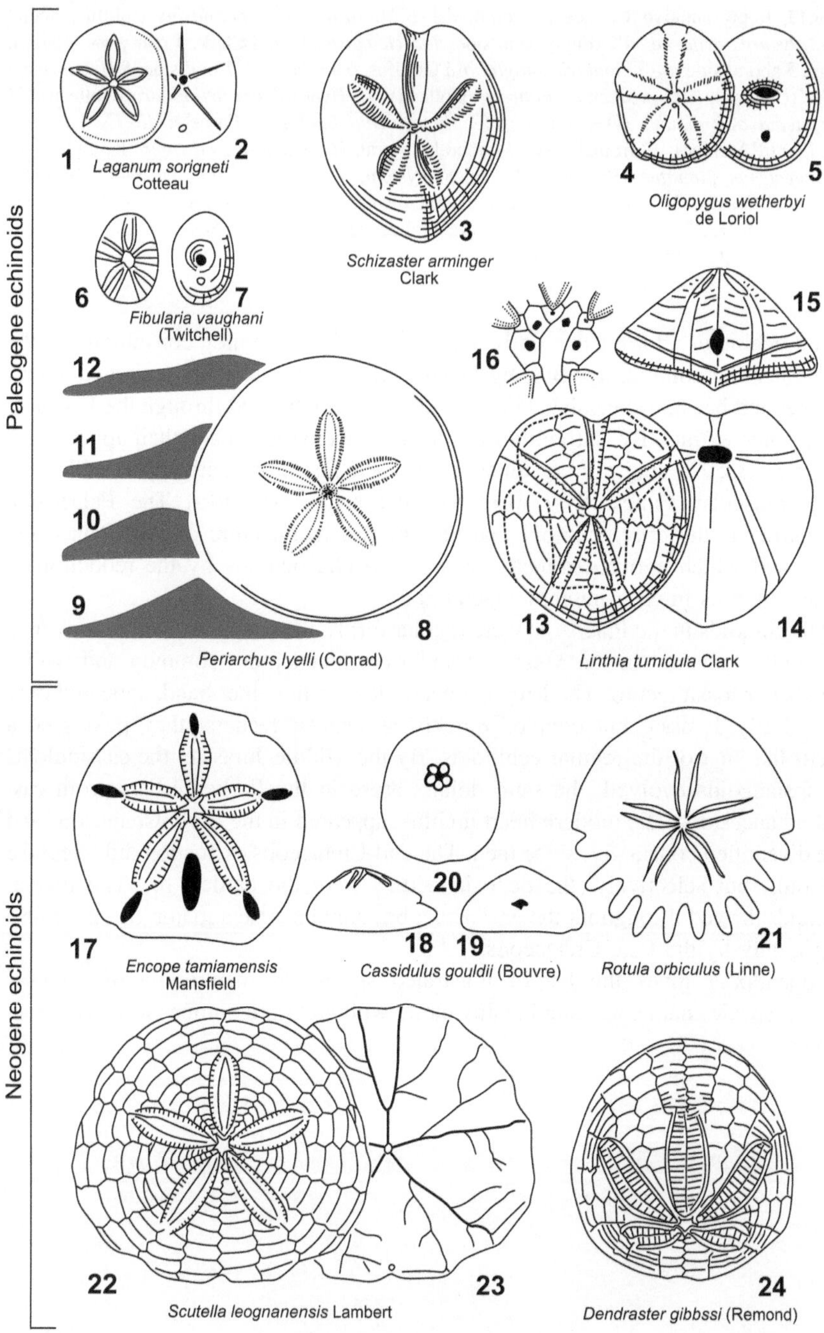
Paleogene echinoids
1
2
Laganum sorigneti Cotteau
3
Schizaster arminger Clark
4
5
Oligopygus wetherbyi de Loriol
6
7
Fibularia vaughani (Twitchell)
12
11
10
9
8
Periarchus lyelli (Conrad)
16
15
13
14
Linthia tumidula Clark
Neogene echinoids
17
Encope tamiamensis Mansfield
20
18
19
Cassidulus gouldii (Bouvre)
21
Rotula orbiculus (Linne)
22
23
Scutella leognanensis Lambert
24
Dendraster gibbssi (Remond)

◀ **Fig. 6.16** Representative Paleogene and Neogene echinoids. *1–16* Paleogene. *1–2 Laganum sorigneti* Cotteau, *1* aboral view, *2* oral view; *3 Schizaster arminger* Clark, aboral view; *4–5 Oligopygus wetherbyi* de Loriol, *4* aboral view, *5* oral view; *6–7 Fibularia vaughani* (Twitchell), *6* aboral view, *7* oral view; *8–12 Periarchus lyelli* (Conrad), *8* aboral view, *9–12* a series of "varieties" (largely geographical) are illustrated by various profiles; *13–16 Linthia tumidula* Clark, *13* aboral view, *14* oral view, *15* end view, *16* apical system; *17–24* Neogene. *17 Encope tamiamensis* Mansfield; *18–20 Cassidulus gouldii* (Bouvre), *18* side view, *19* rear view, *20* oral view; *21 Rotula orbiculus* (Linne); *22–23 Scutella leognanensis* Lambert; *24 Dendraster gibbssi* (Remond)

References

Beaver, H.H., K.E. Caster, J.W. Durham, et al. 1978. *Treatise on invertebrate paleontology. Part S, Echinodermata 1*. The Geological Society of America and the University of Kansas Press.

Bell, B.M. 1976. A study of North American Edrioasteroidea. *New York State Museum, Memoirs* 21: 1–447.

Clausen, S., and A.B. Smith. 2005. Palaeoanatomy and biological affinities of a Cambrian deuterostome (Stylophora), *Nature* 438(7066): 351–354.

Clarkson, E.N.K. 1993. *Invertebrate palaeontology and evolution*. London: Chapman and Hall. 434 p.

Duncan, P.M. 1889. A revision of the genera and great groups of the Echinoidea. *Journal of the Linnean Society, Zoology* 23: 1–311.

Gilliland, P. 1993. The skeletal morphology, systematics and evolutionary history of holothurians. *Special Papers in Palaeontology* 47: 1–147.

Guensburg, T.E., and J. Sprinkle. 1994. Revised phylogeny and functional interpretation of the Edrioasteroidea based on new taxa from the Early and Middle Ordovician of Western Utah. *Fieldiana Geology* 29: 1–41.

Jagt, J.W.A. 2000a. Late Cretaceous Early Palaeogene echinoderms and the K/T boundary in the southeast Netherlandsand northeast Belgium. Part 3, Ophiuroids. *Scripta Geologica* 121: 1–179.

Jagt, J.W.A. 2000b. Late Cretaceous Early Palaeogene echinoderms and the K/T boundary in the southeast Netherlands and northeast Belgium. Part 5, Asteroids. *Scripta Geologica* 121: 377–503.

Kier, P.M. 1965. Evolutionary trends in Paleozoic echinoids. *Journal of Paleontology* 39: 436–446.

Kier, P.M. 1966. Noncidaroid Paleozoic echinoids. In *Treatise on Invertebrate Paleontology*, ed. R.C. Moore. Boulder: University of Kansas Press and Geological Society of America.

Littlewood, D.T.J., A.B. Smith, K.A. Clough, and R.H. Ensom. 1997. The interrelationships of the echinoderm classes: morphological and molecular evidence. *Biological Journal of the Linnean Society* 61: 409–438.

Mooi, R. 1990. Living cassiduloids (Echinodermata: Echinoidea): a key and annotated list. *Proceedings of the Biological Society of Washington* 103: 63–68.

Robison, R.A., and J. Sprinkle. 1969. Ctenocystoidea: new class of primitive echinoderms. *Science* 166(3912): 1512–1514.

Smith, A.B. 1984. *Echinoid Palaeobiology*. London: George Allen and Unwin.

Smith, A.B. 2001. Probing the cassiduloid origins of clypeasteroid echinoids using stratigraphically restricted parsimony analysis. *Paleobiology* 27: 392–404.

Smith, A.B., K. Peterson, D.T.J. Littlewood, and G.A. Wray. 2004. From bilateral symmetry to pentaradiality: the phylogeny of hemichordates and Echinodermata. In *Assembling the Tree of Life*, ed. J. Cracraft, and M. Donoghue. Oxford: Oxford University Press.

Sprinkle, J. 1980. An overview of the fossil record. In: Broadhead, T.W. and Waters, J.A. (Eds.), *Echinoderms, Notes for a Short Course*, University of Tennessee Department of Geological Sciences, Studies in Geology 3: 15–26.
Sprinkle, J. 1982. Echinoderm faunas from the bromide formation (middle Ordovician) of Oklahoma. *The University of Kansas Paleontological Contributions Monograph* 1: 1–369.
Waters, J.A., and C.G. Maples. 1997. Geobiology of echinoderms. *The Paleontological Society Papers* 3: 1–35.

Chapter 7
Graptolites

7.1 Introduction

The Graptolites (*grapto* = write, *lithos* = stone) are very small (<1 cm) extinct planktonic and colonial forms (inference based on shape of colonies, theca, and their recovery from deep "bottom" deposits). They secrete a chitino-phosphatic exoskeleton, and, are hence, grouped under Hemichordates. Most graptolites are found flattened and carbonized in black shales and mudstones, in deep offshore fine-grained sedimentary rocks (Frazier and Schwimme 1987; Benton and Harper 2009). Thus, owning to their unique process of carbonization and highly compressible nature (most fossils being flat), they are also very difficult to study (Prothero 2013; Maletz 2015). However, there are two major graptolite orders that have significance due to their higher preservational potential, Graptoloidea and Dendroidea. Graptoloidea are pelagic, with one basic type of thecae, and the sicula is the initial part of the colony, whereas the Dendroidea are benthic, with two types of theca (autotheca and bitheca), and a stolon. The Dendroidea are stratigraphically less important as they occur attached to the sea floor, they are only important locally.

7.2 General Morphology

Graptolite morphology is considered either at the level of theca (i.e., at the individual level) or that of a rhabdosome (i.e., at the level of a colony; [Fig. 7.1(1–3)]. The graptolite skeleton comprises of rows or lines of small tubes or cups called thecae [Fig. 7.1(1–3)]. It is in these theca that an individual lives called Zooid and each individual is linked together by a common canal [Fig. 7.1(9)]. Overall, the skeleton bears a "saw-tooth-like" appearance [Fig. 7.1(1–2)]. The rows or lines of small thecae may occur on opposing sides of a stipe (biserial), i.e., long blade-like

S. Jain, *Fundamentals of Invertebrate Palaeontology*, Springer Geology,
DOI 10.1007/978-81-322-3658-0_7

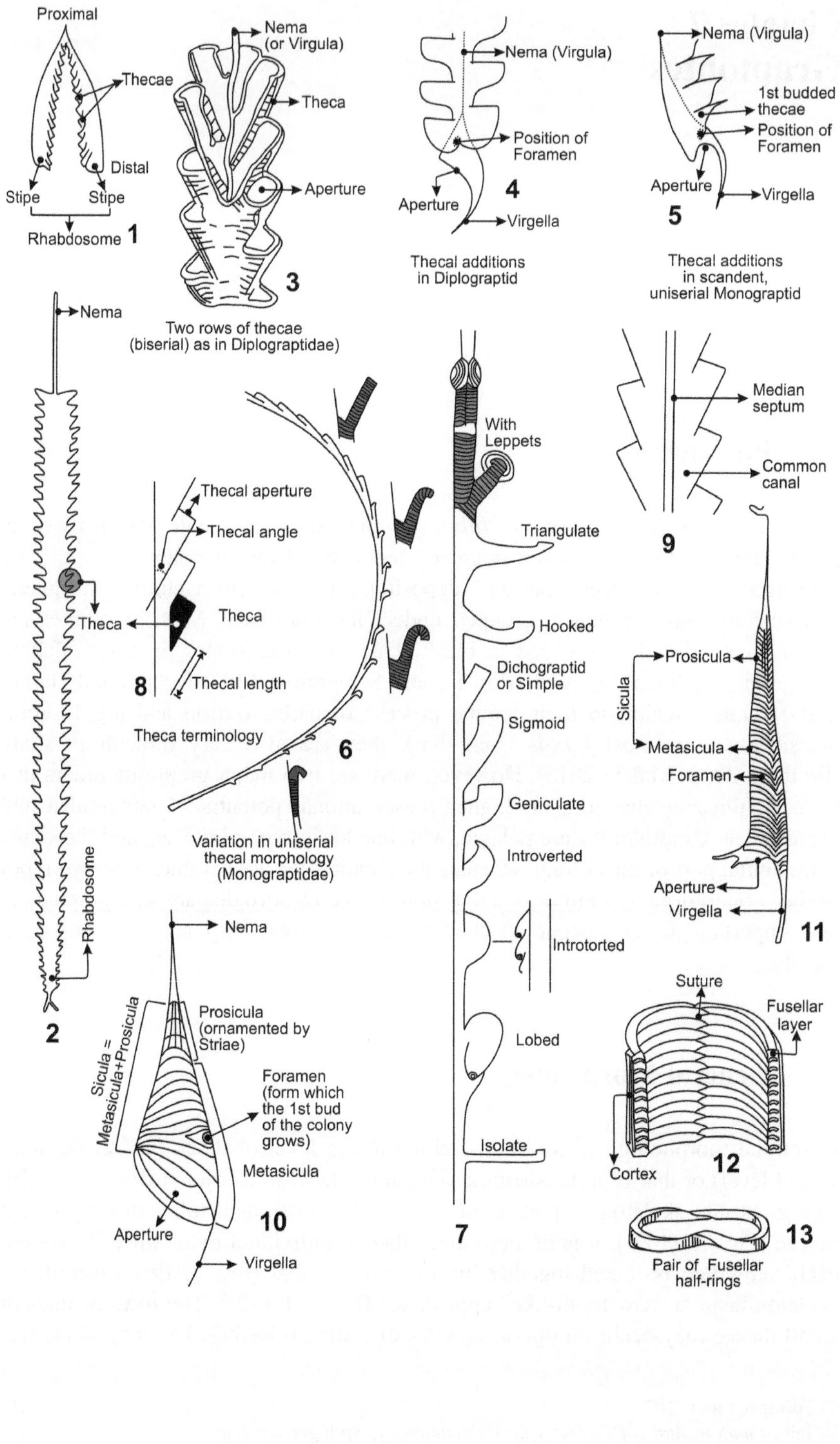
Proximal
Thecae
Distal
Stipe
Stipe
Rhabdosome 1
Nema
(or Virgula)
Theca
Aperture
3
Two rows of thecae
(biserial) as in Diplograptidae)
Nema (Virgula)
Position of
Foramen
Aperture
4
Virgella
Thecal additions
in Diplograptid
Nema (Virgula)
1st budded
thecae
Position of
Foramen
Aperture
Virgella
5
Thecal additions
in scandent,
uniserial Monograptid
Nema
Theca
Rhabdosome
2
Thecal aperture
Thecal angle
Theca
Thecal length
8
Theca terminology
6
Variation in uniserial
thecal morphology
(Monograptidae)
With
Leppets
Triangulate
Hooked
Dichograptid
or Simple
Sigmoid
Geniculate
Introverted
Introtorted
Lobed
Isolate
7
Median
septum
Common
canal
9
Sicula
Prosicula
Metasicula
Foramen
Aperture
Virgella
11
Nema
Sicula =
Metasicula+Prosicula
Prosicula
(ornamented by
Striae)
Foramen
(form which
the 1st bud
of the colony
grows)
Metasicula
Aperture
Virgella
10
Suture
Fusellar
layer
Cortex
12
13
Pair of Fusellar
half-rings

◀ **Fig. 7.1** Graptolite morphology. *1*, *2* Graptolite rhabdosome (skeleton) showing thecae; *3* Biserial (two rows) thecae of Diplogratidae (*Dendrograptus* (*Orthograptus*) *gracilis* Hall); *4* Thecal additions in a Diplogratid; *5* Thecal additions in a uniserial Monograptid; *6* Variation in thecal morphology (Monograptidae); *7* Types of thecae; *8* Thecal terminology; *9* Common canal in thecae (longitudinal section); *10*, *11* Internal structure of theca; *11* sicula of *D.* (*O.*) *gracilis* Hall; *12*, *13* Wall structure showing Fusellar rings and half-rings

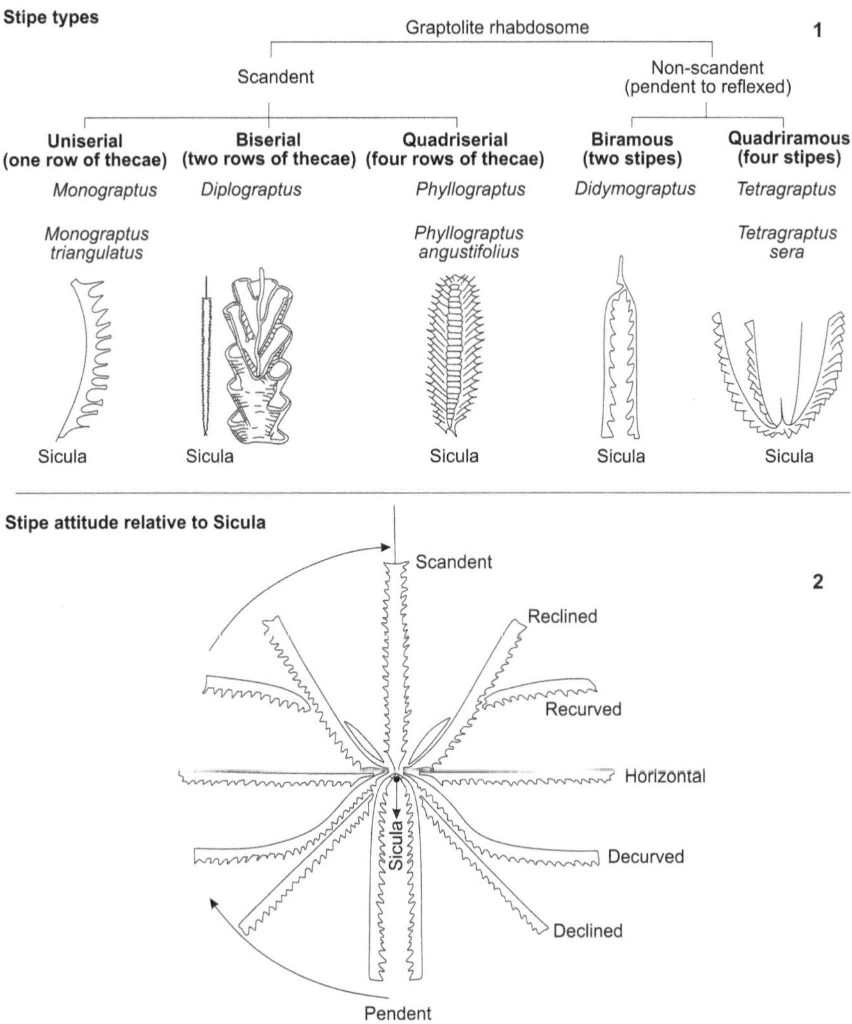

Fig. 7.2 Graptolite stipes. *1* Stipe types; *2* stipe attitude. The branches of the graptoloids show an evolutionary trend from a position in which they hang downwards from nema and sicula, through intermediate positions, to the scandent type of growth, in which the stipes grow upwards along the nema

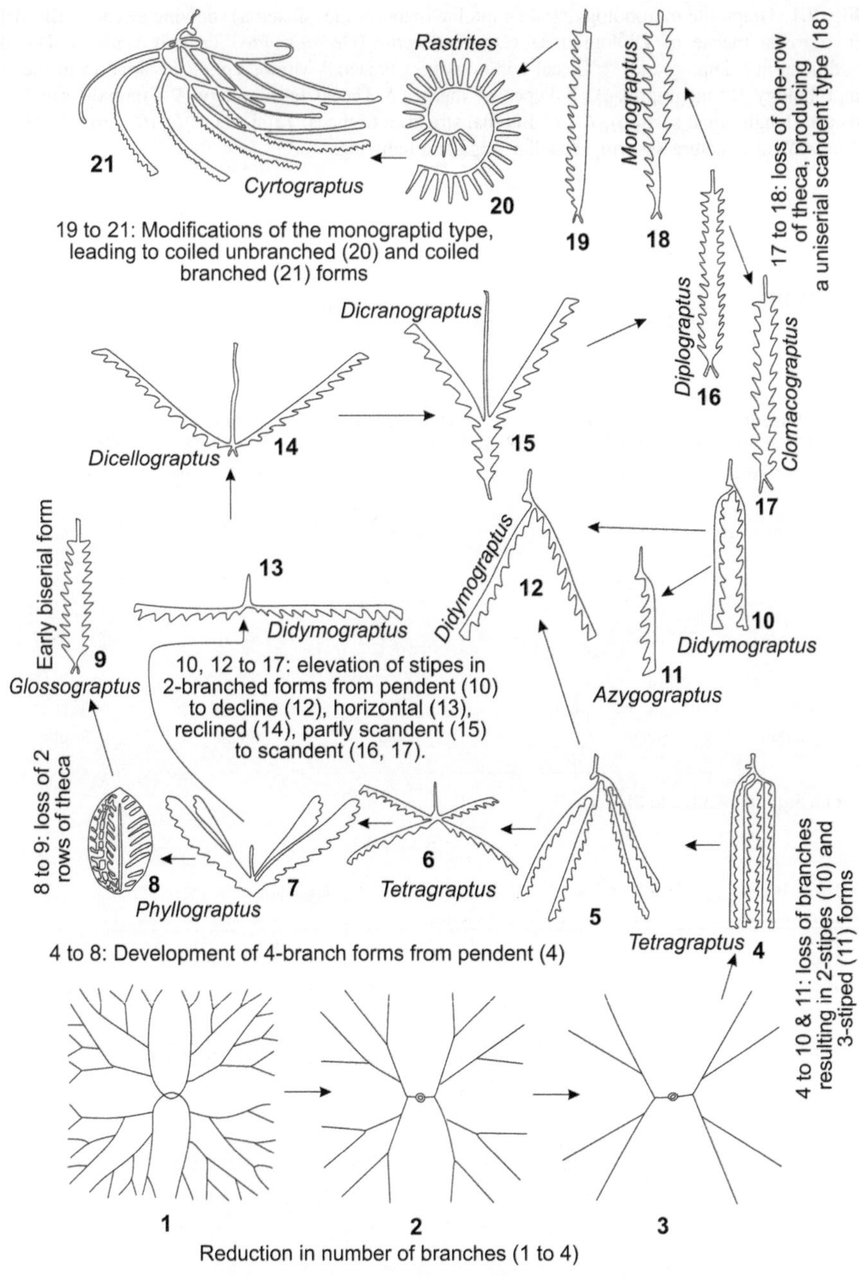
Rastrites
Monograptus
17 to 18: loss of one-row of theca, producing a uniserial scandent type (18)
21
Cyrtograptus
20
19
18
19 to 21: Modifications of the monograptid type, leading to coiled unbranched (20) and coiled branched (21) forms
Dicranograptus
Diplograptus
16
Clomacograptus
14
15
Dicellograptus
17
Early biserial form
13
Didymograptus
12
9
Didymograptus
10
Glossograptus
10, 12 to 17: elevation of stipes in 2-branched forms from pendent (10) to decline (12), horizontal (13), reclined (14), partly scandent (15) to scandent (16, 17).
11
Didymograptus
Azygograptus
8 to 9: loss of 2 rows of theca
8
7
6
Tetragraptus
5
Phyllograptus
4 to 8: Development of 4-branch forms from pendent (4)
Tetragraptus
4
4 to 10 & 11: loss of branches resulting in 2-stipes (10) and 3-stiped (11) forms
1
2
3
Reduction in number of branches (1 to 4)

◀ **Fig. 7.3** Evolution of the graptoloid colony. The diagram illustrates the following trends. *1–4* Reduction in number of branches from many to four; *4–8* development of four-branched forms pendent (*4*), to declined (*5*), horizontal (*6*), recurved (*7*), and scandent (*8*). *8*, *9* By loss of two rows of thecae, production of early biserial forms (*9*); *4*, *10*, *11* Loss of branches, resulting in two-stiped (*10*), one-stiped (*11*) forms; *10*, *12–17* Elevation of stipes in two-branched forms, leading from pendent (*10*), to declined (*12*), horizontal (*13*), reclined (*14*), partly scandent (*15*), to scandent (*16*, *17*); *16–18* loss of one row of thecae, producing a uniserial scandent type form; *19–21* Modification of the monograptid type leading to coiled unbranched (*20*) and coiled branched (*21*) forms

colonies, or be arranged along one side (uniserial) ([Fig. 7.1(2–4, 5–8)] and, respectively). The graptoloids consist of one or more stipes of theca—usually there are 1, 2, 4, or 8 stipes (= uniramous, biramous, and multiramous, respectively) (Fig. 7.2). The entire colony is called a Rhabdosome [Fig. 7.1(1, 2)]. The size of a graptolite colony is often less than one centimeter in length.

At the theca level [Fig. 7.1(3)], variations in thecal orientation and shape and in the presence of lappets and their shape [Fig. 7.1(6, 7)] are noted. However, as these change through an individual's ontogeny, the older and younger theca may differ also within one rhabdosome [Fig. 7.1(6)].

All of these features (i.e., shapes of thecae, and the number and attitude of stipes) are used in the classification and identification of evolutionary relations (Fig. 7.3). The monograptid thecal changes have been used to define successive faunas for stratigraphical correlation—the Early Llandovery (Early Silurian) monograptids have straight or gently curving thecae, followed by simple and triangulate thecae; Late Llandovery is marked by lobate and hooked thecae. The terms used for thecal description are illustrated in Fig. 7.1.

The cavity of the first formed theca connects to that of the sicula through a foramen [Fig. 7.1(10)]. The first formed part of the graptolite is the sicula [Fig. 7.1 (10, 11)], a conical tube with its aperture pointing downwards and terminating at its apex in a long hollow rod-like nema or virgula [Fig. 7.1(10, 11)]. The sicula is divided into two parts—the lower matasicula and the upper prosicula [Fig. 7.1(10, 11)]. The prosicula are ornamented by longitudinal and spiral striae [Fig. 7.1(10, 11)]. The metasicula is ornamented with well-marked rings representing growth increments called fusellae [Fig. 7.1(12, 13)]. The fusellar tissue consists of thin half-rings or complete rings of skeletal material stacked one above the other and uniting along zigzag sutures [Fig. 7.1(12)]. The fusellar half-rings [Fig. 7.1(13)] are joined to a stout rod (or a projecting spine) called the virgella [Fig. 7.1(10)].

In Dendroidea, *Dendrograptus* represents the most representative structure (Fig. 7.4; see also Koszlowshi 1948; Clarkson 1983). The inverted conical sicula stands upright into the sea with its expanded apical base as a holdfast [Fig. 7.4(1, 2)]. Half-way up the sicula arises the stolotheca [Fig. 7.4(3–6)]. It forms a continuous closed chain all the way up the rhabdosome. Two types of thecae arise at equally spaced nodal points—the larger stolotheca and the smaller and narrower, bitheca [Fig. 7.4(4)]. The arrangement of stolotheca and bitheca follow the Winman Rule of alternating triads; the bitheca usually arise at alternate nodal points on opposite sides of the main stem and are carried distal to the autothecae [Fig. 7.4(4–6)]. Unlike in

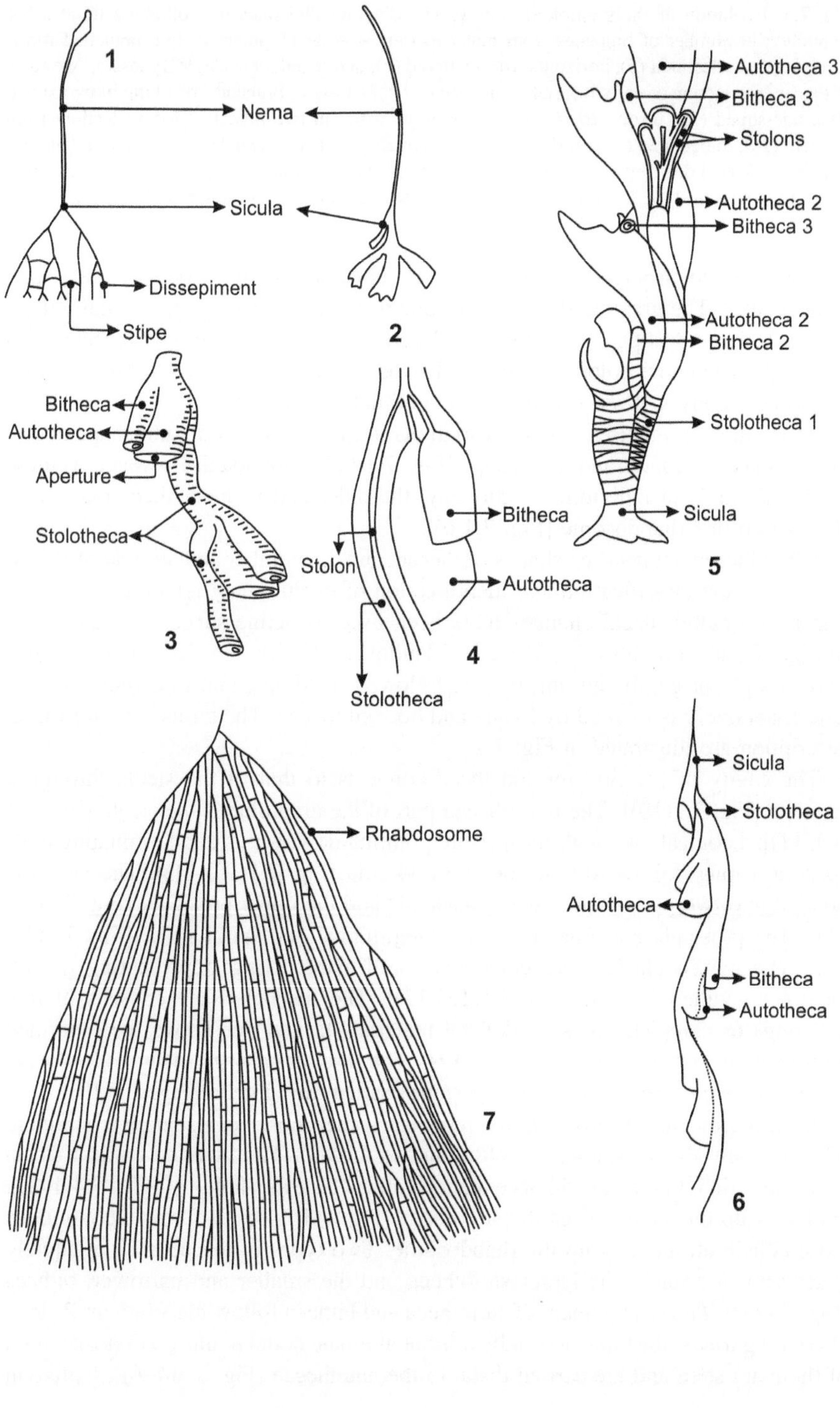
1
Nema
Sicula
Dissepiment
Stipe
2
Autotheca 3
Bitheca 3
Stolons
Autotheca 2
Bitheca 3
Autotheca 2
Bitheca 2
Stolotheca 1
Sicula
5
Bitheca
Autotheca
Aperture
Stolotheca
3
Stolon
Bitheca
Autotheca
Stolotheca
4
Rhabdosome
7
Sicula
Stolotheca
Autotheca
Bitheca
Autotheca
6

◀ **Fig. 7.4** Dendroid morphology. *1*, *2* Dissepiments: these are transverse bars that cross link branching stipes, *1* beginning of a colony, *2* the proximal portion of the colony; *3*, *4* Relationship of thecae with stolon, *3* portion of a branch, *4* enlarged view of *3*; *5* morphology of the proximal end of a dendroid (reverse view), *Dendrograptus communis* Kozlowski (modified after Clarkson 1983); *6* budding in a dendroid graptolite; *7* dendroid rhabdosome

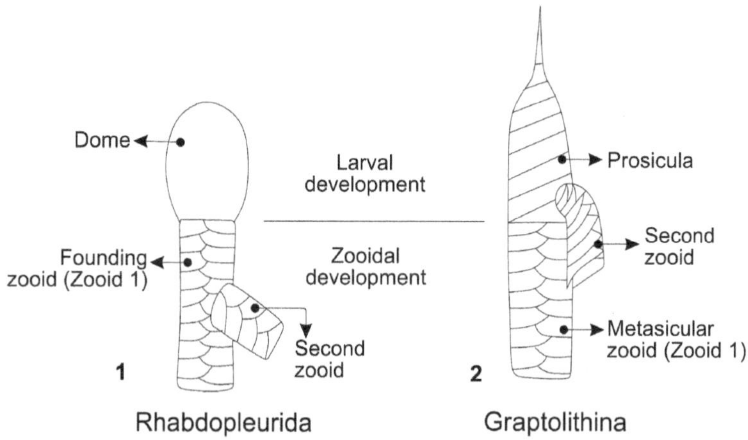

Fig. 7.5 Morphological comparison between modern Hemichordate, *Rhabdopleura* (1) and a graptoloid (after Maletz et al. 2005)

Graptoloidea, in Dendroidea, along the length of stolotheca, runs a tubular thread called stolon [Fig. 7.4(3)], a rudimentary nerve cord that places the graptolites under Hemichordates.

7.3 Taxonomic Relationships of Graptolites

The morphological resemblance and the phosphatic composition of graptolites colonies place them close to the extinct Hemichordates. However, no firm evidence of their biological relationship exists as there are no known living representatives of graptolites to compare with.

7.3.1 Chordates

Graptolites are closely related to Chordates based on the possessing of three characters, (a) a cartilage strengthening/supporting dorsal rod called notochord; this is firm but flexible, (b) a hollow nerve cord and (c) phrangeal gill slits. The Tunicates, "sea squirts" (chordates), also resemble graptolites and share the above three chordate characteristics. Tunicates are attached marine filter feeders with a superficial resemblance to sponges. Their larval stages are mobile and resemble a tadpole; upon maturing, the tail with its notochord is reabsorbed; the animal attaches to the bottom, and begins its filter feeding adult life. The resemblance between the lanclets (small streamlined fish-like chordate) and juvenile tunicates is remarkable, and many believe that the earliest true chordates may have descended from a tunicate-like ancestor.

7.3.2 *Hemichordates*

Rhabdopleura, a modern Hemichordate, has a colonial lifestyle (with a "stalk" and a theca) somewhat similar in appearance to the stipes of a graptolite (see Fig. 7.5; see also Rigby 1994; Maletz et al. 2005). *Rhabdopleura* also possess thecal walls with a fibrous-like pattern known as the Fusellar fabric. The graptolite zooid soft parts are essentially unknown; so it's internal anatomy (whether it possesses pharyngeal gill slits or not) is not known. The Graptolites, due to their phosphatic composition and similarities with modern Hemichordates like *Rhabdopleura*, are considered to be more like the hemichordates than any other animal group. Hence, most workers classify graptolites as a Class of Phylum Hemichordata (see Table 7.1).

7.4 Classification

The graptolite classification is complex (see Bulman 1955, 1970; Clarkson 1983), and hampered by poor preservational record and convergent evolution, i.e., their morphological characters have evolved more than once in different stocks. Hence, species and genera based on these characters could contain graptolites from varied linages (i.e., polyphyletic). Therefore, a classification based on several characters is preferred (Fig. 7.6). A more recent alternate classification of the Pterobranchia is given in Table 7.1 (after Maletz 2014). Recent systematic work suggests that the group is equivalent to modern pterobranchs (Mitchell et al. 2013).

7.5 Geological History

The Graptolites first evolved in Middle Cambrian (Siberian Platform; Obut 1974) and became extinct in either the Late Carboniferous or Early Permian. By Late Cambrian they had become quite common; being most abundant and diverse in

Table 7.1 Graptolite classification (after Maletz 2014)

Phylum Hemichordata. Bateson (1885)
Class Enteropneusta, Gegenbaur (1870)
Class Planctosphaeroidea, van der Horst (1936)
Class Pterobranchia, Lankester (1877)
Subclass Cephalodiscida, Fowler (1892)
Family Cephalodiscidae, Harmer (1905)
Subclass Graptolithina, Bronn (1849)
Incertae sedis Family Rhabdopleuridae, Harmer (1905)
Incertae sedis Family Cysticamaridae, Bulman (1955)
Incertae sedis Family Wimanicrustidae, Bulman (1970)
Incertae sedis Family Dithecodendridae, Obut (1964)
Incertae sedis Family Cyclograptidae, Bulman (1938)
Order Dendroidea, Nicholson (1872b)
Family Dendrograptidae, Roemer (1897) in Frech (1897)
Family Acanthograptidae, Bulman (1938)
Family Mastigograptidae, Bates and Urbanek (2002)
Order Graptoloidea, Lapworth (1875) in Hopkinson and Lapworth (1875)
Suborder Graptodendroidina, Mu and Lin (1981) in Lin (1981)
Family Anisograptidae, Bulman (1950)
Suborder Sinograpta, Maletz et al. (2009)
Family Sigmagraptidae, Cooper and Fortey (1982)
Family Sinograptidae, Mu (1957)
Family Abrograptidae, Mu (1958)
Suborder Dichograptina, Lapworth (1873b)
Family Dichograptidae, Lapworth (1873a, b)
Family Didymograptidae, Mu (1950)
Family Pterograptidae, Mu (1950)
Family Tetragraptidae, Freeh (1897)
Suborder Glossograptina, Jaanusson (1960)
Family Isograptidae, Harris (1933)
Family Glossograptidae, Lapworth (1873b)
Suborder Axonophora, Freeh (1897)
Infraorder Diplograptina, Lapworth (1880)
Family Diplograptidae, Lapworth (1873b)
Subfamily Diplograptinae, Lapworth (1873b)
Subfamily Orthograptinae, Mitchell (1987)
Family Lasiograptidae, Lapworth (1880)
Family Climacograptidae, Freeh (1897)
Family Dicranograptidae, Lapworth (1873b)
Subfamily Dicranograptinae, Lapworth (1873b)

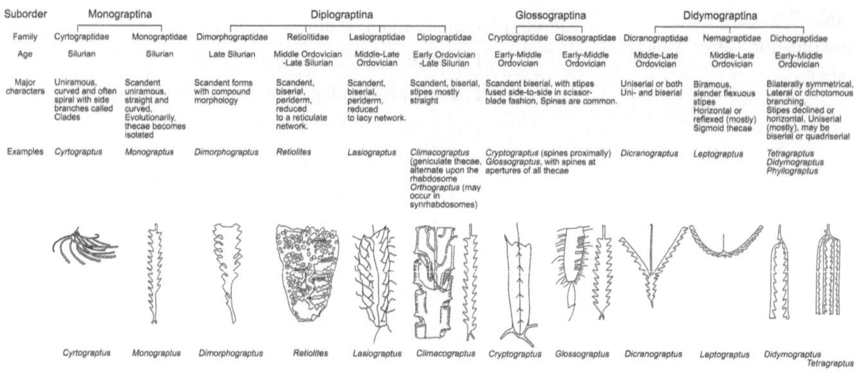

Fig. 7.6 Graptolite classification. The graptolite classification is complex and hampered by poor preservational record and convergent evolution (Polyphyletic). Therefore, a classification based on several characters is preferred. Recent systematic work suggest that the group is equivalent to modern pterobranchs (Mitchell et al. 2013), see also Table 7.2

earlier Paleozoic times. The graptolites evolved from Dendroids in Late Cambrian to Early Devonian times involving changes in the (a) decrease in number of stipes; (b) change in attitude of stipes; (c) increase in complexity of the thecae, and (d) change in the positioning of the thecae. The sessile types like *Dictyonema* [Fig. 7.7(1)], appeared first.

Graptolites are excellent index fossils for Ordovician and Silurian rocks (see Tables 7.2 and 7.3) because of their wide geographic distribution (due to their

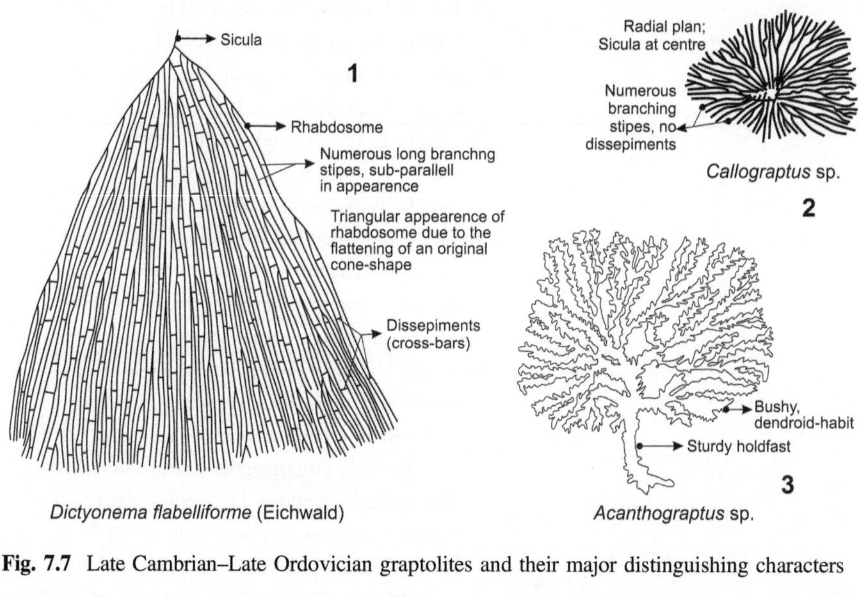

Fig. 7.7 Late Cambrian–Late Ordovician graptolites and their major distinguishing characters

Table 7.2 Ordovician index graptolites. Graptolites are extinct marine colonial animals that thrived from Middle Cambrian to Silurian. They reached their greatest diversity during the Ordovician and are important index fossils for dating Paleozoic rocks

Ma	Period	Epoch	Age/Stage	Australasia Graptolite Zones		Standard Graptolite Zones
	Ordovician	Late	Hirnantian	Bo 5		Normalograptus? persculptus
445				Bo 4		Normalograptus? extraordinarius
			Katian	Bo 3		Paraorthograptus pacificus
				Bo 2		pre-pacificus
450				Bo 1		Climacograptus? uncinatus
				Ea 4		Dicellograptus gravis
				Ea 3		Dicranograptus kirki
				Ea 2		Diplacanthograptus spiniferus
455				Ea 1		Diplacanthograptus lanceolatus
			Sandbian	Gi 2		Orthograptus calcaratus
460				Gi 1		Nemagraptus gracilis
		Middle	Darriwilian	Da 4	b	Archiclimacograptus riddellensis
					a	
465				Da 3		Pseudoclimacograptus decoratus
				Da 2		Undulograptus intersitus
				Da 1		Undulograptus austrodentatus
			Dapingian	Ya 2		Cardiograptus morsus
				Ya 1		Oncograptus upsilon
				Ca 3		Isograptus victoriae maximus
470				Ca 2		Isograptus victoriae victoriae
				Ca 1		Isograptus victoriae lunatus
		Early	Floian	Ch 2		Isograptus primulus
				Ch1		Didymograptus protobifidus
				Be 4		(Pendeograptus fruticosus)
				Be 3		
475				Be 2		
				Be 1		Tetragraptus approximatus
				La 3		
480			Tremadocian	La 2	b	Araneograptus pulchellus
485					a	Aorograptus victoriae
				La 1	b	Psigraptus jacksoni
					a	Anisograptus
488				Pre-La 1		-

Table 7.3 Silurian index graptolites

Ma	Period	Epoch	Age/Stage	Standard Graptolite Zones
414				
415	**Devonian**			Monograptus bouceki-transgrediens-perneri
				Monograptus branikensis-lochkovensis
	Silurian	Pridoli		Monograptus parultimus-ultimatus
				Monograptus formosus
				Neocullograptus kozlowskii-Polonograptus podoliensis
420		Ludlow	Ludfordian	Saetograptus leintwardinensis
				Lobograptus scanicus
			Gorstian	Neodiversograptus nilssoni
		Wenlock	Homerian	Colonograptus praedeubeli-deubeli
				Pristiograptus parvus-Gothograptus nassa
425				Cyrtograptus lundgreni
				Cyrtograptus rigidus-perneri
			Sheinwoodian	Monograptus riccartonensis-belophorus-antennularis
		Llandovery	Telychian	Cyrtograptus centrifugus-murchisoni
				Cyrtograptus lapworthi-insectus
430				Oktavites spiralis Interval zone
				Monograptus griestoniensis-crenulata
				Monograptus crispus
				Spirograptus turriculatus
435				Spirograptus guerichi
			Aeronian	Stimulograptus sedgwickii
				Lituigraptus convolutus
				Monograptus argenteus - leptotheca
				Demirastrites triangulatus-pectinatus
440			Rhuddanian	Coronograptus cyphus
				Orthograptus vesiculosis
				Parakidograptus acuminatus
				Akidograptus ascensus
444				

planktonic nature), restriction to latitudal belts (due to their eurythermic nature), ability to thrive in both deep and shallow waters (due to their epipelagic nature), and their short stratigraphic range (Sadler et al. 2009).

However, there are some problems too—graptolites are usually preserved in highly compressed forms, hence, correct species identification is sometimes difficult, and they are rarely preserved in coarse-grained rocks, hampering correlation between areas of different facies. Thus, correlation based on faunal units (successive faunas), i.e., based on the first appearance of a new species, and is the best possible option for global correlation. Faunal data suggests that the earliest Ordovician (Tremdoc) is marked by the "*anisograptid fauna*" consists

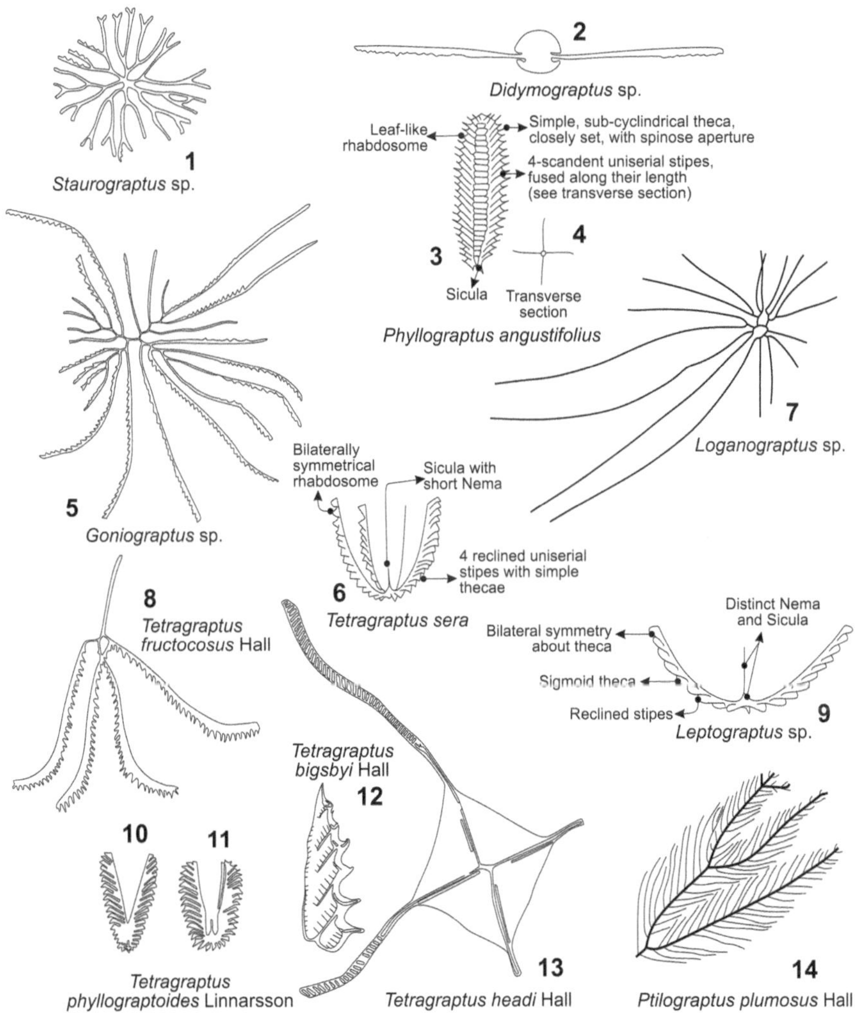

Fig. 7.8 Early Ordovician graptolites and their major distinguishing characters

predominantly of anisograptids (graptoloids with bitheca and multiramous). The succeeding "dichograptid fauna" (Arenig) are largely known from Order Dichograptina; though they lack bitheca, they are multiramous. The most common forms in the "*diplograptid fauna*" (in the rest of the Ordovician) which have fewer stipes (commonly biramous), and are biserial. Finally, the "*monograptid fauna*" (in Silurian to Early Devonian times) is dominated by uniramous and uniserial forms.

Representative graptolite species are illustrated in Figs. 7.7, 7.8, 7.9, 7.10.

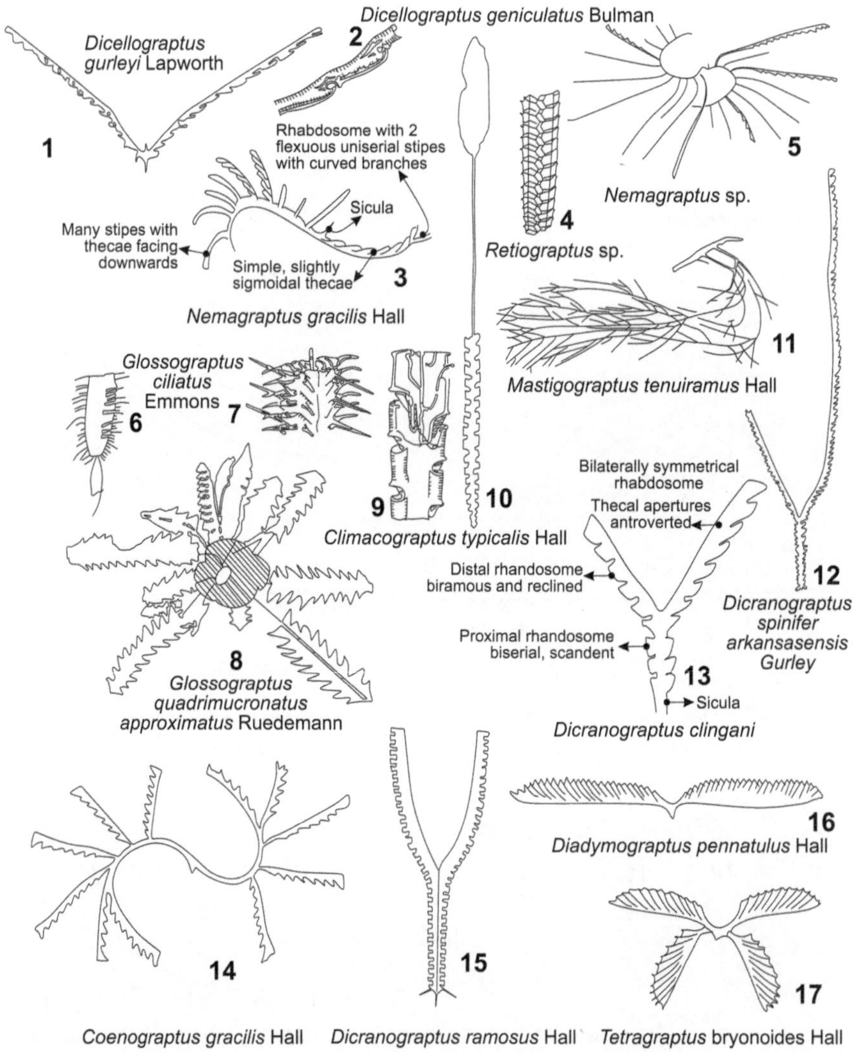

Fig. 7.9 Middle Ordovician graptolites and their major distinguishing characters

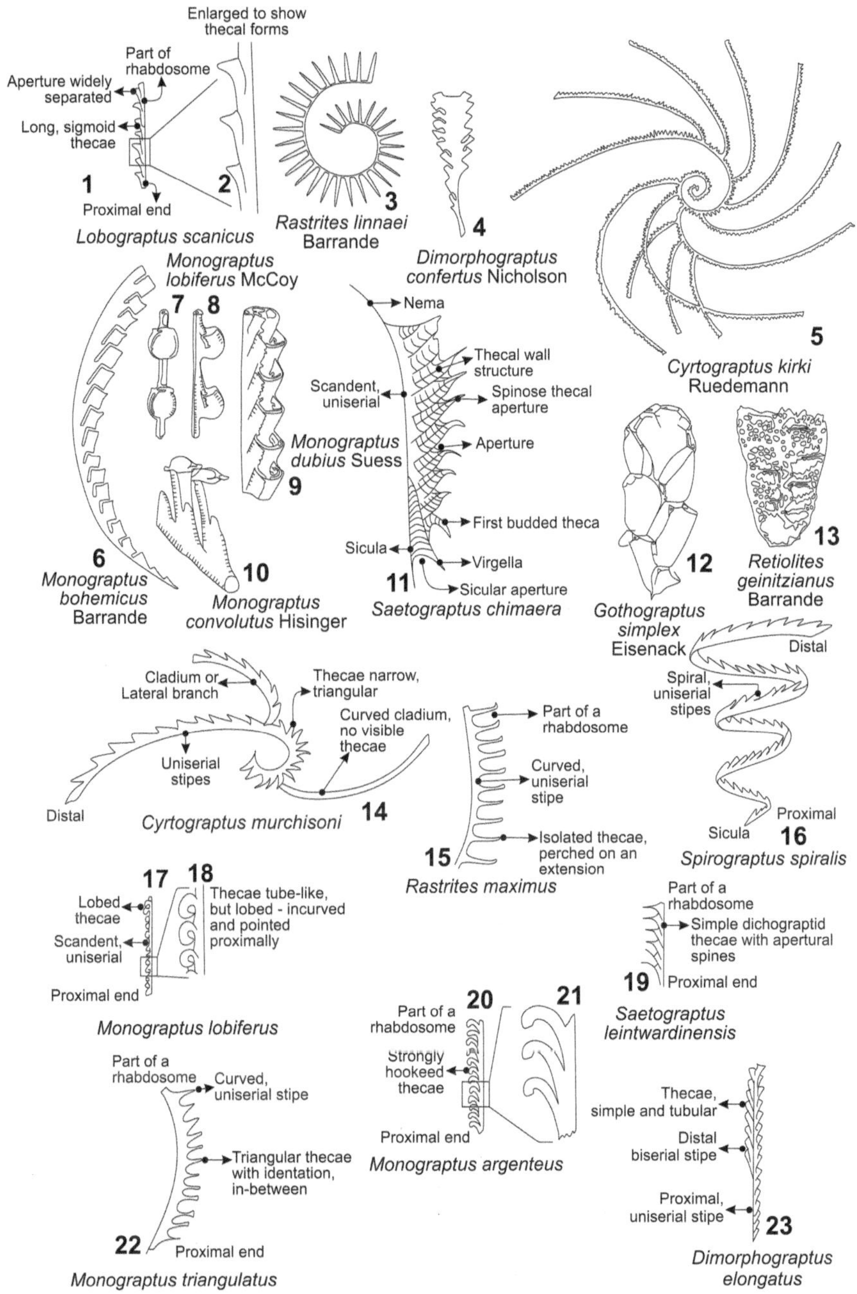

Fig. 7.10 Silurian graptolites and their major distinguishing characters

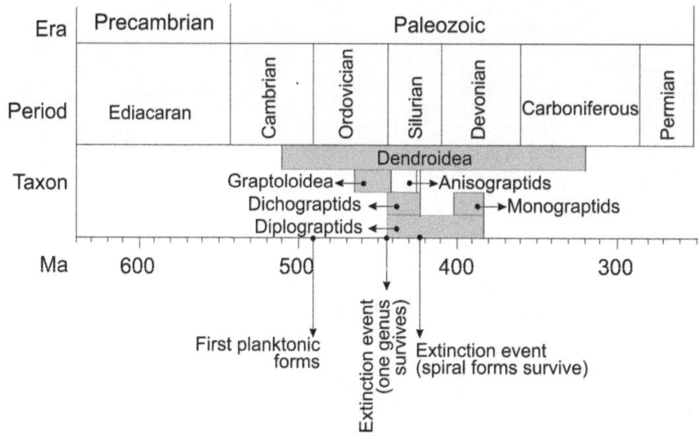

Fig. 7.11 Geological history of graptolites

Appendix 1 gives the list of illustrated specimens mentioning the chapter number, species name, age, and locality along with its figure number within the said chapter (Fig. 7.11).

References

Bates, D.E.B., and A. Urbanek. 2002. The ultrastructure, development, and systematic position of the graptolite genus Mastigograptus. *Acta Palaeontologica Polonica* 47(3): 445–458.

Bateson, W. 1885. The later stages in the development of Balanoglossus kowalevskii, with a suggestion as to the affinities of the Enteropneusta. *Quarterly Journal of Microscopical Science* 25: 81–122.

Benton, M., and D.A.T. Harper. 2009. *Introduction to Paleobiology and the Fossil Record*, 2nd ed., 608 pp. New York: Wiley.

Bronn, H. G. 1849. Handbuch der Geschichte der Natur. Dritter Band, Zweite Abtheilung. II. Theil: Organisches Leben (Schluß). Index palaeontologicus oder Ueberblick der bis jetzt bekannten fossilen Organismen, 1106 pp.

Bulman, O.M.B. 1938. Graptolithina. In *Handbuch der Paläozoologie*, vol. 2D, ed. O.H. Schindewolf, 1–92. Berlin: Borntraeger.

Bulman, O.M.B. 1950. Graptolites from the Dictyonema Shales of Quebec. *Quarterly Journal of the Geological Society of London* 106: 63–99.

Bulman, O.M.B. 1955. Graptolithina. In *Treatise on Invertebrate Paleontology, Part V*, edited by R.C. Moore. Lawrence: Geological Society of America and University of Kansas Press.

Bulman, O.M.B. 1970. Graptolithina. In *Treatise on invertebrate paleontology, part V*, 2nd ed, ed. C. Teichert. Lawrence: Geological Society of America and University of Kansas Press.

Clarkson, E.N.K. 1983. Invertebrate palaeontology and evolution, 452 pp. Blackwell Science.

Cooper, R.A., and R.A. Fortey. 1982. The Ordovician graptolites of Spitsbergen. *Bulletin of the British Museum (Natural History), Geology Series*, 36(3):157–302.

Fowler, G.H. 1892. The morphology of Rhabdopleura normani Allm., 293–297. In Festschrift zum 70ten Geburstag Rudolf Leuckarts, 293 pp.

Frazier, W.J., and D.R. Schwimme. 1987. *Regional stratigraphy of North America*, 744 pp. Berlin: Springer.

Frech, F. 1897. Lethaea geognostica oder Beschreibung und Abbildung für die Gebirgs-Formationen bezeichnendsten Versteinerungen. Herausgegeben von einer Vereinigung von Palaeontologen, 1. Teil - Lethaea Palaeozoica, 544–684 pp.

Gegenbaur, C. 1870. Grundzüge der vergleichenden Anatomie. Zweite, umgearbeitete Auflage. 892 pp.

Harmer, S.F. 1905. The Pterobranchia of the Siboga-Expedition with an account of other species. *Siboga Expedition Monograph* 26: 1–133.

Harris, W.J. 1933. *Isograptus caduceus* and its allies in Victoria. *Proceedings of the Royal Society of Victoria, New Series* 46: 79–114.

Hopkinson, J., and C. Lapworth. 1875. Descriptions of the graptolites of the Arenig and Llandeilo rocks of St. David's. *Quarterly Journal of the Geological Society* 31: 631–672.

Jaanusson, V. 1960. Graptoloids from the Ontikan and Viruan (Ordov.) limestones of Estonia and Sweden. *Bulletin of the Geological Institutions of the University of Uppsala*, 38(3–4):289–366.

Kozlowski, R. 1948. Les graptolithes et quelques nouveaux groupes d'animaux du Tremadoc de la Pologne, Palaeontologia Polon 3: 235 p.

Lankester, E.R. 1877. Notes on the embryology and classification of the animal kingdom; comprising a revision of speculations relative to the origin and significance of the germlayers. *Quarterly Journal of Microscopical Science, New Series* 17: 339–454.

Lapworth, C. 1873a. Notes on the British graptolites and their allies. 1. On an improved classification of the Rhabdophora, part 1. *Geological Magazine* 10: 500–504.

Lapworth, C. 1873b. Notes on the British graptolites and their allies. 1. On an improved classification of the Rhabdophora, part 2. *Geological Magazine* 10: 555–560.

Lapworth, C. 1880. On the geological distribution of the Rhabdophora. Part III. *Annals and Magazine of Natural History, Series*, 5(6):185–207.

Lin, Y.K. 1981. New materials of graptodendroids with special reference to the classification of Graptodendroidea. *Bulletin of Nanjing Institute of Geology and Palaeontology, Academia Sinica* 3: 242–262.

Maletz, J. 2014. The classification of the Pterobranchia (Cephalodiscida and Graptolithina). *Bulletin of Geosciences* 89(3): 477–540.

Maletz, J. 2015. Graptolite reconstructions and interpretations. *Paläontologische Zeitschrift* 89(3): 271–286.

Maletz, J., M. Steiner, and O. Fatka. 2005. Middle Cambrian pterobranchs and the question: what is a graptolite? *Lethaia* 38: 73–85.

Maletz, J., J. Carlucci, and C.E. Mitchell. 2009. Graptoloid cladistics, taxonomy and phylogeny. *Bulletin of Geosciences* 84(1): 7–19.

Mitchell, C.E. 1987. Evolution and phylogenetic classification of the Diplograptacea. *Palaeontology* 30(2): 353–405.

Mitchell, C.E., M.J. Melchin, C.B. Cameron, and J. Maletz. 2013. Phylogenetic analysis reveals that *Rhabdopleura* is an extant graptolite. *Lethaia* 46: 34–56.

Mu, A.T. 1950. On the evolution and classification of graptolites. *Geological Review* 15: 171–183.

Mu, A.T. 1957. Some new or little known graptolites from the Ningkuo Shale (Lower Ordovician) of Changshan, Western Chekiang. *Acta Palaeontologica Sinica* 5(3): 369–437.

Mu, A.T. 1958. *Abrograptus*, a new graptolite genus from the Hulo Shale (Middle Ordovician) of Kiangshan, western Chekiang. *Acta Palaeontologica Sinica* 6(3): 259–265.

Nicholson, H.A. 1872b. A monograph of the British Graptolitidae, 144 pp. Blackwood and Sons.

Obut, A.M. 1964. Podtip Stomochordata. Stomokhordovye. In *Osnovy paleontogii: Echinodermata, Hemichordata, Pogonophora, Chaetognatha*, ed. Y.A. Orlov, 279–337. Moscow: Nedra Press.

Obut, A.M. 1974. New graptolites for the Middle Cambrian of the Siberian Platform. In Graptolite studies in honour of O. M. B. Bulman. Special papers in Palaeontology No. 13, 9–13.

Prothero, D.R. 2013. *Bringing fossils to life: an introduction to paleobiology*, 672 pp. Columbia University Press.

Rigby, S. 1994. Hemichordate skeletal growth, shared patterns in *Rhabdopleura* and graptoloids. *Lethaia* 27: 317–324.

Sadler, P.M., R.A. Cooper, and M.J. Melchin. 2009. High-resolution, early Paleozoic (Ordovician-Silurian) time scales. *Geological Society of America Bulletin* 121: 887–906.

Van Der Horst, C.J. 1936. Planctosphaera and Tornaria. *Quarterly Journal of Microscopical Science* 78: 605–613.

Chapter 8
Brachiopods

8.1 Introduction

Brachiopods are the oldest known shelly invertebrate fossils. *Askepasma saproconcha* Topper, a Paterinida, is the oldest known brachiopod coming from a pre-trilobitic strata (Terreneuvian, Cambrian Stage 2, lower Atdabanian; ~526–530 Ma) within the Early Cambrian succession from South Australia (Topper et al. 2013).

Brachiopods are exclusively marine solitary organisms that live on the ocean floor, inhabiting a wide range of water depths from very shallow waters of rocky shorelines to ocean floors (at places three-and-a-half miles beneath the ocean surface). Spatially, they are found from the warm tropical waters of the Caribbean to the cold Antarctic seas. Around 300 species (~120 genera) are present today and over 30,000 fossil species (~900 alone in the Devonian) have been described, so far. Some living species today appear much like their fossil ancestors such as the modern *Lingula unguis* Linne that looks almost identical to its Paleozoic ancestor of 400 Ma ago. Brachiopods are also called "*Lamp shells*" based on their close resemblance with the shape of an oil lamp used in ancient Greece and Rome (to mostly Mediterranean terebratulid species *Magellania*).

Brachiopods have an excellent geological record especially throughout the Phanerozoic (Cambrian–Quaternary) and are also amongst the most successful benthic invertebrates of the Paleozoic (Cambrian–Permian). Hence, the Paleozoic is often referred as the "*Age of Brachiopods*" with several orders dominating the shallow shelf environments; giving way grudgingly to the rapidly diversifying pelecypods in the Mesozoic (Gould and Calloway 1980; Sepkoski 1996) (see Fig. 8.1). As a phylum, brachiopods show a great variety of changes in the form and function through time, and hence are important for biostratigraphic, paleoecologic, and evolutionary studies marked by characteristic and dominant assemblages through time (Table 8.1). Most Brachiopods are small (<7 cm) with the largest living species having a shell length of about 100 mm (4 in.). However, the

S. Jain, *Fundamentals of Invertebrate Palaeontology*, Springer Geology,
DOI 10.1007/978-81-322-3658-0_8

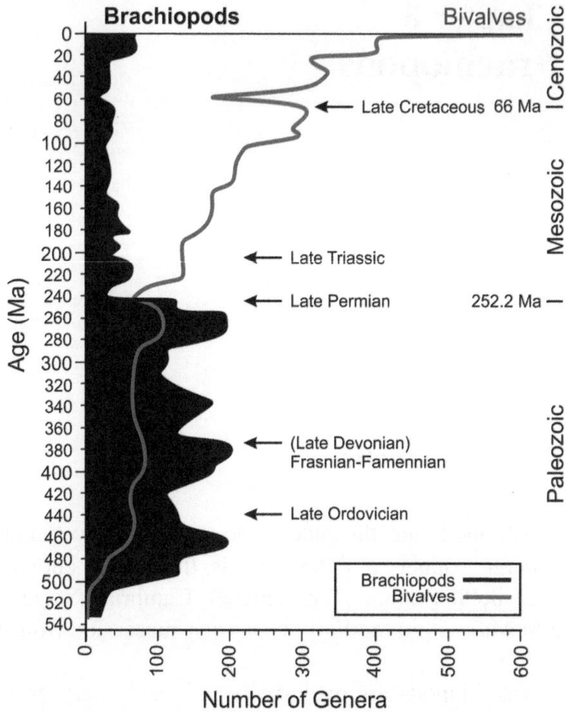

Fig. 8.1 Generic diversity of Brachiopods and Pelecypods in the Phanerozoic (modified from Gould and Calloway 1980; Sepkoski 1996). The Brachiopods are among the most successful benthic invertebrates of the Paleozoic (Cambrian-Permian) so much so that the Paleozoic is often termed as the "*Age of Brachiopods*" with several orders dominating; giving way grudgingly to the rapidly diversifying Pelecypods of the Mesozoic (see Chap. 4). *Arrow* marks major extinction events in brachiopod geological history (see Sect. 8.5 for details)

Table 8.1 Dominant brachiopod assemblages through time

Dominant assemblage	Age
Abundance of Richthofenids and Oldhaminids	Permian
Abundance of Productids	Pennsylvanian (Late Carboniferous) and Permian
Abundance of *Spirifers* and Productids	Mississippian (Early Carboniferous)
Abundance of *Spirifers*	Devonian
Abundance of Pentamerids	Silurian
Flat Orthids (*Resserella* or *Hebertella*)	Ordovician
Abundance of "D-shaped" Strophomenids	Ordovician
Lingulid-Orthid assemblage	Cambrian

Early Carboniferous (Mississippian) *Gigantoproductus giganteus* Martin is the largest of them all with a shell length of over 300 mm (12 in.). Another big form, but somewhat smaller than *Gigantoproductus*, is another Early Mississippian brachiopod *Delepinea*. In general, brachiopod shells with width less than 20 mm (0.8 in.) are considered "small," those between 20 and 50 mm (0.8–2 in.) are "medium" sized and those greater than 50 mm (>2 in.) are large.

8.2 The Shell

Brachiopod terminology is complex with a very elaborate shell description (see also Clarkson 1993). Hence, only major features are covered and illustrated here. Other minor features are enumerated under Terminology, Sect. 8.3. Table 8.2 summarizes the outline of the chapter description that unfolds.

The two mineralized valves that enclose most of the soft body parts constitute the brachiopod shell (Fig. 8.2). Brachiopods extract minerals from the sea water. The mineralized shell is made up of ~50 % calcium carbonate ($CaCO_3$) or phosphate and the other 50 % is composed of chitin or proteins with varying amounts of organic material. The Late Cretaceous *Crania* is the only genus to have chitin as the choice for building its shell (see Lee and Brunton 1986; Emig 2009; Moore 1965). The Paterinates, the earliest brachiopods, possess an organophosphatic shell, the Linguliformeans have phosphatic material combined into their shell fabric, the Craniiformean's shell consists of organocarbonate and in Rhynchonelliformeans it is of low-magnesian calcite.

Table 8.2 A broad outline of the description followed in this chapter

• Two valves: pedicle (ventral) and brachial (dorsal)
• Shell orientation and dimensions
• General shell features
– Brachiopod shell shape
– Fold and sulcus
– Lateral profile
– Size of the two valves
– Position of the umbo
– Strophic versus non-strophic
– Shell ornamentation
• Hinge features: teeth and socket; hinge line
• Features of the anterior part of shell
– Commissure and commissural plane
– Pedicle valve: umbo, interarea, delthyrium, and deltidial plates
Types of deltidial plates
– Brachial valve: notothyrium, chilidium, chilidial plates and Listrium
– Position of the Pedicle foramen relative to the beak ridge
• Brachiopod feeding mechanism: Lophophore, Brachidium, Spiralia and Jugum
Types of Brachidium
• Brachiopod musculature: Diductors, Adductors and Adjusters
• Pallial markings (Vascular markings)
• Shell structure: Punctuate, Pseudopunctate and Impunctate

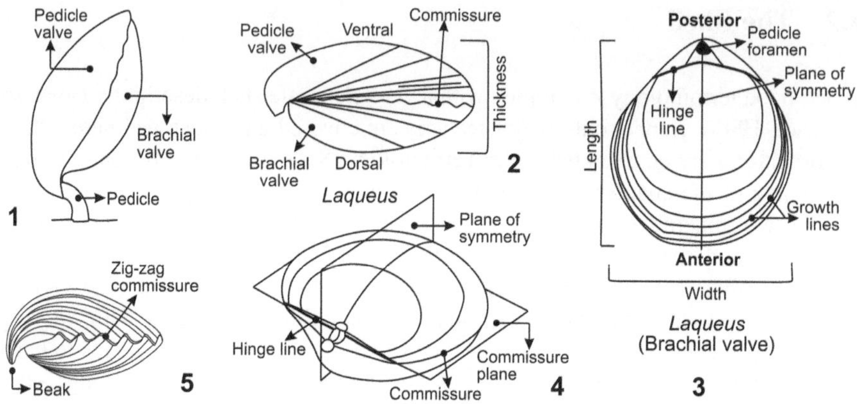

Fig. 8.2 Basic brachiopod shell terminology

The two valves of brachiopods are of unequal size; the smaller half is called the dorsal or brachial valve and the larger half is called the ventral or pedicle valve; the latter hosts the edicle [Fig. 8.2(1, 2)]. Unlike pelecypods (Chap. 4) that have inequilateral valves with a symmetry of the upper and lower valves, the brachiopod valves are symmetrical about a median plane thus, giving it a left–right symmetry [Fig. 8.2(3, 4)]. Hence, brachiopods are bilaterally symmetrical about a plane drawn perpendicular to the line of contact of the two closed valves, i.e., each valve is bilaterally symmetrical and the shell as a whole is asymmetrical [Fig. 8.2(3, 4)].

The brachiopod shell, while describing, is always oriented with the posterior margin (the hinge line) placed above, and the anterior margin, below [Fig. 8.2(3)]. A line drawn from the beak to the anterior margin describes its length [Fig. 8.2(3)]; and one at right angles to the same, in the direction of right and left, is its width [Fig. 8.2(3)]; a third line drawn perpendicularly to the other two, and passing through the centers of the valves, measures its thickness [Fig. 8.2(2)]. In some, the posterior margin of the pedicle valve is convex, and curved to form a beak [Fig. 8.2 (5)] which may be pointed, or perforated by a round opening, called the pedicle foramen [Fig. 8.2(3)]; the pedicle protrudes from the latter [Fig. 8.2(1)].

At the anterior end of the brachiopod shell, the Commissure forms a line of junction between the two valves [Fig. 8.2(2)]; its plane of symmetry is called the commissural plane which divides the shell into two unequal halves [Fig. 8.2(4)]. In some brachiopod lineages (as in Rhynchonellida) a zigzag commissure is noted. This "zig-zag" pattern increases the amount of water filtered and also protects the delicate lophophore by not allowing entry to overly large-sized particles [Fig. 8.2 (5)]. Three major types of commissure are noted and are illustrated in Fig. 8.3(1, 3).

The shape of the brachiopod shell varied; the major types are illustrated in Fig. 8.3(4–16). Additionally, the two brachiopod valves also join together in different ways; the anterior margin of one valve is frequently indented by a median Sinus, and the other (in the Pedicle valve) usually exhibits a corresponding Fold, or elevation [Fig. 8.3(17–30)]. Their lateral profile also varies from being smooth to

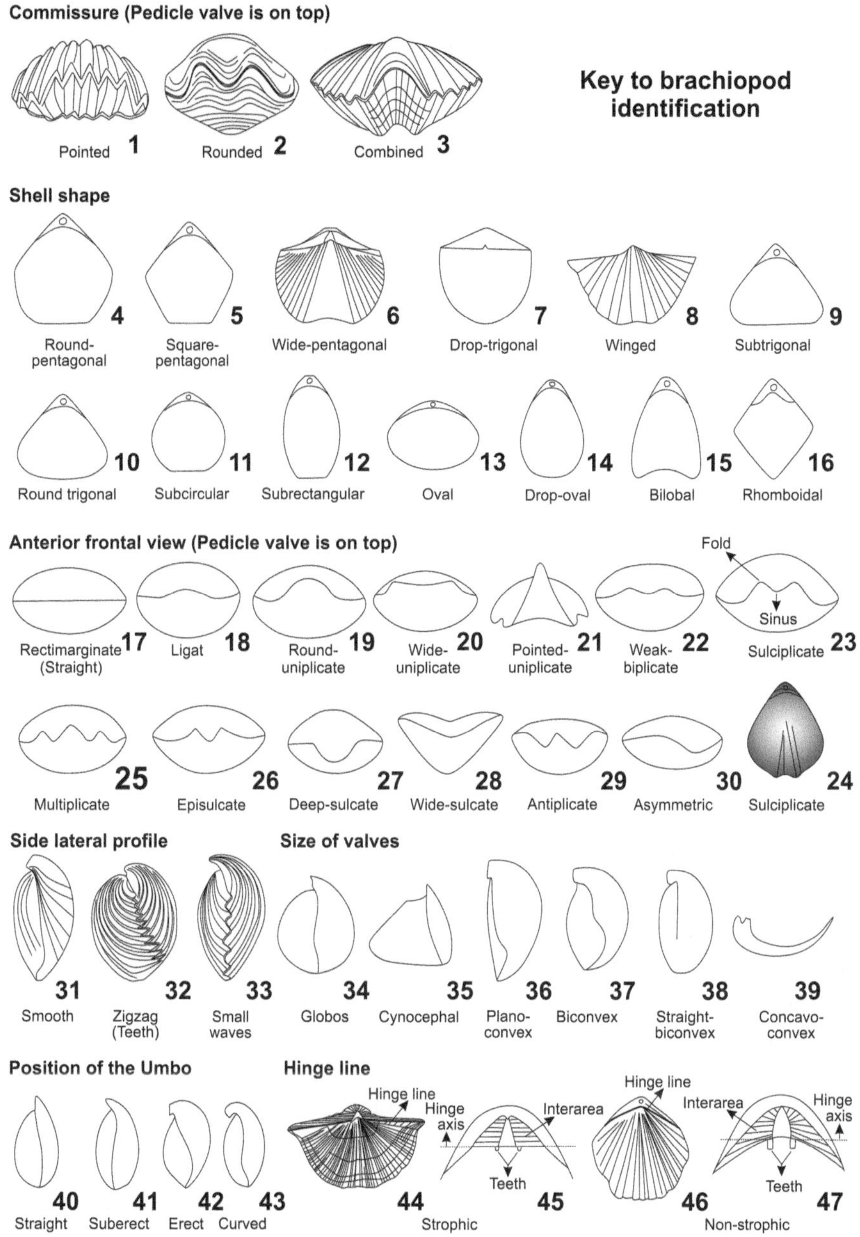

Fig. 8.3 Key to brachiopod identification. Seven parameters are used to identify brachiopods, namely type of commissure, shell shape, type of frontal (anterior) shell view, type of shell lateral profile, size of valves, position of umbo and type of hinge line. These are basic morphologies but in nature there may also exist several combinations of these

zigzag to wavy [Fig. 8.3(31–33)]. The size of the two valves (brachial and pedicle) also differs [Fig. 8.3(34–39)]. The position of the umbo varies from being straight to curved [Fig. 8.3(40–43)] and last, the hinge line (the posterior margin of the shell parallel to the hinge axis) can be either straight (strophic) or curved (astrophic) [Fig. 8.3(44–47)].

In the earliest growth stages of the shell, it is invariably smooth, and may remain so throughout life, but the greater number of shells develops radial striae, ribs, or undulations, and these are usually crossed by concentric growth lines, or lamellar, which are sometimes of great width, or may extended into spines (see Fig. 8.4). Fundamentally, the ornamentation on the brachiopod shell is a combination of radiating ribs and concentric growth lines [Fig. 8.4(1)] resulting in varied patterns of ornamentation [Fig. 8.4(2–16)]. Additional ornamentation terminology is enumerated under Sect. 8.3.3

On the basis of shell structure, the brachiopods have traditionally been divided into two taxonomic groups—Articulata and Inarticulata. The Articulata are characterized by the presence of teeth, sockets, and a definite hinge (the place where the two valves meet at their posterior end) (see [Fig. 8.5(1–4)]). The hinge area found in them is important for taxonomic classification and is a diagnostic feature of certain brachiopod groups. On the other hand, the Inarticulates have no teeth or hinge and the pedicle, when present, passes between the two valves of the shell.

Between the two valves, the posterior portion of the hinge structure is the Interarea [Fig. 8.5(1)]. Its left and right portions are often closed off by plates, thereby leaving a triangular opening through which the pedicle protrudes. In the brachial valve, the opening (a notch instead of the round pedicle foramen) in the interarea, is called the Notothyrium and Delthyrium for the pedicle valve [Fig. 8.5 (2–4)]. The pedicle foramen (the opening for the pedicle; [Fig. 8.5(2, 5, 6)] may be enclosed on the anterior end by a single plate; it is then called a Xenidium [Fig. 8.5 (6)] and when done by a pair of plates, the structure is called a Deltidium [Fig. 8.5 (13)], and the plates are then called Deltidium plates [Fig. 8.5(14)]. Varied types of deltidial plates are illustrated in [Fig. 8.5(8–12)]. If plates enclose the notch (Notothyrium/Delthyrium), it is called Chilidium [Fig. 8.5(15)] and the plates are called Chilidial plates [Fig. 8.5(16)]. The Chilidium is a convex plate which often covers the cardinal process of the dorsal valve [Fig. 8.5(15)]. The cardinal area is a term applied to the flattened or curved triangular area which is observed frequently between hinge line and beak ([Fig. 8.5(4)]; the cardinal process encloses it). It is well developed in the pedicle valve rather than in the dorsal valve, and is bisected medially by the triangular delthyrium [Fig. 8.5(2)]. Listrium is a plate closing the progressive track of the pedicle opening or pedicle cleft, posterior to the apex of the pedicle valve [Fig. 8.5(21)], as is commonly noted in Discinids. The position of the pedicle foramen relative to the beak ridge also varies [Fig. 8.5(26–31)]. Other relevant terminology of the brachiopod shell are enumerated under Sect. 8.3.5 and illustrated in Fig. 8.5.

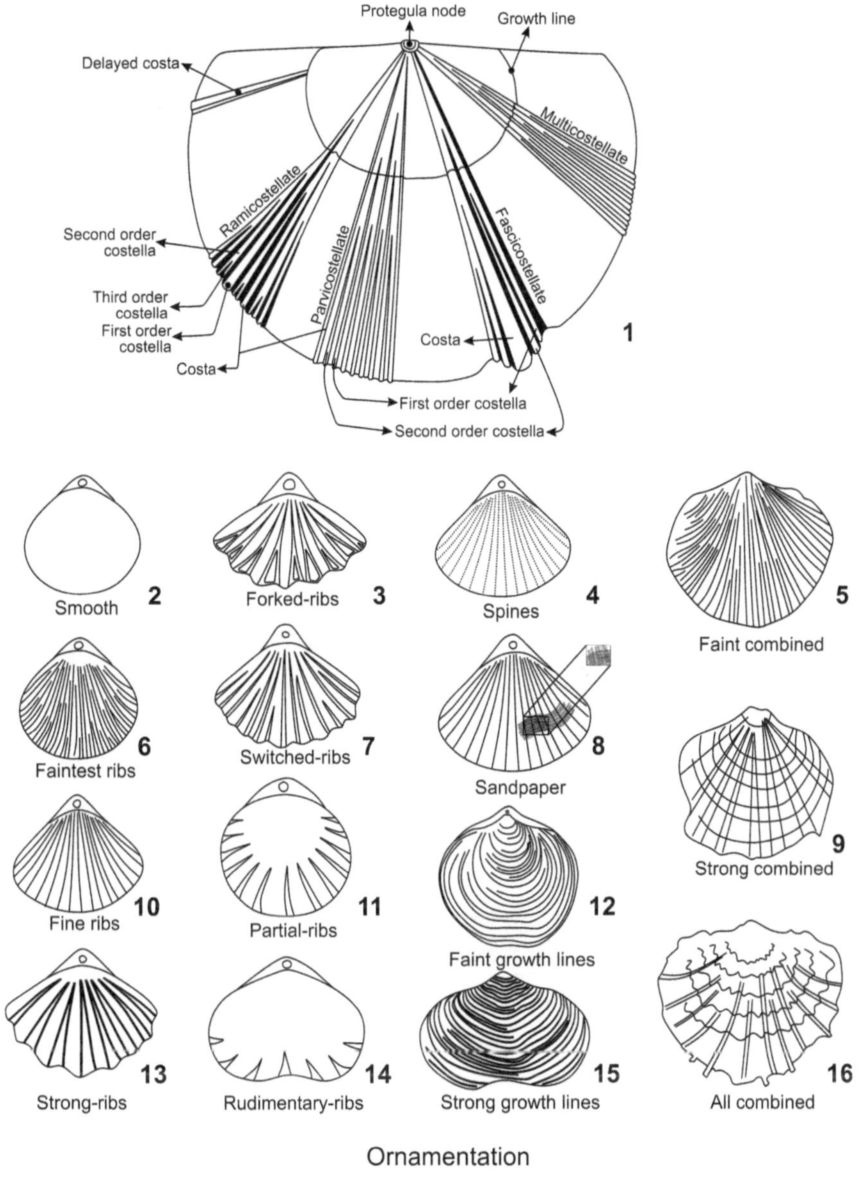

Fig. 8.4 Key to brachiopod ornamentation. These are the basic types of ornamentation but in nature there may also exist several combinations of these

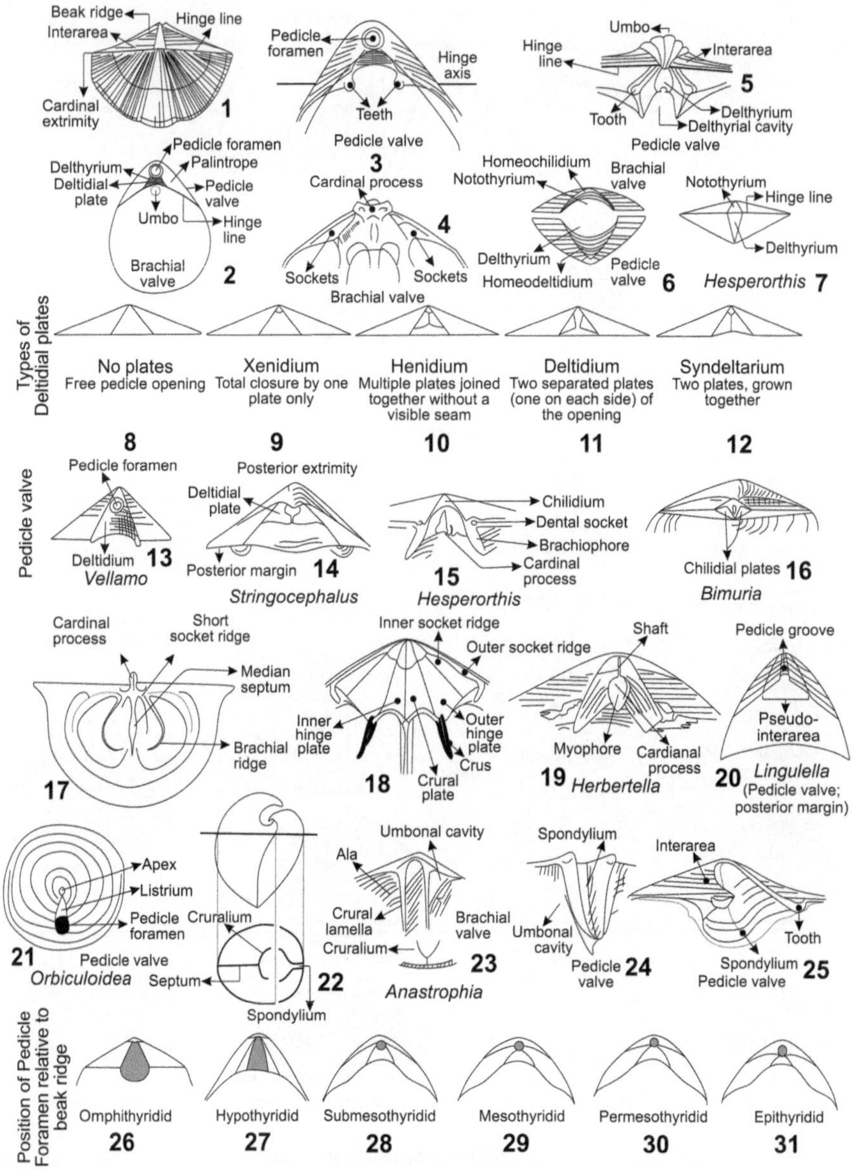

Fig. 8.5 Basic brachiopod shell terminology used in the text. Brachiopod shell terminology is complex and very elaborate. Minor features are not illustrated. For explanation see text and refer to Sect. 8.3—Terminology

Brachiopods are filter feeders and collect food particles through a ciliated organ called the Lophophore (see [Fig. 8.6(1)]; also noted in Recent Terebratulids). Lophophore is a feeding structure and is a characteristic feature of the phylum, consisting of a pair of ciliated, twisted projections that create water currents and then filters out the microscopic food particles [Fig. 8.6(1)]. It is symmetrically placed around the mouth, and is suspended from the anterior body wall; it may also be attached to the dorsal mantle and occupies the mantle cavity (see [Fig. 8.6(1)]).

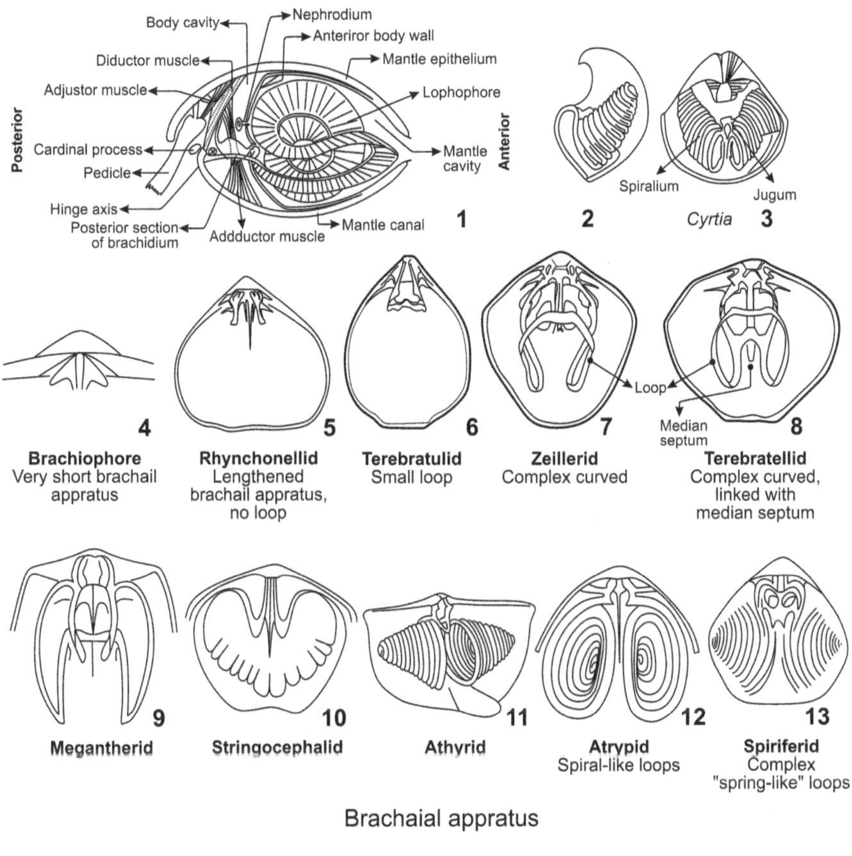

Fig. 8.6 The lophophore and types of brachidium. The lophophore is a feeding structure and the calcareous support of the lophophore is the brachidium which consists of an elaborate long calcareous loop system (*4–13*). The pair of spirally coiled lamellae is called the Spiralia (*2*) and the medially placed structure of secondary shell connecting two primary lamellae of spiralia is called the Jugum (*3*). The shape of the brachidia is highly variable; from a short stout set of supports called the brachiophors (*4*) to a conical spiral pointing laterally as in spiriferids (*13*)

The calcareous support of the Lophophore is called the Brachidium, an elaborate long calcareous loop system; the pair of spirally coiled lamellae supporting the lophophore is called Spiralia (see [Fig. 8.6(2, 3)]). Jugum is the medially placed structure of secondary shell connecting two primary lamellae of spiralia [Fig. 8.6 (3)]. The shape of the Brachidia is highly variable; from a short stout set of supports called the Brachiophors [Fig. 8.6(4)] to a conical spiral pointing laterally as in Spiriferids [Fig. 8.6(13)]. Various types are illustrated in Fig. 8.6(3–12). The brachidia are of great importance in brachiopod systematics.

The brachiopod uses muscles to open and close its shell [Fig. 8.7(1, 2)]. In Articulates, there are three sets of muscles [Fig. 8.7(3, 4)]—the diductors, which by contraction opens the valves; the adductors, which by contraction closes the valves; and the pedicle muscles, or adjustors that by contraction withdraws the pedicle [Fig. 8.7(1, 2)]. The points of attachment (i.e., points of insertion and origin) of these muscles leave more or less distinct impressions (scars) in the valves [Fig. 8.7 (1, 2, 5)]. The number and form of these scars provide important diagnostic characters. The two adductor muscles, each divided dorsally, are commonly present to produce a single pair of scars located between diductor impressions in the pedicle valve [Fig. 8.7(5)] and two pairs (Anterior and Posterior adductors) in the brachial valve [Fig. 8.7(1, 2)]. The Diductor muscles open the shell. The Adjustor muscles constitute two pairs of muscles in many articulated brachiopods; they provide movement to pedicle or shell and arise from the proximal end of the pedicle valve [Fig. 8.7(5)]. The diductors, or opening muscles, originate at the anterior ventral edge of the visceral area, and on either side of the median ridge; the scars of these muscles being usually the largest and deepest of any in the animal [Fig. 8.7(5, 6)]. They taper rapidly in crossing the interior cavity, and their small extremities are attached to the anterior portion of the cardinal process. Myophore is the differentiated site of the diductor muscle attachment on the cardinal process [Fig. 8.5(18)].

The Pallial markings (also called Vascular markings) (Fig. 8.8) are impressions of mantle canals on the shell's interior. The mantle canals are flattened tube-like extensions of the body cavity into the mantle.

The brachiopod shell in thin section has two layers; the inner, made up of inclined fibers of calcite and the outer, a lamellar layer, with low angle calcite crystals (Fig. 8.9). If small tubes or pores penetrate the shell, the shell is called punctuate [Fig. 8.9(1, 4)]. If there are perpendicular rods of calcite in the fibrous inner layer, it is called Pseudopunctate (as in Strophominids) [Fig. 8.9(3, 6)]. If the shell is solid it is called impunctate [Fig. 8.9(2, 5)]. Both punctate and impunctate types have little taxonomic relevance. A puncta (pl., punctae) is a perforation penetrating the shell to connect with the periostracum (the organic external layer of shell) [Fig. 8.9(1, 4)]. Punctate shells have punctae (sing. Puncta: [Fig. 8.9(1)]).

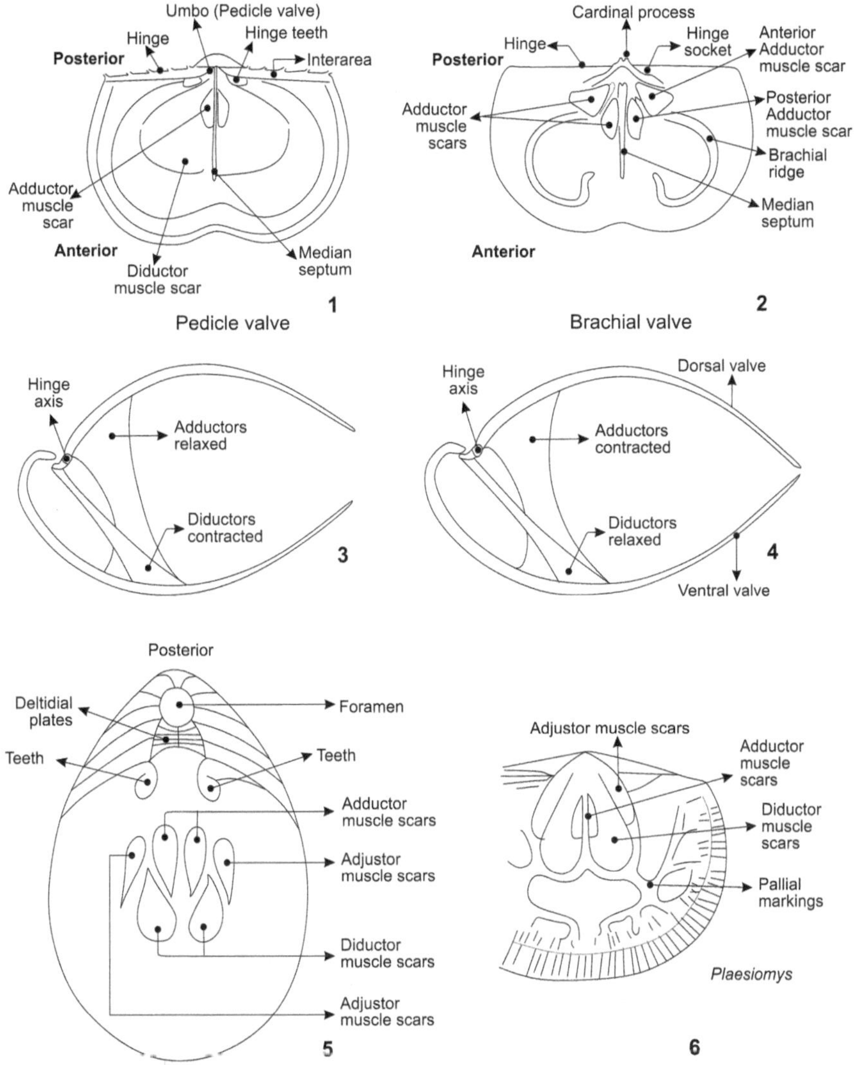

Fig. 8.7 Muscle attachments and scars. *1–2* General overview of muscle attachments and scars of both pedicle and brachial valves. *3–4* Muscles used for closing and opening of valves. *5* Details of muscle attachments and scars. *6* Arrangement of muscle scars as noted in *Plaesiomys* Hall and Clarke, a Middle to Late Ordovician brachiopod genus (Ohio, USA)

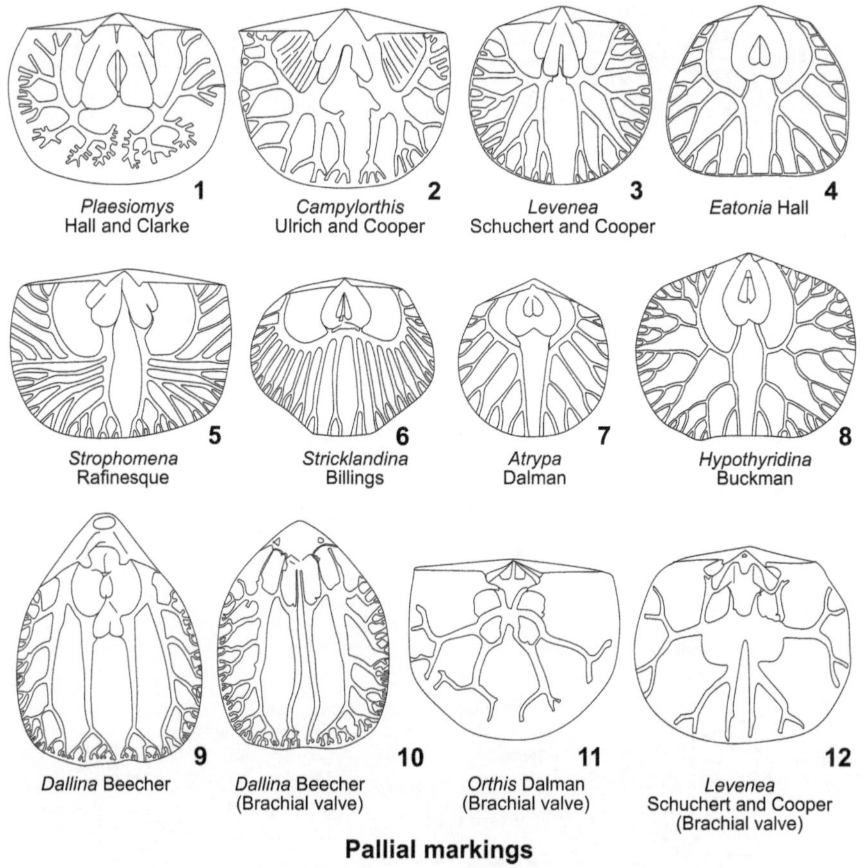

Fig. 8.8 Types of Pallial markings (also called vascular markings) are impressions of mantle canals on the shell's interior. Illustrated examples include: *Plaesiomys* Hall and Clarke, Middle to Late Ordovician (Ohio, USA); *Campylorthis* Ulrich and Cooper, Middle Ordovician (Wisconsin, USA); *Levenea* Schuchert and Cooper, Silurian to Middle Devonian (Tennessee, USA); *Eatonia* Hall, Early Devonian (Maryland, USA); *Strophomena* Rafinesque, Middle to Late Ordovician (Ohio, USA); *Stricklandina* Billings, Early Silurian (North America); *Atrypa* Dalman, Early Silurian to Late Devonian (Nova Scotia, Canada); *Hypothyridina* Buckman, Middle to Late Devonian (Plymouth, England); *Dallina* Beecher, Eocene to? Recent (Japan); *Orthis* Dalman, Early to? Middle Ordovician (Russia)

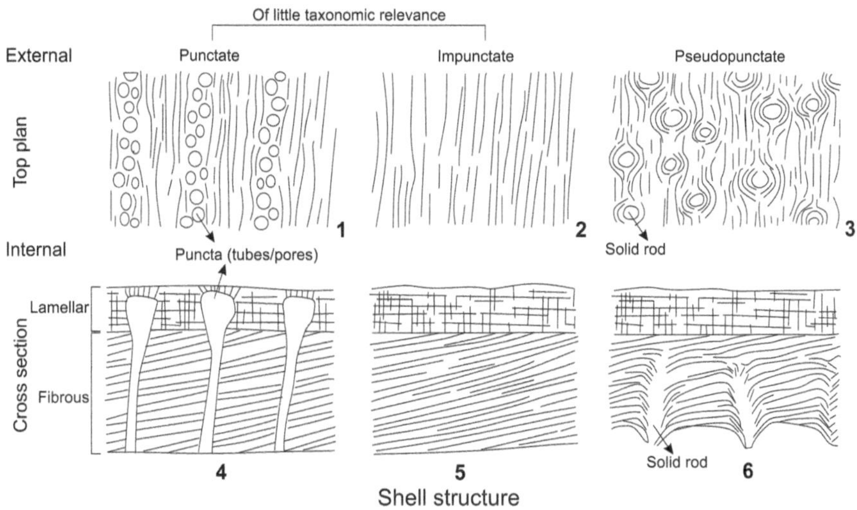

Fig. 8.9 Types of external and internal (cross section) structure of the brachiopod shell

8.3 Terminology

The brachiopod terminology is explained briefly under the following four subheads and illustrated in Figs. 8.2, 8.3, 8.4, 8.5, 8.6, 8.7, 8.8, 8.9, 8.10:

8.3.1 General terms
8.3.2 Shell features
8.3.3 Shell ornamentation
8.3.4 Cardinal area (Posterior region of the shell).

8.3.1 General Terms

8.3.1.1 **Beak**: The commonly pointed extremity of the umbo. It is the valve's apical portion [Fig. 8.2(5)]

8.3.1.2 **Commissure**: A junction line between the margins of valves [Figs. 8.2(2) and 8.3(1–3)]

8.3.1.3 **Dorsal valve**: It is smaller than the pedicle valve and possess a distinctive muscle-scar pattern and Lophophore (as it houses the lophophore, it is also called the Brachial valve) [Fig. 8.2(1, 2)]

8.3.1.4 **Protegula node (= Protegulal node)**: This is the apical portion of the adult shell and the site of Protegulum and of later shell growth [Fig. 8.4 (1)]

8.3.1.5 **Protegulum**: This is the first formed shell of organic material secreted by both mantles [Fig. 8.2(4)]

8.3.1.6 **Umbo**: This is the apical portion of a valve that houses the beak [Fig. 8.2(2)]

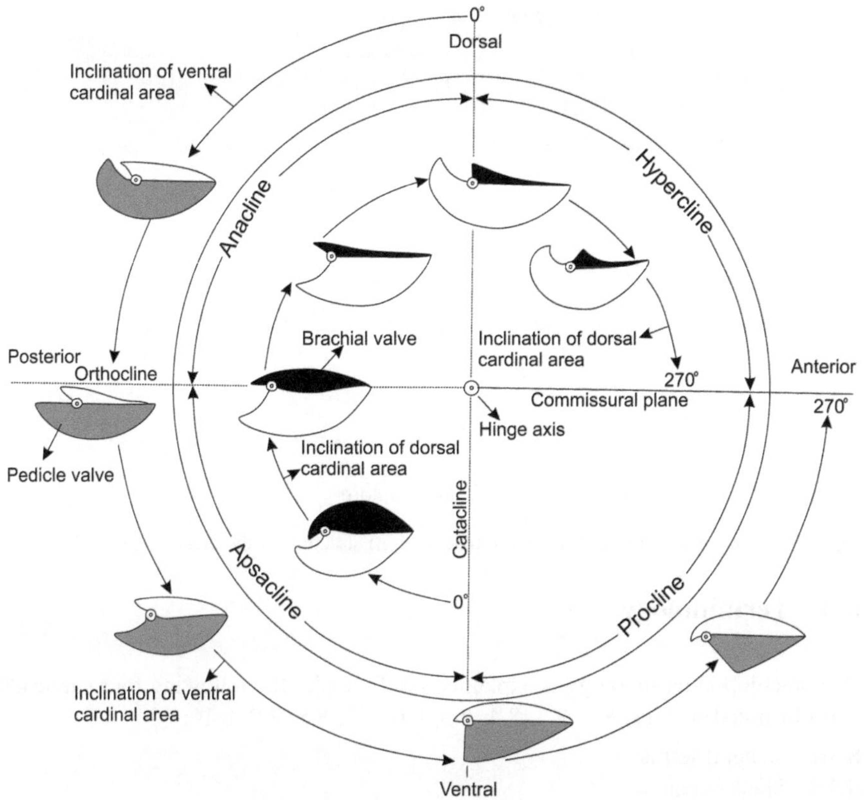

Fig. 8.10 Inclination of cardinal area or of pseudointerarea

8.3.1.7 **Valve**: One of the two separate halves that make up the brachiopod shell [Fig. 8.2(1)]

8.3.1.8 **Ventral valve (= Pedicle valve)**: Larger than the dorsal valve and containing teeth (when present), and the pedicle that commonly emerges from it (as it houses the pedicle, it is also called the pedicle valve) [Fig. 8.2(1)].

8.3.2 Shell Features

8.3.2.1 **Fold**: Major elevation of the valve's surface; externally convex in transverse profile and radial from the umbo [Fig. 8.3(23)]

8.3.2.2 **Growth lines**: Concentric line on outer shell's surface formed when the forward growth of shell temporarily ceased [Fig. 8.2(3)]

8.3.2.3 **Sulcus**: Major depression of the valve's surface; externally concave in transverse profile [Fig. 8.3(23)].

8.3.3 Shell Ornamentation

8.3.3.1 **Costa (pl., costae)**: These are the first-formed radial ridge on the shell's surface; often interchangeably used for any coarse rib [Fig. 8.4(1)]

8.3.3.2 **Costate**: Radially disposed costae [Fig. 8.4(13)]

8.3.3.3 **Costella (pl., costellae)**: Costellae are a fine ribs; Capilla/Capillate are very fine ribs (see [Fig. 8.4(1, 6)]). Costellate possess Costellae. Costella is a radial ridge on the shell's surface originating later than costa bifurcation of costa or costella or formed by the intercalation between earlier formed ribs [Fig. 8.4(1)]

8.3.3.4 **Fascicostellate**: These are Costae and Costellae arranged into bundles [Fig. 8.4(1)]

8.3.3.5 **Parvicostellate**: Numerous Costellae arising by intercalation between widely spaced costae [Fig. 8.4(1)]

8.3.3.6 **Plication**: These are prominent undulation of the commissure. On the shell's interior, it is reflected as a crest directed dorsally [Fig. 8.3(2)] and is often associated with a fold on the dorsal and sulcus on the ventral valve

8.3.3.7 **Ramicostellate**: Having a Costellae on the shell resulting from branching [Fig. 8.4(1)]

8.3.3.8 **Ruga (pl., rugae)**: These are concentric and rarely oblique wrinkling noted on the shell's exterior surface [Fig. 8.4(16)].

8.3.4 Cardinal Area (Posterior Region of the Shell)

8.3.4.1 **Cardinal process**: Blade or variably shaped boss of secondary shell. The cardinal process is positioned medially in the posterior end of the brachial valve. Its function is the separation or attachment of paired diductor muscles (muscles that open valves) [Fig. 8.5(4)]

8.3.4.2 **Cardinalia**: Placed in the posterio-median region of the dorsal valve, these are structures of secondary shell that are associated with the support of lophophore, muscle attachment, and articulation [Fig. 8.5(4)]

8.3.4.3 **Chilidium**: It is a crescentic-shaped plate that covers the top of the notothyrium. When present, it is convex externally and spans for variable distances, ventrally, over the proximal end of the cardinal process and chilidial plates [Fig. 8.5(15, 16)]

8.3.4.4 **Crura (sing., Crus)**: These are paired processes that support the distal end of the lophophore [Fig. 8.5(18)]

8.3.4.5 **Crural plates**: A plate extending from the inner edge of the outer hinge plate or crural base to the floor of the dorsal valve [Fig. 8.5(18)]

8.3.4.6 **Deltidial plates**: These are two plates that are medially placed from the margins of the delthyruim, and partly or completely closing the latter [Fig. 8.5(13, 14)]

8.3.4.7 **Delthyruim**: An aperture bisecting the ventral cardinal area or pseudointerarea. It serves as a pedicle opening [Fig. 8.5(13, 14)]

8.3.4.8 **Hinge line**: It is straight, parallel with hinge axis, and constitutes the posterior margin of the shell. Previously used as synonym for cardinal margin [Figs. 8.3(44, 46); 8.5(1, 7)]

8.3.4.9 **Homeochilidium**: An externally convex triangular plate closing almost all or only the apical part of the notothyrium (best seen in Paterinides) [Fig. 8.5(6)]

8.3.4.10 **Homeodeltidium**: An externally convex triangular plate closing almost all or only the apical part of the delthyrium (best seen in Paterinides) [Fig. 8.5(6)]

8.3.4.11 **Inclination of cardinal area or of pseudointerarea**: These are terms used to describe the condition of valves (dorsal or ventral) when viewing the specimen in a lateral profile with beaks to left, and the dorsal (brachial) valve uppermost. Then, referring the cardinal area to its position within one of four quadrants defined by the commissure plane and plane normal to it and symmetry plane, touching base of cardinal areas (Fig. 8.10). It is Orthocline when the cardinal area is lying on the continuation of the commissure plane. Thence, moving clockwise, if cardinal area in the first quadrant, it is weakly to strongly Apsacline; Catacline is at 90° to the Orthocline; and continuing counterclockwise into the bottom right quadrant, the cardinal area is weakly to strongly Procline

8.3.4.12 **Notothyrium**: A median subtriangular opening bisecting the dorsal cardinal area or pseudointerarea [Fig. 8.5(6, 7)]

8.3.4.13 **Palintrope**: A curved surface of shell, bounded by beak ridges and cardinal margin of astrophic shells [Fig. 8.5(2)]

8.3.4.14 **Pedicle foramen**: Subcircular to circular perforation of the shell through which the pedicle passes [Figs. 8.2(3), 8.5(2, 3, 24–29)]

8.3.4.15 **Pedicle groove**: It divides the ventral pseudointerarea medially thereby allowing passage for the pedicle; the groove is subtriangular in shape [Fig. 8.5(19)]

8.3.4.16 **Pedicle**: It is a cuticle-covered, stalk-like appendage that is variably developed and protrudes from the pedicle valve [Fig. 8.2(1)]

8.3.4.17 **Pseudointerarea**: In some inarticulated brachiopods, it forms the flattened, posterior sector of the shell and is secreted by the posterior sector of the mantle [Fig. 8.5(20)]

8.3.4.18 **Socket**: These are pits for the accommodation of the hinge teeth within the posterior margin of the dorsal valve [Fig. 8.5(4)]

8.3.4.19 **Spondylium**: It is a trough shaped, spoon-like apparatus composed of dental plates found in the pedicle valve (in the beak region); it is a curved calcareous platform for muscle attachment and best noted in Pentamerids [Fig. 8.5(22, 24, 25)].

8.4 Classification

Brachiopods have a long and varied geological history going back to the Early Cambrian. Over 12,000 fossil species and ~350 living species have been recorded, belonging to nearly 6000 genera; a great majority of the modern brachiopods are the Rhynchonelliforms (Articulata, excluding Craniida) (see also Williams et al. 2000).

The brachiopods experienced significant convergent evolution and reversals (where the latest group loses a characteristic feature that it acquired from an intermediate group and reverts back to the characteristics of the older group). Hence, defining higher levels of classification (such as Order) would be premature. Hence, workers have a recommended a bottom–up approach that identifies genera and then places them into intermediate groups (Carlson 2001). But, others are of the view that some characteristic patterns are stable enough to merit higher level classifications, although the fundamental premise, "*what constitutes higher-level classification*" is still a matter of intense debate (see also Carlson 2001).

The "*traditional*" classification was defined in 1869 (see ITIS), but the 1990s saw the establishment of two more approaches (see Table 8.3; see also Carlson 2001). These are

Table 8.3 Traditional classification

Inarticulata[a]	Articulata
No hinge	Hinge present
Complex muscle system	Simple muscle system
Common in Cambrian	Common since Ordovician-Devonian
Shell $Ca_3(PO_4)^2$, chitin	Calcite shell
Digestive system has anus	No anus
Cambrian–recent	Cambrian-recent
~220 genera	~3200 genera
Class inarticulata:	Class articulata:
Most common brachiopods of the Cambrian that eventually gave way to the articulates (the "Paleozoic fauna") during the Ordovician. The shell is composed of chitinophosphate (calcium phosphate embedded in an organic matrix), although in several groups a calcium carbonate skeleton is also secreted. The valves are almost always simple, without complex ribbing, spines, or other types of ornamentation; held together by muscles and body wall. The Cardinal process is absent	They are a diverse and complex class that makes up ~958 % of all the known brachiopods. They remained benthic filter-feeders for most of the Paleozoic. Shell is made of calcium carbonate. They are characterized by an articulated hinge structure in the posterior. In some brachiopods the hinge is either straight (strophic) or curved (astrophic). Teeth in pedicle valve and socket in brachial valve
This class contains five orders, only three are commonly recorded and only two extant orders (Lingulida and Acrotretida) have fossil records that go beyond the Early Paleozoic	Seven orders are recognized

[a]Lingula is the oldest "living fossil"

1. In the "*traditional*" classification, the Articulata have toothed hinges (i.e., hinged valves) between the valves, while the hinges of the Inarticulata are held together only by muscles (i.e., unhinged valves; lacking tooth-and-socket articulation) (see also Moore 1965). In inarticulates the valves are mostly of chitinophosphatic composition while articulates possess calcareous valves (see Table 8.3).
2. A 1990s classification based on shell composition placed Craniida and "articulate" brachiopods within Calciata; the former two have calcitic shells. Similarly, based on common chitin and calcium phosphatic composition, the Lingulida and Discinida were combined under Lingulata (see also Carlson 2001) (see Table 8.4).
3. Another tripartite 1990s classification scheme has Craniida as a separate group—the Craniformea. Here, Linguliformea includes Lingulida and Discinida, and Rhynchonelliformea has Rhynchonellida and Terebratulida (see Table 8.5; Fig. 8.11) (see also Williams et al. 2000; Holmer 2001; Carlson and Leighton 2001; Milsom and Rigby 2009).

Table 8.4 Three high-level classification of brachiopods

<table>
<tr><th>Parameters</th><th colspan="5">Description</th></tr>
<tr><td>Traditional classification</td><td colspan="3">Inarticulata</td><td colspan="2">Articulata</td></tr>
<tr><td>Calciata approach</td><td colspan="2">Lingulata</td><td colspan="3">Calciata</td></tr>
<tr><td>Three-part approach</td><td colspan="2">Linguliformea</td><td>Craniformea</td><td colspan="2">Rhynchonelliformea</td></tr>
<tr><td>Orders</td><td>Lingulida</td><td>Discinida</td><td>Craniida</td><td>Terebratulida</td><td>Rhynchonellida</td></tr>
<tr><td>Hinge</td><td colspan="3">No teeth</td><td colspan="2">Teeth and sockets</td></tr>
<tr><td>Anus</td><td colspan="3">On front of body, attend of U-shaped gut</td><td colspan="2">None</td></tr>
<tr><td rowspan="2">Pedicle</td><td colspan="2">Contains coelom with muscles</td><td>No pedicle</td><td colspan="2">No coelom, muscles where joins</td></tr>
<tr><td>Long burrows</td><td>Short, attached to hard surfaces</td><td>None, cemented to surface</td><td colspan="2">Short, attached to hard surfaces</td></tr>
<tr><td>Periostracum</td><td colspan="2">Glycosaminoglycans and chitin</td><td>Chitin</td><td colspan="2">Proteins</td></tr>
<tr><td>Primary (middle) mineralized layer of shell</td><td colspan="2">Glycosaminoglycans and apatite (calcium phosphate)</td><td colspan="3">Calcite (a form of calcium carbonate)</td></tr>
<tr><td>Inner mineralized layer of shell</td><td colspan="2">Collagen and other proteins, chitinophosphate and apatite</td><td>Calcite</td><td colspan="2">Proteins and calcite</td></tr>
<tr><td>Chaetae around opening of shells</td><td colspan="2">Yes</td><td>No</td><td colspan="2">Yes</td></tr>
<tr><td>Coelom fully divided</td><td colspan="2">Yes</td><td>No</td><td colspan="2">Yes</td></tr>
</table>

Table 8.5 Brachiopod classification (Linguliformea, Craniiformea, and Rhynchonelliformea) (after Holmer 2001; Carlson and Leighton 2001; Williams et al. 2000; Milsom and Rigby 2009)

Subphylum	Order	Major characteristics	Stratigraphical range
Linguliformea	Lingulida	These tongue-shaped (linguliform) to circular in outline suspension feeders are the most conservative brachiopods. *Lingula* (meaning "little tongue") had already appeared in the Tommotian (earliest Cambrian) and is still present, thus, is an excellent of a living fossil. Its tongue-shaped shell (biconvex and oval to squarish in outline) has remained more or less unchanged through time. Its shell is made of calcium phosphate (chitinophosphatic). The valves lack articulation and are separate but, possess a complex set of muscles. A muscular pedicle extrudes out between the valves, either through a foramen, or from the posterior end of the shell	Early Cambrian to Recent
	Acrotretida	The acrotretids, recorded from the Terreneuvian Epoch (earliest Cambrian) are the first hard-shelled organisms, with a fairly constant shell morphology. The shell is characterized by round, convex, and cap-shaped brachial valve and a pedicle valve that is flat and often with a pedicle opening. Its characteristic shell outline and very small size resembles Paterinids but differ in possessing a $CaCO_3$ shell, a distinctly conical shape of the pedicle valve and more particularly by the presence of a minute pedicle foramen, located at or just behind the apex	Early Cambrian to Devonian
	Discinida	These inarticulate, small, long-lived group possess chitinophosphatic shells. The valves are rounded. Brachial valve is convex, cap-shaped and discoid to conical; the pedicle valve is flat. Indenting the posterior margin of the pedicle valve, the pedicle opening is a deep narrow notch. In some, this opening is an enclosed narrow slit located behind the apex	Early Ordovician to Recent
	Siphonotretida	They have subcircular and biconvex valves with spines. They have elongate pedicle foramen	Cambrian to Ordovician
	Paterinida	The tiny Paterinides are representatives of the first evolutionary wave of invertebrates; better known as the Tommotian biota. Morphologically, they are close to the Rhynchonelliforms. However, the Paterinides are a problematic group as they share characters both with inarticulates and articulates. With the inarticulates, they share resemblance in possessing a chitinophosphatic shell, whereas with the articulates, they	Early Cambrian to Late Ordovician

(continued)

Table 8.5 (continued)

Subphylum	Order	Major characteristics	Stratigraphical range
		share their biconvex valves, their essentially straight posterior margin, well-developed pseudointeraea and the medianly arched plates. The Paterinides are characterized by a distinct palintrope on the pedicle valve; the delthyrium is often closed by a homeodeltidium. The paterinides are largely limited to Early–Middle Cambrian, however, some rare forms have also been recorded from the Late Ordovician, as well	
Craniiformea	Craniida	Most Craniidae, a low diversity group, are extinct, but 20 species of this 470-million-year-old lineage have survived. The V*aldiviathyris quenstedti*, is one such living fossil that has remained more or less unchanged for the last 35 Ma. It is also one of the oldest and most long-lived species known to science. Craniidae possess a $CaCO_3$ shell and are attached to the substrate by the pedicle valve; however, they lack the pedicle	Early Ordovician to Recent
	Craniopsida	These are characterized by small oval valves with internal platforms and very distinctive concentric growth lines	Early Cambrian to Early Carboniferous
	Trimerellida	These are characterized by their large size, thick inequivalved aragonitic shell, muscle platforms, and umbonal cavities. Advanced forms have rudimentary articulation. They possess a prominent triangular and transversely striated ventral cardinal area. One of them, the *Trimerella* is an important index fossil for the Silurian	Middle Ordovician to Late Silurian
Rhynchonelliformea	Chileida	These are characterized by inarticulated, calcitic, strophic (wide hinge line) shells. They lack articulatory structures, and umbonal perforation	Early Cambrian-Permian
	Dictyonellida	The valves are biconvex with large umbonal opening covered by a colleplax, a triangular plate—in the umbonal foramen	Ordovician to Permian
	Naukatida	These possess biconvex shells with articulatory structures and an apical foramen	Cambrian
	Obolellida	These resemble Acrotretids. The obolellids are characterized by circular to oval valves, foliated, impunctate, biconvex shell with a very variable disposition of the pedicle, which may emerge from a grooved interarea or a foramen situated either apically or at the anterior to the beak. The obolellids are similar to the lingulids, as they lack hinge, and, in this character, they are also typically inarticulate. Additionally, the obolellids also resemble the inarticulates, in their general external appearance and in the similar	Early to Middle Cambrian

(continued)

Table 8.5 (continued)

Subphylum	Order	Major characteristics	Stratigraphical range
		distribution of muscle scars. As such, the obolellids are a short-lived group restricted to Early Cambrian and whose relationship to the other "inarticulates" is not very well understood	
	Kutorginida	The Kutorginids are the most ancestral (lacking complex morphologic features) of the articulates, and restricted to the Early–Middle Cambrian interval. These ventri-biconvex forms also represent the first evolutionary wave of invertebrates (the Tommotian biota) and are also the earliest order with a calcareous shell. The Kutorginid shell is impunctate with a fibrous microstructure. They possess a strophic (straight) hinge with a prominent beak in the pedicle valve. Similar to the billingsellacean articulates, the Kutorginids also possess a supra-foramen and an interarea (or a pseudointerarea). The delthyrium is very wide and triangular. Both these groups are partially closed by a pseudodeltidium. The brachial valve has a prominent interarea. A plate is present and is similar to the chilidium of other articulates; the plate closes the space beneath the beak	Early to Middle Cambrian
	Orthotetida	First noted in the Ordovician, the Orthotetids reached their acme in the Permian, with the display of bizarre flattened forms like *Leptodus*. The pedicle valve in *Leptodus* somewhat resembles an oyster, and also has an oyster-like behavior (i.e. attachment of its pedicle valve to other shells). The extremely thin brachial valve is flat or slightly concave, fitting neatly over the pedicle valve	Early Ordovician to Late Permian
	Billingsellida	All subsequent articulate brachiopods were derived from this ancestral stock. The group has a laminar secondary shell, very characteristic of articulate brachiopods. However, this fundamental distinction in forms which otherwise, morphologically, are very similar, is quite puzzling	Middle Cambrian to Ordovician
	Strophomenida	These Paleozoic "*petrified butterflies*" were an immensely successful group, with huge morphological diversity. Their shells are concavo–convex, strophic, pseudopunctate with a very well-defined interarea (cardinal area) on one or both valves. Shell remains attached by a pedicle in early growth stages, but becomes free-lying, later in the ontogeny. Costae (fine radiating lines) are present at the surface. The shell, as such, lacks spines	Early Ordovician to Early Carboniferous

(continued)

Table 8.5 (continued)

Subphylum	Order	Major characteristics	Stratigraphical range
	Productida	These cup-shaped brachiopods displayed truly bizarre variations with some forms mimicking even corals. Another characteristic feature is their strongly concavo–convex shells with hollow spines. Some have developed very large spines or have specialized as reef-forming forms. Spines develop either along the posterior margin or are distributed more or less abundantly over other parts of the shell surface. Their unusual convexity of the pedicle valve is another defining character. The anterior portion of both valves are prolonged, thus, producing a feature called Trail. Due to this, the two valves, when closed, are nearly or actually in contact with each other. Hinge line is equal or nearly equal to the width of the shell. Interareas are not conspicuous, even when present in some shells. The productids dominated the Late Paleozoic	Ordovician to Triassic
	Protorthida	These are characterized by well-developed interareas, primitive articulation and ventral free spondylium	Cambrian to Devonian
	Orthida	They are the most ancestral of the articulates. They lack complex morphological features and possess biconvex valves with wide and strophic hinge line, flanked by distinct and well-developed interareas. Most orthids possess an open delthyrium and notothyrium, but in few, they are covered by plates. The cardinalia (the brachial valve interior structures) are usually simple. In orthids, shell is both punctate and Impunctate and shell shape ranges from being subcircular to elliptical. The brachial valve is usually flatter. Shell commonly displays radiating ribs, sulcus and fold structures. Interesting, the basic structure of orthids, throughout the Paleozoic, has remained more or less unchanged. The orthids are very common during the Early Paleozoic; they appeared first in the mid-Early Cambrian and thrived during the Ordovician. The Late Ordovician extinction almost wiped them out, but few managed to survive until the Permian	Early Cambrian to Permian
	Pentamerida	Shell is impunctate with a pentagonal outline. Their robust spondylium (a characteristic feature), a spoon shaped, curved calcareous platform structure at the rear portion of the pedicle valve (for muscle attachment in the beak region) differs from other articulates. Another characteristic feature is their fold and sulcus; these are reversed in regard to other brachiopods; the fold is on the pedicle valve. The Pentamerid valves are strongly biconvex, smooth or finely costate. The hinge line is short; strophic or astrophic. The pentamerids are useful index fossils for the Silurian	Cambrian to Devonian

(continued)

Table 8.5 (continued)

Subphylum	Order	Major characteristics	Stratigraphical range
	Rhynchonellida	These morphologically conservative group, possibly evolved from the pentamerids, radiated in the Middle Ordovician and maintained a relatively constant diversity since then. Most Mesozoic brachiopods are rhynchonellids. Their shell is strongly biconvex (almost globose), usually impunctate, heavily plicate with a very short astrophic hinge line. Externally, the brachial valve is characterized by a fold, and the pedicle valve by a sulcus with strong radiating ribs. The pedicle valve has the beak that usually overlaps that of the brachial valve, in order to allow the shell to open and close. Only the pedicle valve has an interarea, and its delthyrium is usually partially covered by deltidial plates. The internal structure is simple with no complicated support system for the brachidium	Ordovician to Recent
	Atrypida	The Atrypids are characterized by a spiral lophophore and are also the first brachiopods to possess them. Most atrypid shells are smooth with a rounded outline, although some do possess finely costate or medium to coarse plications. Some have alate lamellae or wide "wings," and in some, the lamellae folds into delicate spines. Some possess characteristic ornamentation as in *Atrypa* with concentric growth lines or lamellar outgrowths. The atrypids are characterized by a very inconspicuous interarea on the pedicle valve, an unremarkable beak, a short hinge line and a spiralia whose initial portion bends outward and encloses between them the rest of the spiral cones	Ordovician to Devonian
	Athyridida	The athyrid shells are characteristically smooth with a rounded outline. They are impunctate, and possess prominent beaks but with no perceptible interarea on the pedicle valve. They closely resemble Atrypids as both have smooth shells. However, in spite of their close external resemblance, internally, they differ considerably as their brachidium is more complex; the athyrid brachidium sometimes bears long pointed processes, or loops like the handles of a scissor. Similar is the case with terebratulids (such as with *Dielasma* and *Kingena*) and rhynchonellids (such as with *Spirifer*), with which they only resemble externally (in external morphology). However, with the latter (the Spiriferida proper), the only common character is the spiral brachidium which is also directed laterally outward. The Athyrids thrived during Silurian and Devonian, common early Carboniferous, before they went extinct in the Permian	Ordovician to Jurassic

(continued)

Table 8.5 (continued)

Subphylum	Order	Major characteristics	Stratigraphical range
	Spiriferida	These punctate or impunctate, strophic or astrophic forms have a "winged" appearance due to their very wide hinge line. They have varied morphology, are strongly plicate or costate, and often bear a fold and sulcus. However, their most characteristic feature is a spirally coiled brachidium (a spiral support for the lophophore, the spiralia) that "points" toward the cardinal extremities. The Spiriferids were an important group in Middle and Late Paleozoic times and in the Early Mesozoic. *Spirifer*, *Neospirifer*, *Mucrispirifer* and *Composita* are important index fossils for the Carboniferous–Permian duration	Ordovician to Jurassic
	Thecideida	These are small forms with strophic shells having a complex spiralia including brachial ridges and median septum	Triassic to Recent
	Terebratulida	These punctate, strongly biconvex, smooth or finely costate forms are the most abundant articulate brachiopods today. They are called as "lamp shells" due to their resemblance in shape with ancient oil lamps in Italy and Greece (Mediterranean). The terebratulids possess a prominent beak which usually has a rounded pedicle opening. They also bear a short curved hinge line with no interarea. Besides, Terebratulids, Rhynchonellida is the only other living articulate brachiopod order. However, the rhynchonellids are punctate, bulbous in shape with a simple calcareous loop supporting the lophophore. They are characterized by a very short hinge line, circular or ovoid shell outline with variable surface ornamentation; a single species often shows both smooth and ribbed forms. Additionally, the rhynchonellids have a circular pedicle opening, or foramen, located in the beak. The terebratulids, during early or Middle Silurian times, possibly evolved from the atrypids. Early genera were almost circular to elongate-oval, with smooth or finely costate shells, although, the later forms (during Cretaceous and Tertiary) became coarsely plicate. The terebratulids are recorded since the Early Devonian but became only abundant in the Mesozoic and Cenozoic	Early Devonian to Recent

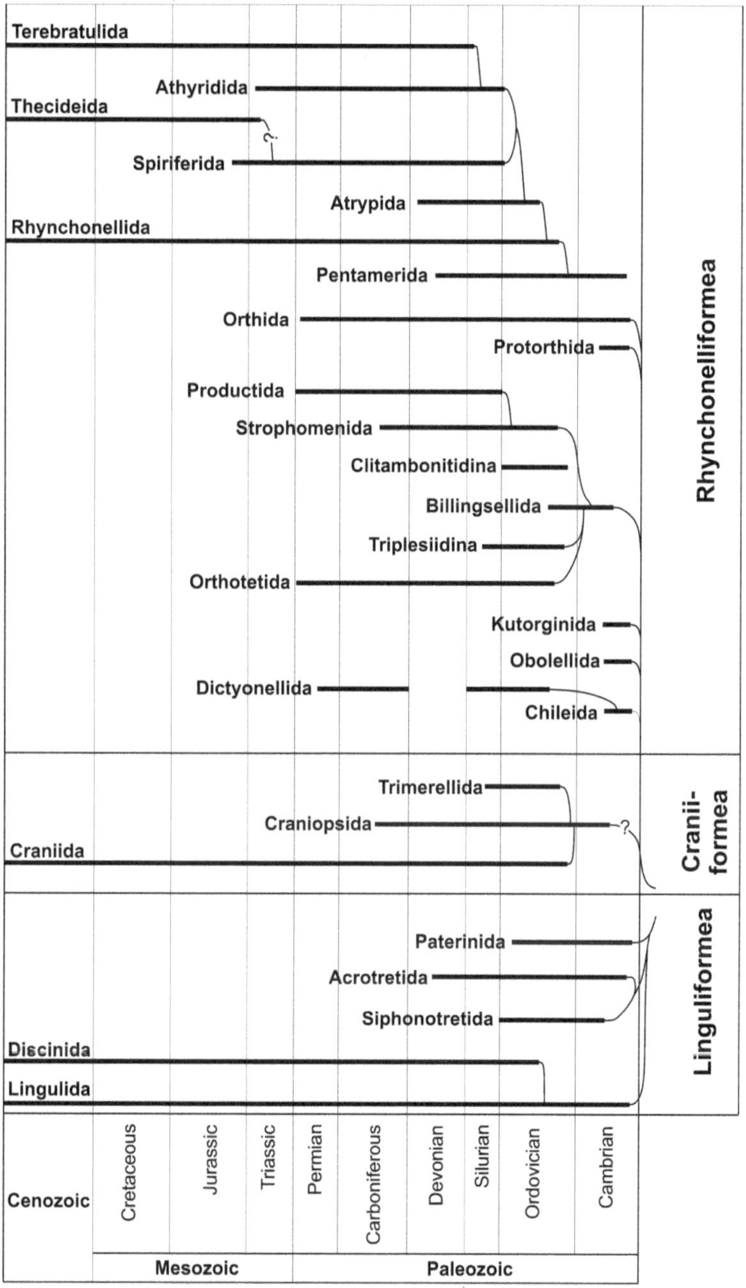

Fig. 8.11 The three-part scheme classification of brachiopods. The current classification places the Craniida in a separate group of its own, the Craniformea. The Lingulida and Discinida are grouped as Linguliformea, and the Rhynchonellida and Terebratulida as Rhynchonelliformea (after various workers; see Williams et al. 2000; Holmer 2001; Carlson and Leighton 2001; Milsom and Rigby 2009)

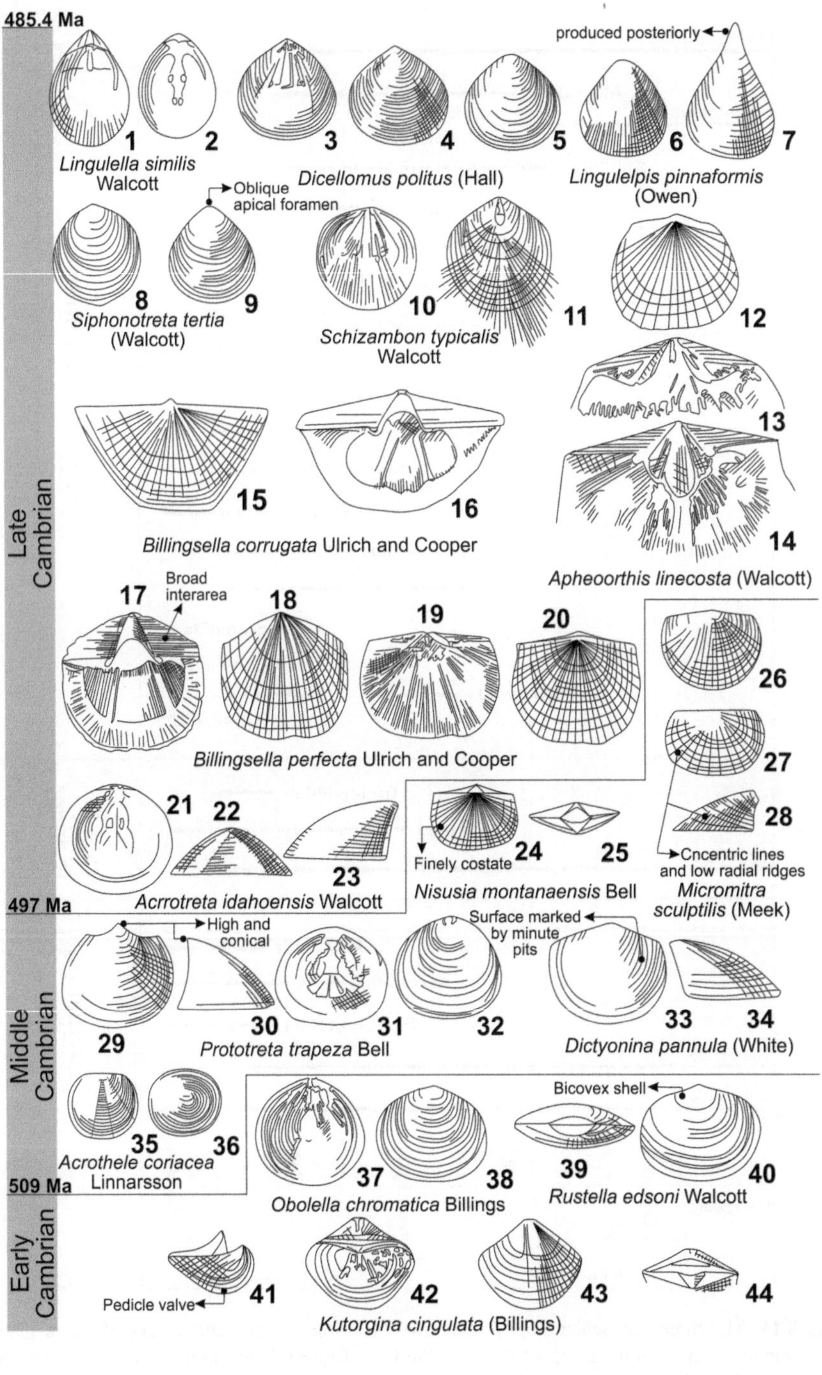
485.4 Ma
produced posteriorly
1 2 3 4 5 6 7
Lingulella similis Walcott
Dicellomus politus (Hall)
Lingulelpis pinnaformis (Owen)
Oblique apical foramen
8 9 10 11 12
Siphonotreta tertia (Walcott)
Schizambon typicalis Walcott
13
15 16
14
Billingsella corrugata Ulrich and Cooper
Apheoorthis linecosta (Walcott)
Late Cambrian
17 Broad interarea
18 19 20
26
27
Billingsella perfecta Ulrich and Cooper
21 22 23
24 25
28
Cncentric lines and low radial ridges
Finely costate
497 Ma
Acrrotreta idahoensis Walcott
Nisusia montanaensis Bell
Micromitra sculptilis (Meek)
High and conical
Surface marked by minute pits
29 30 31 32 33 34
Prototreta trapeza Bell
Dictyonina pannula (White)
Middle Cambrian
Bicovex shell
35 36
Acrothele coriacea Linnarsson
509 Ma
37 38
Obolella chromatica Billings
39 40
Rustella edsoni Walcott
Early Cambrian
Pedicle valve
41 42 43 44
Kutorgina cingulata (Billings)

◀ **Fig. 8.12** Few representatives of Cambrian brachiopods. *1–23* Late Cambrian. *1, 2 Lingulella similis* Walcott, *1* pedicle valve, *2* brachial valve; *3–5 Dicellomus politus* (Hall), *3* and *5* pedicle valve, *4* brachial valve; *6, 7 Lingulelpis pinnaformis* (Owen), *6* brachial valve, *7* pedicle valve produced posteriorly; *8, 9 Siphonotreta tertia* (Walcott), *8* brachial valve, *9* pedicle valve; *10, 11 Schizambon typicalis* Walcott, *10* brachial valve, *11* pedicle valve; *12–14 Apheoorthis linecosta* (Walcott), *12* and *14* pedicle valve, *13* brachial valve, pedicle valve has unclosed delthyrium and a pseudospondylium; *15, 16 Billingsella corrugata* Ulrich and Cooper, *15* interior of a pedicle valve, *16* brachial valve; *17–20 Billingsella perfecta* Ulrich and Cooper, *17, 18* pedicle valve, *19, 20* brachial valve; *21–23 Acrrotreta idahoensis* Walcott, *21* brachial valve, *22, 23* pedicle valve, posterior and side views, respectively. *24–36* Middle Cambrian. *24, 25 Nisusia montanaensis* Bell, *24* brachial valve; *25* posterior view, this is the oldest and well-developed articulate brachiopod; *26–28 Micromitra sculptilis* (Meek), *26* and *28* pedicle valve, *27* brachial valve; *29–32 Prototreta trapeza* Bell, *29, 30* pedicle valve, *31, 32* brachial valve; *33, 34 Dictyonina pannula* (White), pedicle valve; *35, 36 Acrothele coriacea* Linnarsson, *35* brachial valve, *36* pedicle valve. *37–44* Early Cambrian. *37, 38 Obolella chromatica* Billings, *37* brachial valve, *38* pedicle valve; *39, 40 Rustella edsoni* Walcott, *39* posterior view, *40* pedicle valve; *41–44 Kutorgina cingulata* (Billings), *41* Side view, *42* brachial valve, *43* pedicle valve, *44* Interareas

8.5 Geological History

The Earliest Cambrian Tommotian Stage (see Kouchinsky et al. 2001 for Stage details) saw the appearance of the oldest group of brachiopods—the Paterinates (see also Table 8.5). The Paterinates share compositional similarity with linguliformeans, both have an organophosphatic shell. However, the shell structure of the group is quite different, possessing true interareas, delthyria, notothyria, with a functional diductor muscle systems.

The Precambrian is marked by the absence of brachiopods.

The non-articulated groups dominated the Cambrian faunas (Fig. 8.12) in association and as minor constituents by chileids, naukatids, obolellids, kutorginids, billingsellids, protorthids, orthids, and pentamerids (the articulated taxa).

The Ordovician (Figs. 8.13, 8.14, 8.15, 8.16, 8.17) saw the dominance of deltidiodont orthides and strophomenids. The spire (lophophore)-bearing brachiopods reached their acme after the end-Ordovician extinction and by the mid-Palaeozoic (particularly during the Silurian–Devonian: Figs. 8.19, 8.20, 8.21, 8.22, 8.23, 8.24, 8.25, 8.26) in favorable carbonate environments (Fig. 8.18).

Remarkable experimentations were noted in Carboniferous (Mississippian–Pennsylvanian: Figs. 8.27, 8.28, 8.29) and particularly in the Permian (Fig. 8.30) where they mimicked corals [Fig. 8.30(7–9)] or developed extravagant clusters of spines; some reduced their shells, thereby presenting soft tissues to the outside environment. However, this spectacular show of diversity came to an end due to the end-Permian extinction (Fig. 8.1).

In general, for the Mesozoic and Cenozoic brachiopod landscape, the cyrtomatodont rhynchonelliformeans possessing either crurae (rhynchonellids) or loops (terebratulids) dominated.

Over time, the brachiopods experienced five major extinction events, namely Late Ordovician (445–443 Ma), Late Devonian (at the Frasnian–Famennian stage

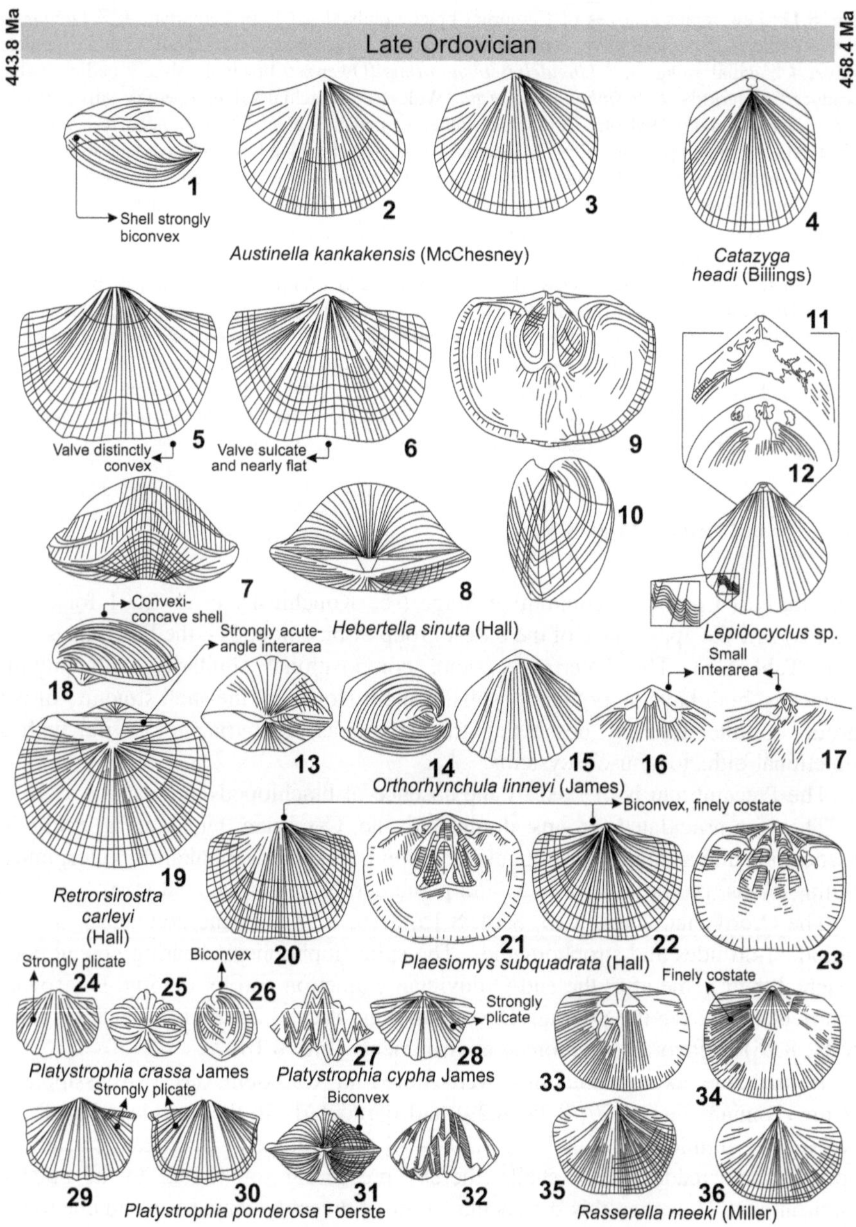
443.8 Ma
Late Ordovician
458.4 Ma
Shell strongly biconvex
Austinella kankakensis (McChesney)
Catazyga headi (Billings)
Valve distinctly convex
Valve sulcate and nearly flat
Convexi-concave shell
Strongly acute-angle interarea
Hebertella sinuta (Hall)
Lepidocyclus sp.
Small interarea
Orthorhynchula linneyi (James)
Biconvex, finely costate
Retrorsirostra carleyi (Hall)
Strongly plicate
Biconvex
Plaesiomys subquadrata (Hall)
Finely costate
Strongly plicate
Platystrophia crassa James
Platystrophia cypha James
Strongly plicate
Biconvex
Platystrophia ponderosa Foerste
Rasserella meeki (Miller)

◀ **Fig. 8.13** Few representatives of Late Ordovician brachiopods. *1–3 Austinella kankakensis* (McChesney), *1* side view, *2* pedicle valve, *3* brachial valve; *4 Catazyga headi* (Billings), brachial valve; *5–10 Hebertella sinuta* (Hall), *5* brachial valve, *6* pedicle valve, *7* anterior view, *8* posterior view, *9* interior of pedicle valve, *10* side view; *11, 12 Lepidocyclus* sp., *11* interior of pedicle valve, *12* interior of brachial valve; *13–17 Orthorhynchula linneyi* (James), *13* posterior view, *14* side view, *15* brachial valve, *16, 17* interior of brachial valve; *18, 19 Retrorsirostra carleyi* (Hall), *18* side view, *19* pedicle valve; *20–23 Plaesiomys subquadrata* (Hall), *20* brachial valve, *21* interior of pedicle valve, *22* pedicle valve, *23* interior of brachial valve; *24–26 Platystrophia crassa* James, *24* brachial valve, *25* pedicle valve, *26* side view; *27, 28 Platystrophia cypha* James, *27* anterior view, *28* brachial valve; *29, 32 Platystrophia ponderosa* Foerste, *29* brachial valve, *30* pedicle valve, *31* posterior view, *32* anterior view; *33–36 Rasserella* meeki (Miller), *33* interior of brachial valve, *34* interior of pedicle valve, *35* brachial valve, *36* pedicle valve

boundary; ~372.2 Ma), Late Permian (252.2 Ma), Late Triassic (202 Ma), and Late Cretaceous (66 Ma) (see Fig. 8.1).

Almost 80 % of the existing brachiopod families were wiped out (in two phases) by the Late Ordovician (late Katian-Hirnantian; about 445–443 Ma) glaciation (see also Rasmussen and Harper 2011). However, there was recovery and subsequent radiation. The deltidiodont groups (such as the orthids and strophomenids), declined, thereafter. On the other hand, the spire-bearing groups (the atrypids, athyridids, and spiriferids, together with the pentamerids), dominated, largely due to the prevailing carbonate environment.

The climatic change at the Frasnian–Famennian stage boundary (~372.2 Ma; Late Devonian) saw the demise of atrypids and pentamerids, and severely affected the orthides and strophomenids. The spiriferids and rhynchonellids due to their preference for deeper water habitats, survived, and staged an impressive recovery, thereafter (see also McGhee 1996; Sokiran 2002). But the most characteristic feature of this stage boundary event was the show of spectacular diversity by recumbent brachiopod megaguilds, largely dominated by the productids.

Most of the ecologically and taxonomically diverse brachiopods saw their demise in the massive Late Permian extinction event (~252.2 Ma) which also led to the disappearance of over 90 % of all living species (see Bottjer et al. 2008). Thereafter, disaster taxa such as lingulids dominated. However, the brachiopod fauna later diversified and were largely dominated by rhynchonellids and terebratulids.

Most of the remaining spiriferids and the last of strophomenids disappeared due the Late Triassic extinction event (~202 Ma). However, the rhynchonellide and terebratulide continued to dominate from the Permian (see Tanner et al. 2004).

Seventy percent of Chalk brachiopod faunas in NW Europe disappeared by the Late Cretaceous event (~66 Ma). However, in the subsequent Danian (Early Paleogene) limestone facies saw radiation that involved many of the preextinction taxa (see also Johansen 1987, 1988).

Appendix 1 gives the list of illustrated specimens mentioning the chapter number, species name, age, and locality along with its figure number within the said chapter.

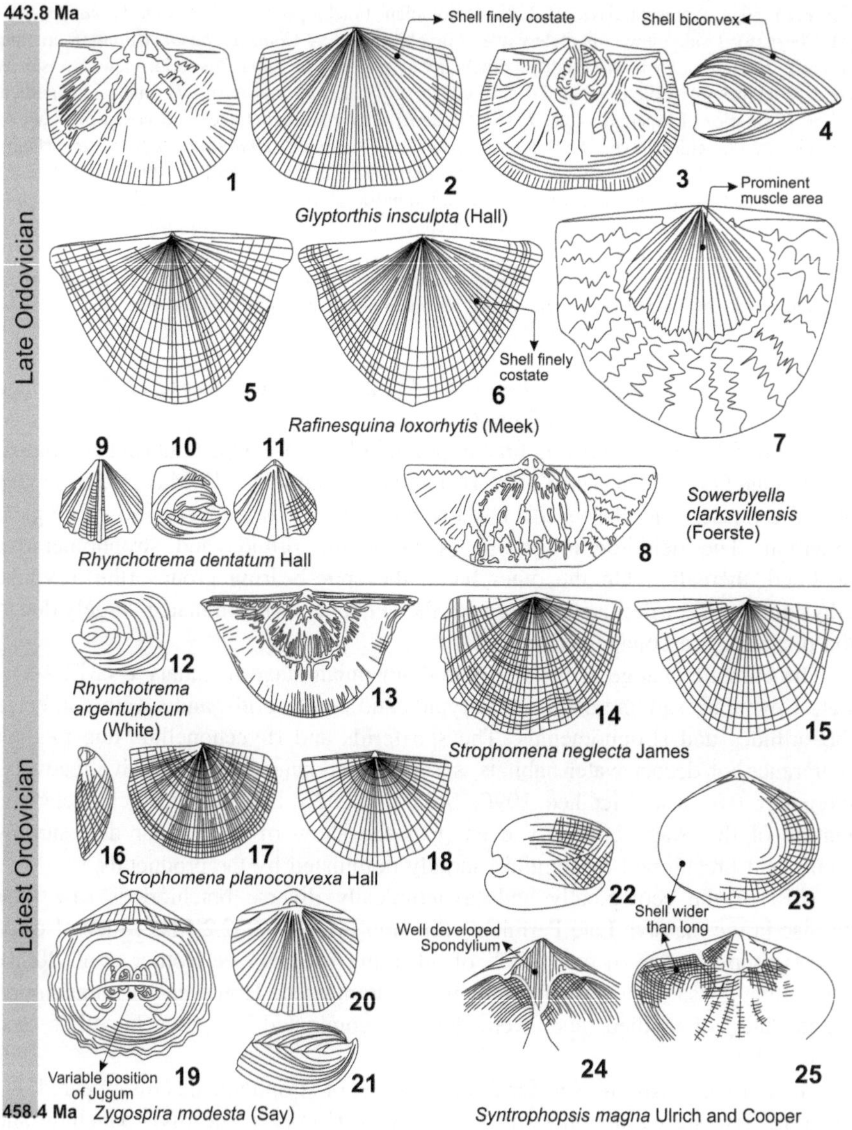

Fig. 8.14 Few representatives of Late Ordovician brachiopods. *1–11* Latest Ordovician (Richmondian). *1–4 Glyptorthis insculpta* (Hall), *1* interior of brachial valve, *2* brachial valve, *3* interior of pedicle valve, *4* side view; *5–7 Rafinesquina loxorhytis* (Meek), *5* brachial valve, *6* pedicle valve, *7* interior of pedicle valve; *8–10 Rhynchotrema dentatum* Hall, *8* pedicle valve, *9* side view, *10* brachial valve; *11 Sowerbyella clarksvillensis* (Foerste), interior of brachial valve. *12–23* Late Ordovician. *12 Rhynchotrema argenturbicum* (White), side view; *13–15 Strophomena neglecta* James, *13* interior of pedicle valve, *14* pedicle valve, *15* brachial valve; *16–18 Strophomena planoconvexa* Hall, *16* side view, *17* pedicle valve, *18* brachial valve; *19–21 Zygospira modesta* (Say), *19* interior of brachial valve, *20* brachial valve, *21* side view; *22–25 Syntrophopsis magna* Ulrich and Cooper, *22* side view, *23* brachial valve, *24* interior of pedicle valve, *25* interior of brachial valve

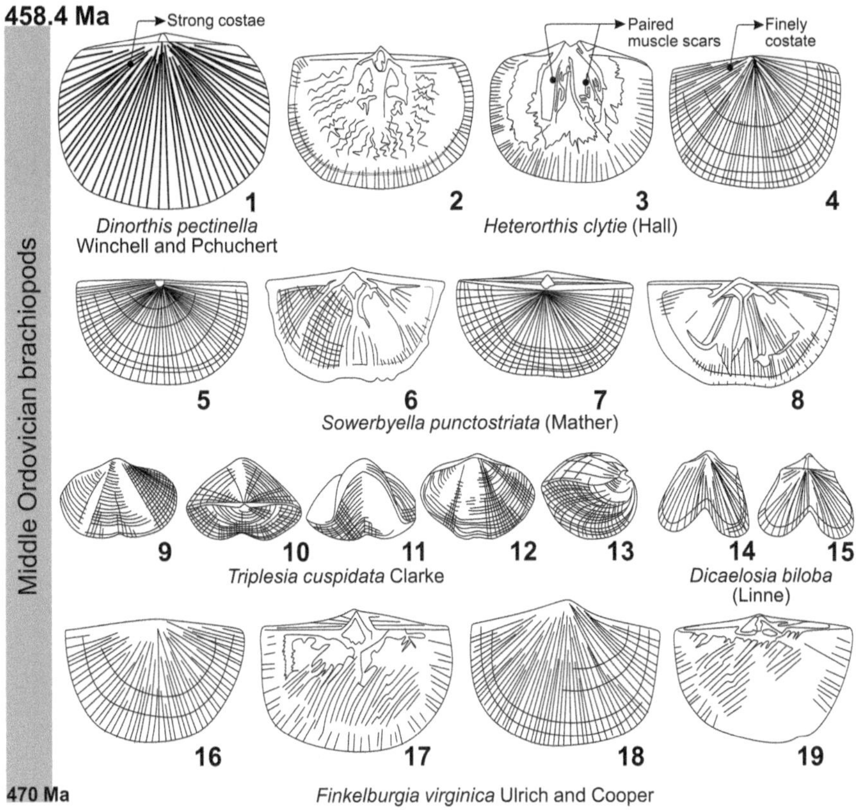

Fig. 8.15 Few representatives of Ordovician brachiopods. *1–8* Latest Ordovician (Trentonian). *1 Dinorthis pectinella* Winchell and Pchuchert, pedicle valve; *2–4 Heterorthis clytie* (Hall), *2* interior of brachial valve, *3* interior of pedicle valve, *4* brachial valve; *5–8 Sowerbyella punctostriata* (Mather), *5* brachial valve, *6* interior of pedicle valve, *7* brachial valve, *8* interior of brachial valve; *9–15* Middle Ordovician. *9–13 Triplesia cuspidata* Clarke; *14*, *15 Dicaelosia biloba* (Linne), *14* pedicle valve, *15* brachial valve; *16–19* Late Cambrian to Early Ordovician, *Finkelburgia virginica* Ulrich and Cooper, *16* brachial valve, *17* interior of pedicle valve, *18* pedicle valve, *19* brachial valve

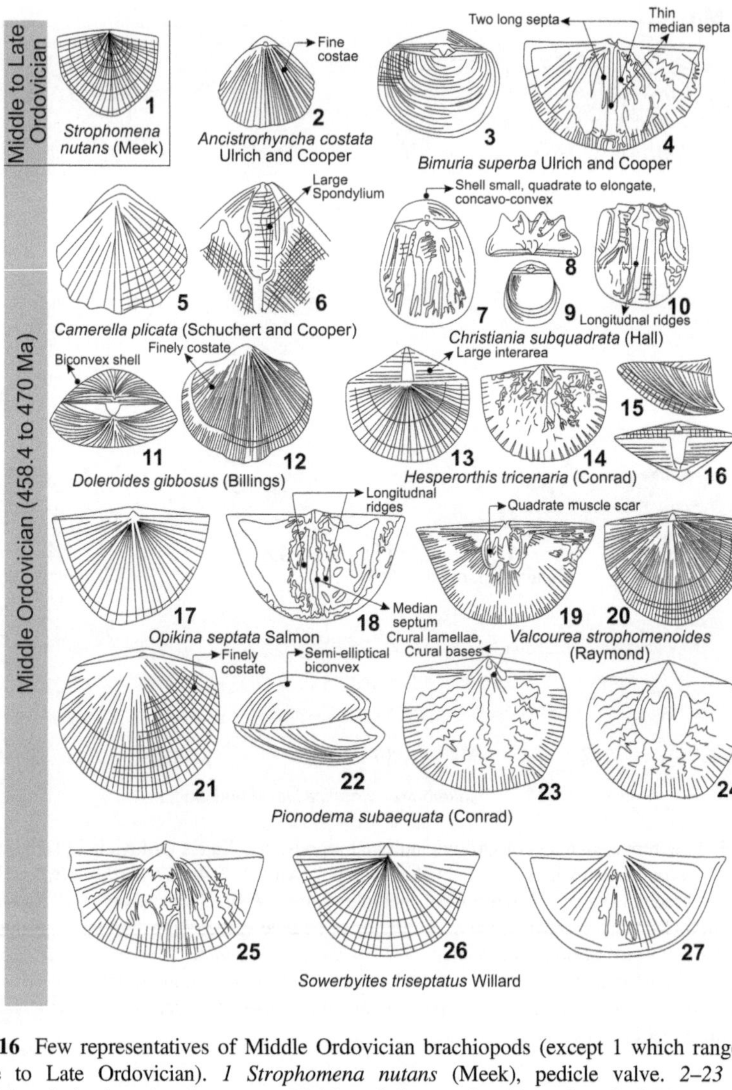

Fig. 8.16 Few representatives of Middle Ordovician brachiopods (except 1 which ranges from Middle to Late Ordovician). *1 Strophomena nutans* (Meek), pedicle valve. *2–23* Middle Ordovician. *2 Ancistrorhyncha costata* Ulrich and Cooper, brachial valve; *3*, *4 Bimuria superba* Ulrich and Cooper, *3* brachial valve, *4* interior of brachial valve; *5*, *6 Camerella plicata* (Schuchert and Cooper), *5* brachial valve, *6* interior of pedicle valve; *7–10 Christiania subquadrata* (Hall), *7* pedicle valve, *8* posterior view of brachial valve; *9* interior of brachial valve, *10* brachial valve; *11*, *12 Doleroides gibbosus* (Billings), *11* posterior view, *12* pedicle valve; *13–16 Hesperorthis tricenaria* (Conrad), *13* brachial valve, *14* interior of brachial valve, *15* side view, *16* posterior view; *17*, *18 Opikina septata* Salmon, *17* brachial valve, *18* interior of brachial valve; *19*, *20 Valcourea strophomenoides* (Raymond), *19* interior of pedicle valve, *20* brachial valve; *21–24 Pionodema subaequata* (Conrad); *25–27 Sowerbyites triseptatus* Willard, *25* interior of pedicle valve, *26* brachial valve, *27* interior of brachial valve

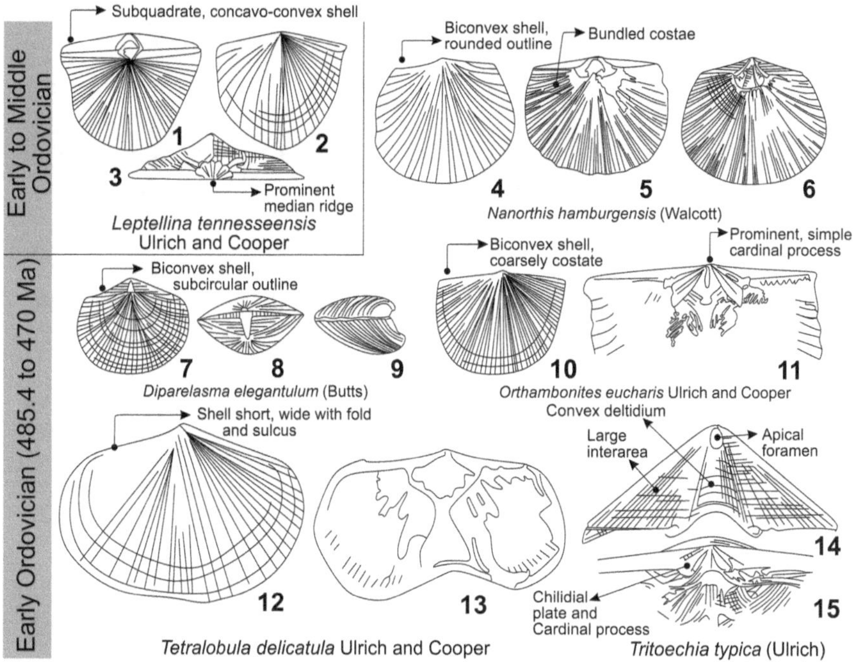

Fig. 8.17 Few representatives of Early Ordovician brachiopods (except *1–3* which ranges from Early to Middle Ordovician). *1–3 Leptellina tennesseensis* Ulrich and Cooper, *1* brachial valve, *2* pedicle valve, *3* posterior view of the interior of brachial valve; *4–6 Nanorthis hamburgensis* (Walcott), *4* brachial valve, *5* interior of brachial valve, *6* interior of pedicle valve; *7–9 Diparelasma elegantulum* (Butts), *7* brachial valve, *8* posterior view, *9* side view; *10, 11 Orthambonites eucharis* Ulrich and Cooper, *10* brachial valve, *11* interior of brachial valve; *12, 13 Tetralobula delicatula* Ulrich and Cooper, *12* interior of pedicle valve, *13* pedicle valve; *14, 15 Tritoechia typica* (Ulrich), *14* Interarea of pedicle valve, *15* Interarea of brachial valve

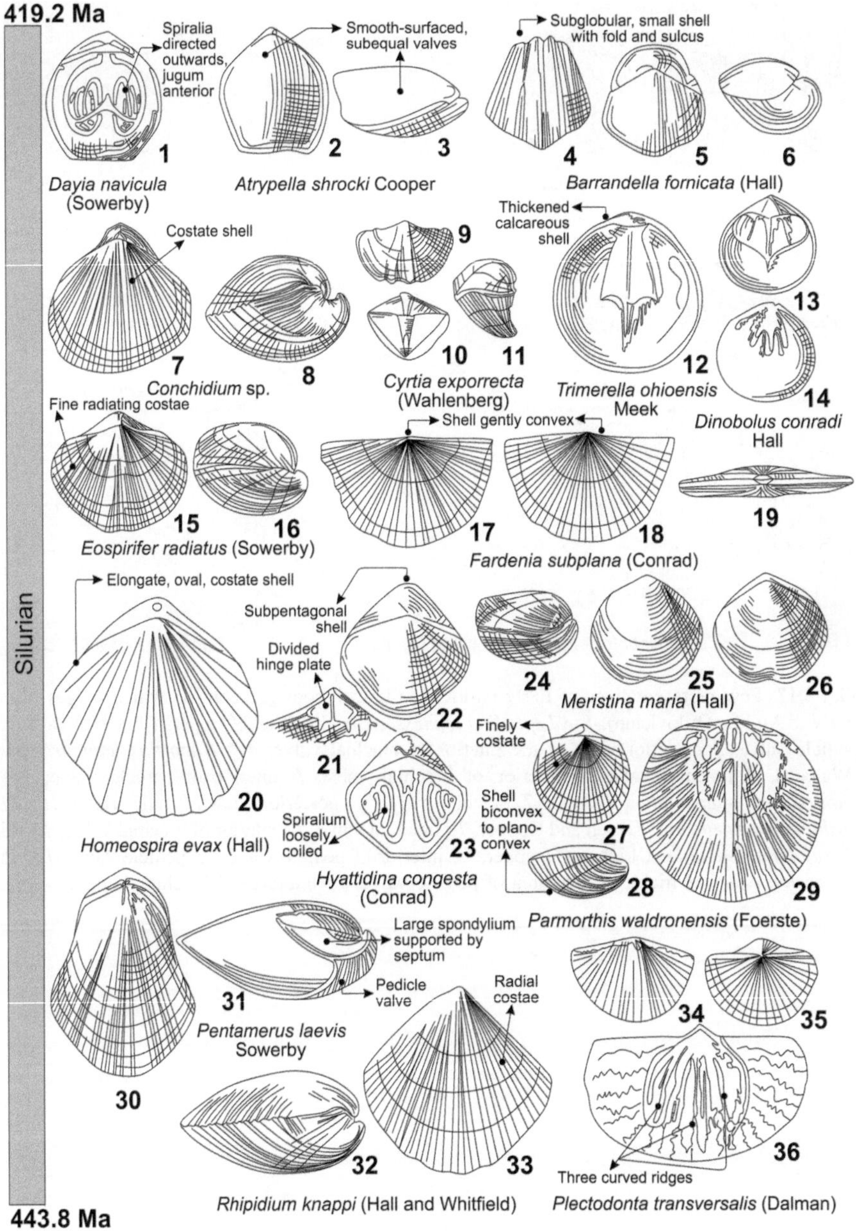
419.2 Ma
Spiralia directed outwards, jugum anterior
Smooth-surfaced, subequal valves
Subglobular, small shell with fold and sulcus
1
2
3
4
5
6
Dayia navicula (Sowerby)
Atrypella shrocki Cooper
Barrandella fornicata (Hall)
Costate shell
Thickened calcareous shell
9
13
7
8
10
11
12
14
Conchidium sp.
Cyrtia exporrecta (Wahlenberg)
Trimerella ohioensis Meek
Dinobolus conradi Hall
Fine radiating costae
Shell gently convex
15
16
17
18
19
Eospirifer radiatus (Sowerby)
Fardenia subplana (Conrad)
Silurian
Elongate, oval, costate shell
Subpentagonal shell
Divided hinge plate
24
25
26
Meristina maria (Hall)
22
21
Finely costate
20
Spiralium loosely coiled
23
Shell biconvex to plano-convex
27
28
29
Homeospira evax (Hall)
Hyattidina congesta (Conrad)
Parmorthis waldronensis (Foerste)
Large spondylium supported by septum
Pedicle valve
31
Radial costae
34
35
Pentamerus laevis Sowerby
30
32
33
36
Three curved ridges
Rhipidium knappi (Hall and Whitfield)
Plectodonta transversalis (Dalman)
443.8 Ma

◀ **Fig. 8.18** Few representatives of Middle Silurian brachiopods (except *1* which is Late Silurian). *1 Dayia navicula* (Sowerby), brachial valve; *2–36* Middle Silurian. *2, 3 Atrypella shrocki* Cooper [Middle Silurian (Niagaran)], *2* side view, *3* brachial valve; *4–6 Barrandella fornicata* (Hall), *4* pedicle valve, *5* side view, *6* brachial valve; *7, 8 Conchidium* sp., *7* brachial valve, *8* side view; *9–11 Cyrtia exporrecta* (Wahlenberg), *9* brachial valve, *10* posterior view, *11* side view; *12 Trimerella ohioensis* Meek, brachial valve: *13, 14 Dinobolus conradi* Hall, *13* interior of pedicle valve, *14* pedicle valve; *15, 16 Eospirifer radiatus* (Sowerby), *15* brachial valve, *16* side view; *17–19 Fardenia subplana* (Conrad), *17* brachial valve, *18* pedicle valve, *19* posterior view; *20 Homeospira evax* (Hall), brachial valve; *21–23 Hyattidina congesta* (Conrad), *21* Divided hinge plate of brachial valve, *22* brachial valve, *23* spiralium of brachial valve; *24–26 Meristina maria* (Hall), *24* side view, *25* pedicle valve, *26* brachial valve; *27–29 Parmorthis waldronensis* (Foerste), *27* brachial valve, *28* side view, *29* interior of brachial valve; *30, 31 Pentamerus laevis* Sowerby, *30* brachial valve, *31* cross section with pedicle valve below; *32, 33 Rhipidium knappi* (Hall and Whitfield), *32* side view, *33* brachial valve; *34–36 Plectodonta transversalis* (Dalman), *34* pedicle valve, *35* brachial valve, *36* interior of brachial valve

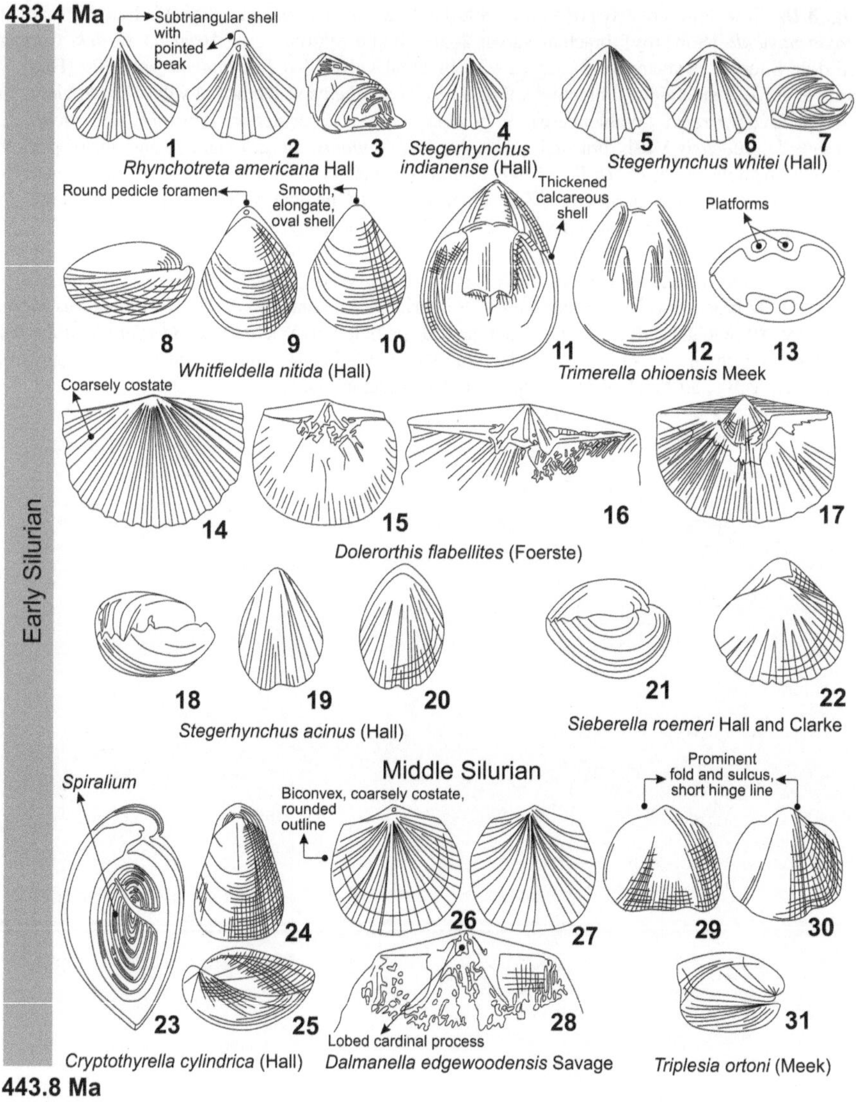

Fig. 8.19 Few representatives of Silurian brachiopods. *1–22* Middle Silurian. *1–3 Rhynchotreta americana* Hall, *1* pedicle valve, *2* brachial valve, *3* side view; *4 Stegerhynchus indianense* (Hall), pedicle valve; *5–7 Stegerhynchus whitei* (Hall), *5* pedicle valve, *6* brachial valve, *7* side view; *8–10 Whitfieldella nitida* (Hall), *8* side view, *9* brachial valve, *10* pedicle valve; *11–13 Trimerella ohioensis* Meek (Middle Silurian, Niagaran), *11* interior of pedicle valve, *12* pedicle valve, *13* transverse section of pedicle valve; *14–17 Dolerorthis flabellites* (Foerste) (Middle Silurian, Niagaran), *14* pedicle valve, *15*, *16* interior of brachial valve, *17* pedicle valve; *18–20 Stegerhynchus acinus* (Hall), *18* side view, *19* pedicle valve, *20* brachial valve; *21*, *22 Sieberella roemeri* Hall and Clarke, *21* side view, *22* brachial valve; *23–30* Early Silurian. *23–25 Cryptothyrella cylindrica* (Hall), *23* section showing spiralium, *24* brachial valve, *25* side view; *26–28 Dalmanella edgewoodensis* Savage, *26* brachial valve, *27* pedicle valve, *28* interior of brachial valve; *29–31 Triplesia ortoni* (Meek), *29* pedicle valve, *30* brachial valve, *31* side view

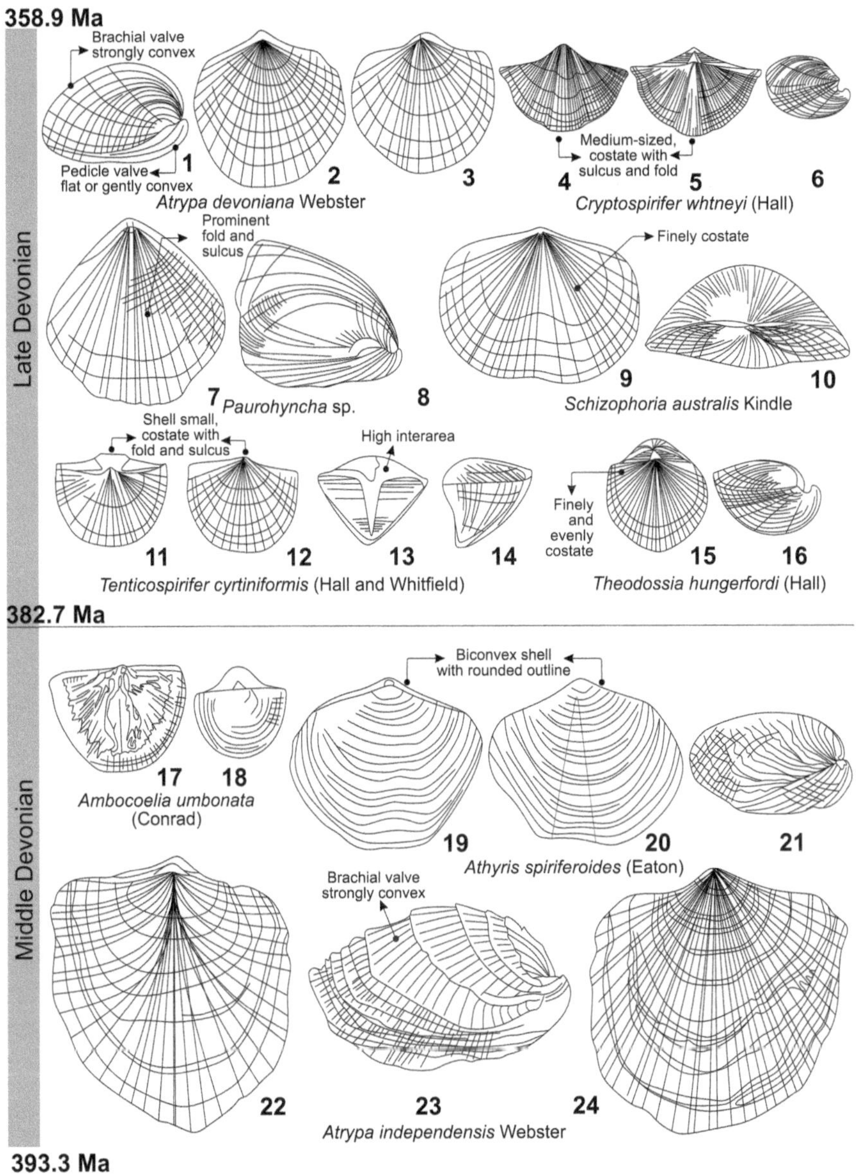

Fig. 8.20 Few representatives of Devonian brachiopods. *1–16* Late Devonian. *1–3 Atrypa devoniana* Webster, *1* side view, *2* pedicle valve, *3* brachial valve; *4–6 Cryptospirifer whtneyi* (Hall), *4* pedicle valve, *5* brachial valve, *6* side view; *7, 8 Paurohyncha* sp., *7* brachial valve, *8* side view; *9, 10 Schizophoria australis* Kindle, *9* pedicle valve, *10* posterior view; *11–14 Tenticospirifer cyrtiniformis* (Hall and Whitfield), *11* pedicle valve, *12* brachial valve, *13* posterior view, *14* side view; *15, 16 Theodossia hungerfordi* (Hall), *15* brachial valve, *16* side view; *17–24* Middle Devonian. *17, 18 Ambocoelia umbonata* (Conrad), *17* interior of brachial valve, *18* brachial valve; *19–21 Athyris spiriferoides* (Eaton), *19* brachial valve, *20* pedicle valve, *21* side view; *22–24 Atrypa independensis* Webster, *22* brachial valve, *23* side view, *24* pedicle valve

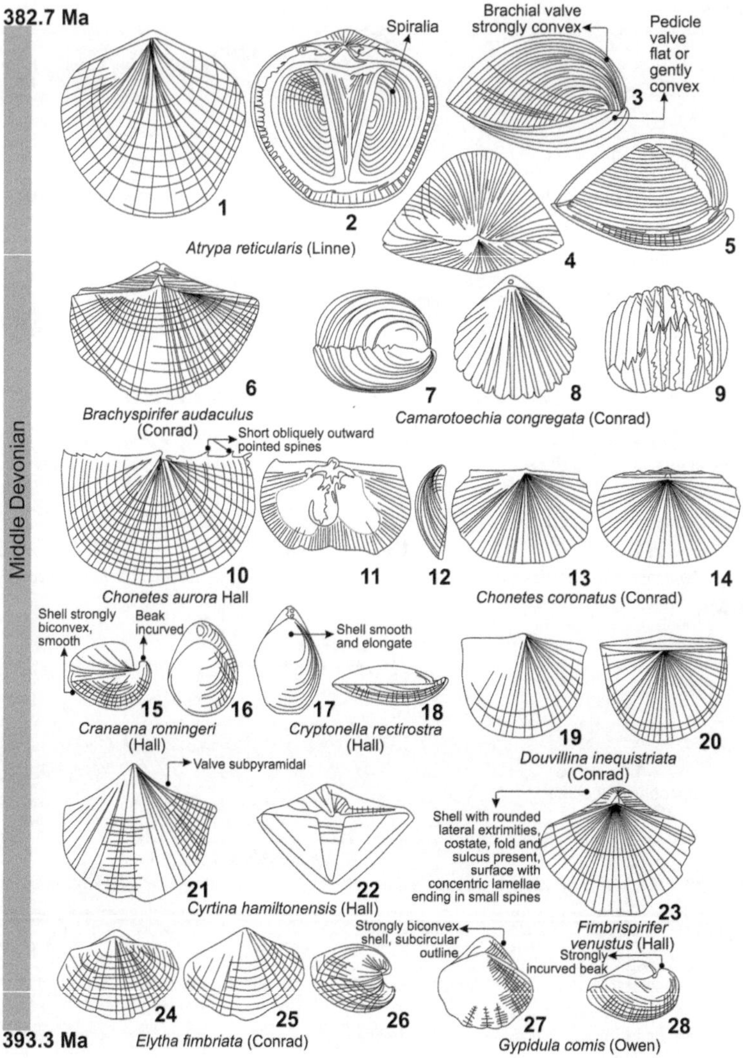

Fig. 8.21 Few representatives of Middle Devonian brachiopods. *1–5 Atrypa reticularis* (Linne), *1* brachial valve, *2* interior of brachial valve showing the spiralia, *3* side view, *4* posterior view, *5* sectional view; *6 Brachyspirifer audaculus* (Conrad), brachial valve; *7–9 Camarotoechia congregata* (Conrad), *7* side view, *8* brachial valve, *9* anterior view; *10 Chonetes aurora* Hall, pedicle valve; *11–14 Chonetes coronatus* (Conrad), *11* interior of brachial valve, *12* side view, *13* pedicle valve, *14* brachial valve; *15*, *16 Cranaena romingeri* (Hall), *15* side view, *16* brachial valve; *17*, *18 Cryptonella rectirostra* (Hall), *17* brachial valve, *18* side view; *19*, *20 Douvillina inequistriata* (Conrad), *19* pedicle valve, *20* brachial valve; *21*, *22 Cyrtina hamiltonensis* (Hall), *21* pedicle valve, *22* posterior view; *23 Fimbrispirifer venustus* (Hall), brachial valve; *24–26 Elytha fimbriata* (Conrad), *24* brachial valve, *25* pedicle valve, *26* side view; *27*, *28 Gypidula comis* (Owen), *27* brachial valve, *28* side view

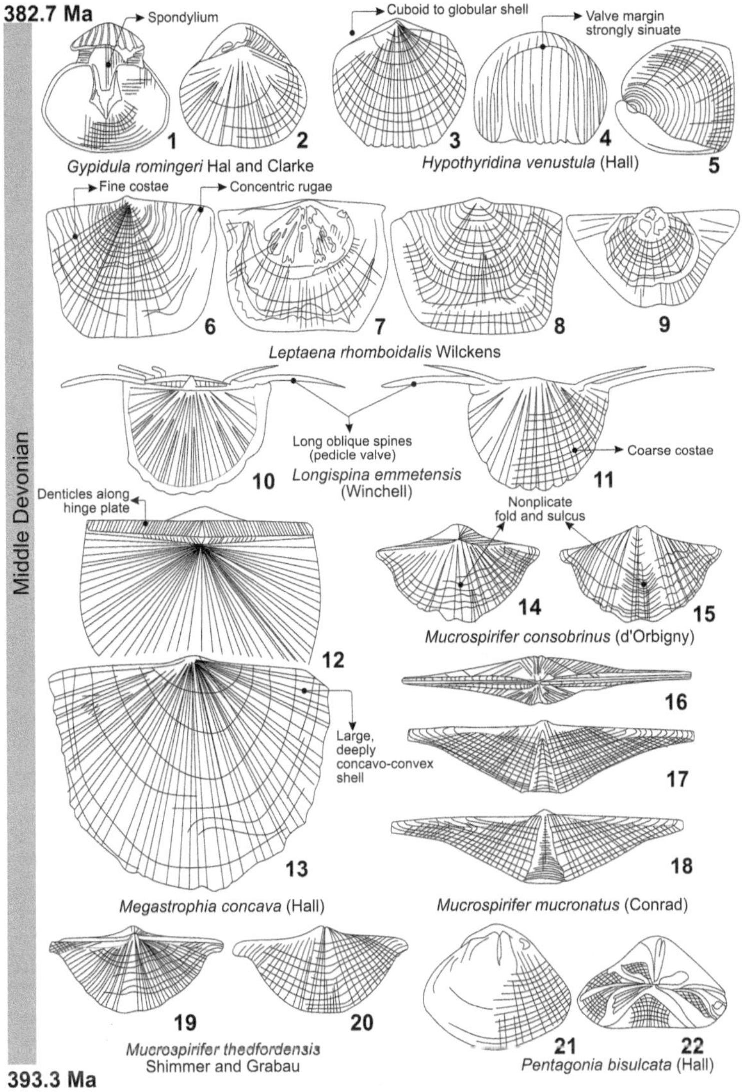

Fig. 8.22 Few representatives of Middle Devonian brachiopods. *1, 2 Gypidula romingeri* Hal and Clarke, *1* interior of brachial valve, *2* brachial valve; *3–5 Hypothyridina venustula* (Hall), *3* brachial valve, *4* anterior view, *5* side view; *6–9 Leptaena rhomboidalis* Wilckens, *6* brachial valve, *7* interior of pedicle valve, *8* pedicle valve, *9* interior of brachial valve; *10, 11 Longispina emmetensis* (Winchell), *10* brachial valve, *11* pedicle valve; *12, 13 Megastrophia concava* (Hall), *12* brachial valve, *13* pedicle valve; *14, 15 Mucrospirifer consobrinus* (d'Orbigny), *14* brachial valve, *15* pedicle valve; *16–18 Mucrospirifer mucronatus* (Conrad), *16* posterior view, *17* pedicle valve, *18* brachial valve; *19, 20 Mucrospirifer thedfordensis* Shimmer and Grabau, *19* brachial valve, *20* pedicle valve; *21, 22 Pentagonia bisulcata* (Hall), *21* brachial valve, *22* posterior view

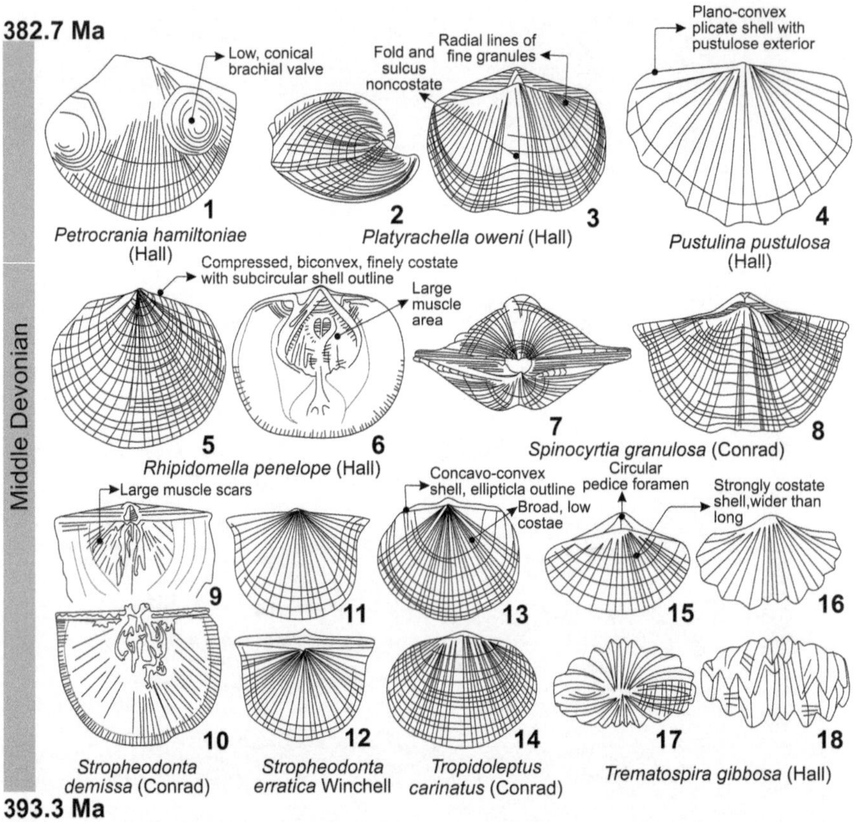

Fig. 8.23 Few representatives of Middle Devonian brachiopods. *1 Petrocrania hamiltoniae* (Hall), two specimens attached to a larger *Stropheodonta* shell, brachial valve; *2*, *3 Platyrachella oweni* (Hall), *2* side view, *3* brachial valve; *4 Pustulina pustulosa* (Hall), brachial valve; *5*, *6 Rhipidomella penelope* (Hall), *5* pedicle valve, *6* interior of pedicle valve; *7*, *8 Spinocyrtia granulosa* (Conrad), *7* posterior view, *8* brachial valve; *9*, *10 Stropheodonta demissa* (Conrad), *9* interior of pedicle valve, *10* interior of brachial valve; *11*, *12 Stropheodonta erratica* Winchell; *13*, *14 Tropidoleptus carinatus* (Conrad), *13* brachial valve, *14* pedicle valve; *15–18 Trematospira gibbosa* (Hall), *15* brachial valve, *16* pedicle valve, *17* posterior view, *18* anterior view

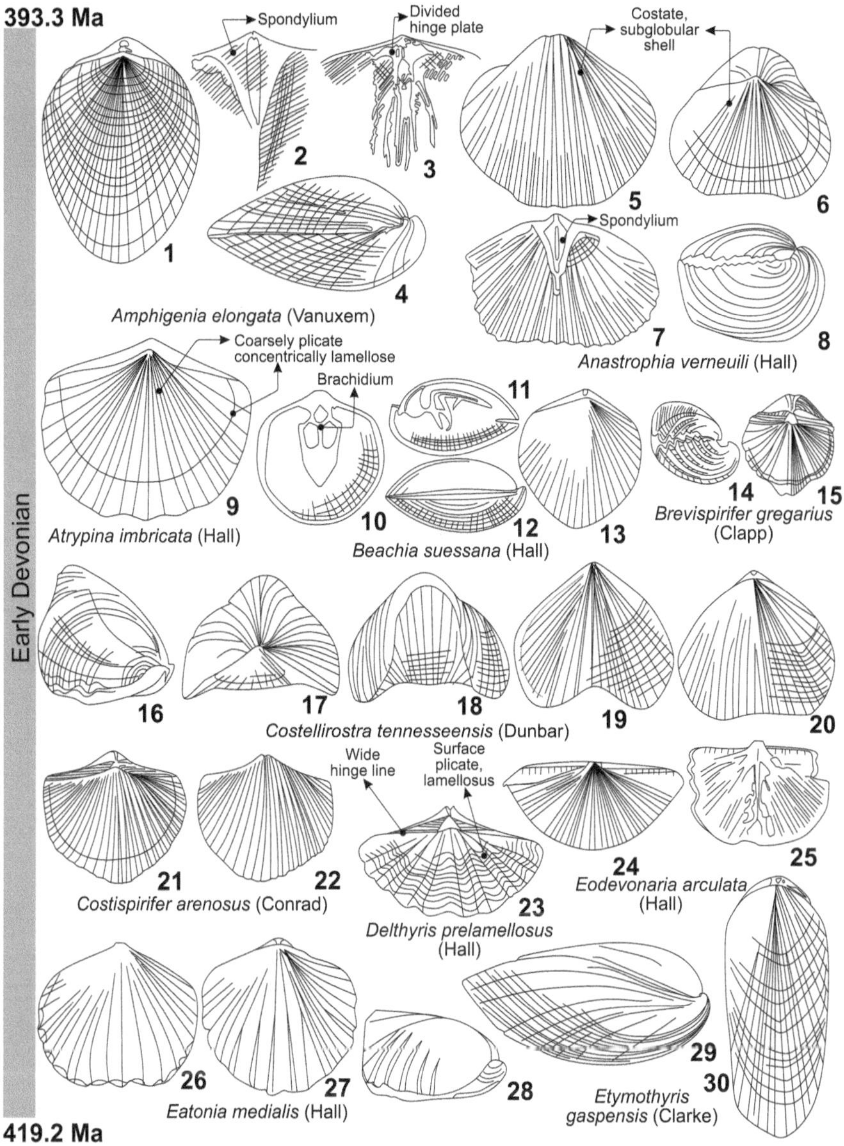

Fig. 8.24 Few representatives of Early Devonian brachiopods. *1–4 Amphigenia elongata* (Vanuxem), *1* brachial valve, *2* interior of pedicle valve, *3* interior of brachial valve, *4* side view; *5–8 Anastrophia verneuili* (Hall), *5* pedicle valve, *6* brachial valve, *7* interior of pedicle valve, *8* side view; *9 Atrypina imbricata* (Hall), brachial valve; *10–13 Beachia suessana* (Hall), *10* interior of brachial valve, *11* sectional view, *12* side view, *13* brachial valve; *14, 15 Brevispirifer gregarius* (Clapp), *14* side view, *15* brachial valve; *16–20 Costellirostra tennesseensis* (Dunbar), *16* side view, *17* posterior view, *18* anterior view, *19* pedicle valve, *20* brachial valve; *21, 22 Costispirifer arenosus* (Conrad), *21* brachial valve, *22* pedicle valve; *23 Delthyris prelamellosus* (Hall), brachial valve; *24, 25 Eodevonaria arculata* (Hall), *24* pedicle valve, *25* interior of brachial valve; *26–28 Eatonia medialis* (Hall), *26* pedicle valve, *27* brachial valve, *28* side view; *29, 30 Etymothyris gaspensis* (Clarke), *29* side view, *30* brachial valve

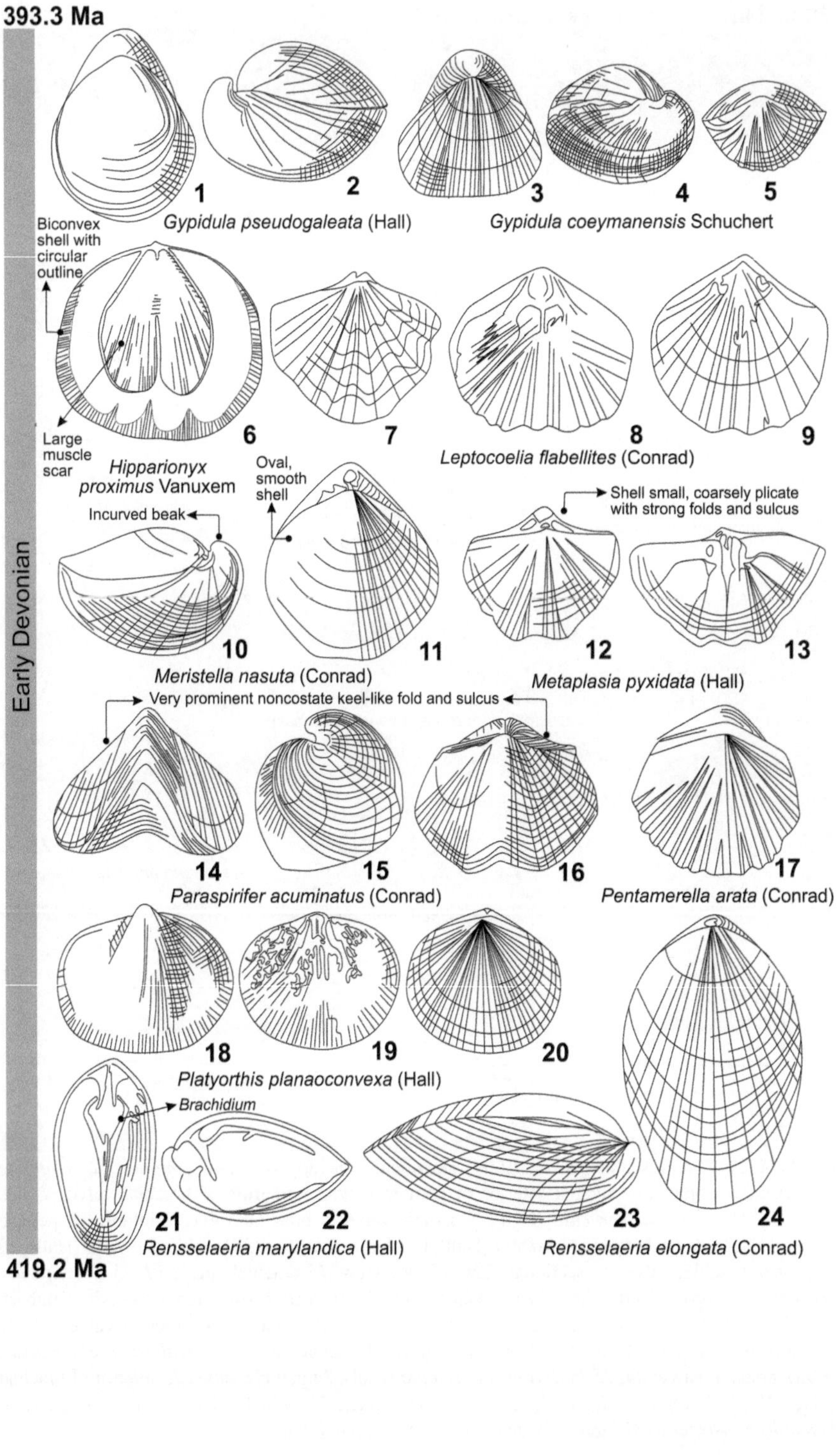

393.3 Ma
Early Devonian
419.2 Ma
1
2
3
4
5
Gypidula pseudogaleata (Hall)
Gypidula coeymanensis Schuchert
Biconvex shell with circular outline
Large muscle scar
6
7
8
9
Hipparionyx proximus Vanuxem
Leptocoelia flabellites (Conrad)
Oval, smooth shell
Incurved beak
Shell small, coarsely plicate with strong folds and sulcus
10
11
12
13
Meristella nasuta (Conrad)
Metaplasia pyxidata (Hall)
Very prominent noncostate keel-like fold and sulcus
14
15
16
17
Paraspirifer acuminatus (Conrad)
Pentamerella arata (Conrad)
18
19
20
Platyorthis planaoconvexa (Hall)
Brachidium
21
22
23
24
Rensselaeria marylandica (Hall)
Rensselaeria elongata (Conrad)

◀ **Fig. 8.25** Few representatives of Early Devonian brachiopods. *1, 2 Gypidula pseudogaleata* (Hall), *1* brachial valve, *2* side view; *3–5 Gypidula coeymanensis* Schuchert, *3* brachial valve, *4* side view, *5* posterior view; *6 Hipparionyx proximus* Vanuxem, interior of pedicle valve; *7–9 Leptocoelia flabellites* (Conrad), *7* brachial valve, *8* interior of brachial valve, *9* interior of pedicle valve; *10, 11 Meristella nasuta* (Conrad), *10* side view, *11* brachial valve; *12, 13 Metaplasia pyxidata* (Hall). *12* brachial valve, *13* interior view; *14–16 Paraspirifer acuminatus* (Conrad), *14* anterior view, *15* side view, *16* brachial valve; *17 Pentamerella arata* (Conrad), brachial valve; *18–20 Platyorthis planoconvex* (Hall), *18* pedicle valve, *19* interior of brachial valve, *20* brachial valve; *21, 22 Rensselaeria marylandica* (Hall), *21* interior of brachial valve, *22* sectional view; *23, 24 Rensselaeria elongata* (Conrad), *23* side view, *24* brachial valve

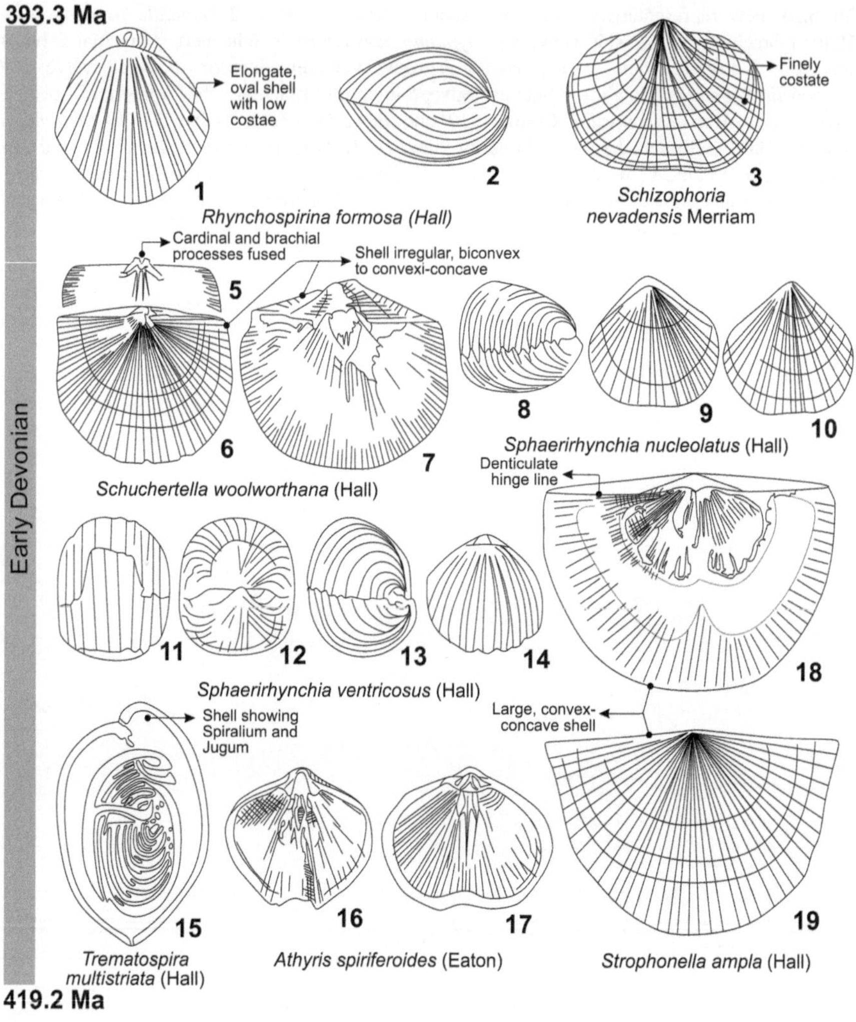

Fig. 8.26 Few representatives of Early Devonian brachiopods. *1*, *2 Rhynchospirina formosa* (Hall), *1* brachial valve, *2* side view; *3 Schizophoria nevadensis* Merriam, pedicle valve; *5–7 Schuchertella woolworthana* (Hall), *5* brachial valve, *6* pedicle valve, *7* interior of pedicle valve; *8–10 Sphaerirhynchia nucleolatus* (Hall), *8* side view, *9* brachial valve, *10* pedicle valve; *11–14 Sphaerirhynchia ventricosus* (Hall), *11* anterior view, *12* posterior view, *13* side view, *14* brachial valve; *15 Trematospira multistriata* (Hall), sectional view; *16*, *17 Athyris spiriferoides* (Eaton), *16* interior of pedicle valve, *17* interior of brachial valve; *18*, *19 Strophonella ampla* (Hall), *18* interior of pedicle valve, *19* brachial valve

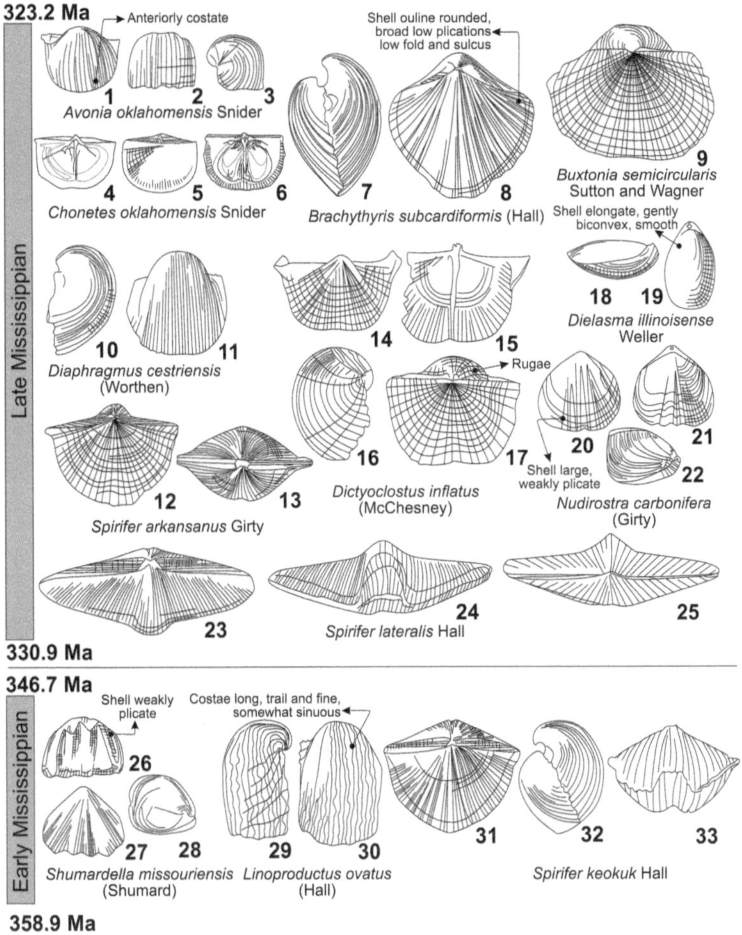

Fig. 8.27 Few representatives of Mississippian brachiopods. *1–25* Late Mississippian. *1–3 Avonia oklahomensis* Snider, *1* pedicle valve, *2* anterior view, *3* side view; *4–6 Chonetes oklahomensis* Snider, *4* interior of pedicle valve, *5* brachial valve, *6* interior of brachial valve; *7, 8 Brachythyris subcardiformis* (Hall), *7* side view, *8* brachial valve; *9 Buxtonia semicircularis* Sutton and Wagner (Late Mississippian, Chesteran), brachial valve; *10, 11 Diaphragmus cestriensis* (Worthen) (Late Mississippian, Chesteran), *10* side view, *11* pedicle valve; *12, 13 Spirifer arkansanus* Girty, *12* brachial valve, *13* posterior view; *14–17 Dictyoclostus inflatus* (McChesney) (Late Mississippian, Chesteran), *14* posterior view, *15* interior of brachial valve, *16* side view, *17* brachial valve; *18, 19 Dielasma illinoisense* Weller (Late Mississippian, Chesteran), *18* side view, *19* brachial valve; *20–22 Nudirostra carbonifera* (Girty), *20* pedicle valve, *21* brachial valve, *22* side view; *23–25 Spirifer lateralis* Hall (Late Mississippian, Meramecian), *23* brachial valve, *24* anterior view, *25* posterior view; *26–33* Early Mississippian. *26–28 Shumardella missouriensis* (Shumard), *26* anterior view, *27* pedicle valve, *28* side view; *29, 30 Linoproductus ovatus* (Hall) (Early Mississippian, Osagian), *29* side view, *30* pedicle valve; *31–33 Spirifer keokuk* Hall (Early Mississippian, Osagian), *31* brachial valve, *32* side view, *33* anterior view

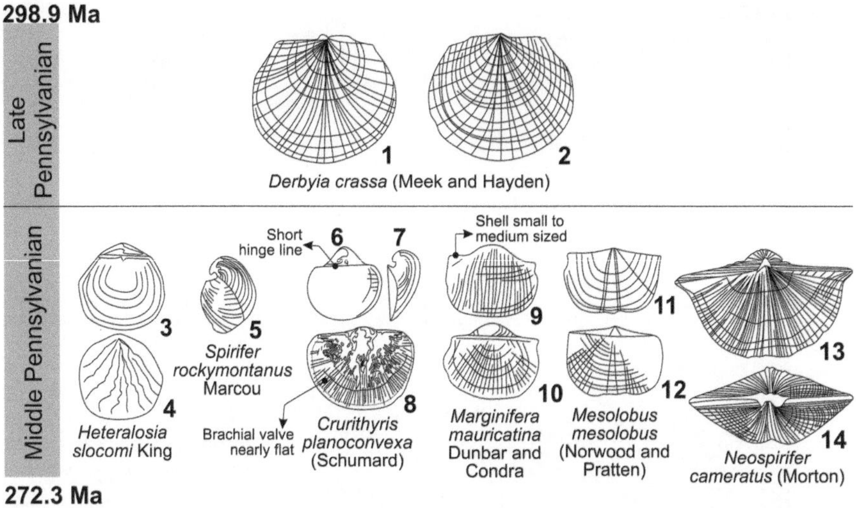

Fig. 8.28 Few representatives of Middle Pennsylvanian and Permian brachiopods. *1*, *2* Late Pennsylvanian, *Derbyia crassa* (Meek and Hayden) (Late Pennsylvanian, Virgilian), *1* brachial valve, *2* pedicle valve; *3–14* Middle Pennsylvanian. *3*, *4 Heteralosia slocomi* King, *3* brachial valve, *4* pedicle valve; *5 Spirifer rockymontanus* Marcou, side view; *6–8 Crurithyris planoconvexa* (Schumard), *6* brachial valve, *7* side view, *8* interior of brachial valve; *9*, *10 Marginifera mauricatina* Dunbar and Condra, *9* pedicle valve, *10* brachial valve; *11*, *12 Mesolobus mesolobus* (Norwood and Pratten); *13*, *14 Neospirifer cameratus* (Morton), *13* brachial valve, *14* posterior view

Fig. 8.29 Few representatives of Late Pennsylvanian brachiopods. *1–3 Chonetina granulifer* Owen, *1* pedicle valve, *2* brachial valve, *3* interior of brachial valve; *4–6 Dielasma bovidense* (Morton), *4* brachial valve, *5* side view, *6* pedicle valve; *7–10 Dictyoclostus americanus* Dunbar and Condra, *7* brachial valve, *8* pedicle valve, *9* posterior view, *10* side view, *11* interior of brachial valve; *12*, *13 Echinoconchus semipunctatus* (Shepard), *12* brachial valve, *13* pedicle valve; *14*, *15 Marginifera lasallensis* (Worthen), *14* pedicle valve, *15* brachial valve; *16–18 Juresania nebrascensis* (Owen), *16* interior of brachial valve, *17* side view, *18* brachial valve; *19–21 Meekella striatocostata* (Cox), *19* brachial valve, *20* posterior view, *21* side view; *22*, *23 Juresania symmetrica* (McChesney), *22* pedicle valve, *23* brachial valve; *24*, *25 Spiriferella texanus* (Meek), *24* brachial valve, *25* side view; *26*, *27*: *Chonetina flemingi* (Norwood and Pratten) (Late Pennsylvanian, Missourian), *26* pedicle valve, *27* brachial valve; *28–31 Derbyia cymbula* Hall and Clarke (Late Pennsylvanian, Virgilian), *28* interior of pedicle valve, *29* posterior view of brachial valve, *30* posterior view, *31* brachial valve; *32–34 Nudirostra carbonifera* (Girty) (Late Pennsylvanian, Missourian), *32* pedicle valve, *33* brachial valve, *34* side view; *35*, *36 Linoproductus oklahomae* Dunbar and Condra (Late Pennsylvanian, Missourian), *35* pedicle valve, *36* brachial valve; *37–43 Enteletes hemiplicatus* (Hall) (Late Pennsylvanian, Missourian), *37* side view, *38* brachial valve, *39* pedicle valve, *40* interior of brachial valve, *41* sectional view normal to commissure, *42*, *43* interior of pedicle valve

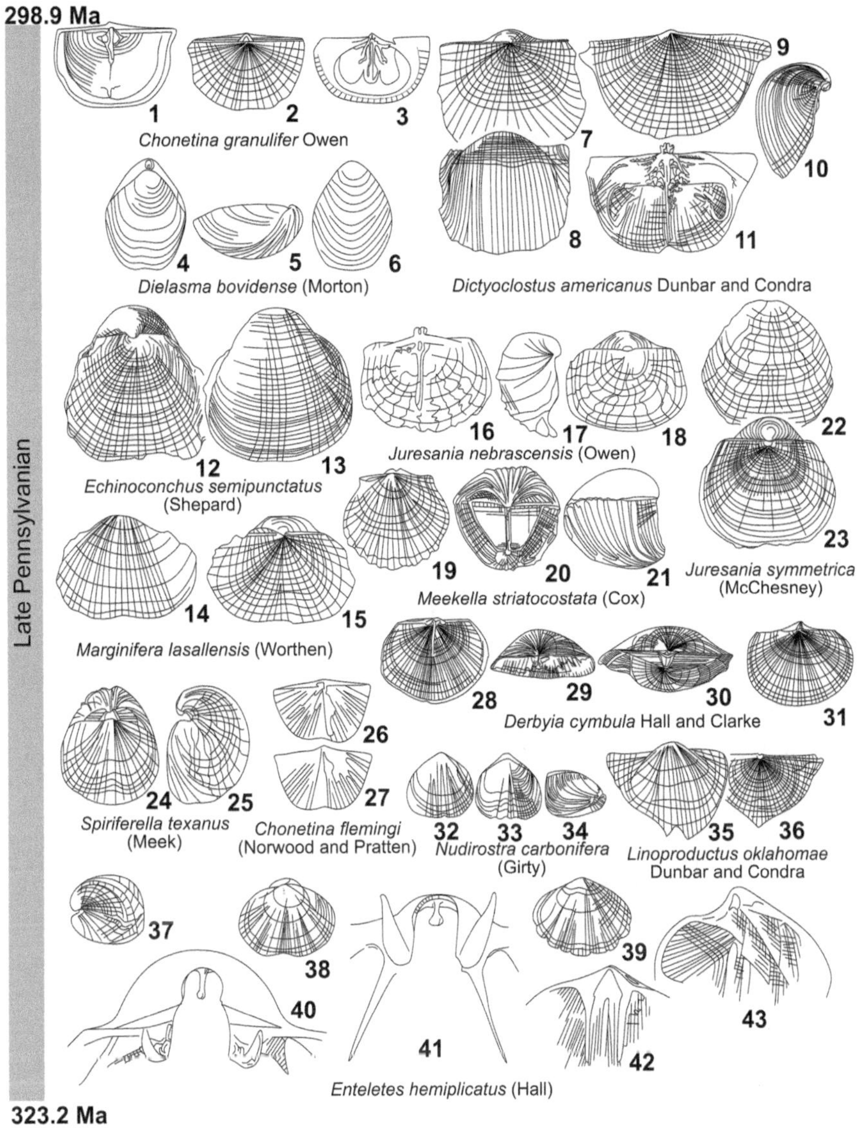
298.9 Ma
Late Pennsylvanian
1 2 3
Chonetina granulifer Owen
4 5 6
Dielasma bovidense (Morton)
7 8 9 10 11
Dictyoclostus americanus Dunbar and Condra
12 13
Echinoconchus semipunctatus (Shepard)
14 15
Marginifera lasallensis (Worthen)
16 17 18
Juresania nebrascensis (Owen)
19 20 21
Meekella striatocostata (Cox)
22 23
Juresania symmetrica (McChesney)
24 25
Spiriferella texanus (Meek)
26 27
Chonetina flemingi (Norwood and Pratten)
28 29 30 31
Derbyia cymbula Hall and Clarke
32 33 34
Nudirostra carbonifera (Girty)
35 36
Linoproductus oklahomae Dunbar and Condra
37 38 39 40 41 42 43
Enteletes hemiplicatus (Hall)
323.2 Ma

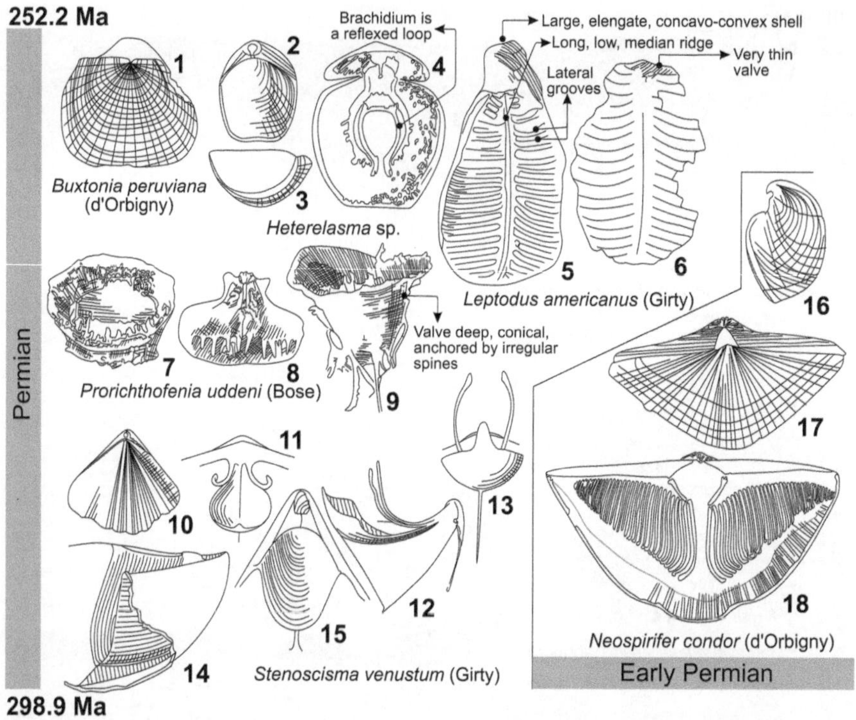

Fig. 8.30 Few representatives of Permian brachiopods. *1 Buxtonia peruviana* (d'Orbigny), brachial valve; *2–4 Heterelasma* sp., *2* brachial valve, *3* side view, *4* interior of brachial valve; *5, 6 Leptodus americanus* (Girty), *5* interior of pedicle valve, *6* interior of brachial valve; *7–9 Prorichthofenia uddeni* (Bose), *7* oblique view of brachial valve, *8* interior of brachial valve, *9* side view of pedicle valve; *10–15 Stenoscisma venustum* (Girty), *10* brachial valve, *11–13* interior of brachial beak, *14, 15* spondylium; *16–18* Early Permian, *Neospirifer condor* (d'Orbigny), *16* side view, *17* brachial valve, *18* interior of brachial valve showing spiralia

References

Bottjer, D.J., M.E. Clapham, M.L. Fraiser, and C.M. Powers. 2008. Understanding mechanisms for the end-Permian mass extinction and the protracted Early Triassic aftermath and recovery. *GSA Today* 18(9): 4–10. doi:10.1130/GSATG8A.1.

Carlson, S.J. 2001. Ghosts of the past, present, and future in brachiopod systematics. *Journal of Paleontology* 75(6): 1109–1111.

Carlson, S.J., and L.R. Leighton. 2001. The phylogeny and classification of Rhynchonelliformea, 27–52. In *Brachiopods ancient and modern: a tribute to G. Arthur Cooper*, *The Paleontological Society, Papers*, edited by White and Allmon. Vol. 7, 257 pp.

Clarkson, E.N.K. 1993. *Invertebrate palaeontology and evolution*, 4th ed. Cambridge: Blackwell Science. 468 pp.

Emig, C.C. 2009. *Nummulus brattenburgensis* and *Crania craniolaris* (Brachiopoda, Craniidae). Carnets de Géologie (Notebooks on Geology), 2009/08 (CG2009_A08).

Gould, S.J., and C.B. Calloway. 1980. Clams and brachiopods: ships that pass in the night. *Paleobiology* 6: 383–396.

Holmer, L.E. 2001. Phylogeny and Classification: Linguiliformea and Craniiformea, p. 11–26. In *Brachiopods Ancient and Modern: A Tribute to G. Arthur Cooper. Paleontological Society Papers 7*, edited by White and Allmon. New Haven: Yale University Reprographics.

ITIS. Brachiopoda (Integrated Taxonomic Information System). Retrieved from http://www.itis.gov. Accessed January 16, 2016.

Johansen, M.B. 1987. Brachiopods from the Maastrichtian-Danian boundary sequence at Nye Kløv, Jylland, Denmark. *Fossils and Strata* 20: 1–100.

Johansen, M.B. 1988. Brachiopod extinctions in the Upper Cretaceous to lowermost Tertiary chalk of Northwest Europe. In *Paleontology and evolution: extinction events*, edited by Lamolda, M. A., E.G. Kauffman, and O.H. Walliser. Revista Española de Paleontologia, numero Extraordinario, 41–56.

Kouchinsky, A., S. Bengtson, V.V. Missarzhevsky, S. Pelechaty, P. Torssander, and A.K. Val'kov. 2001. Carbon isotope stratigraphy and the problem of a pre-Tommotian stage in Siberia. *Geological Magazine* 138(4). doi:10.1017/S0016756801005684.

Lee, D.E., and C.H.C. Brunton. 1986. *Neocrania* n.gen., and a revision of cretaceous-recent brachiopod genera in the family Craniidae. *Bulletin of the British Museum, Natural History. Geology* 40(4): 141–160.

McGhee, G.R. 1996. *The Late Devonian mass extinction. The Frasnian-Famennian crisis*, 303 pp. New York: Columbia University Press.

Milsom, C., and S. Rigby. 2009. *Brachiopods. Fossils at a glance*. New York: Wiley. 37.

Moore, R.C. 1965. *Brachiopoda. Treatise on Invertebrate Paleontology. Part H.*, Vol. I. Geological Society of America, University of Kansas Press.

Rasmussen, C.M.Ø., and D.A.T. Harper. 2011. Did the amalgamation of continents drive the end Ordovician mass extinctions? *Palaeogeography, Palaeoclimatology, Palaeoecology* 311: 48–62.

Sepkoski, J. 1996. Competition in Macroevolution: the double wedge revisited. In *Evolutionary paleobiology*, edited by Jablonski, D. et al., 211–255. University of Chicago Press.

Sokiran, E.V. 2002. Frasnian-Famennian extinction and recovery of rhynchonellid brachiopods from the East European Platform. *Acta Palaeontologica Polonica* 47(2): 339–354.

Stearn, C.W., and R.L. Carroll. 1989. *Paleontology: the record of life*, 453 pp. New York: Wiley.

Tanner, L.H., S.G. Lucas, and M.G. Chapman. 2004. Assessing the record and causes of Late Triassic extinctions. *Earth-Science Reviews* 65: 103–139.

Topper, T.P., L.E. Holmer, C.B. Skovsted, G.A. Brock, U. Balthasar, C.M. Larsson, S. Petterson Stolk, and D.A.T. Harper. 2013. The oldest brachiopods from the lower Cambrian of South Australia. *Acta Palaeontologica Polonica* 58(1): 93–109.

Williams, A., C.H.C. Carlson, S.J. Brunton. 2000. Outline of suprafamilial classification and authorship. In *Brachiopoda. Treatise on invertebrate paleontology*, edited by Williams, A., C. H.C. Carlson, and S.J. Brunton. New York: Geological Society of America and The University of Kansas.

Chapter 9
Class Gastropoda

9.1 Introduction

Gastropods are one of the main divisions of phylum Mollusca and include organisms that bear a coiled [Fig. 9.1(1)] or uncoiled (Vermiform or worm-like; [Fig. 9.1(2)] calcareous shell; few others, such as slugs, have no hard parts. Gastropods, originally, were exclusively marine but during Mesozoic and Cenozoic times, became adapted for life in fresh waters, and finally also enabled them to invade dry land. However, a large majority of the group still maintained existence in the sea. Taken together, recent and fossil species of gastropods considerably outnumber all other species of mollusks combined.

Gastropods essentially live on shallow sea bottoms, but few have been dredged from the ocean depths of more than 3 miles ($\sim$4.8 km). Still others swim in near-surface waters of the open oceans far from land. Terrestrial snails are the only mollusks that have acquired lungs, and are able to move out of water bodies. They can climb trees and ascend mountains to an elevation of 18,000 ft ($\sim$5486 m) above mean sea level. On the whole, gastropods are inactive animals, well characterized by the adjective "sluggish," a word derived from slug. They depend on their shell and on their retiring habits to protect them from enemies.

9.2 Shell

The average size of a gastropod shell is $\sim$25 mm (1 in.) in length or diameter, though, fully grown adults range from 0.5 mm to $\sim$60 cm (0.002 to $\sim$2 ft).

S. Jain, *Fundamentals of Invertebrate Palaeontology*, Springer Geology,
DOI 10.1007/978-81-322-3658-0_9

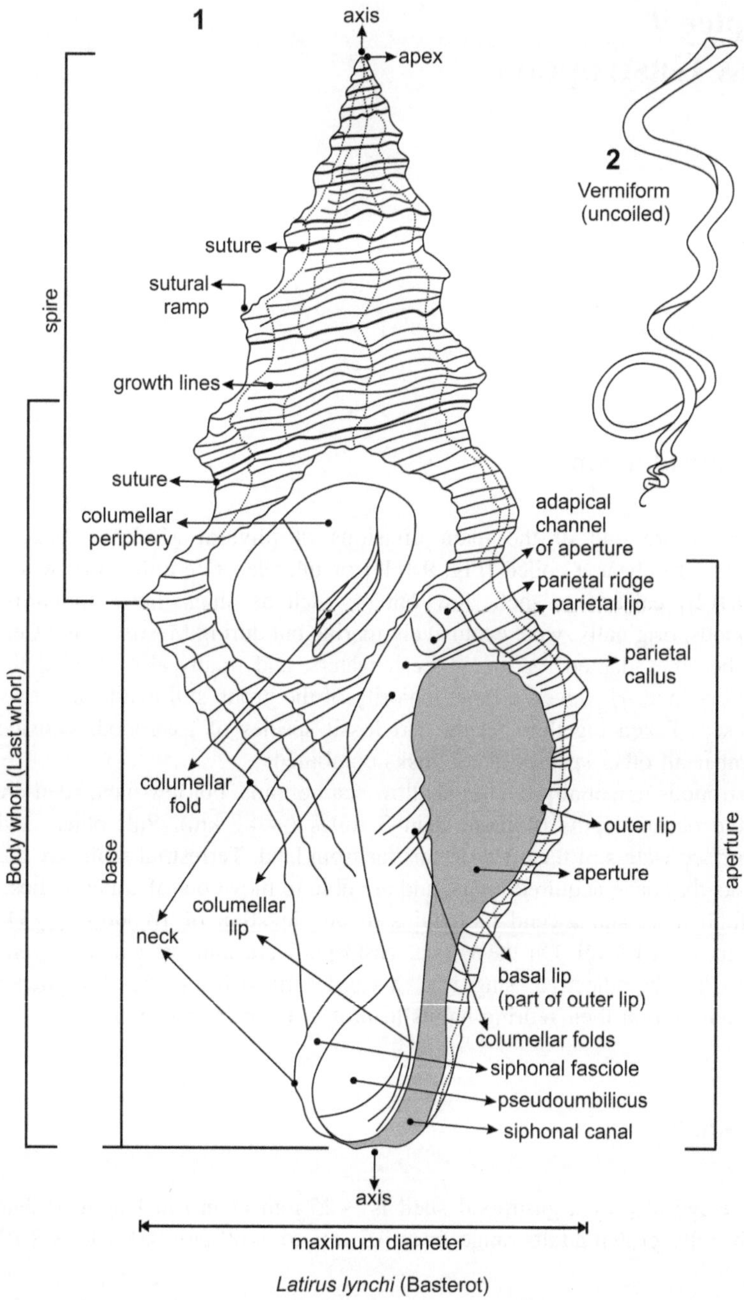

Fig. 9.1 Gastropod terminology

9.2.1 *Uncoiled Shells*

Uncoiled gastropods are conispiral forms (shells with coil that lie nearly or exactly in a plane) during their early ontogeny but later change the parameters of shell growth so that the youngest whorls cease to maintain contact [Fig. 9.1(2)].

9.2.2 *Coiled Shells*

Gastropods generally possess coiled shells, and in most, the plan of coiling gives rise to or is associated with pronounced asymmetry of the body [Fig. 9.1(1)]. A few gastropods have curved shells which define only part of a single 360° volution. Although such shells can be said to exhibit only a tendency toward coiling, either incipient or vestigial, they may be associated with those having a distinctly coiled form.

9.3 Types of Coiling

Gastropod shells typically have a spirally coiled structure, and the spiral is invariably one or other of two types: planispiral and conispiral (Fig. 9.2). The simplest and seemingly most primitive type of gastropod coiling is that in which the mid-line of the curved tubular shell lies entirely in a single plane—this is Planispiral [Fig. 9.2(1–3)]. The initial turns are close to the point of origin, and the later ones are progressively farther out from the center, following the course of a logarithmic spiral, but the mid-line does not swerve from a plane. The half of a shell lying on one side of the plane of coiling is the mirror image of the other half [Fig. 9.2(3)]. Such coiling is illustrated by *Bellerophon* and it is the normal type of coiling among cephalopods; only few living gastropods have planispiral shells. Planispiral coiling is thus morphologically intermediate between depressed hyperstrophic coils and the raised orthostrophic coils [see Fig. 9.2(4–9)].

Another much more common pattern of spiral coiling in gastropods is marked by the deviation from a plane [Figs. 9.2(4–6)]. A component of shift away from a plane produces a form similar to that of a wire wound around a cone, from the apex to its base. Such a spirally wound wire, representing the mid-line of the coiled gastropod shell, defines a Conispiral form [Figs. 9.2(4–6)]. Nearly all snails have gently to steeply sloping conispiral patterns.

Some shells coil in a depressed cone and are designated as Hyperstrophic (*hyper* = ultra; *stroph* = turn) [Figs. 9.2(4–6)]; the normal type of conispiral shell is termed as Orthostrophic (*ortho* = erect) [Figs. 9.2(7–9)].

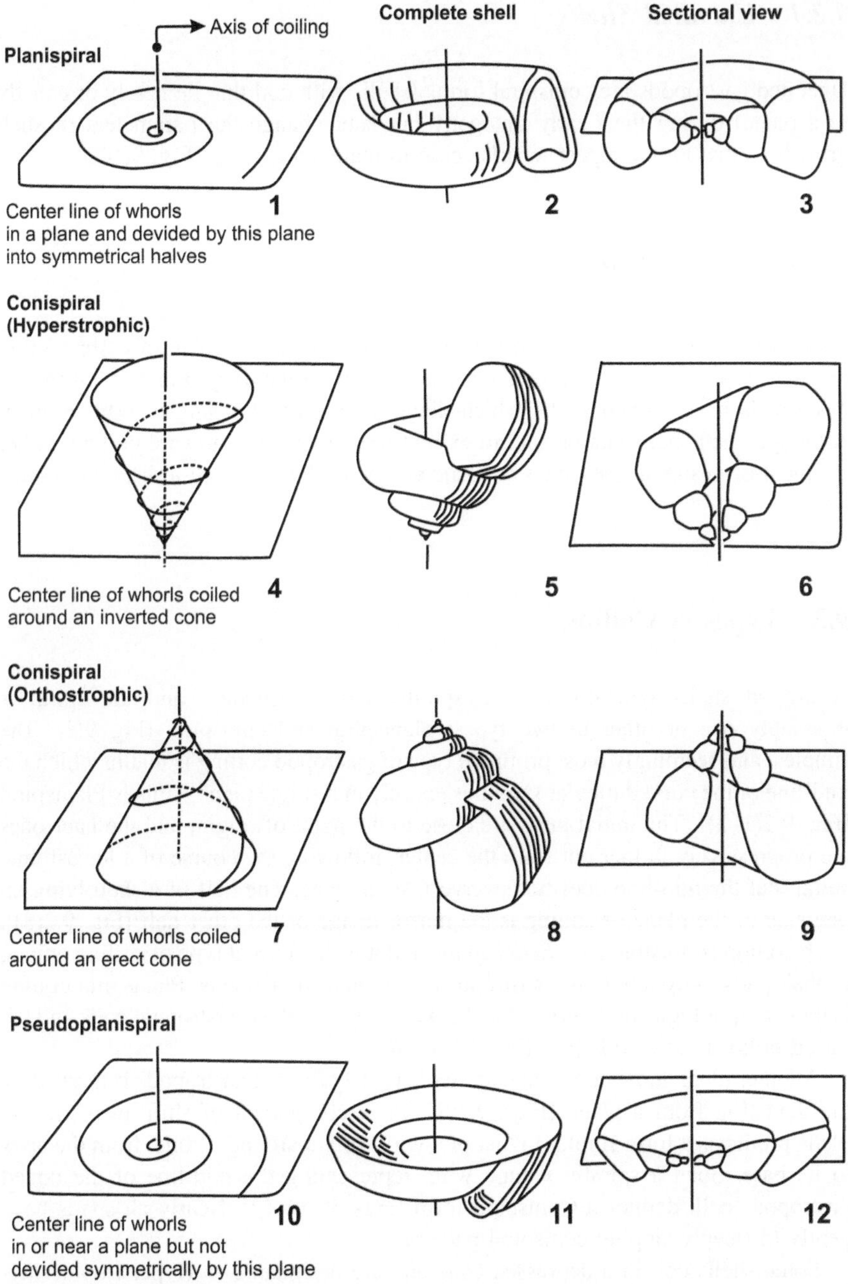

Fig. 9.2 Coiling in gastropods

Fig. 9.3 Gastropod shell orientation

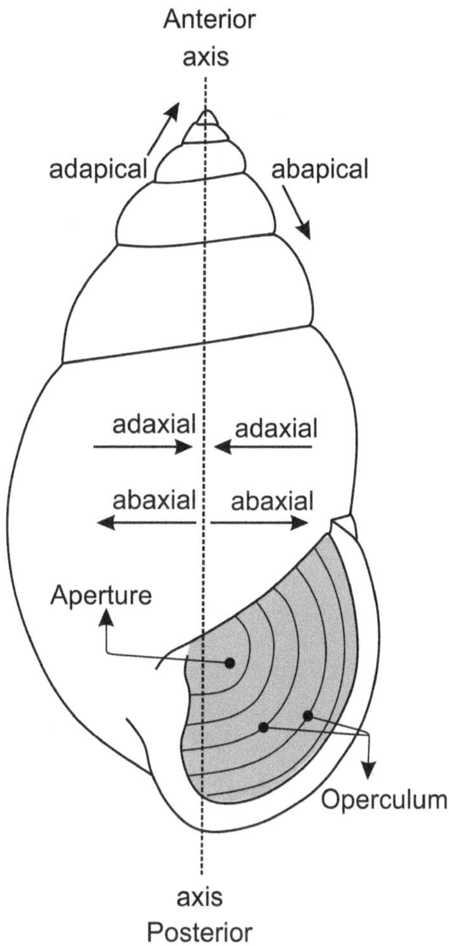

Some conispiral shells are so flat that the mid-line of all turns of the coil may lie nearly or exactly in a plane. Such shells are not planispiral, because one side is not the mirror image of the other; they are Pseudoplanispiral [Figs. 9.2(10–12)].

The axis of coiling, in all types, is an imaginary line drawn through the origin of the spiral in such direction that successive points outward along the spiral mid-line of the coiled shell have a gradually (logarithmically) increased distance from any selected point on the axis [Fig. 9.2(1, 4) and 9.2(10)]. It is the line around which the shell seems to coil. In planispiral shells, the axis is normal to the plane of coiling [Fig. 9.2(1)] and in conispiral shells, it runs from the apex to the center of the base of the cone [Fig. 9.2(4) and (7)].

9.4 Shell Orientation

Coiled gastropod shells are oriented in a consistent manner for purposes of description and illustration. Planispiral shells ordinarily are placed with the plane of coiling in a vertical position and the axis horizontal (Fig. 9.3). Use of the prefixes "ad" and "ab" provides a directional terminology for describing gastropod shells; "ad" means toward, and "ab" means away from (Fig. 9.3). Thus, *adapical* means toward the apex, and *abapical* means away from the apex.

Planispiral shells, like that of *Bellerophon* [Fig. 9.2(1–3)], presumably were carried by the snails which inhabited them with the plane of coiling vertical and the aperture directed downward. The outer part of the aperture, farthest from the center of coiling, is interpreted as the most posterior and an opposite point toward the head of the animal, the most anterior (Fig. 9.3).

Conispiral shells are placed with the axis of coiling vertical and with the origin of coiling (apex of the shell) uppermost [Fig. 9.2(7–9)]. These shells normally are borne by living snails with the aperture directed forward and downward and the point of the coiled shell backward. Thus, posterior means in the direction of the aperture, as used for these shells, and anterior is toward the point of the cone (Fig. 9.3). As such, shells are oriented in illustrations; the anterior extremity is uppermost and the posterior part lowermost (Fig. 9.3). Some terminology to describe shell orientation is given in Fig. 9.3.

9.5 Terminology

The general terms describing the features of a gastropod shell are briefly described below and illustrated in Figs. 9.1, 9.2, 9.3 and 9.4.

9.5.1 **Anterior**: It is toward the point of the shell cone (Fig. 9.4).

9.5.2 **Aperture:** The opening of a gastropod shell (Figs. 9.1 and 9.3).

9.5.3 **Apex:** The tip of the spire of a shell (Figs. 9.1 and 9.3).

9.5.4 **Axis:** An imaginary line that runs straight from the anterior to the posterior side of the shell (Figs. 9.1 and 9.3).

9.5.5 **Basal fasciole:** A shell corrugation marked by curved growth lines which define the shift in position of the siphonal notch is termed Basal fasciole [Fig. 9.4(2)].

9.5.6 **Base:** It is the extremity opposite to the Apex [Fig. 9.4(1)].

9.5.7 **Body whorl:** This is where the organism lives [Figs. 9.1 and 9.4(1)].

9.5.8 **Callus:** A localized thickened part of the inductura outside the aperture, which may partly or entirely seal the umbilicus [Fig. 9.4(12)].

9.5.9 **Canal:** Semi-tubular anterior extension of aperture, enclosing siphon; at least slightly open alongside, not closed like a pipe [Figs. 9.1 and 9.4(3)].

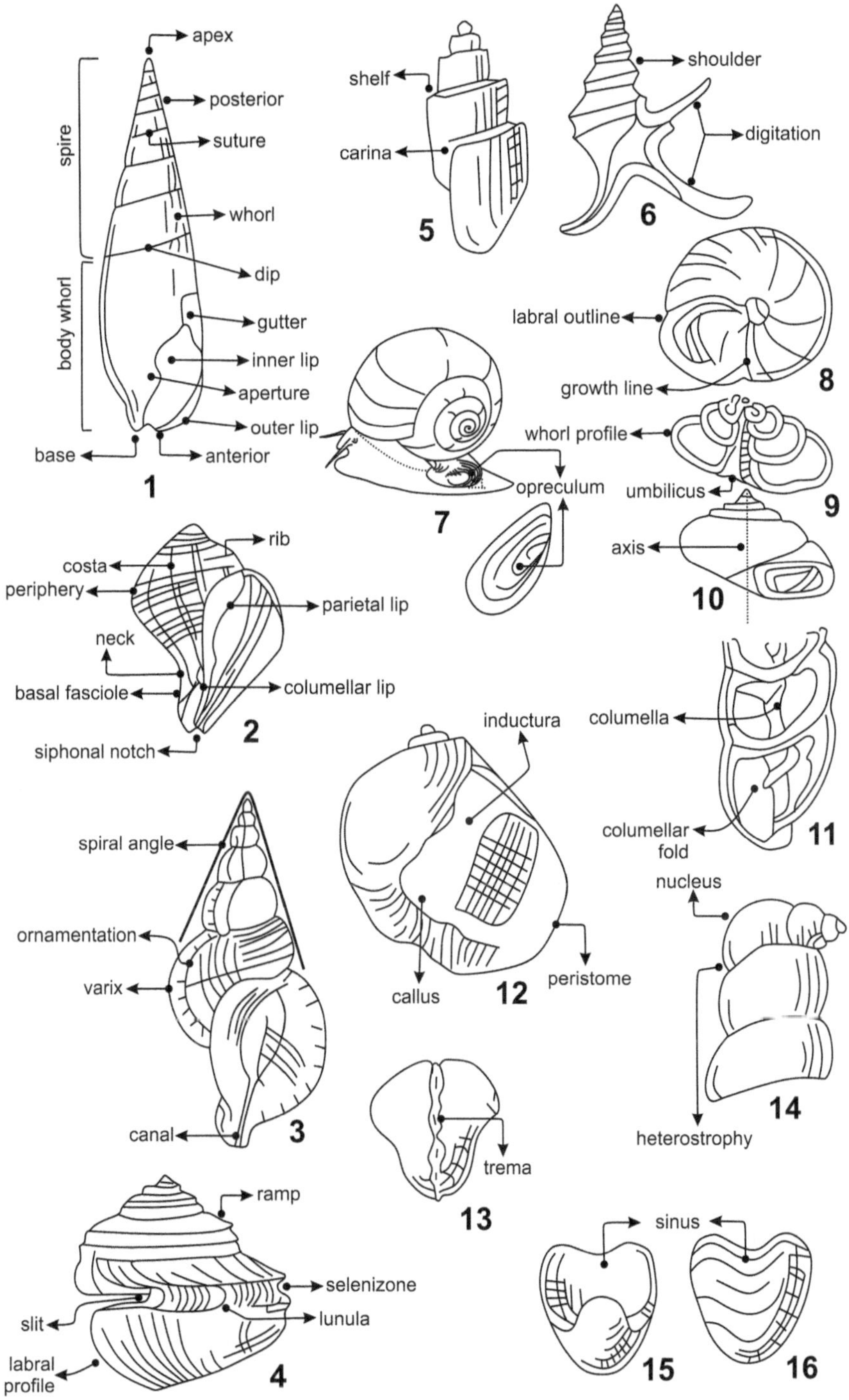

Fig. 9.4 Gastropod terminology

9.5.10 **Carina:** Spiral keel on exterior of whorl, generally at edge of shelf. Sharply accentuated angulation between the ramp or shelf and the surface next to it [Fig. 9.4(5)].

9.5.11 **Columella:** Solid or perforate pillar formed by inner walls of a conispiral shell. Shells in which the inner edges of the whorls touch the axis of coiling have a solid rod-like center, which is termed Columella (Fig. 9.4(11)]; see also Fig. 9.1). Shells having a narrow umbilicus are said to possess a Perforate columella.

9.5.12 **Columellar fold:** Spiral elevation on columella produced by localized thickening of shell [Fig. 9.4(11); see also Fig. 9.1]. Columella may be smooth or bear spirally arranged ridges. These are Columellar folds.

9.5.13 **Columellar lip:** Part of inner lip of aperture adjoining columella [Fig. 9.4 (2); see also Fig. 9.1].

9.5.14 **Costa:** Coarse threadlike thickening of shell running spirally or axially [Fig. 9.4(2)].

9.5.15 **Digitation:** Finger-like outward projection of outer lip of aperture [Fig. 9.4 (6)]. The outer lip of some gastropods carries projections, those which extend inward being called teeth, and periodically formed outward extensions being known as digitations. A few species are distinguished by remarkably long and slender digitations, which denote unusual narrow and attenuated temporary protrusions of the shell-secreting mantle in building them.

9.5.16 **Dip:** Deviation of suture from a plane normal to the shell axis [Fig. 9.4(1)].

9.5.17 **Growth lines:** Marking on shell parallel to apertural margin, denoting a former position of aperture (Fig. 9.1).

9.5.18 **Gutter:** Groove or canal at posterior extremity of aperture, in some gastropods marking location of anal outlet. An ex-current passageway is marked at the posterior edge of the aperture in some gastropods by a notch or groove (gutter). This also may indicate the position of the anus [Fig. 9.4 (1)].

9.5.19 **Heterostrophy (*hetero* = different; *strophy* = turning):** Abrupt change in the type of coiling between nucleus and later-formed part of shell; this involves change in position of the axis of coiling. Best noted in the Late Triassic–Early Cretaceous genus *Bandellina* (see Bandel 1995, 1996 and 2002) [Fig. 9.4(14)].

9.5.20 **Inductura:** Layer of lamellar shell material made by the mantle along inner lip of aperture or extending over shell surface beyond outer lip (next-to-last whorl), characterized by smooth surface; includes callus [Fig. 9.4(12)].

9.5.21 **Inner lip:** Margin of aperture adjacent to the next-to-last whorl; may include parietal lip and columellar lip [Fig. 9.4(1)].

9.5.22 **Labral outline:** Shape of outer lip in view normal to aperture [Fig. 9.4(8)].

9.5.23 **Labral profile:** Shape of outer lip in view parallel to edge of aperture [Fig. 9.4(4)].

9.5.24 Lunula (plLunulae): Crescentic growth line on selenizone. The extremity of the slit, on the side away from the aperture, is narrowly rounded, and as the shell margin is built forward the shape of the base of the slit is recorded by successive sharp-curved growth lines [Fig. 9.4(4)].

9.5.25 Neck: Constricted anterior part of body whorl of some gastropods, exclusive of canal (Fig. 9.1).

9.5.26 Nucleus: Embryonic gastropod shell, commonly consisting of one to four whorls.

9.5.27 Operculum: Horny or horny-calcareous plate carried on posterior part of foot, used to close aperture when gastropod withdraws into its shell (Fig. 9.3).

9.5.28 Ornamentation: Raised or depressed markings of shell surface other than growth lines (Fig. 9.1).

9.5.29 Outer lip: Edge of aperture on side away from next-to-last whorl. The outer side of the aperture, away from the shell, is the outer lip (Fig. 9.1).

9.5.30 Parietal lip: Part of inner lip which adjoins next-to-last whorl (Fig. 9.1).

9.5.31 Periphery: Part of whorl farthest from shell axis; the exterior of the shell farthest removed from the axis of coiling (Fig. 9.1).

9.5.32 Peristome: Margin of whole aperture (Fig. 9.1).

9.5.33 Posterior: Direction backward from head of gastropod, in spiral shells toward apex (Figs. 9.1 and 9.3).

9.5.34 Ramp: Sloping surface of a whorl next below a suture (= sutural ramp; Fig. 9.1; see also [Fig. 9.4(4)]).

9.5.35 Rib: Well-marked linear elevation of shell surface, larger and broader than a costa [Fig. 9.4(2)].

9.5.36 Selenizone (Slit band): Sharply defined band parallel to coiling of whorls, which bears crescentic growth lines denoting a notch or slit in outer lip. The belt of lunulae, which commonly is defined by bordering ridges, constitutes the selenizone or so-called slit band [Fig. 9.4(4)].

9.5.37 Shelf: Horizontal or sub-horizontal part of whorl surface next to a suture, bordered on side toward periphery of whorl by a sharp angulation or by a carina (Fig. 9.1).

9.5.38 Shoulder: Salient angulation of a whorl parallel to coiling. An angulation between the ramp or shelf and the surface next to it constitutes a Shoulder (Fig. 9.1).

9.5.39 Sinus: Reentrant in outer lip with nonparallel sides. Other snails, including especially many Paleozoic planispiral and conispiral genera, have an indentation of varying depth in the outer lip which is inferred to have served the same function as the gutter—a passageway for waste from the anus and for water after it had bathed the gills. This indentation is called a sinus, if it is comparatively broad, and a slit (see below), if it is narrow, parallel-sided, and more or less deep. In a few fossil snails, the slit is excessively deep, reaching more than half-way around the body whorl [Fig. 9.4(15–16)].

9.5.40 **Siphonal notch:** Reentrant at junction of outer and columellar lips occupied by siphon. At the anterior extremity of the aperture, where outer and inner lips meet, many advanced types of gastropods bear a groove (siphonal notch) which may be extended along a shell outgrowth (canal) that holds the siphon [Fig. 9.4(2)].

9.5.41 **Slit:** More or less deep, parallel-sided reentrant in outer lip, which gives rise to a selenizone [Fig. 9.4(4)].

9.5.42 **Spire:** Coiled gastropod shell exclusive of body whorl (Fig. 9.1).

9.5.43 **Spiral angle:** Angle formed by lines tangent to two or more whorls on opposite sides of shell; inasmuch as lines tangent to all whorls of the spire may define a curve, spiral angle is commonly determined by drawing straight-line tangents to lowermost whorls of spire [Fig. 9.4(3)].

9.5.44 **Suture:** Spiral line of junction between surfaces of any two whorls; includes external sutures on outer side of shell and umbilical sutures within umbilicus. Sutures can be of different types such as flush, impressed, or channeled, and their inclination (Dip) may be a significant shell character (Fig. 9.1).

9.5.45 **Trema (pl. tremata):** Perforation of shell, generally formed by periodic closure of a slit, but occurring also at apex of some cap-shaped shells. In some shells, like that of the modern abalone (*Haliotis*), the slit is discontinuous, and unsettled openings (tremata, sing. trema) are left behind; these function as outlets for water and waste from the shell interior [Fig. 9.4(13)].

9.5.46 **Umbilicus:** Central cavity of a shell formed by walls on inner sides of whorls; most common is basal umbilicus of orthostrophic conispiral shells but also included are apical umbilicus of convolute and hyperstrophic shells and lateral umbilici of planispiral (isotrophic) shells. All shells in which the inner sides of the whorls lie outside the axis of coiling possess a narrow or wide open space within the encircling whorls. This is an Umbilicus [Fig. 9.4(9)].

9.5.47 **Varix:** Ridge, flange, or row of spines parallel to growth lines and marking modification of shell at former position of aperture [Fig. 9.4(3)]. On parts of the body whorl and spire behind the aperture, this may be shown distinctly by the growth lines and in some shells also by flange-like ridges called varices. A varix marks a former location of the aperture where a pause in forward growth of the shell was accompanied by construction of special apertural features. These may include digitations which have to be cut away by resorption when growth of the body whorl brings the functional aperture around to the position of the abandoned one. Varices on successive whorls may be aligned, indicating rather remarkable regularity in pulsatory building of the shell, or they may be offset, indicating unevenly distributed pauses in growth.

9.5.48 **Whorl:** Single complete loop of a spiral shell. It is a 360° volution (a complete turn) of the shell (Fig. 9.1).

9.5.49 Whorl profile: Transverse contour of surface of a whorl in a plane intersecting the axis of coiling; differs from labral profile and labral outline (Fig. 9.1).

9.6 Shell Form

All shell forms are illustrated in Fig. 9.5. However, these twenty five types of shell forms are by no means an end-all, as minor variations and varied combinations of these would exist.

9.7 Classification

Gastropod classification is problematic for 3 reasons: the group possess huge morphological and anatomical variability; the use of limited number of distinguishing morphological characters (in initial classification schemes) such as shape of the shell, position of the mantle cavity, or the arrangement of various organs (such as gills or head) and the recent knowledge of deep-sea faunas associated with hydrothermal vents that revealed new gastropod groups with unusual anatomical features.

The German zoologist, Johannes Thiele, integrated earlier classifications and identified three subclasses, Prosobranchia, Opisthobranchia, and Pulmonata (Table 9.1). The Prosobranchia were further divided into three orders, Archaeogastropoda, Mesogastropoda, and Neogastropoda. Thiele's system was used by Clarkson (1993) and is given in Table 9.2.

However, the application of new methods such as transmission electron microscopy (TEM), discovery of deep-sea hydrothermal vent faunas, and the analyses with more morphological and developmental characters have necessitated the need for a new classification scheme based on phylogenetic analysis using morphological characters (Ponder and Lindberg 1997) (Fig. 9.6). The Patellogastropoda (Docoglossa, Cyclobranchia) represents the sister group to all other living gastropods. The former along with their coiled ancestors form a subclass, Eogastropoda. Others and their ancestors are placed in the subclass, Orthogastropoda, which has four main groups of living gastropods—Neritimorpha, Archaeogastropoda, Caenogastropoda, and Heterobranchia (Fig. 9.6). The latter two are sister groups, united under Apogastropoda. The extant Caenogastropoda has two orders, Architaenioglossa and Sorbeoconcha. Terrestrial Cyclophoroidea

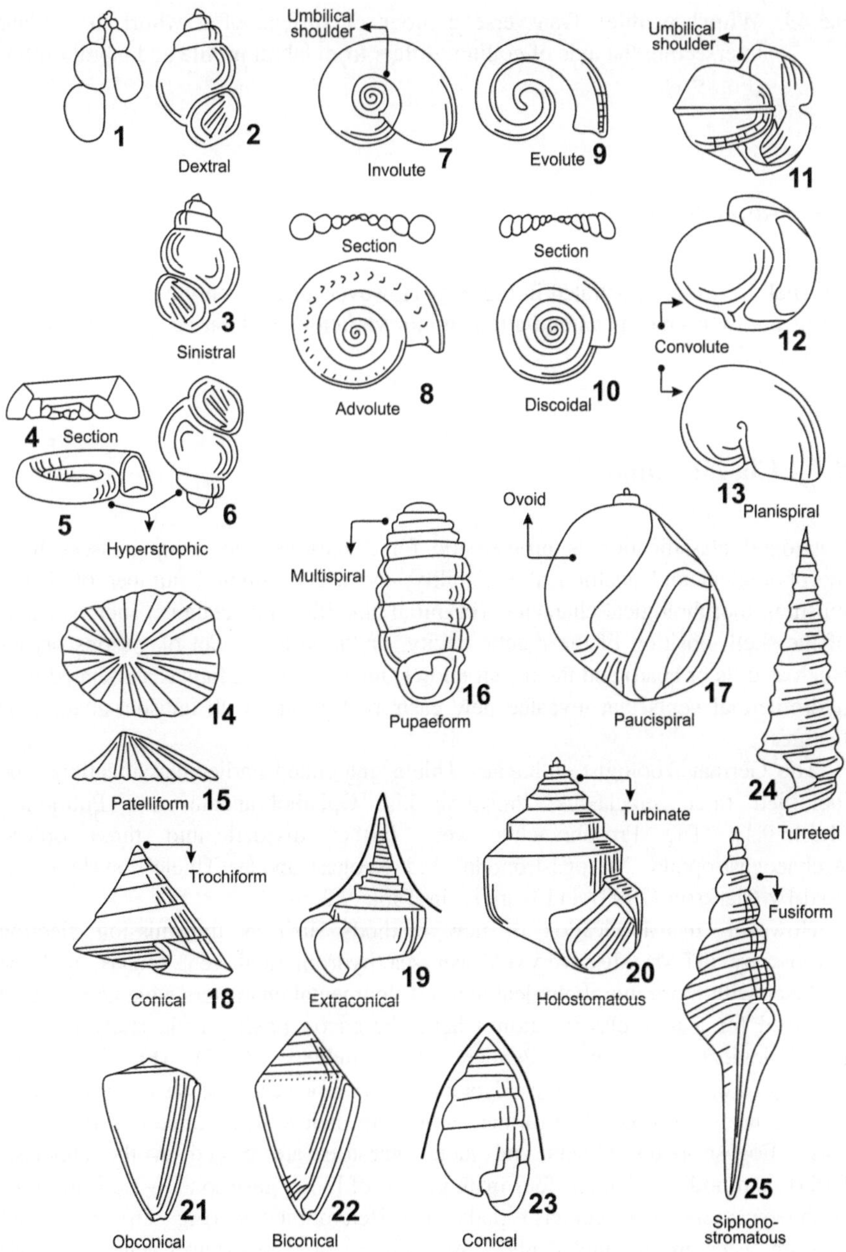

Fig. 9.5 Gastropod shell forms

Table 9.1 The German zoologist Johannes Thiele's classical three-partite classification

Class	Subclass	Order	Age
Gastropoda	Prosobranchiata	Early Cambrian—Recent	
		Mesogastropoda	Ordovician—Recent
		Neogastropoda	Cretaceous—Recent
		Archaeogastropoda	Cambrian to Recent
	Opisthobranchiata	Cretaceous—Recent	
		Pleurocoela	Mississippian-Recent
		Pteropoda	Cretaceous-Recent
		Sacoglossa	Recent
	Pulmonata	Pennsylvanian—Recent	
		Basommatophora	Jurassic-Recent
		Stylommatophora	Pennsylvanian-Recent

and fresh water Ampullarioidea form the Architaenioglossa. The mostly marine Sorbeoconcha represents a highly diverse group.

The Heterobranchia includes Thiele's "Opisthobranchia" and "Pulmonata," as well as some "prosobranch" groups, such as the Valvatoidea and Architectonicoidea. The majority of the lower Heterobranchia (Opisthobranchia or sea slugs) are also marine gastropods. The higher Heterobranchia (Pulmonata) form a dominant group of terrestrial gastropods, but also occur in fresh water environments. In several classifications, Pulmonata have been divided into three orders, Systellommatophora (terrestrial gastropods), Basommatophora, and Eupulmonata. The ancient marine Basommatophora have been separated into the Archaeopulmonata and the fresh water Basommatophora into the Brachiopulmonata. The classification followed here is given in Table 9.2.

Figures 7.22 illustrate gastropods through time mentioning their major distinguishing characters.

9.7.1 *Class Amphigastropoda*

The shells are symmetrical, non-coiled or planispirally coiled. The mantle cavity and gills are located in posterior; primitive position (*amphi* = on both sides, refers to symmetry). The Amphigastropods range from Cambrian to Permian (possibly Triassic). They include the superfamily Tryblidiacea and possibly also the Bellerophontacea, which together contain nine families. Both super families appeared in the Cambrian but attained peak development during the early Palaeozoic.

Table 9.2 Gastropod classification as followed here (after Clarkson 1993)

Class Amphigastropoda
Tryblidiids
Bellerophontids
Class Prosobranchia
Order Aspidobranchia
Order Pectinibranchia
Order Archaeogastropoda (or Aspidobranchs)
Suborder Rhipidoglossa
Suborder Docoglossa
Suborder Macluritids
Suborder Euomphalids
Suborder Pleurotomariids
Patellids and Cocculinids
Trochonematids and Trochids
Loxonematids and Subulitids
Naiads
Order Mesogastropoda
Superfamily Cerithiids
Superfamily Epitoniids
Superfamily Strornbids
Superfamily Tonnids
Superfamily Naticids
Superfamily Cypraeids
Superfamily Pyrarnidellids
Superfamily Calyptraeids and Hipponicids
Superfamily Nerineids
Superfamily Rissoids
Superfamily Cyclophorids, Valvatids, and Littoriaids
Order Neogastropoda
Suborder Muricids
Suborder Buccinids
Suborder Volutids
Suborder Conids
Class Opisthobranchia
Order Pleurocoela
Order Pteropoda
Order Sacoglossa
Class Pulmonata
Order Basommatophora Order Stylommatophora

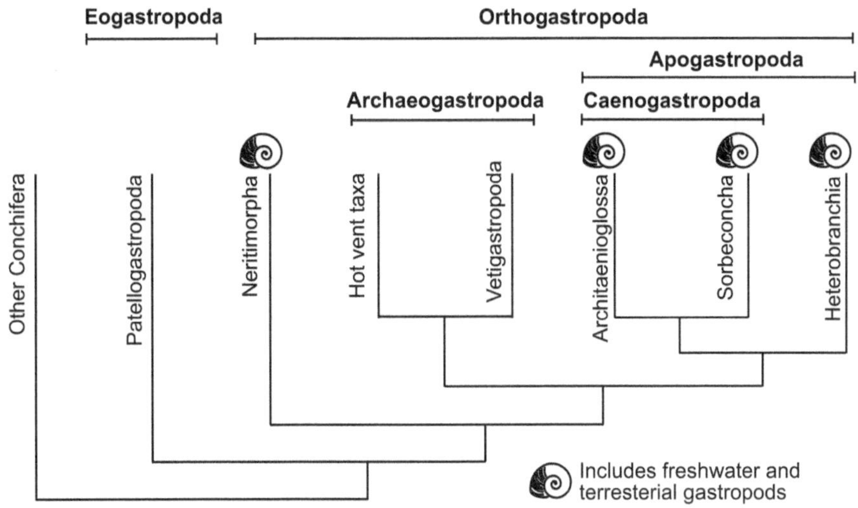

Fig. 9.6 Gastropod classification. Recent classification scheme of living gastropods, illustrating their phylogenetic relationships and the distribution of fresh water and terrestrial groups

9.7.1.1 Tryblidiids

These include shells ranging from high conical non-coiled shapes to very low, partly coiled forms. All are symmetrical, and the tip of the coiled shells is pointed forward. They range from Cambrian to Devonian.

9.7.1.2 Bellerophontids

The bellerophontids are more abundant and advanced than the Tryblidiids. Some have an evolute, incompletely coiled shell form, whereas others are advolute, slightly to strongly involute, or distinctly convolute. The advolute and involute shells have a more or less open umbilicus on each side, but the convolute types lack umbilici and only the body whorl is visible from the exterior. Generally, a slit is present at the mid-point of the outer lip, and a selenizone can be traced along the mid-line of the last-formed whorl by the sharply curved pattern of growth lines (lunulae) and commonly by the slightly raised or depressed surface along the band. Different species exhibit a considerable range of spiral and transverse markings. The occurrence of both together produces a reticulate pattern. Peak development of the bellerophontids is from Ordovician to Devonian.

9.7.2 Class Prosobranchia

These have cap-shaped or conispiral shell with mantle cavity, gills in anterior position and a neural loop twisted like a figure of eight (*proso* = forward; *branchia* = gills). They range from Cambrian to Recent and are an important constituent of the post-Palaeozoic faunas. More than half of all gastropods are prosobranchs; there are two living orders.

9.7.2.1 Aspidobranchia (Aspido, Shield; Branchia, Gills)

These have one or two gills which bear double rows of respiratory leaflets, and the heart contains two auricles.

9.7.2.2 Pectinibranchia (*Pectin* = Comb)

These have only one gill which bears a single row of leaflets, and the heart contains one auricle. The Pectinibranch snails include both Mesogastropoda and Neogastropoda. This class is also known as Streptoneura (*strepto* = twisted; *neura* = nerve cord), in reference to the figure-eight pattern of their nerve cord.

9.7.2.3 Order Archaeogastropoda

The primitive form have two unequal gills with double rows of leaflets but advanced forms have only one gill (*archaeo* = ancient). They range from Cambrian to Recent. Living Archaeogastropods (or Aspidobranchs) are divided into two suborders on the basis of the structure of the radula:

9.7.2.3.1 Rhipidoglossa (*rhipido*, fan; *glossa*, tongue):

In which the radula has rows of very numerous teeth that diverge like ribs of a fan.

9.7.2.3.2 Docoglossa (*doco*, bar):

In which the radula has rows made up of a few strong teeth.

These suborders can be recognized fairly well among fossils by their shell characters, but because the radulas are not preserved, hence many fossil genera cannot be classified definitely in terms of the suborders.

9.7.2.3.3 Macluritids:

These important, but short-lived group (Early Ordovician—Devonian) is typified by Genus *Maclurites*. Most of the shells are large. They are very low conispiral coils having advolute or slightly involute whorls. One side of the shell is nearly flat, and the opposite one bears a concavity formed by the wide umbilicus.

9.7.2.3.4 Euomphalids:

Thes low-to-flat-spired snails, some hyperstrophic like the Macluritids, are grouped under superfamily Euomphalacea. They are very abundant in the Palaeozoic and are useful index fossils. Euomphalids (Early Ordovician—Cretaceous) peaked twice—in Devonian and Triassic by disappeared by the Cretaceous.

9.7.2.3.5 Pleurotomariids:

Genus Pleurotomaria (Late Cambrian—Recent) has a conispiral shell characterized by its moderately wide spiral angle, fairly even sides, and a somewhat flattened base, but its most characteristic feature is a prominent slit, accompanied by a well-defined selenizone or slit band. The selenizone extends around the outer part of the body whorl and can be seen on spiral whorls near the sutures between whorls. Most genera are described from the Silurian. Modern snails classed among the Pleurotomariacea are characterized chiefly by slit-bearing, cap-shaped shells.

9.7.2.3.5.1 Patellids and Cocculinids

These are cap-shaped shells, most of which lack a sign of coiling in the adult. They are moderately common in the Mesozoic and Cenozoic and absent from the Palaeozoic. Living genera are more than twice as numerous as those genera known in Tertiary or Mesozoic times.

9.7.2.3.5.2 Trochonematids and Trochids

These were probably derived from Pleurotomariid ancestors. Trochonematids are common in Devonian and other Palaeozoic systems. They mostly have a pleurotomarian shape but lack the slit and slit band. The first-mentioned group is mainly Palaeozoic (Early Ordovician) but ranges to the close of Mesozoic, whereas the second group is exclusively post-Palaeozoic and mostly Cenozoic.

9.7.2.3.5.3 Loxonematids and Subulitids

These are distinguished by their high and narrow spire. Both make an appearance in the Ordovician and persist throughout the Mesozoic; the Subulitids are the longer lasting superfamily. The Loxonematids show a gradual increase of genera with peak diversity in the Triassic, then followed by a quick decline.

9.7.2.3.5.4 Naiads

The superfamily Neritacea consists mostly of ovoid shells but contains some cap-shaped forms. Deposits of callus on the inner lip and other special features of the aperture distinguish many genera, especially the later ones. The neritids are a minor constituent of the Palaeozoic gastropod faunas but have increased since Permian to a present-day maximum.

9.7.2.4 Order Mesogastropoda

These have only one gill with a single row of leaflets (*Meso* = intermediate). They range from Middle Ordovician to Recent. The Mesogastropoda include considerably more than half of the order, defined by characters of the gill and heart, which is called Pectinibranchia (*pectini* = comb) because of the single row of gill leaflets, or Monotocardia because of the single auricle of the heart. Living mesogastropods are all classed on the basis of their radula structure in the suborder Taenioglossa (*taenio* = band), which typically has seven teeth in each row. If we neglect half a dozen genera from Pennsylvanian and Permian rocks, which are rather doubtfully placed with mesogastropods, then this order would wholly be post-Palaeozoic in distribution. It includes most known Mesozoic snails, but the number of Cenozoic genera is far greater than those of Mesozoic age.

9.7.2.4.1 Cerithiids

Cerithiacea is one of the leading superfamily groups, characterized by very numerous whorls and the tall, turreted form of the shell. The aperture of many, but not all, is siphonostomatous, although the long canal is not developed. Among genera which are especially numerous and important as fossils are *Cerithium* (Cretaceous-Recent), including well-ornamented high-spired shells, and *Turritella* (also Cretaceous-Recent but chiefly Tertiary), containing a host of exceptionally tall, slender, many-whorled shells. The aberrant cerithiid genus, *Vermicularia* (Cretaceous to Recent), starts growth like *Turritella* but soon builds its narrow shell tube irregularly in almost any direction. Another member of the superfamily is *Vermetus* (Pliocene to Recent), which is attached by the apical part of its shell in a fixed position with the aperture pointed upward; individuals of some species grow closely packed together in the form of gastropod “reefs.”

9.7.2.4.2 Epitoniids

The superfamily Epitoniacea somewhat resembles the cerithiids, but the shells have round holostomatous apertures and the whorls commonly are marked by oblique frill-like expansions. Many species have very distinctive, rather delicate shells. They are smaller on the average than the cerithiids, and they are less numerous. A few genera occur in Cretaceous rocks, but they are mainly Tertiary and Recent group.

9.7.2.4.3 Strornbids

The Strombacea are an important superfamily of marine mesogastropods, which is represented by some of the larger, more colorful shells belonging to genus *Strombus* found on Florida (USA) and other sea beaches. The exterior generally bears axial and revolving ribs and nodes. The spire is moderately elevated in turret form, and the aperture is long, with flaring lips in the adult. The group first appeared in the Triassic, but numerous, distinctive fossils are noted from Jurassic, Cretaceous, and Tertiary formations. Some of these, such as *Aporrhais*,

Drepanocheilus, and *Anehura*, from the Jurassic and Cretaceous, have very striking digitations of the adult outer lip, which make them very distinguishable. Even more unique in its display of long curved digitations is *Pterocera*, a modern strombid, which is known as a fossil only in some Pleistocene deposits.

9.7.2.4.4 Tonnids

These late-appearing, highly organized marine mesogastropods resemble strombids in many ways are grouped in the superfamily Tonnacea. Most of them have a strongly siphonostomatous aperture, and they are rather strongly sculptured. The shells are medium to large, with the body whorl predominating greatly over the spire. Several genera are characterized by the strong development of varices, which tend to be aligned on the whorls.

9.7.2.4.5 Naticids

These are a conservative assemblage of ovoid to globose marine snails having nearly smooth shells that mostly is only an inch or two in diameter. They span from Triassic to Recent. Many of them bear a prominent callus deposits that modifies the configuration of the inner lip and tends to close the umbilicus. The naticid shells are very common in some Tertiary formations.

9.7.2.4.6 Cypraeids

The brilliantly polished, mostly very smooth-surfaced "cowry shells" that bear no visible spire as adults and possess a long, narrow apertural slit on one side are members of Cypraeacea. They occur as fossils in Cretaceous and Tertiary rocks but have not been found in older deposits. They are exclusively marine gastropods of world-wide distribution that have their peak development at the present time.

9.7.2.4.7 Pyrarnidellids

The high-spired marine snails called pyramidellids closely parallel the cypraeids in number of genera represented by fossils and in geologic distribution, but they are far less conspicuous because a majority of them are less than 0.5 in. in height. Indeed, the adults of many species range from 0.04 to 0.10 in., being appropriately classifiable as microfossils.

9.7.2.4.8 Calyptraeids and Hipponicids

These are two quantitatively less prominent assemblages of marine mesogastropods (Cretaceous to Recent), but they are structurally distinctive, especially in the development of internal platforms for muscle attachment. Also, a few genera, such as *Calyptraea* and *Crepidula*, the latter commonly known as the "slipper shell," are well represented.

9.7.2.4.9 Nerineids

The superfamily Nerineacea includes Jurassic and Cretaceous gastropods, many of which draw attention on account of their exceptional form, for the height of the shell may be more than 20 times the diameter of the body whorl. The spiral angle is so small that the sides of the spire are nearly parallel. The most distinctive feature of the nerineids, however, is their internals that can be seen only in broken shells or by making sections. When cut in half longitudinally, the inner walls of the whorls are found to bear projections which run spirally from the aperture to the apex, or nearly to it. Some shells are umbilicate, whereas others have a strong columella. The spiral ridges occur in both, and they produce varying cross sections of the whorl interiors from genus to genus.

9.7.2.4.10 Rissoids

A numerically important but otherwise rather unimpressive division of the mesogastropods comprises the Rissoacea. They include a few fresh water forms, but most of them live in the sea. The shells are small, and a large majority of them have a height-range of 0.5–1 in. They have slightly rounded conical outlines and a holostomatous aperture which generally lacks a rim. The shell surface is smooth or moderately sculptured. A few genera have been recorded from Jurassic and Cretaceous rocks. The main development of the group is in Tertiary and Recent.

9.7.2.4.11 Cyclophorids, Valvatids, and Littoriaids

Superfamilies composed mainly of fresh water mesogastropods, but include some that are at home in brackish waters or live along coasts in the zone between high and low tides, are grouped together here. They have low conical to nearly flat coiled shells which are smooth or moderately sculptured, mostly with axially disposed ribs. The aperture is Holostomatous. Except few doubtfully genera of Pennsylvanian and Permian, the group is wholly post-Palaeozoic, having maximum development in Tertiary and Recent.

9.7.2.5 Order Neogastropoda

Neogastropoda (order; *neo* = new, recent) have a gill structure like that of the Mesogastropoda but have a siphonostomatous aperture, and typically provided with a well-developed canal. The range from Cretaceous to Recent and are less diverse than Mesogastropods. The Neogastropods comprise the more progressive, highly specialized Pectinibranchs, and as represented by living genera, are equivalent to the suborder Stenoglossa (*steno* = narrow), as they have a radula in which only one to three teeth occur in each row. Fossil Neogastropods are grouped in four superfamilies: Muricacea, Buccinacea, Volutacea, and Conacea.

9.7.2.5.1 Muricids

This group of Neogastropods is distinguished especially by the prominence of the canal that extends forward from the aperture and by the absence of columellar folds. The surface is strongly sculptured, and many shells bear knobby or spinose varices. Although Muricids were first noted in the Cretaceous, the majority of forms come from Tertiary.

9.7.2.5.2 Buccinids

The shells are mostly spindle-shaped and less strongly sculptured than the Muricids. The Buccinids have a long or short canal, and generally, the columella lacks folds. They range from the Cretaceous to Recent.

9.7.2.5.3 Volutids

These are characterized by egg-shaped to fusiform shells that mostly somewhat sculptured; some may bear fairly strong axial or spiral ribs and a few have knobby spines along the shoulders of the whorls. A well-developed canal may be present, or the anterior edge of the aperture merely contains a siphonal notch. The columella is generally marked by spiral folds. Many common living marine snails belong in this group. *Athleta*, *Cancellaria*, *Harpa*, *Liopeplum*, *Oliva*, *Olivella*, *Volutoderma*, and *Volutomorpha* are important constituents of various Cretaceous and Tertiary faunas.

9.7.2.5.4 Conids

The shell form of *Conus*, which gives this superfamily its name, is readily distinguished from that of other Neogastropods, because the elongate body whorl tapers evenly, with straight or gently curved sides, from the widest part of the shell, at the shoulder of the body whorl, to the anterior extremity. The spire may be moderately elevated or nearly flat, but it has a conical form, and the shell as a whole may be described as biconical or obconical. The aperture is slit-like. The conids occur rarely in the Cretaceous, but their number expands greatly in the Tertiary.

9.7.3 Class Opisthobranchia

Shell reduced in size, commonly internal or absent; mantle cavity and gill, where present, in rear position as a result of twisting back from prosobranch condition, gill commonly absent, being replaced by respiratory structure developed in the mantle or entire outer surface; neural loop not crossed in figure-eight (*opistho* = backward). The Opisthobranchs ranges from Pennsylvanian to Recent are exclusively marine form, and are much less important in the fossil record when compared with other classes. They possess diverse shell-less forms; only two groups have hard parts and hence, capable of preservation, the Pleurocoela (that live in shallow seas, like most Prosobranchs), and the Pteropoda (open-ocean pelagic snails).

9.7.3.1 Order Pleurocoela

Shell, mantle cavity, and gill present. They range from Mississippian to Recent. The Pleurocoela have conispiral shells; some are subglobular or ovoid in outline, others rather high conical with a rounded base, and still others biconical or obconical. A few have thick shells. Some are very delicate and ill-suited for preservation.

A few genera, such as the fairly common and widely distributed *Actaeonina*, are recorded from Mississippian to Recent. However, a large majority of them are restricted to the Mesozoic and Cenozoic.

9.7.3.2 Order Pteropoda

Slender conical shell present or absent; lack distinct head, foot modified as paired wing-like fins for swimming; gills present (*ptero* = wing; *pod* = foot). They range from?Cambrian-?Permian; Cretaceous-Recent. The Pteropods (wing-footed swimming gastropods), are generally elongated and bilaterally symmetrical. Only part of this group is provided with a shell, but the majority are naked. Both kinds swarm in many portions of the open sea. The calcareous covering of shelled forms is very thin and may be transparent. They are mostly very small, less than a half inch in length, and shaped like a very narrow straight-sided cone, but a few are spirally coiled. Some have flattened shells provided with lateral keels, and the aperture may be covered by a thin operculum. Abundance of pteropod shells in deep-sea deposits is the basis for calling them pteropod ooze. A widely distributed pteropod-like fossil in Cambrian rocks, and recorded at many places in Ordovician and Silurian strata, is termed *Hyolithes*. The genus occurs sparingly in younger Palaeozoic deposits as high as Permian. The shell has an elliptical or subtriangular cross section and narrows from the aperture to a sharp point. The surface is smooth or marked by fine longitudinal striations. An operculum fits over the aperture. That this fossil is really a gastropod has been doubted, for conceivably it belongs to some other, entirely extinct group of invertebrates. Fossils which are identified certainly as pteropods range from Cretaceous through Pleistocene, and these are a good deal smaller than most specimens of *Hyolithes*. Discovery of a Middle Cambrian *Hyolithes* in the Burgess shale of western Canada, which not only has the operculum joined to the aperture but shows two symmetrical impressions projecting laterally in the position of the paired wing foot of the pteropods, suggests that this ancient fossil belongs to Pteropods. If this is true, the range of the order is Cambrian to Recent. However, the Burgess Shale fossils do not at all prove the pteropod affinities of *Hyolithes*.

9.7.3.3 Order Sacoglossa

Shell lacking except in larval stage; no gills; unknown as fossils. Stratigraphic range is Recent.

9.7.4 Class Pulmonata

Mostly shell-bearing but lacking an operculum; neural cod not in form of figure of eight; mantle cavity modified as an air-breathing lung (*pulmo* = lung). Next to the

Prosobranchs, Pulmonate gastropods are the most numerous. Although they range from Pennsylvanian to Recent, they are quite common in Tertiary deposits. They are adapted to fresh water and terrestrial modes of living, and hence supplement marine invertebrates in furnishing paleontological materials for zonation and correlation of the rock column, inasmuch as they occur where marine shells are lacking. Pleistocene stratigraphy of the non-glaciated part of the Great Plains (USA) has been worked out reliably in recent years mainly because it has been found possible to differentiate and correlate widely distributed Pulmonate gastropod faunas within the sediments.

9.7.4.1 Order Basommatophora

Fresh water Pulmonates having eyes at the base of posterior tentacles (*bas* = base; *ommato* = eye; *phora* = carry). They range from Jurassic to Recent. This group is invariably provided with shells that range in shape from moderately elevated cones with rounded bases to discoidal forms, all classifiable as conispiral. A few left-handed genera, such as *Physa*, are very common, but most shells are dextral (right-handed). The aperture is holostomatous, and the inner lip is smooth or bears spiral folds. Callus deposits near the umbilical opening, if present, are inconspicuous or lacking, and accordingly most shells are either phaneromphalous or anomphalous, the latter possessing a columella. Genera represented by fossil shells belong to Basommatophora; this group assignment is based either by studying the soft parts of living representatives, or by the comparison of shell with that of some living member of the order.

9.7.4.2 Order Stylommatophora

These terrestrial Pulmonates have eyes at the tip of posterior tentacles (*stylo* = stalk). They range from Pennsylvanian to Recent. These land snails include many shell-less forms; others have thin to moderately thick calcareous conispiral shells that range from very low-spired to steep-sided, high-spired forms. Majority forms have many whorls, generally but not always, more than among basommatophorans. They are small to medium in size, averaging about 0.5 in. in length or width. The aperture, which mostly has a rather evenly rounded outer lip, has a thin peristome or bears a thickened rim. Tooth-like projections of the inner lip, and in some shells, in the outer lip also, modify the appearance of aperture. Pupidae are characterized by such features, and because of their close resemblance of some Pennsylvanian non-marine gastropods to modern *Pupa*, these Palaeozoic fossils are judged to belong among the Stylommatophorans.

9.8 Geological History and Distribution

The geological range of the Gastropoda is from Early Cambrian to Present. The oldest gastropod, *Aldanella attleborensis* (Shaler and Foerste), comes from basal Tommotian ($\sim$525–521 Ma; see Peng et al. 2012 for chronology) in northern Siberia and Newfoundland strata; it is an mm-sized, low trochospiral, and dextrally coiled shell (Figs. 9.1, 9.10). Gastropods are not common in the Cambrian, but at least nine families of amphigastropods and archaeogastropod prosobranchs are noted that include non-coiled cap-shaped forms, and planispiral, orthostrophic conispiral, and hyperstrophic conispiral shells (Fig. 9.7; Cambrian forms). The

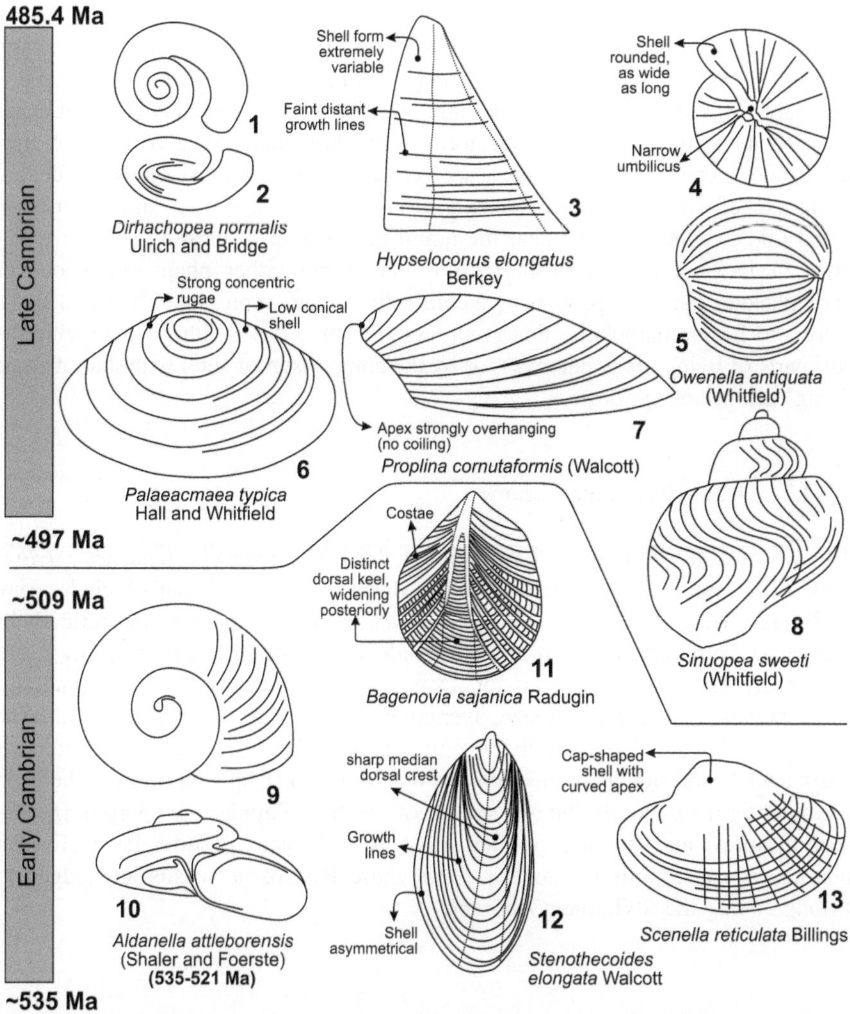

Fig. 9.7 Early to Late Cambrian gastropods and their major distinguishing characters

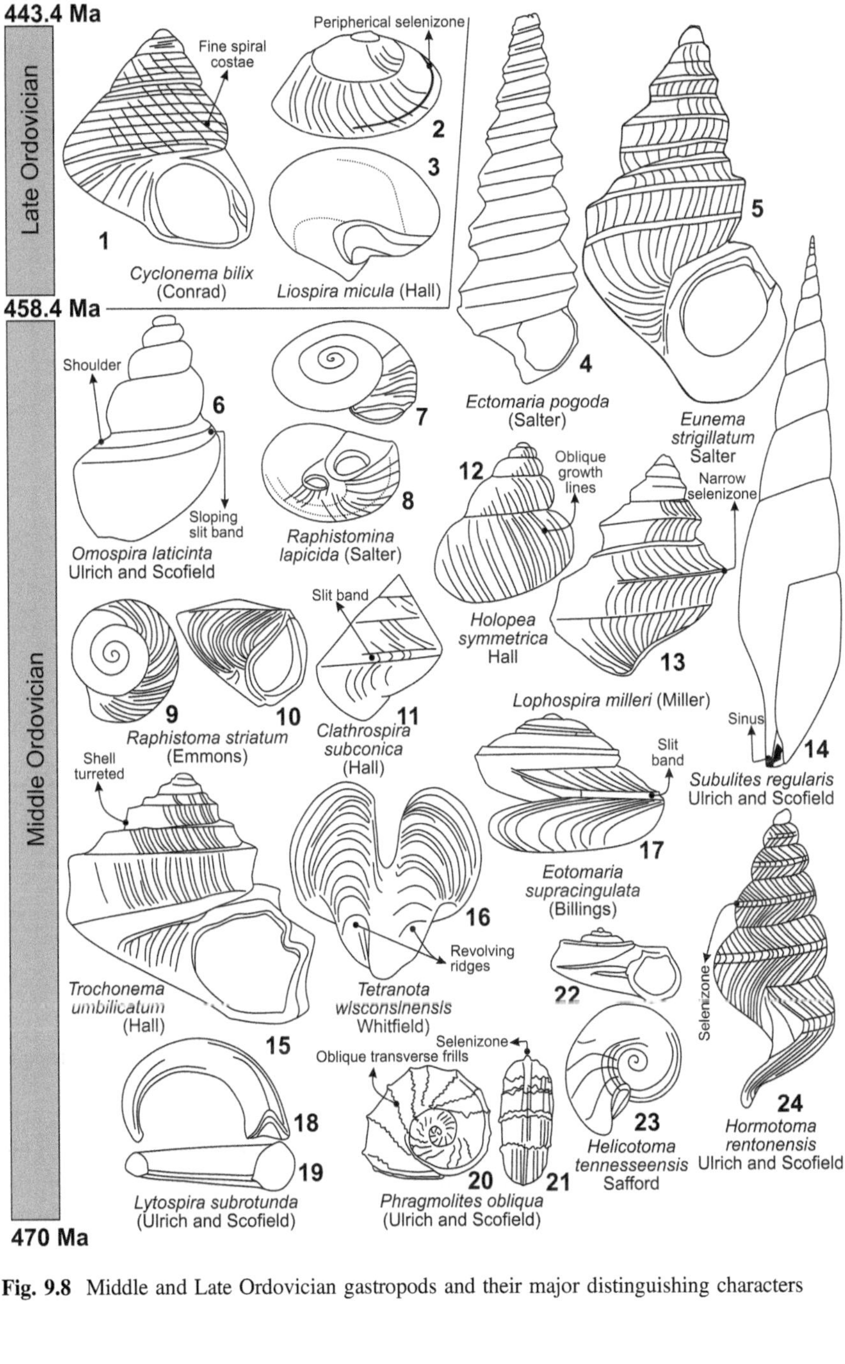

Fig. 9.8 Middle and Late Ordovician gastropods and their major distinguishing characters

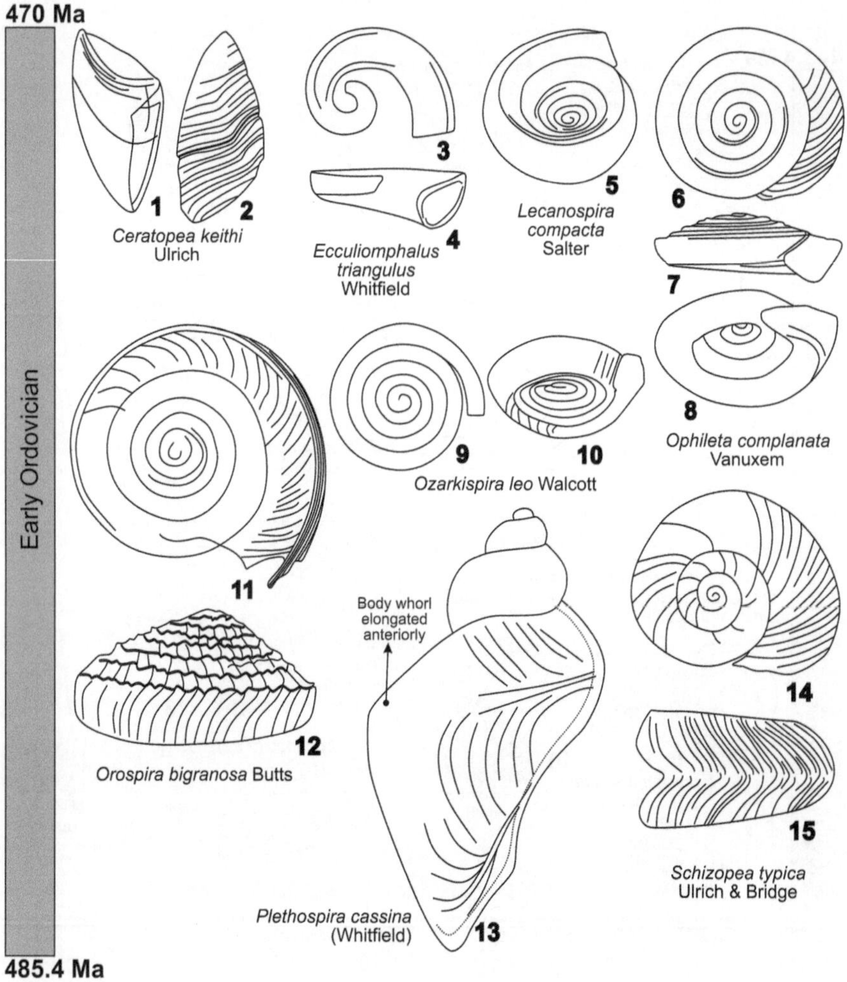

Fig. 9.9 Early Ordovician gastropods and their major distinguishing characters

beginning of the amphigastropods and all main divisions of the early Prosobranchs are recognized within the Cambrian strata, itself.

Ordovician records a great diversity; forms belonging to families that previously were in the Cambrian and 14 additional ones. Nearly all of these are confined to the Palaeozoic rocks. Eight additional families of archaeogastropods appeared in the Silurian strata, two in the Devonian, one in the Mississippian, and three in the Permian (Figs. 9.8, 9.9, 9.10, 9.11, 9.12, 9.13, 9.14, 9.15, 9.16, 9.17, 9.18, 9.19, 9.20, 9.21 and 9.22).

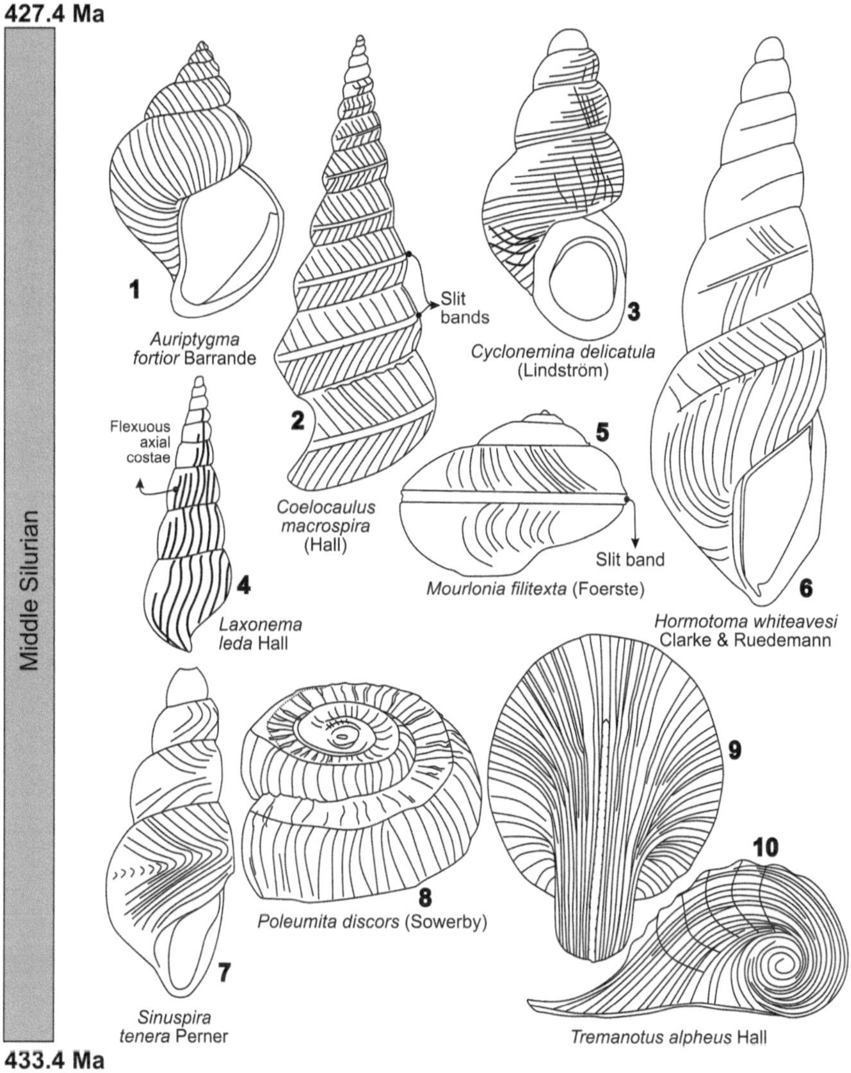

Fig. 9.10 Middle Silurian gastropods and their major distinguishing characters

The Mesogastropods range through Mesozoic rocks but increase in the Cenozoic, whereas Neogastropods are confined to Cretaceous and post-Cretaceous deposits. All but a few superfamily groups of these two orders are more abundant today than at any time in the past.

The Opisthobranch gastropods are recorded from Mississippian to Recent. If the narrow conical Hyolithids are pteropods, which is very doubtful, the Opisthobranch group is as old as Cambrian. The Opisthobranchs are far outranked by Prosobranchs in paleontological importance.

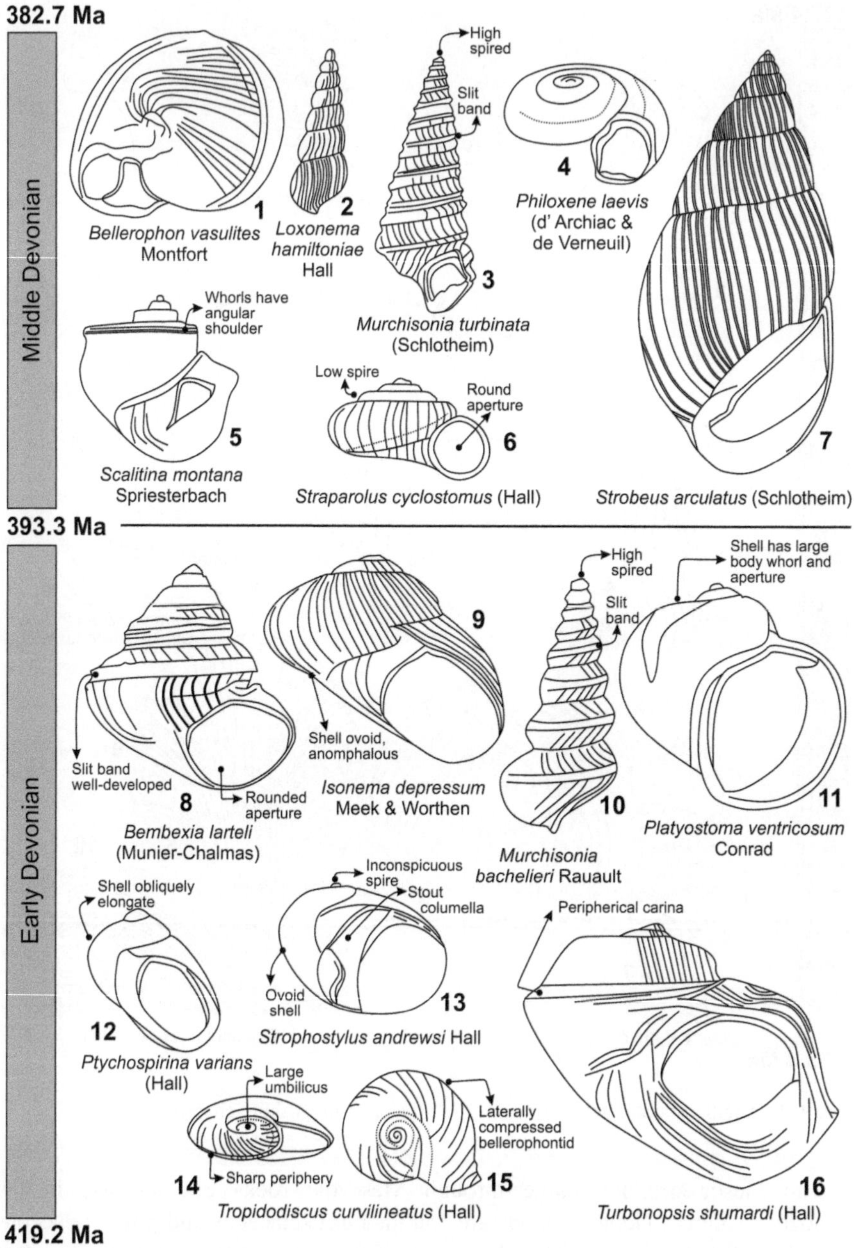

Fig. 9.11 Early and Middle Devonian gastropods and their major distinguishing characters

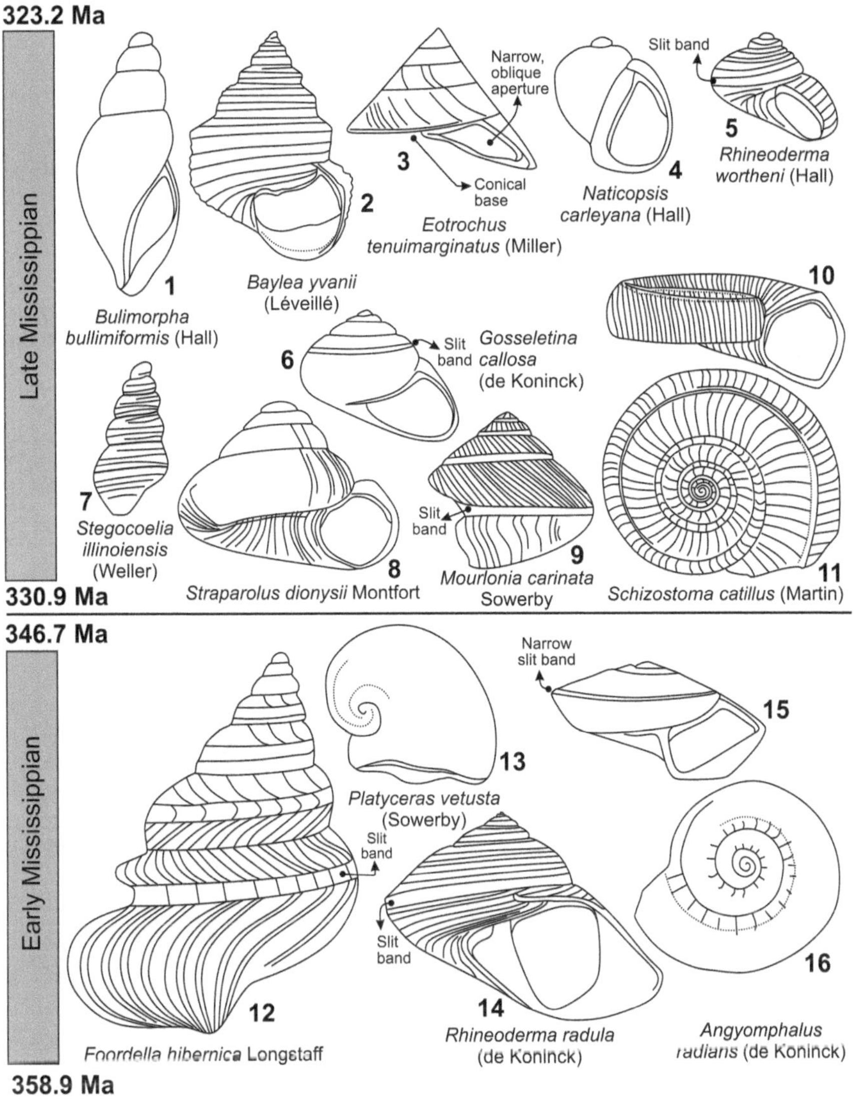

Fig. 9.12 Early and Late Mississippian gastropods and their major distinguishing characters

The Pulmonate gastropods are late arrivals, due to their evolutionary advancement in being adapted to life on land. A few forms belonging to this class seem to be represented among the Pennsylvanian fossils, but the main geological record comes from the Cenozoic. Modern faunas include nearly 1000 genera of Pulmonates. Thus, they seem to be at the peak of their development.

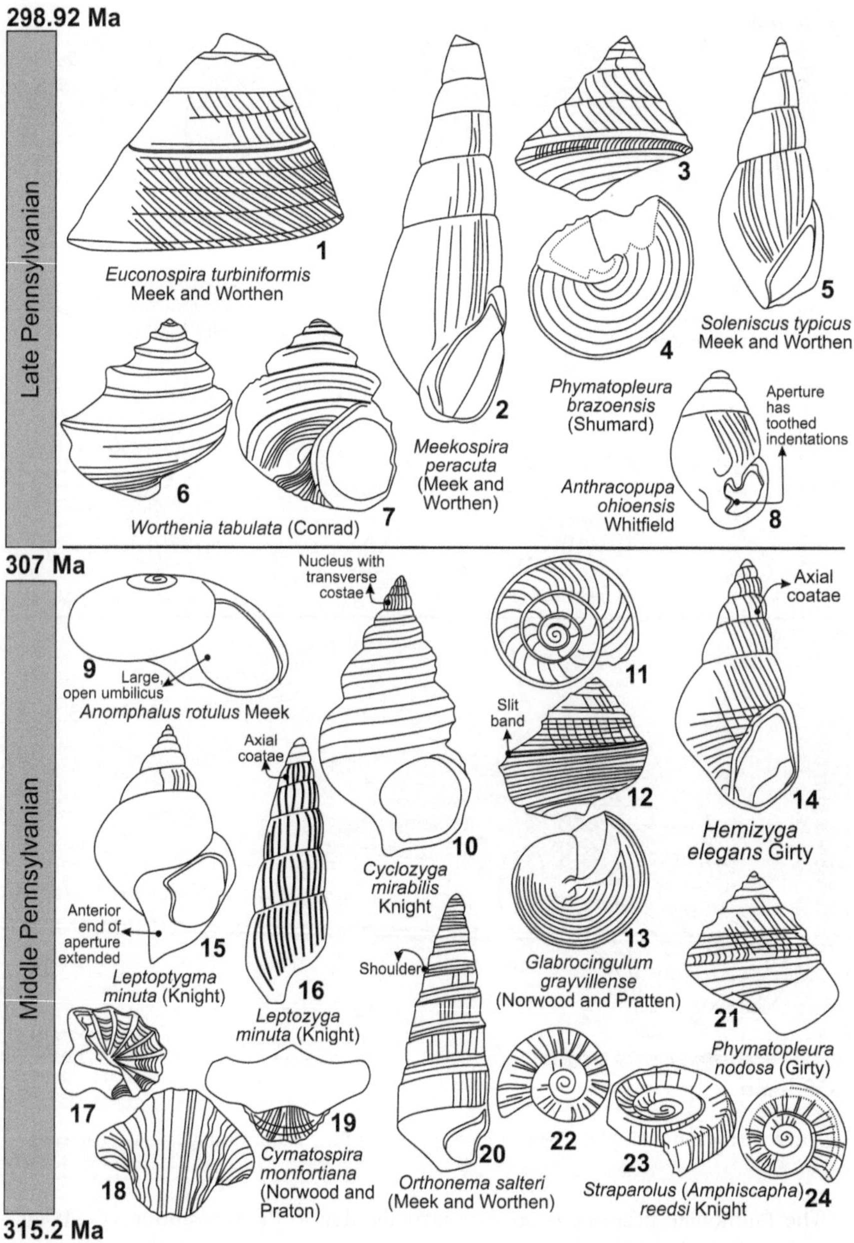

Fig. 9.13 Middle and Late Pennsylvanian gastropods and their major distinguishing characters

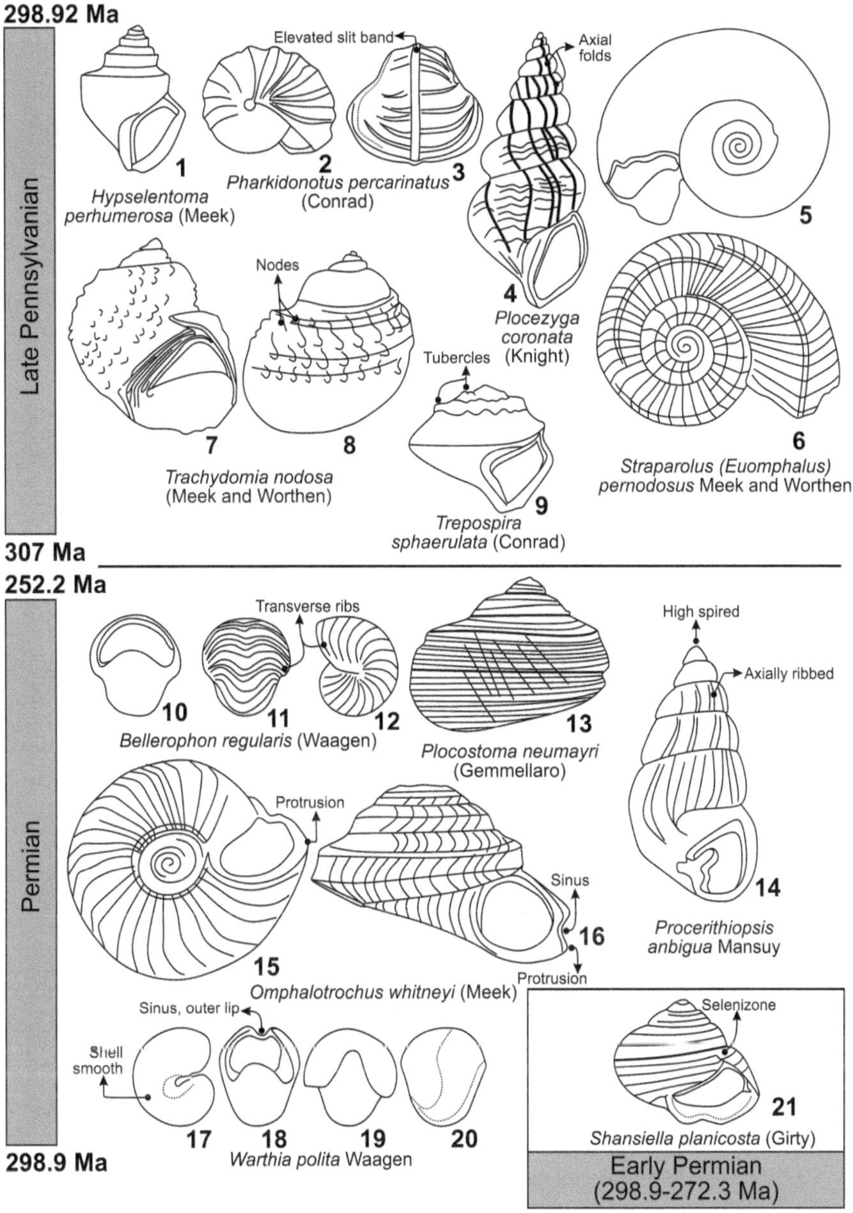

Fig. 9.14 Late Pennsylvanian and Permian gastropods and their major distinguishing characters

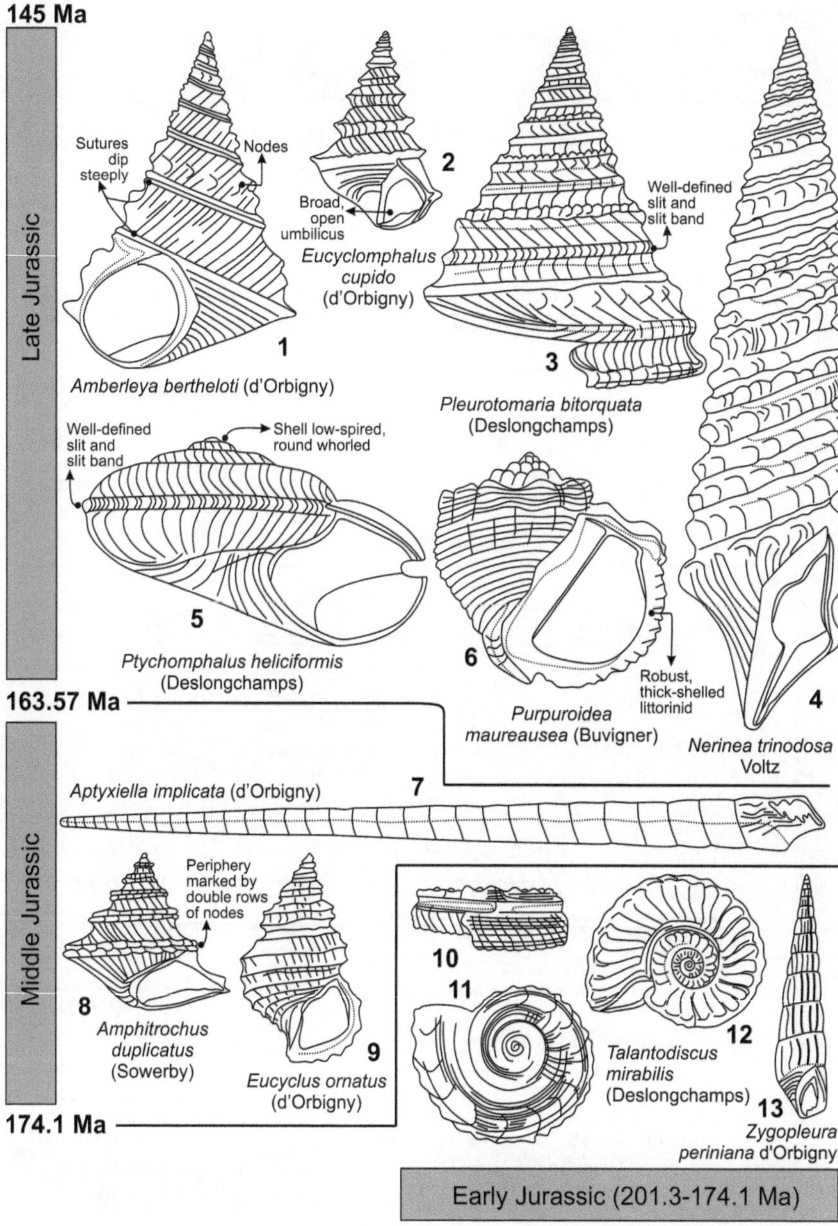

Fig. 9.15 Early, Middle and Late Jurassic gastropods and their major distinguishing characters

Appendix 1 gives the list of illustrated specimens mentioning the chapter number, species name, age and locality along with its figure number within the said chapter.

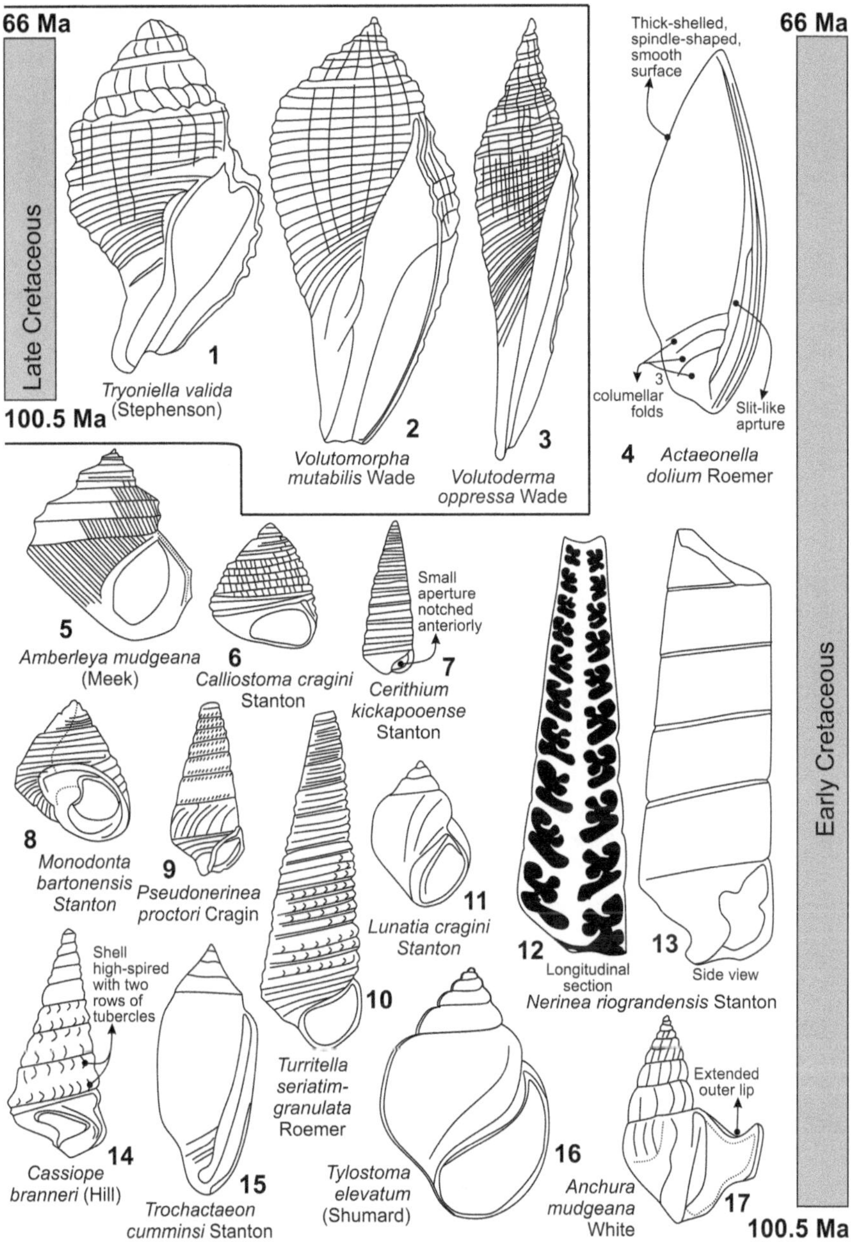

Fig. 9.16 Early and Late Cretaceous gastropods and their major distinguishing characters

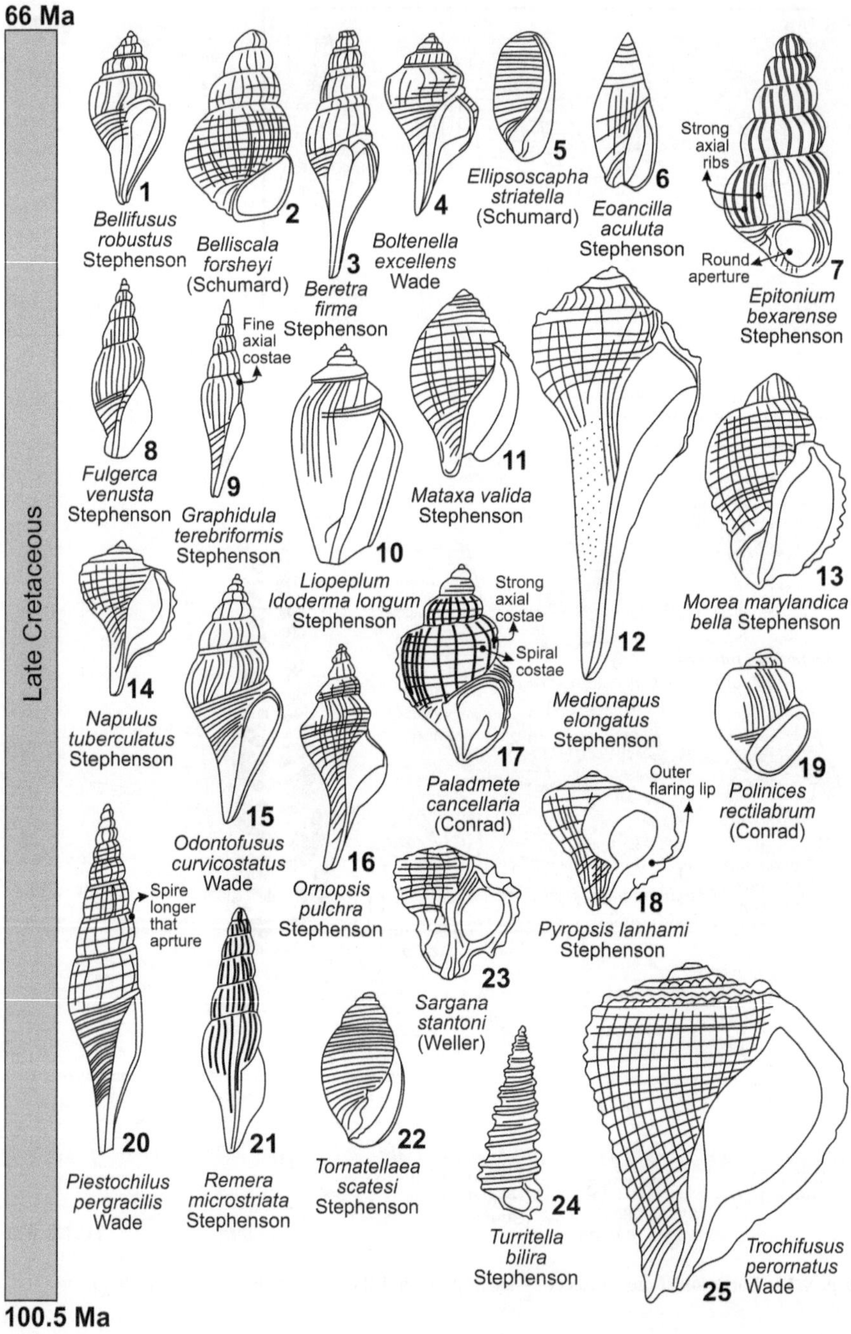

Fig. 9.17 Late Cretaceous gastropods and their major distinguishing characters

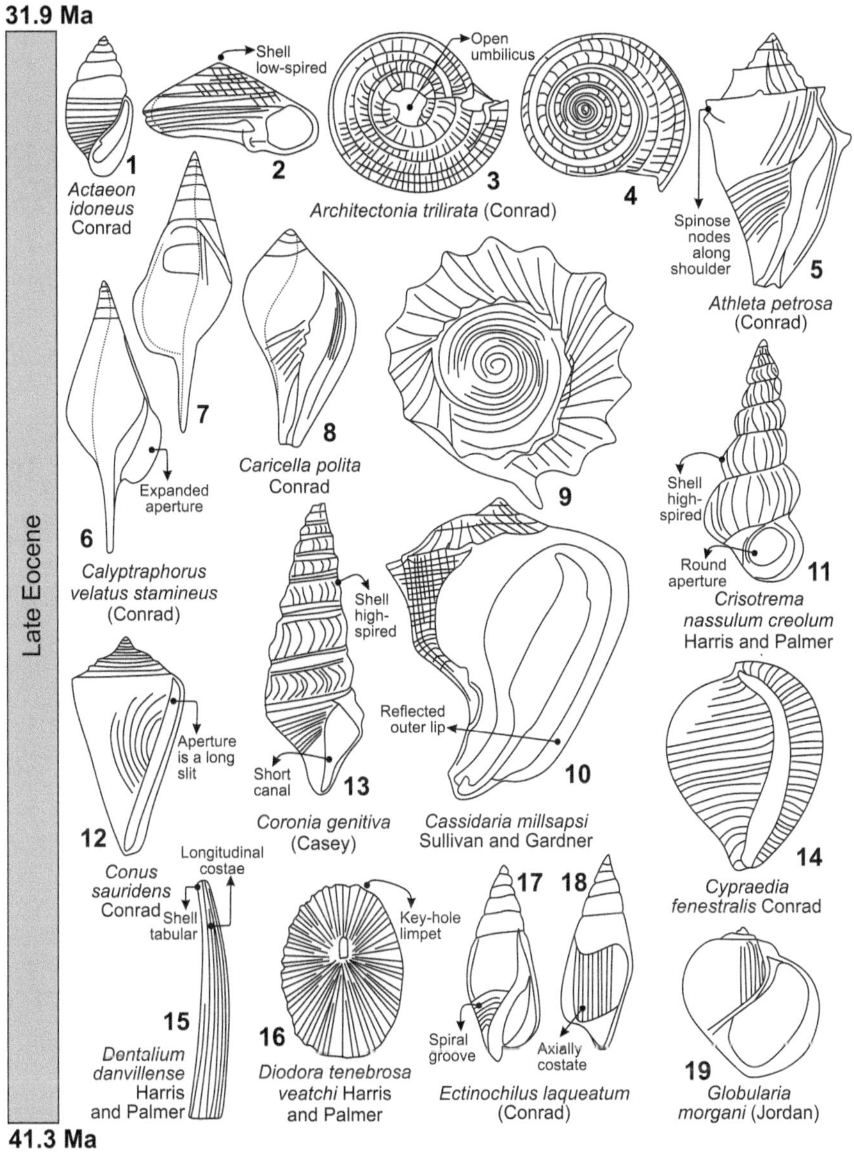

Fig. 9.18 Late Eocene gastropods and their major distinguishing characters

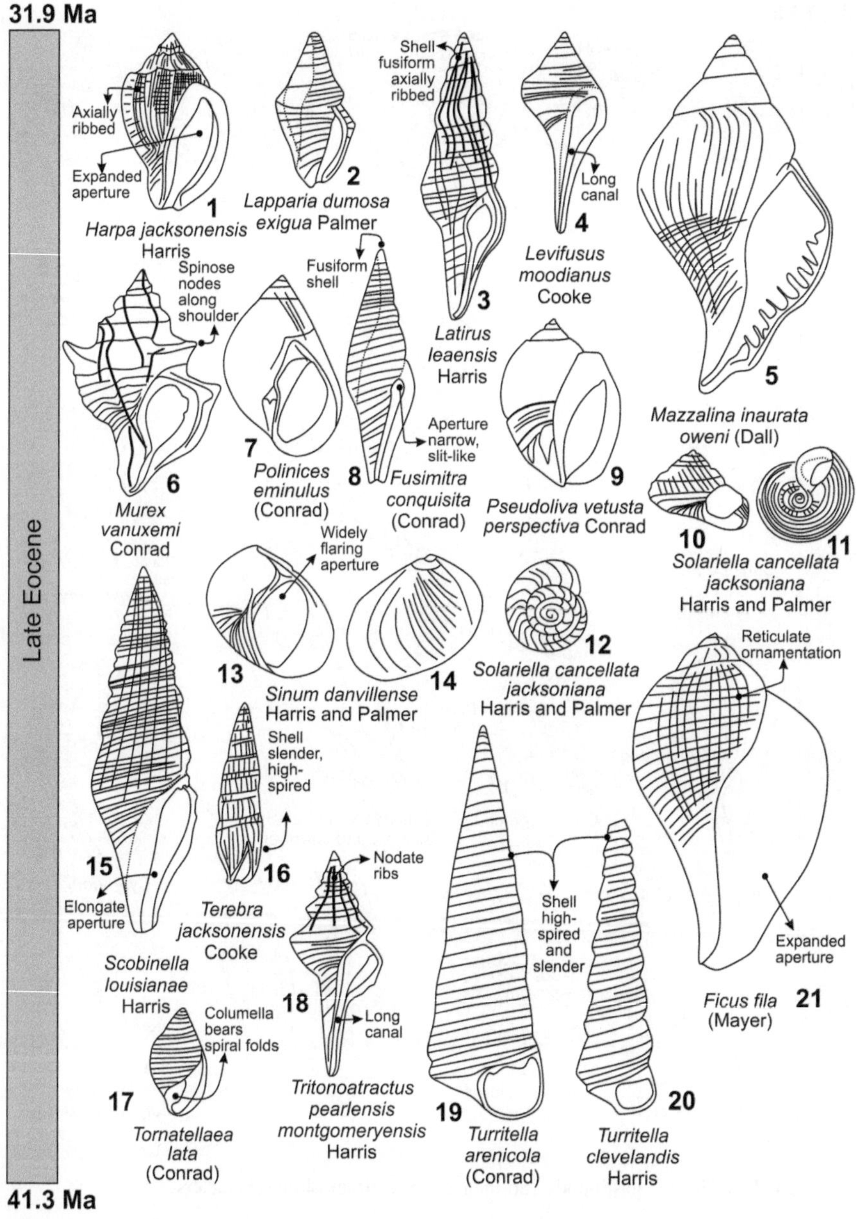

Fig. 9.19 Late Eocene gastropods and their major distinguishing characters

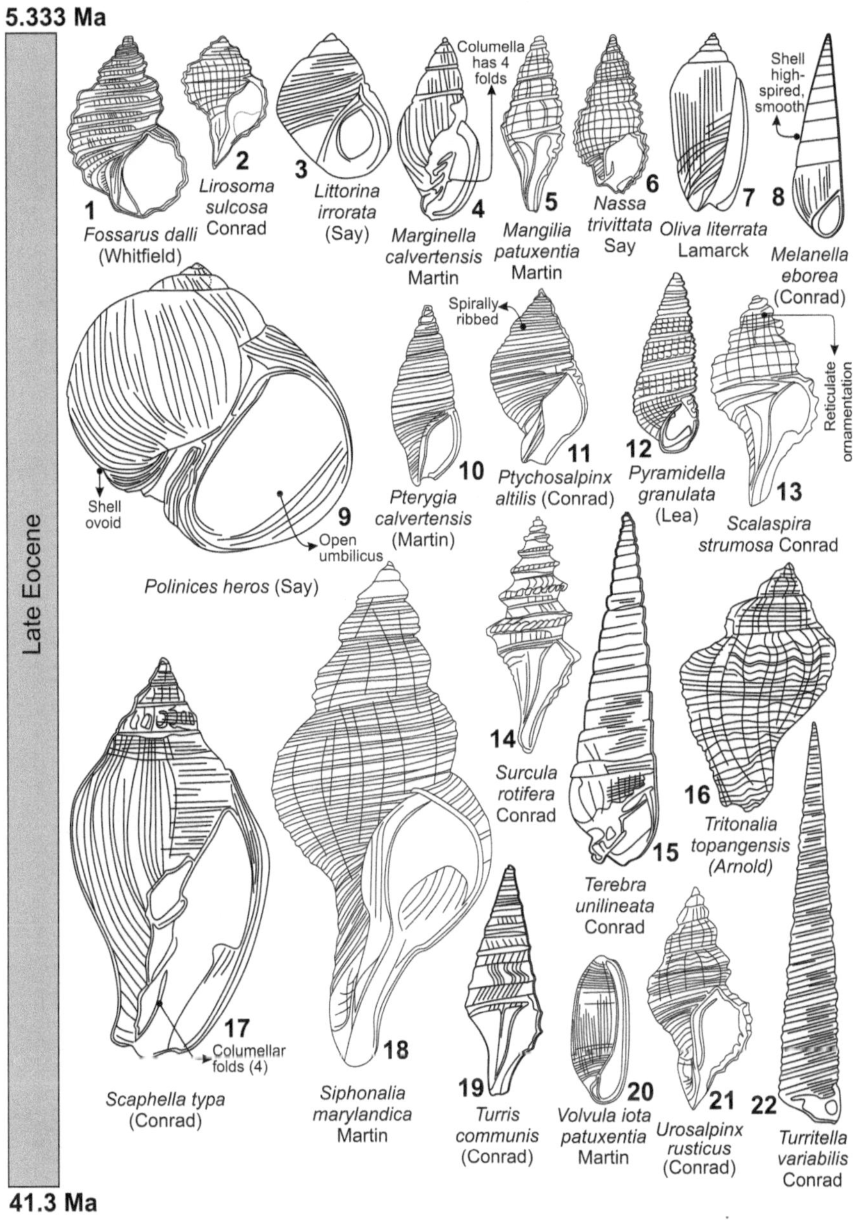

Fig. 9.20 Late Eocene gastropods and their major distinguishing characters

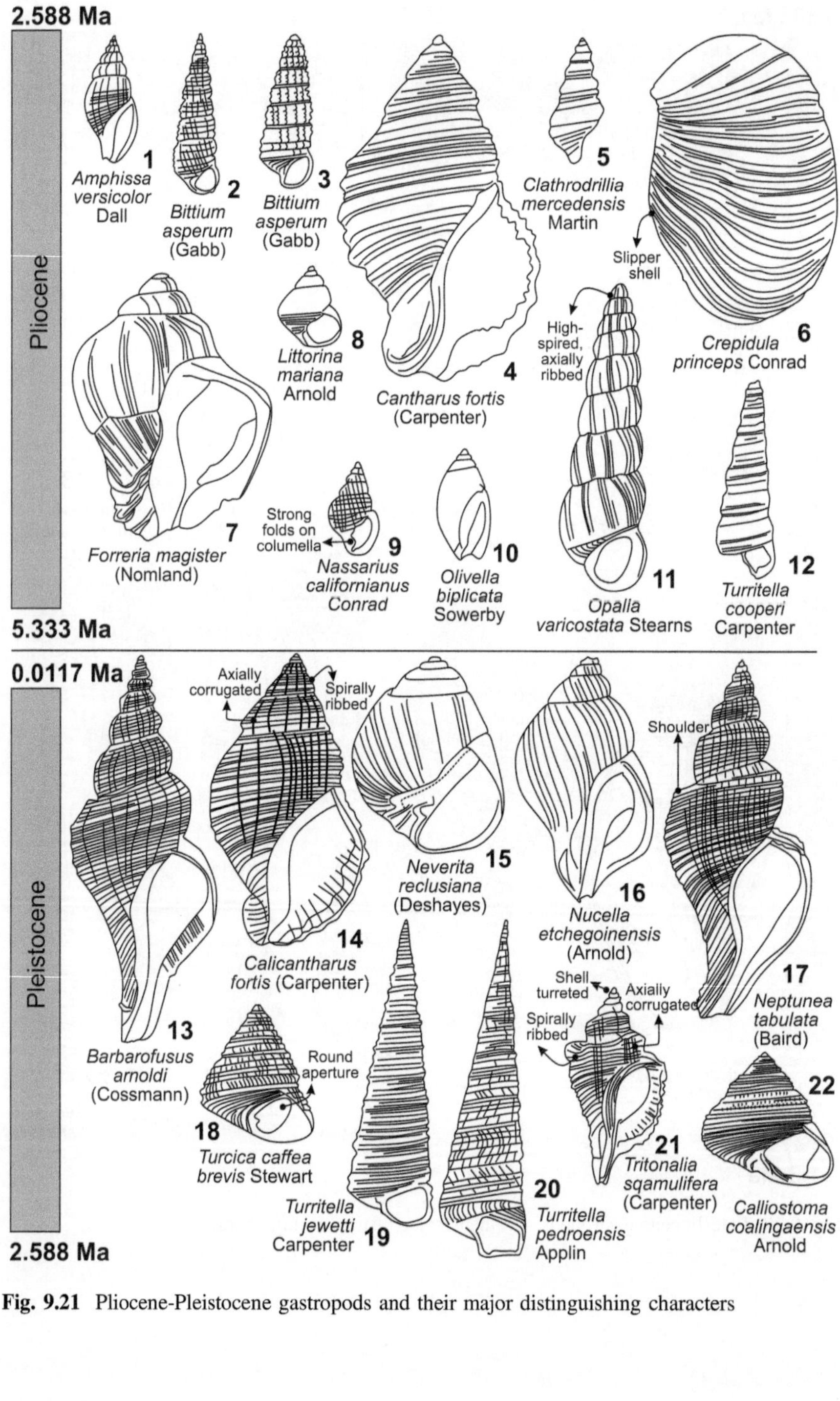

Fig. 9.21 Pliocene-Pleistocene gastropods and their major distinguishing characters

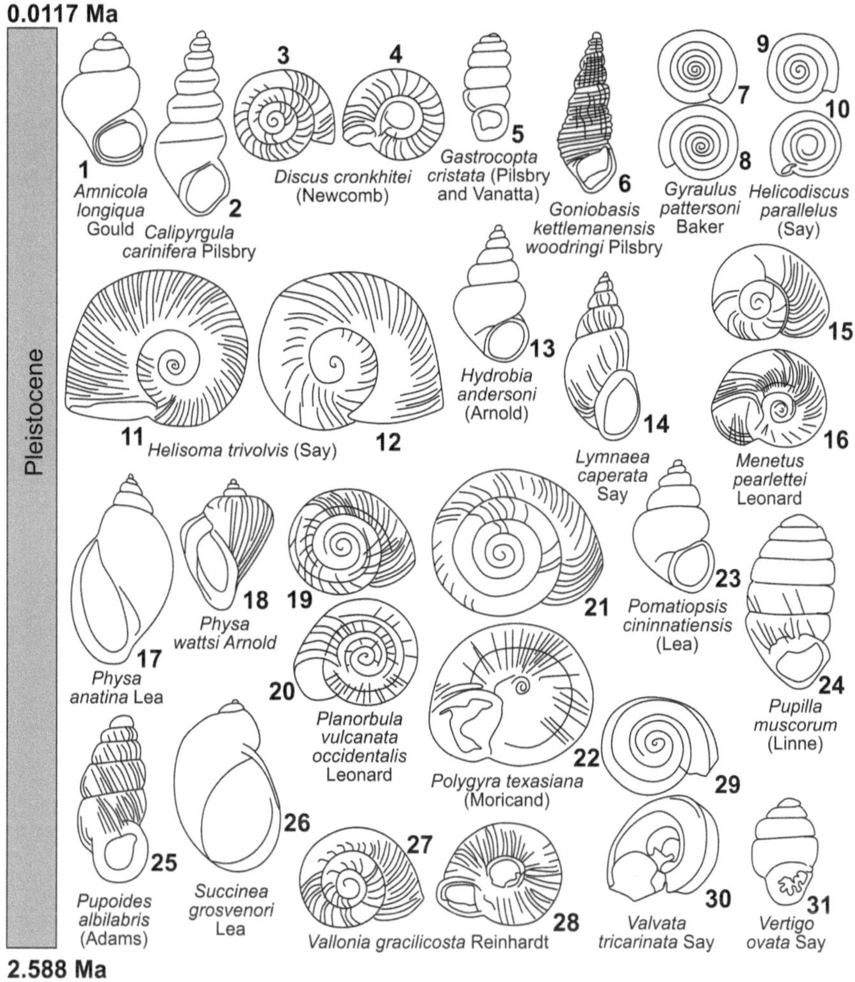

Fig. 9.22 Pleistocene gastropods and their major distinguishing characters

References

Bandel, K. 1995. Mathildoidea (Gastropoda, Heterostropha) from the Late Triassic St Cassian Formation. *Scripta Geologica* 111: 1–83.

Bandel, K. 1996. Some heterostrophic gastropods from Triassic St. Cassian Formation with a discussion on the classification of the Allogastropoda. *Paläontologische Zeitschrift* 70: 325–365.

Bandel, K. 2002. About the Heterostropha (Gastropoda) from the Carboniferous and Permian. *Mitteilungen aus dem Geologisch-Paläontologischen Institut der Universität Hamburg* 86: 45–80.

Bowles, E. 1939. Eocene and Paleocene Turritellidae of the Atlantic and Gulf Coastal Plain of North America: *Jour. Paleontology* 13: 267–336.

Clarkson, E.N.K. 1993. *Invertebrate palaeontology and evolution*. London: Chapman and Hall. 434 p.

Dall, W.H. 1909. Contributions to the tertiary paleontology of the pacific coast, I, The Miocene of Astoria and Coos Bay, Oregon: *U.S. Geological Survey Professional Paper* 59: 1–278.

Peng, S.C., Babcock, L.E., Zuo, J.X., Zhu, X.J., Lin, H.L., Yang, X.F., Oi, Y.P., Bagnoli, G., Wang, L.W. 2012. Global Standard Stratotype-Section and Point (GSSP) for the Base of the Jiangshanian Stage (Cambrian: Furongian) at Duibian, Jiangshan, Zhejiang, Southeast China. *Episodes* 35: 462–477.

Ponder, W.F., and D.R. Lindberg. 1997. Towards a phylogeny of gastropod molluscs: an analysis using morphological characters. *Zoological Journal of the Linnean Society* 119: 83–256.

Chapter 10
Corals

10.1 Introduction

Corals belong to Phylum Cnidaria (*Ni-dá-ri-a*: corals, sea anemones, hydra, jelly-fish, sea fans, and the Portuguese man-of-war) that contain stinging cells [Cnidoblasts: variously referred to as Nematocyte or Cnidocyte; Fig. 10.1(1)] commonly located in tentacles. The tentacles are the extension of the Ectoderm (=outer Epidermis). A cnidoblast (stinging cell) is an explosive cell containing a secretory organelle or cnida for prey capture and as a mechanism of defense from predators. Although, many corals appear to be a single organism (a corallite), are in fact a colony of many individual (corallum), and yet with genetically identical coral polyps [Fig. 10.1(2)].

The Cnidarians have two basic body plans: swimming medusae and sessile polyps [Fig. 10.1(3)], both are radially symmetrical with mouths surrounded by tentacles that bear cnidocytes. The polyp is somewhat cylindrical [Fig. 10.1(3)] with a body wall surrounding the Enteron [Fig. 10.1(3)], and an oral disk (or surface) in which a central mouth is surrounded by one or more circlets of tentacles, consisting of a basal disk [Fig. 10.1(4, 5)] commonly attached to a substrate. The body wall of a polyp is divided into an inner Gastrodermis (Endoderm) and an outer Epidermis (Ectoderm), and separated by a third layer, the gelatinous Mesogloea [Fig. 10.1(3)]. Coral polyps have a ciliated Pharynx [Fig. 10.1(5)], a tube extending from the mouth down into the Enteron [Fig. 10.1(3)]. Radiating vertical partitions in the enteron are called Mesenteries [Fig. 10.1(4, 5)] and these are attached to the body wall of the polyp and the oral disk. Some of the mesenteries are also attached at their inner ends to the pharynx and these are said to be complete; others are not, and are accordingly called incomplete; mesenteries consist of an infold of gastro-dermis enclosing the mesogloea [Fig. 10.1(3–5)]. The position and number of mesenterial pairs can be deduced in fossil corals as skeletal partitions (septa) grow upward in positions corresponding to the space between the mesenteries of each pair. The arrangement of early mesenteries in the larva defines a plane of symmetry

S. Jain, *Fundamentals of Invertebrate Palaeontology*, Springer Geology,
DOI 10.1007/978-81-322-3658-0_10

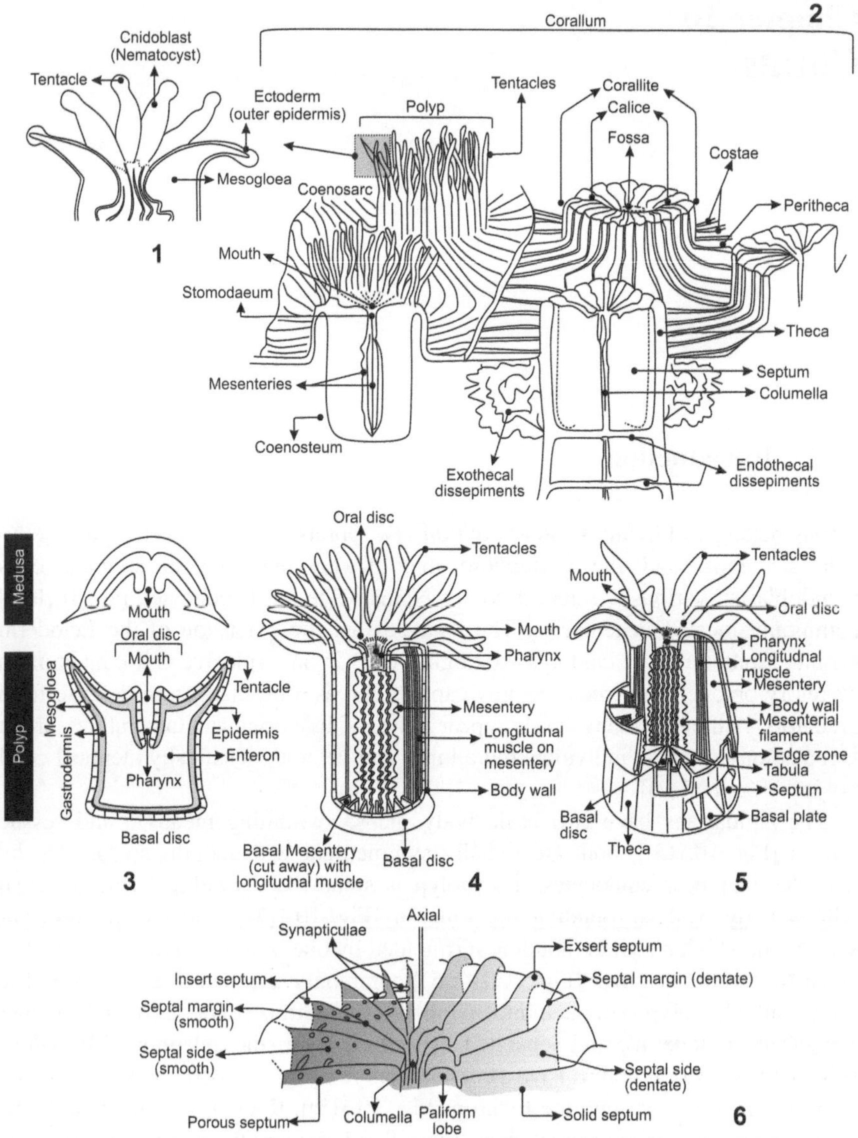

Fig. 10.1 Coral morphology. *1* Corals contain stinging cells called Cnidoblasts (often referred as Nematocyte or Cnidocyte) are located in tentacles. The tentacles are extension of the ectoderm, the outer epidermis. *2* The internal structure of a Corallum. *3* The two basic body plans of a Cnidarian—the swimming medusae and sessile polyp. *4* Sectional view of anthozoan polyp showing the pharynx and radially arranged mesenteries that characterize anthozoan polyps. *5* Sectional view of a young solitary Scleractinian coral. *6* Top view of a corallite showing septae

by which all other adult structures are oriented. Subsequent mesenteries appear in couples symmetrically about the plane in a definite sequence, position, and number that forms the basis of classification into different subclasses and orders (see also Hill 1956, 1981; Wells 1956; Moore 1956; Oliver and Coates 1987; Clarkson 1998).

10.2 Classification

Class Anthozoa are exclusively marine solitary or colonial polyps that differ from other Cnidarians in lacking the medusoid stage (see Table 10.1). Geologically the anthozoans are important as their polyp produces calcified skeletons that are preserved as fossils. The anthozoans are divided into two subclasses, Octocorallia and Zoantharia. The Octocorallia (octocorals or soft corals) have a poor fossil record but are well represented in modern seas. The Zoantharia are common as fossils and are divided into three orders; the Tabulata, Rugosa, and Scleractinia (see also Table 10.1). Most of the fossils of these orders are referred informally as "corals." Class Anthozoa is divided into three subclasses—Ceriantipatharia, Octocorallia, and Zoantharia. Only Zoantharia are considered here (see also Table 10.1). Their general characteristics are given in Fig. 10.2.

10.2.1 Subclass Zoantharia ("Corals"; Cambrian-Recent)

This exclusively marine subclass includes corals (with skeletons) and sea anemones (without skeletons). Most Zoantharia are colonial and reef builders, living as deep as 90 m in the present oceans (Piper 2007). However, some solitary forms live as deep as 1000 m whereas rare occurrences are even noted at the depth of trenches ($\sim$6000 m) (Broadhurst and Simpson 1972). Zoantharia prefer warm shallow seas and are attached to pebbles or boulders when the current is strong or to different substrates such as sand patches when the current is weak. There are eight extinct orders based on differences in their skeletal structure, septal arrangement, and mode of colony formation (Oliver 1980, 1996); all have calcitic skeletons, except for Kilbuchophyllida and Numidiaphyllida that were probably aragonitic (Scrutton 1997).

Three orders are detailed here namely Rugosa, Tabulata, and Scleractinia (see also Table 10.1). The epithecal wall, radial septa, and tabulae are central to a coral structure and its classification; these vary in number and in prominence, thus, separating one order from the other (see Fig. 10.2).

Each coral skeleton, the Corallum [Figs. 10.1(2) and 10.3(1)], is composed of relatively few characters. Common to all corals is Epitheca, the outer wall [Fig. 10.3(1)] which is often irregular in form and has a rough (rugose) texture, usually with a series of incremental growth bands on its surface [Fig. 10.2(1)].

Table 10.1 Simplified classification of Phylum Cnidaria

Phylum Cnidaria
- Tissue-grade organisms
- diploblastic metazoa with radial, biradial or radial bilateral symmetry lacking a through gut or anus; tentaculate, equipped with specialized stinging cells (cnidoblasts or nematocysts).
- Generally polymorphic.

  Class Hydrozoa
  - Radial or tetrameral symmetry
  - polypoidal stage dominant
  - typically colonial and polymorphic
  - hydrocorallines (fire coral) Tertiary to Recent (Milliporina and Stylasterina)
  - trachylines (entirely medusoid) ?Jurassic to Recent
  - siphonophores (complex medusoid colonies lacking hard parts), Ordovician to Recent
  - actinulids

  Class Scyphozoa
  - predominantly marine and medusoid
  - radial tetrameral symmetry
  - thick mesoglea
  - fission and strobilation
  - no velum
  - Ediacaran to Recent

  Class Anthozoa
  - marine
  - no medusoid stage, always possess a stomodaeum

    Subclass Ceriantipatheria
    - no fossil record

    Subclass Octocoralla
    - horny branching skeletons, commonly spiculate or with calcified core, anastomosing branches, includes gorgonians, eight tentacles on polyps.
    - Pennatulacea are also present in Ediacaran faunas, first spicules are Ordovician, good gorgoneans first appeared in Cretaceous.

    Subclass Zooantheria
    - includes unmineralized anemones

      Order Tabulata
      - Calcite corallum, invariably colonial, individual corallites small, prominent tabulae, septae reduced or commonly absent
      - Early Middle Ordovician through Permian

      Order Rugosa or Tetracorallia
      - Calcite corallum, solitary or colonial, septal insertion in quadrants. commonly forming a biradial symmetry, epitheca normally present and commonly rugose
      - Middle Ordovician (Blackriveran) to Permian
      - proseptum divided into cardinal (C) and counter (K) septae; alar (A) septae are next inserted adjacent to the cardinal septum; counter-lateral (KL) septae are then inserted next to the counter septum
      - fossulae occur because of lack of metaseptae inserted in later growth stages adjacent to cardinal or alar septae.

      Order Scleractinia or Hexacorallia
      - Aragonite corallum, solitary or colonial, septal insertion between directing mesentaries in multiples of six
      - Middle Triassic to Recent
      - Probably evolved from unmineralized Zoantherian after the Permo-Triassic extinction event
      - hermatypic forms with Zooxanthellae
      - reef ecology hinges around Zooxanthellae (largely governed by photic zone, temp constraints, latitudinal restrictions, etc.)

Shaded portion is detailed in this book (see text for explanation)

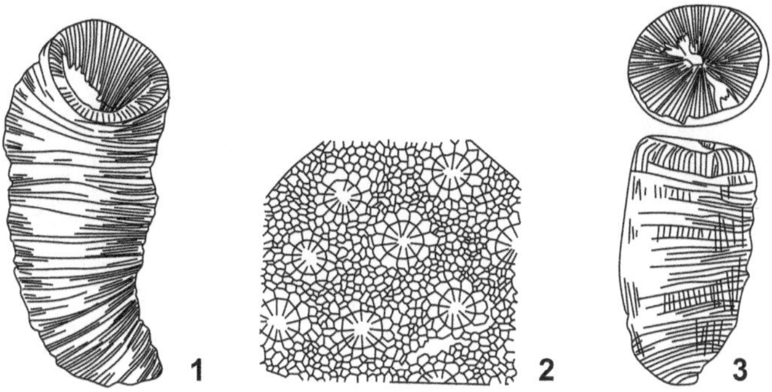

1 *Heliophyllum halli* Edwards and Haime

2 *Heliolites interstinctus* (Linne)

3 *Montlivaltia nattheimensis* Milaschewitsch

Coral group / **Features**	**Rugose**	**Tabulate**	**Scleractinia**
Growth mode	colonial and solitary	colonial	colonial and solitary
Speta	6 prosepta; later septas in only 4 spaces	septa weak or absent	6 prosepta; later septas in all 6 spaces
Tabulae	usual	well-developed	absent
Skeletal material	calcite	calcite	aragonite
Stability	poor	poor	good with basal plate
Range	Ordovician-Permian	Ordovician-Permian	Triassic-Recent
Colonial integration	Low	High	Low to very high

Fig. 10.2 Identification key for the three main groups of hard corals showing their distinctive morphology

These growth bands are added by the polyp which secretes the shell at its junction with the top surface of the skeleton, the Calice [Figs. 10.1(2) and 10.3(1)]. Each c the polyp and the skeleton. In solitary (individual) corals an individual calice takes up the whole of the corallum surface [Figs. 10.1(2) and 10.3(1)]. In colonial (compound) corals there may be many corallites, each with its own calice serving one of the many polyps of the colony [Fig. 10.1(1)].

In all the three coral orders, some of the interior structural elements within the coral skeleton are not developed; only Tabula and Septa are (singular: Tabula and Septum, respectively). The Tabulae [Figs. 10.2 and 10.3(1, 6, 12)] approximate to the horizontal that are added as the coral grows upwards; they represent the development of successive calices with growth of the coral. Septa [Fig. 10.3(2–5)] are vertical walls that are radially arranged so that they extend inwards from the inner wall of the epitheca [Figs. 10.3(1, 2)]. Septa supports mesenteries, the folded inner surface of the enteron [Fig. 10.1(4, 5)]. With continued coral growth, more septa are added radially [Fig. 10.3(3–5)]. The pattern of septal additions is important to the classification of a coral type (Fig. 10.2).

The most complex of all corals, the Rugose corals [Fig. 10.2(1)], have two additional structural components to their skeleton—Axial complex and Dissepiments [Fig. 10.3(5, 6)]. The Axial complex is characteristic and comprises

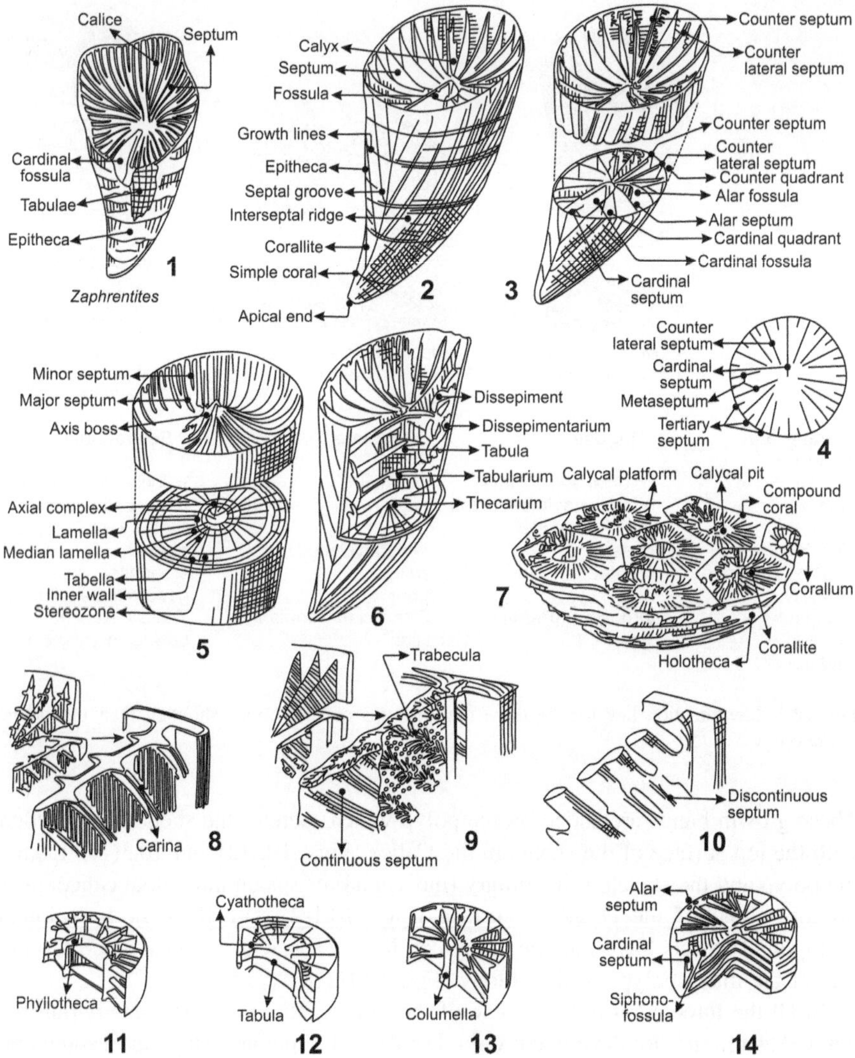

Fig. 10.3 Rugose coral morphology

of an array of features, ranging from a single column of calcite [the Columella; Fig. 10.3(13)] to a series of net-like and web-like structural elements [Fig. 10.3(5)]. The Axial complex (sometimes also mentioned as the Axial structure) generally occupies the central axis of the corallum [Fig. 10.3(5)]. The Dissepiments are generally restricted to the outermost part of the shell's interior, close to the interior of the outer wall, the epitheca [Fig. 10.3(6)]. They serve as a junction between the tabulae and the interior of the epitheca, thereby improving the fit of the polyp to the calice [Fig. 10.3(6)].

Coral shapes are highly variable (Figs. 10.4 and 10.5), as also their modes of colony formation [Fig. 10.4(23–30)]. The corallites can be arranged in certain pattern, thus aiding in their classification. Major types of colony formation are enumerated in Table 10.2 and illustrated in Figs. 10.4 and 10.5.

As mentioned above, rugose, tabulate and scleractinian corals contribute to the fossil record. Both rugose and tabulate corals are characteristic of the Palaeozoic but were wiped out in the end-Permian extinction; to be replaced by scleractinians, sometime in the mid-Triassic. The scleractinians continued to proliferate to the present day. The general characteristics of these three fossil groups are briefly given below.

10.2.1.1 Rugose Corals

The rugose corals:

- are extinct, solitary, and colonial and are relatively complex.
- have calcite corallum, tabulae, septa, dissepiments, and an axial complex.
- have dissepiments organized into a special area called the Dissepimentarium.
- have a variable axial complex.
- have septal insertion in quadrants.
- have biradial symmetry; the epitheca is generally present and is mostly rugose.
- have proseptum divided into cardinal (C) and counter (K) septae; the Alar (A) septae are next that are inserted adjacent to the cardinal septum. The counter-lateral (KL) septae are then inserted next to the counter septum.
- have fossulae due to the absence of metaseptae; the latter are inserted in later growth stages adjacent to the cardinal or alar septae.
- have septae that are inserted in four specific areas, forming gaps called Fossulae, with the development of a bilateral symmetry.

The Rugose corals are also called “Tetracoralla” or “Horn corals” due to their unique horn-shaped chamber with a wrinkled or rugose wall or epitheca [Figs. 10.2 (1) and 10.3(1)]. The solitary ones are generally horn-shaped and hence, called “Horn corals” (such as the Carboniferous genus *Zaphrentites*). The colonial types commonly possess hexagonal corallites. These individual corallites are either fused together forming massive colonies (compound coralla) or remain separate to form fasciculate colonies. The solitary corals range in diameter from few mm in length to 14 cm, with a height of almost a meter. Some colonies are as big as 4 m in diameter.

Internally, the rugose coral skeleton is composed of tabulae and septae, but the skeleton characteristically possesses an axial complex, and a zone of dissepiments (dissepimentarium) around the outer edge of each corallite (Fig. 10.3). It is perhaps these two characteristics, more than any other, which helps distinguish the rugose corals from the other two skeletal types (Fig. 10.2). The axial complex is unique to rugose corals [Fig. 10.3(5, 6)] and consists of modifications of the major septa to form a column [the columella; see Fig. 10.3(13)] as in *Siphonodendron*, or

Table 10.2 The corallites can be arranged in certain patterns, thus aiding in their classification

Major modes of colony formation	Characteristics
Cerioid	The corallites are juxtaposed, but each corallite retains its own wall [Fig. 10.4(29)]; massive corals that have corallites sharing common walls
Dendroid	The corallites branch from each other in a dendritic pattern [Fig. 10.4(29)]
Fasciculate	The corallites are cylindrical but not in contact. They may be dendroid [with irregular branches; Fig. 10.5(20)] or phaceloid [with more or less subparallel corallites with connecting processes; Fig. 10.5(18)]
Flabelloid	The corallites are arranged in a single series, in long meandering rows or valleys that share a common base, but the walls (or ridges) of adjacent valleys are not connected [Fig. 10.5(12)]
Flabello-meandroid (=flabellate)	Corallites are arranged in long meandering rows with a common base; the walls may be partially fused [Figs. 10.4(28) and 10.5(13)]
Hydrophoroid	Coral that possess cone-shaped protuberances between corallites [Figs. 10.4(24) and 10.5(17)]
Meandroid	The corallites are arranged in multiple series; the adjacent valleys share the same ridge [Figs. 10.4(23) and 10.5(14, 15)]
Phaceloid	The corallites separated by a void space; the walls are distinct and separated by coenosteum [Figs. 10.4(26) and 10.5(18)]
Plocoid	These are short stalked and isolated corallites, and separated by coenosteum [Figs. 10.4(25) and 10.5(10)]
Solitary	The corallum is formed by a single corallite (one individual) [Figs. 10.4(2–6), 10.5(1, 2)]
Subplocoid	The corallites are sometimes separated by the coenosteum; each corallite possess its own wall
Thamnasterioid	The septa of adjacent corallites are confluent and often twisted or sinuous in form; plating coral with no walls surrounding corallites [Figs. 10.4(27) and 10.5(16)]

modifications of the tabulae to form a broad axial zone, as in *Aulophyllum* [see Fig. 10.3(5)]. Dissepiments are found in Scleractinians, but in rugose corals these are concentrated into a Dissepimentarium [Fig. 10.3(6)].

Each rugose corallum has a bilateral symmetry imposed by the mechanism of septal insertion through the growth or ontogeny of a single individual [Fig. 10.6(1)] (see also Nield and Tucker 1985; Doyle 1996; Clarkson 1998). As the coral grows, septa are inserted regularly at four specific points. The first is a single septum, the proseptum, which divides the corallite into two. With additional growth, the proseptum is divided into two: the cardinal (C) and counter (K) prosepta [Fig. 10.6 (1.1)]. Continued growth is noted along a set pattern with additional septa insertion; first, on either side of the cardinal septum [Alar septa; A; Fig. 10.6(1.2)], and second, on either side of the counter septum [counter-lateral septa; KL; Fig. 10.6 (1.3)]. Further insertions of smaller septa are then made adjacent to each of the four alar and counter-lateral septa together [Fig. 10.6(1.4)]. Insertion of further septa

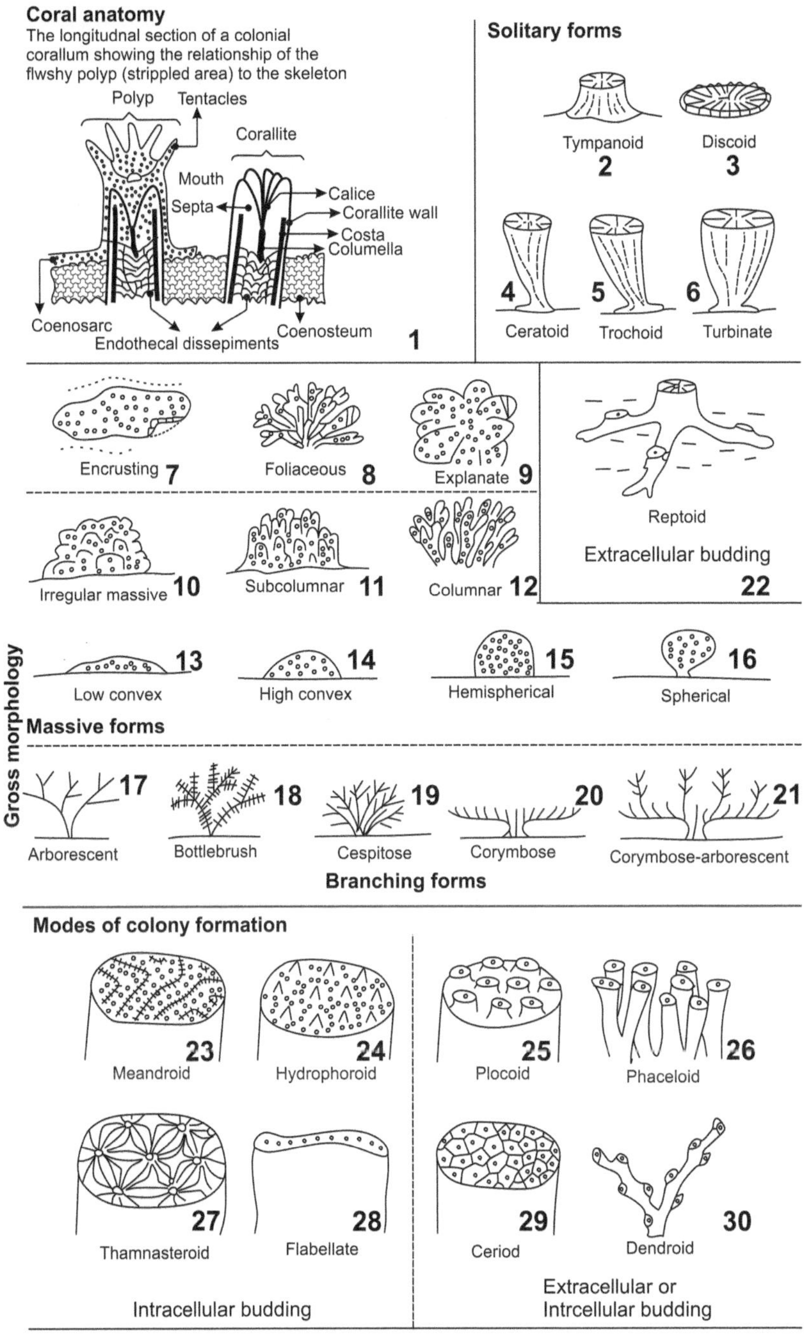

Fig. 10.4 Coral shapes (*1–22*) and modes of colony formation (*23–31*). The corallites can be arranged in certain pattern, thus, aiding in their classification

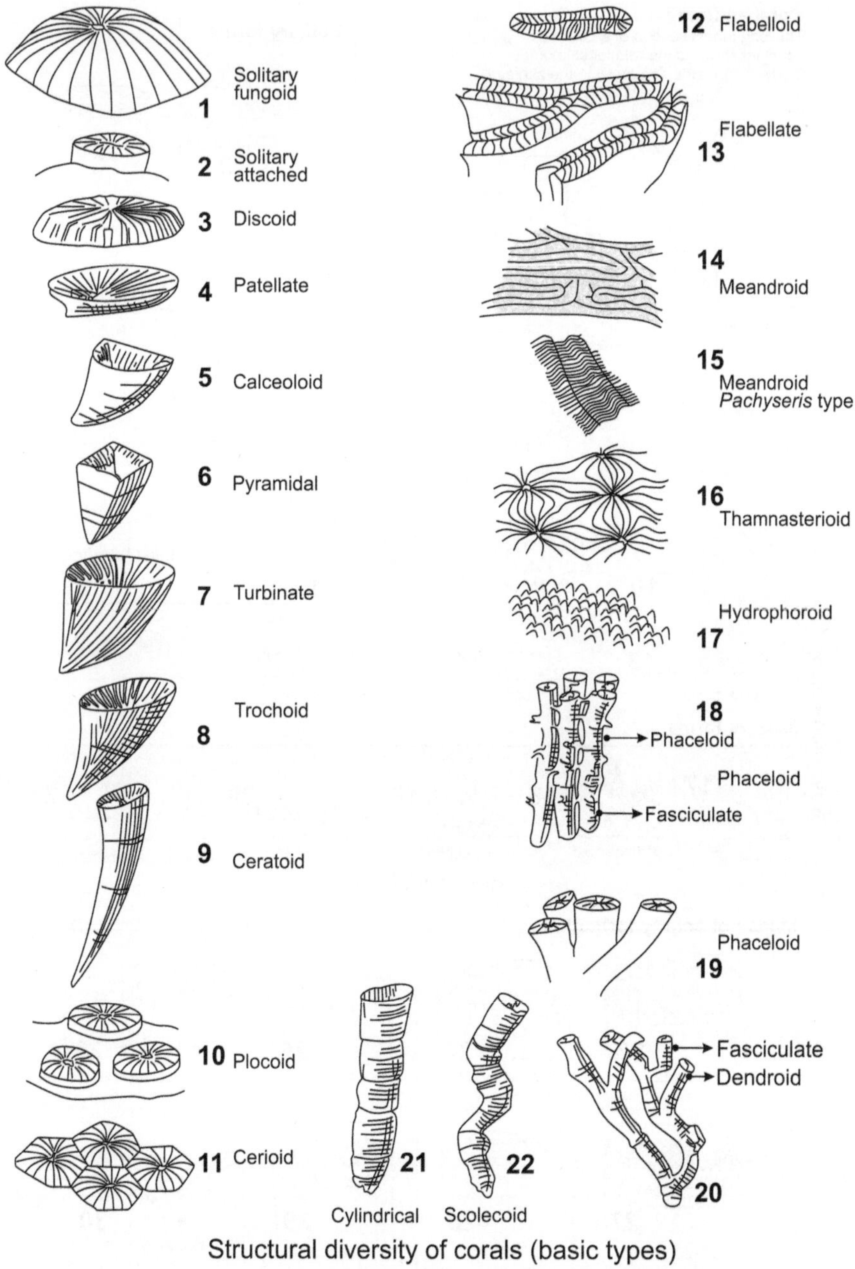

Fig. 10.5 The structural diversity of Corals (shapes continued from Fig. 10.4)

forces a readjustment of their position to make room, and crowding together in the four insertion points eventually leads to the creation of areas which are free from septa, the Fossulae [Fig. 10.6(1.5, 1.6)]. Although it is often difficult to identify fossulae in many rugose corals, the most noticeable is the Cardinal fossula adjacent to the cardinal septum [Fig. 10.6(1.6–1.7)], particularly well developed in the Carboniferous genus *Zaphrentites* (see also Nield and Tucker 1985; Doyle 1996; Clarkson 1998). This pattern of septal insertion, leading to a strong bilateral symmetry is unique to rugose corals, and distinguishes them from the scleractinians which have a radial pattern of septal insertion [see Fig. 10.6(1.2)].

Thus, the main items of classificatory importance among the rugose corals are (a) details of septal pattern, (b) presence or absence of tabulae, dissepiments, and carinae, (c) presence or absence of an epitheca and (d) shape of the corallite (see also Nield and Tucker 1985; Doyle 1996; Clarkson 1998).

10.2.1.2 Tabulate Corals

The tabulate corals:

- are always compound and relatively simple.
- have variable colony shapes, most commonly with polygonal or rounded corallites surrounded by a dense calcareous mass (brain-coral-like); with individual corallites joined in a kind of chain, forming the "fence" (picket fence-like); or forming simple branching corallites.
- possess calcite corallum that is mostly colonial; individual corallites are small.
- possess tabulae that are prominent; the septae are reduced or commonly absent.
- possess only tabulae as major internal shell features.

The Tabulate corals are characterized by the reduction of septa and the relative prominence of the horizontal elements in their structure (the tabulae). The individual corallites are small and open (mm sized), but their colonies are often a meter (m) across. The closely packed corallites (=the "honeycomb" arrangement) and the presence of mural pores (that in some corals [*Favosites*; see Middle Silurian corals; Fig. 10.12(11, 12)]) indicate that some degree of colonial integration was present in tabulate corals. Even in *Syringopora* [see Middle Silurian coral; Fig. 10.12(13)], a fasciculate form, cross-linking tubes have been recorded.

10.2.1.3 Scleractinian Corals

The Scleractinian corals:

- are compound or solitary; compound corals may be massive (brain-coral-like); arranged in linear series; branching; or composed of subparallel, cylindrical corallites
- possess aragonitic corallum

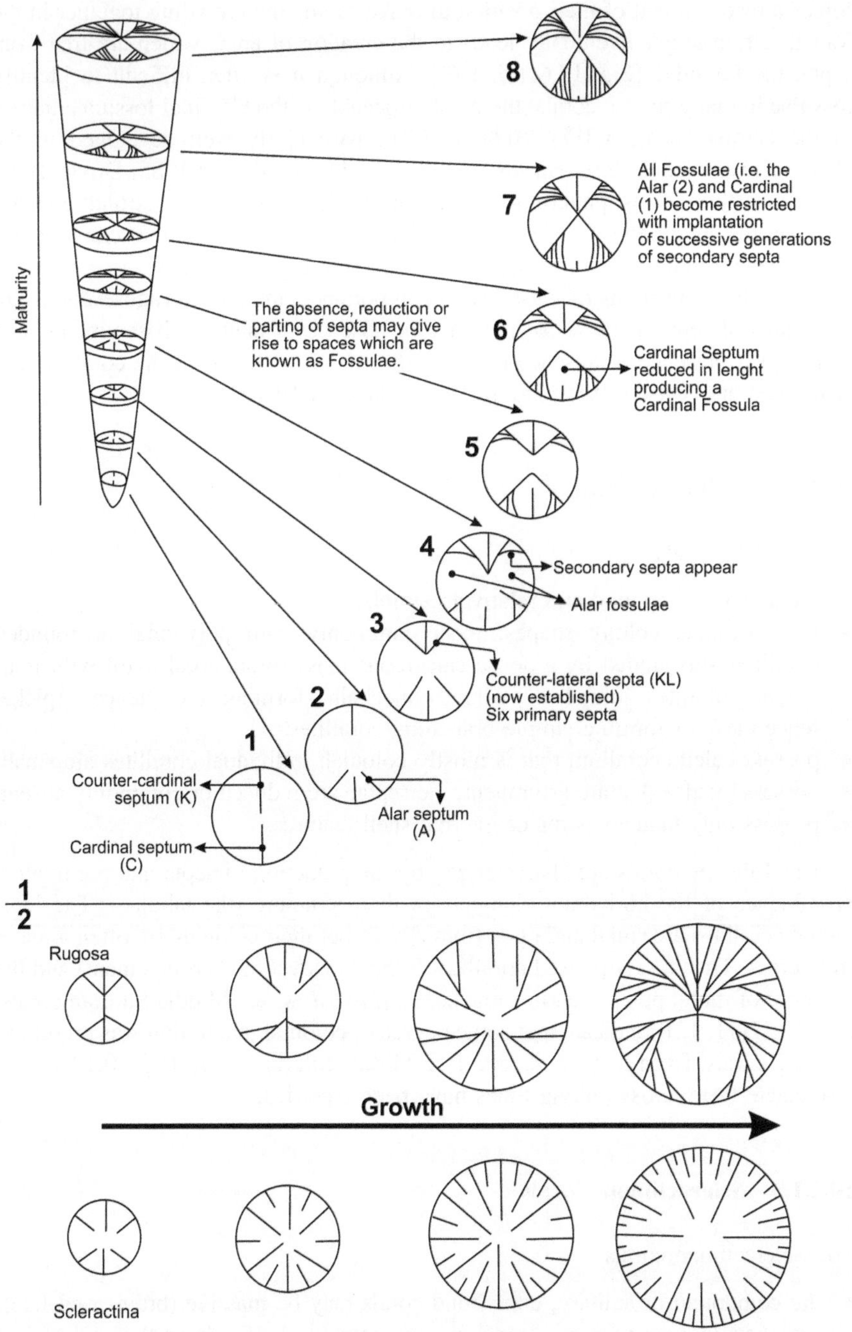
8
7
All Fossulae (i.e. the Alar (2) and Cardinal (1) become restricted with implantation of successive generations of secondary septa
Matrurity
6
The absence, reduction or parting of septa may give rise to spaces which are known as Fossulae.
Cardinal Septum reduced in lenght producing a Cardinal Fossula
5
4
Secondary septa appear
Alar fossulae
3
Counter-lateral septa (KL) (now established) Six primary septa
2
1
Counter-cardinal septum (K)
Alar septum (A)
Cardinal septum (C)
1
2
Rugosa
Growth
Scleractina

◀ **Fig. 10.6** Septal growth. *1* As growth proceeds, the metasepta are introduced between the prosepta in a regular fashion. The process of insertion can be investigated by taking serial sections along the corallum. Here, we see the six primary septa. Note how the metasepta are then inserted at four separate points, coming to lie on either side of the cardinal septum as well as on the "counter" side of each alar septum. Detailed analysis of similar sequences is vital to the fine taxonomy of these fossil corals. The early portion of the corallite consists of a nonseptate cup. Septa (vertical partitions) are added two at a time until a total of six (designated Protosepta) is present. The six protosepta are identified as Cardinal (C), Counter (K), Alar (two in number; A), and Counter-lateral (two in number; KL). Secondary septa are later added in cycles of four, one septum in each of the quadrants on either side of the cardinal septum, secondary septa appear after progressively branching off both sides of the cardinal septum. In the counter quadrants, between the alar and counter-lateral septa, secondary septa are added by branching off the counter side only of the alar septa. Counter-lateral septa lose their distinctive appearance and come to resemble the secondary septa. The primary and the secondary septa are termed Major septa. Minor septa are short septa between major septa that do not follow the rules of insertion outlined above. Most rugose corals will have an equal number of major and minor septa. *2* Comparative septal growth between Rogose and Scleractinian corals (see text for details)

- possess tabulae and septa, but no dissepimentarium or axial complex
- have septa which are inserted radially in regular groups of six, with no fossulae.
- are hermatypic forms with Zooxanthellae; their reef ecology hinges around Zooxanthellae (photic zone, temp constraints, latitudinal restrictions, etc.)
- evolved from the unmineralized Zoantherian, post the Permo-Triassic extinction event

The Scleractinians (Fig. 10.7), like rugose corals (from which they probably evolved), can be either solitary or compound; the former have a less complex skeleton (see also Nield and Tucker 1985; Doyle 1996; Clarkson 1998). The scleractinians possess an aragonite skeleton whereas the rugosan's skeleton is made of calcite. The scleractinians possess tabulae and septae together with dissepiments, but lack the axial complex or dissepimentarium (Fig. 10.7). Septal insertion is less complex than in Rugosa. The scleractinians are often referred to as "hexacorals" because the septal insertion is regular and in multiples of six which results in a radially symmetrical corallite [Fig. 10.6(1.2)]. Like the Rugosa, they also have six prosepta in the calyx. Subsequent metaseptal insertions, are, however, in multiples of six [Fig. 10.6(1.2)]. The septa are closely associated with messentaries of the endoderm. Solitary polyps may be large (up to 25 mm across), but in compound forms they average about 1–3 mm in diameter. Both corallite morphology and the coenosteum among them are distinguishing characters for assigning species names (see also Nield and Tucker 1985; Doyle 1996; Clarkson 1998).

The scleractinians owe their success partly to being able to cement themselves down, and partly to their ability to fuse polyps. But the major element in their success has been their symbiotic relationship with nonmotile Dinoflagellates. The Dinoflagellates are photosynthesizing microorganisms that live in the endodermal cells and are usually referred to as Zooxanthellae. Being plant-like, these manufacture food from CO_2 and sunlight, and they probably donate some of the products to their hosts. But the dinoflagellate and the coral both require phosphorus to grow,

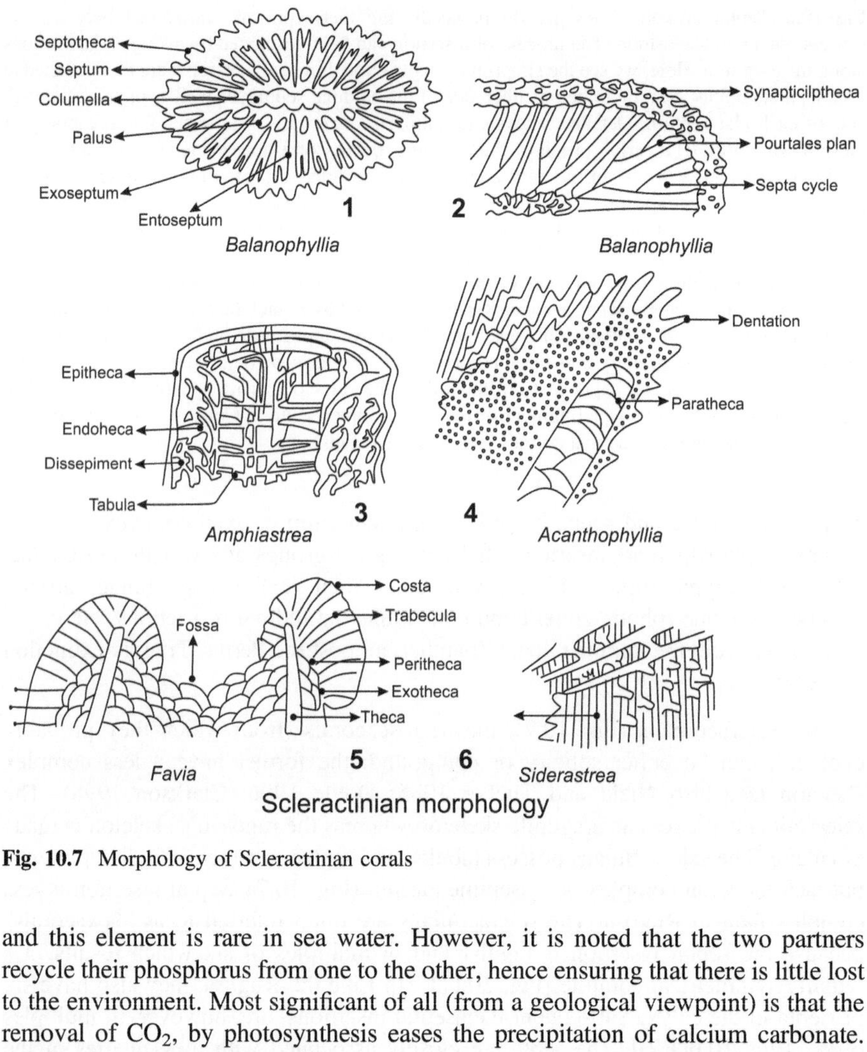

Fig. 10.7 Morphology of Scleractinian corals

and this element is rare in sea water. However, it is noted that the two partners recycle their phosphorus from one to the other, hence ensuring that there is little lost to the environment. Most significant of all (from a geological viewpoint) is that the removal of CO_2, by photosynthesis eases the precipitation of calcium carbonate. Scleractinia can actually calcify ten times more quickly by day than by night, when photosynthesis cannot occur. But this symbiosis only works if the coral lives in the photic zone of the upper ocean. In fact, it is found that corals which grow in reefs (the hermatypic corals) are also those in which the symbionts are found (Fig. 10.8). Hermatypic corals are ecologically restricted to waters of precise salinity, temperature and clarity in tropical seas. Ahermatypic corals (which are non-reef-forming and lack zooxanthellae) are much less particular. They can exist at depths of 6000 m and survive temperatures as low as 1 °C. They are mostly solitary, and some even live in icy but clean waters.

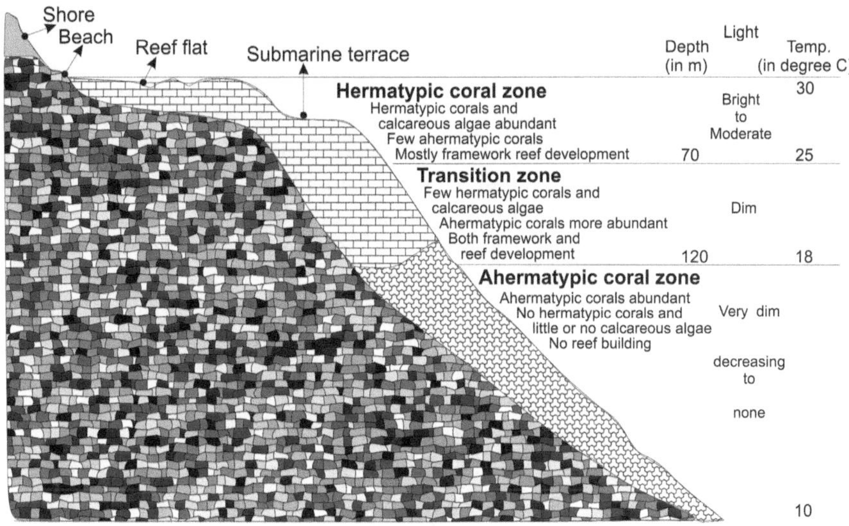

Fig. 10.8 Distribution of Hermatypic and Ahermatypic corals on the basis of depth, light and temperature

10.3 Terminology

A brief alphabetically arranged glossary of morphological terms employed for description of Rugosa, Scleractinia, and Tabulata is given below. Note that as many terms are applied to 2 or more of these ordinal divisions of the Zoantharia, wherever need the classification of terms is indicated as Rugosa, Scleractinia, and Tabulata.

10.3.1 **Ahermatypic**: Not reef-forming corals (Fig. 10.8).

10.3.2 **Alar fossula**: Relatively prominent interseptal space developed in position of an alar septum or adjoining it on side toward counter septum [Rugosa; Fig. 10.6(1.4)].

10.3.3 **Alar septum (symbol, A)**: One of two protosepta located about midway between cardinal and counter septa, distinguished by insertion of newly formed metasepta on side facing counter septum [Rugosa; Figs. 10.3(3, 4) and 10.6(1.2)].

10.3.4 **Apical end**: It is the pointed proximal extremity of a corallite where growth begins [Fig. 10.3(2)].

10.3.5 **Axial boss**: It is the central prominence in calyx formed by an axial structure [Fig. 10.3(5)].

10.3.6 **Axial structure**: A collective term used for various longitudinal structures (be it a solid or spongy rodlike columella or an axial vortex) in the axial region of a corallite, typically consisting of or in various combinations: adaxially extended septa, septal elements more or less detached from septa proper (e.g., median plate), pali, paliform lobes, a styliform axial rod, tabular modifications, dissepimental structures, and tabellae [Rugosa, Scleractinia; Figs. 10.1(5) and 10.3(5, 13)].

10.3.7 **Axial**: With reference to corallite oral-aboral axis [Fig. 10.1(6)].

10.3.8 **Basal disk**: Aboral fleshy part of coral polyp, typically subcircular in outline (see oral disk, basal plate) (Scleractinia) [Fig. 10.1(3–5)].

10.3.9 **Basal plate**: Thin, initially formed part of corallite from which septa begin to be built upward [Scleractinia; Fig. 10.1(5)].

10.3.10 **Calice**: A cup-shaped depression on the corallite's oral surface [Rugosa, Scleractinia; Figs. 10.1(2) and 10.3(1)].

10.3.11 **Calycal platform**: Part of calice floor having a sub-horizontal plane or outwardly sloping (everted) form; generally surrounds a calicular pit [Rugosa; Fig. 10.3(7)].

10.3.12 **Calyx**: A cup shapes structure in which the polyp sits and secreted by the lower portion of the polyp [Fig. 10.3(2)].

10.3.13 **Cardinal fossula**: Relatively prominent interseptal space developed in position of cardinal septa (See closed fossula, open fossula, also other types alar, counter) [Rugosa; Figs. 10.3(1, 3) and 10.6(1.6)].

10.3.14 **Cardinal quadrant**: It is the part of thecarium between cardinal septum and either of alar septa [Fig. 10.3(3)].

10.3.15 **Cardinal septum** (**symbol**, **C**): Protoseptum in plane of bilateral symmetry of a corailite, distinguished from other protosepta by insertion of newly formed metasepta adjacent to it on both sides (Sec other types: alar. counter, counter-lateral) [Rugosa; Fig. 10.3(3–4) and 10.6(1.6)].

10.3.16 **Carina**: Flange-like elevation on side of septum formed by thickened trabeeula [Rugosa; Fig. 10.3(8)].

10.3.17 **Ceratoid**: Very slenderly conical, horn-shaped solitary corallite [Rugosa, Scleractinia; Figs. 10.4(4) and 10.5(9)].

10.3.18 **Cerioid**: Massive corallum in which walls of adjacent polygonal corallites are closely united (Rugosa, Scleractinia, Tabulata) [Figs. 10.4(29) and 10.5(11)].

10.3.19 **Coenenchyme** (**=Coenosarc**): It is a collective term for both eoenosteum and coeoosarc; in compound anthozoans, it is the mesogloea that surrounds and unites the polyps [Scleractinia, Tabulata; Fig. 10.1(2)].

10.3.20 **Coenosarc**: This is a common soft tissue connecting coral in a colony [Fig. 10.1(2)].

10.3.21 **Coenosteum** (**-a**) [**or peritheca** (**-ae**)]: The skeleton between corallites within a colony [Scleractinia; Fig. 10.1(2)].

10.3.22 **Coenosteum**: These are skeletal deposits formed between individual corallites of a colony [Scleractinia; Fig. 10.1(2)].

10.3.23 **Coensarc** (**=peritheca**): This is the living axial part of a coral colony [Scleractinia; Fig. 10.1(2)].

10.3.24 **Columella (-ae)**: A central axial (solid or non-solid calcareous) structure within a corallite formed by various modifications of inner edges of septa; commonly projects into calice in the form of a calicular boss [Rugosa, Scleractinia; Figs. 10.1(2), 10.3(13) and 10.7(1)].

10.3.25 **Corallite**: This is the exoskeleton formed by an individual coral polyp within a colony [Rugosa, Scleractinia, Tabulata; Figs. 10.1(2), 10.3(2, 7) and 10.4(1)].

10.3.26 **Corallum**: Exoskeleton of a coral colony or solitary coral [Rugosa, Scleractinia, Tabulata; Figs. 10.1(2), 10.3(2, 7) and 10.4(1)].

10.3.27 **Costa (-ae)**: extension of a septum on the outer side of a corallite wall; a rib or rib-like structure [Scleractinia; Fig. 10.1(2)].

10.3.28 **Counter septum (symbol, K)**: Protoseptum opposite cardinal septum in position (see other types: alar, cardinal, counter-lateral) [Rugosa; Figs. 10.3(3) and 10.6(1.1)]

10.3.29 **Counter-Lateral septum (symbol, KL)**: One of 2 protosepta that adjoin counter septum on either side (see other types: alar, cardinal, counter) [Rugosa; Fig. 10.6(1.3)].

10.3.30 **Cyanotheca**: Inner wall formed by sharp deflections and union of tabulae

10.3.31 **Dissepiment**: Small domed plate forming a cyst-like enclosure in peripheral region of a corallite [Rugosa, Scleractinia, Tabulata; Fig. 10.3(6)].

10.3.32 **Dissepimentarium**: Peripheral zone of corallite interior occupied by dissepiments [Rugosa, Scleractinia; Fig. 10.3(6)].

10.3.33 **Ectoderm**: Outer layer of oral and basal disks, tentacles, and column wall of coral polyp [Fig. 10.1(2)].

10.3.34 **Edge zone**: Fold of body wall of coral polyp extending over edge of wall [Scleractinia; Fig. 10.1(5)].

10.3.35 **Endoderm**: Inner layer of outer body walls of coral polyp and occurring as a double lamina in mesenteries [see ectoderm, mesogloea; Fig. 10.1 (3)].

10.3.36 **Endotheca**: Collective term for dissepiments inside corallite wall [Scleractinia; Fig. 10.1(2)].

10.3.37 **Enteron**: It is the intestine or gut of an organism [Fig. 10.1(3)].

10.3.38 **Epidermis**: External epithelium of a coral polyp and coenenchyme derived from the ectoderm [Fig. 10.1(3)].

10.3.39 **Epitheca**: Sheath of skeletal tissue laterally surrounding a corallite comprising extension of basal plate [Rugosa, Scleractinia, Tabulata; Fig. 10.3(1)].

10.3.40 **Fossa**: [Fig. 10.7(5)].

10.3.41 **Fossula**: Interseptal space distinguished by its unusual shape and size (see types; alar, cardinal, counter, closed, open) [Rugosa; Figs. 10.3(6)].

10.3.42 **Gastrodermis**: The tissue lining the stomach [Fig. 10.1(3)].

10.3.43 **Hermatypic**: Reef-forming corals (Fig. 10.8).

10.3.44 **Holotheca**: It is the wrinkled lamina deposited by colonial corals to cover base of corallum [Fig. 10.3(7)].

10.3.45 **Interseptal ridge**: The external surface of the theca may be either smooth, or show longitudinal grooves and ridges. Grooves correspond in position to the septa and are called septal grooves (in Rugosa corals). The bulging out of the polyp wall between the septa forms a ridge, called the Interseptal ridges. The longitudinal ridges are extensions of the septa through the wall, and the grooves correspondingly indicate interseptal positions (in Hexacoralla) [Fig. 10.3(2)].

10.3.46 **Lamella**: These are short longitudinal plates within the axial region [Fig. 10.3(5)].

10.3.47 **Major septum**: One of the protosepta or metasepta (see minor septum) [Rugosa, Scleractinia; Fig. 10.3(4)].

10.3.48 **Massive**: Corallum composed of corallites closely in contact with one another (Rugosa, Scleractinia, Tabulata; Fig. 10.4).

10.3.49 **Mesentery**: Fleshy radially disposed lamina (a fold of the peritoneum) attached to inner surface of oral disk and column wall of a coral polyp [Scleractinia; Fig. 10.1(4, 5)].

10.3.50 **Mesogloea**: Noncellular jellylike middle layer of outer walls and mesenteries of coral polyps [see ectoderm, endoderm; Fig. 10.3(3)].

10.3.51 **Metaseptum**: One of the main septa of a corallite other than protosepta, generally distinguished by their extension axially much beyond that of minor septa (see major septum) [Rugosa, Scleractinia; Fig. 10.3(4)].

10.3.52 **Minor septum**: One of the relatively short septa that commonly are inserted between adjacent major septa [Rugosa, Scleractinia; Fig. 10.3 (4)].

10.3.53 **Mural pore**: Circular or oval small hole in wall between adjoining corallites, as in some tabulates [Tabulata; Fig. 10.12(11, 12)].

10.3.54 **Nematocyst**: Stinging or adhesive body characteristic of cnidarians [Fig. 10.1(1)].

10.3.55 **Oral disk**: It is that part of the polyp through the center of which the mouth opens, including peristomal tissue and tentacles. [Fig. 10.1(5)].

10.3.56 **Peritheca**: the living tissue surrounding or between corallites (=coenosarc) [Scleractinia; Fig. 10.3(1)].

10.3.57 **Pharynx**: [Fig. 10.1(4, 5)].

10.3.58 **Polyp**: It is the fundamental structural unit of an anthozoan [Fig. 10.1(1)]. It consisting of a sac-like cylindrical body, a basal (aboral) disk and an oral disk bearing mouth and tentacles [Fig. 10.1(1)].

10.3.59 **Septal cycle**: All septa belonging to a single stage in ontogeny of corallite as determined by order of appearance of septal groups, 6 protosepta comprising first cycle and later-formed exosepta and entosepta in constantly arranged succession being introduced in sextants (Scleractinia; Fig. 10.6).

10.3.60 **Septum** (**-a**): These are radially arranged vertical partitions within a corallite. They can be exsert, insert or even, with respect to the corallite wall and occurring between or within mesenterial pairs (Rugosa, Scleractinia, Tabulata) [Figs. 10.1(2, 4, 5), 10.3(2) and 10.7(1)].

10.3.61 **Solitary**: Corallite of polyp not forming part of a colony [Rugosa, Scleractinia; Fig. 10.7(1, 9)].

10.3.62 **Stereozone**: Peripheral or subperipheral in position, this is an area of dense skeletal deposits in a corallite (Rugosa, Scleractinia, Tabulata) [Fig. 10.3(5)].

10.3.63 **Stomodaeum**: Esophagus-like tubular passageway or pharynx leading from mouth of coral polyp to gastrovascular cavity [Fig. 10.1(2)].

10.3.64 **Synapticulum** (**-ae**): Small rods or bar connecting opposed faces of adjacent septa and perforating mesenteries between them [Scleractinia; Fig. 10.1(6)].

10.3.65 **Tabella**: Small subhorizontally disposed plate in central part of corallite forming part of an incomplete tabula (Rugosa, Tabulata) [Fig. 10.3(5)].

10.3.66 **Tabula** (**-ae**): Transverse partition of corallite, nearly plane, or upwardly convex or concave, extending to outer walls or occupying only central part of corallite. These are horizontal partitions that allow for upward growth of a polyp by isolating the surface from the underlying calcium carbonate skeleton [Rugosa, Scleractinia, Tabulata; Figs. 10.1(5), 10.3(1, 6) and 10.7(3)].

10.3.67 **Tabularium**: Axial part of the interior of a corallite in which tabulae are developed [Rugosa, Scleractinia, Tabulata; Fig. 10.3(6)].

10.3.68 **Tentacle**: Movable tubular extension of soft integument rising from oral disk of coral polyp, closed terminally at tip, commonly simple but rarely forked [Fig. 10.1(1–4)].

10.3.69 **Trabecula**: These are pillar of radiating calcareous fibers comprising skeletal element in the structure of septum and related components [Rugosa, Scleractinia, Tabulata; Figs. 10.3(9) and 10.7(5)].

10.3.70 **Transverse division**: Formation of new coral polyps by separation of parent by splitting into two parts transverse to oral-aboral axis (Scleractinia) [Fig. 10.1(6)].

10.3.71 **Wall [or theca** (**-ae**)]: Skeletal deposit inclosing column of polyp and uniting outer edges of sepia; it is variously formed in different corallites (see septotheca, paratheca, synapticulotheca) [Scleractinia; Figs. 10.3(1), 10.1(5) and 10.7(1, 2, 4)].

10.3.72 **Zooxanthella**: Symbiotic unicellular yellow-brown protistan in endoderm of hermatypic coral polyps.

10.4 Coral Reefs

Coral reefs proliferate in the warm waters (25–30 °C) of the mid-latitudes (between 30°N and 25°S) where the water is strongly illuminated by sunlight. In areas, where much of the sunlight is filtered out (deep waters between 50 and 100 m), the number of reef-building coral species is greatly reduced, and at greater depths (>100 m), most reef builders disappear (Fig. 10.8).

Corals are divided into two major ecological groups. Some corals contain numerous tiny, microscopic, brownish colored spheres (single-celled symbiotic algae called zooxanthellae) in their endodermal tissues that require sunlight for photosynthesis. Some corals do not these and hence, disappear in absence of sunlight. This basic difference in their light requirements, categorizes them into hermatypic or reef-building corals and ahermatypic or Non-reef-building corals (Fig. 10.8), respectively.

The hermatypic types grow faster and deposit mineralized skeletal materials at a much more rapid rate than the ahermatypic types. Although these two groups are the primary constructors of the reef framework, many other attached organisms contribute their skeletal materials, as well. Some of these secondary framework constructors include hydrozoan corals in the orders Milleporina and Stylasterina, *Heliopora,* and *Tubipora* in the anthozoan orders Coenothecalia and Stolonifera, encrusting foraminifers, bryozoans, attached gastropod and bivalve molluscs, calcareous sponges, and a few ahermatypic scleractinian corals. Also included are coralline algae that cements various corals together with compounds of calcium, and tube worms and molluscs that donate their hard skeletons (Cousteau 1985).

Reef-building (or hermatypic) corals, especially those belonging to order Scleractinia ("Stony corals") are largely responsible for establishing the framework of reefs (Fig. 10.9). They are important today and were also in the past due of their extraordinary ability to calcify. The tiny coral polyps produce calcium carbonate ($CaCO_3$) that form reefs. Interestingly, huge $CaCO_3$ deposits, some with important hydrocarbon reserves, owe their origin to ancient coral reefs. The rate at which $CaCO_3$ is laid down varies from species to species; stony coral colony can increase in height or length (and hence, in $CaCO_3$) by as much as 10 cm a year (similar to the growth of a human hair), whereas, other corals, like the dome and plate species are more bulky and may only grow 0.3–2 cm per year (Ross 2007).

The Tabulate corals have contributed to the Palaeozoic reef formation (Fig. 10.9) and were accompanied by the stromatoporoids, to which they were subordinate in reefs that grew in vigorous, shallow waters (Fig. 10.9). Deeper-water reefs tend to have a greater proportion of tabulates. Forms such as *Halysites* [see Fig. 10.12(1)] were characteristic of areas with very high sedimentation; their mode of

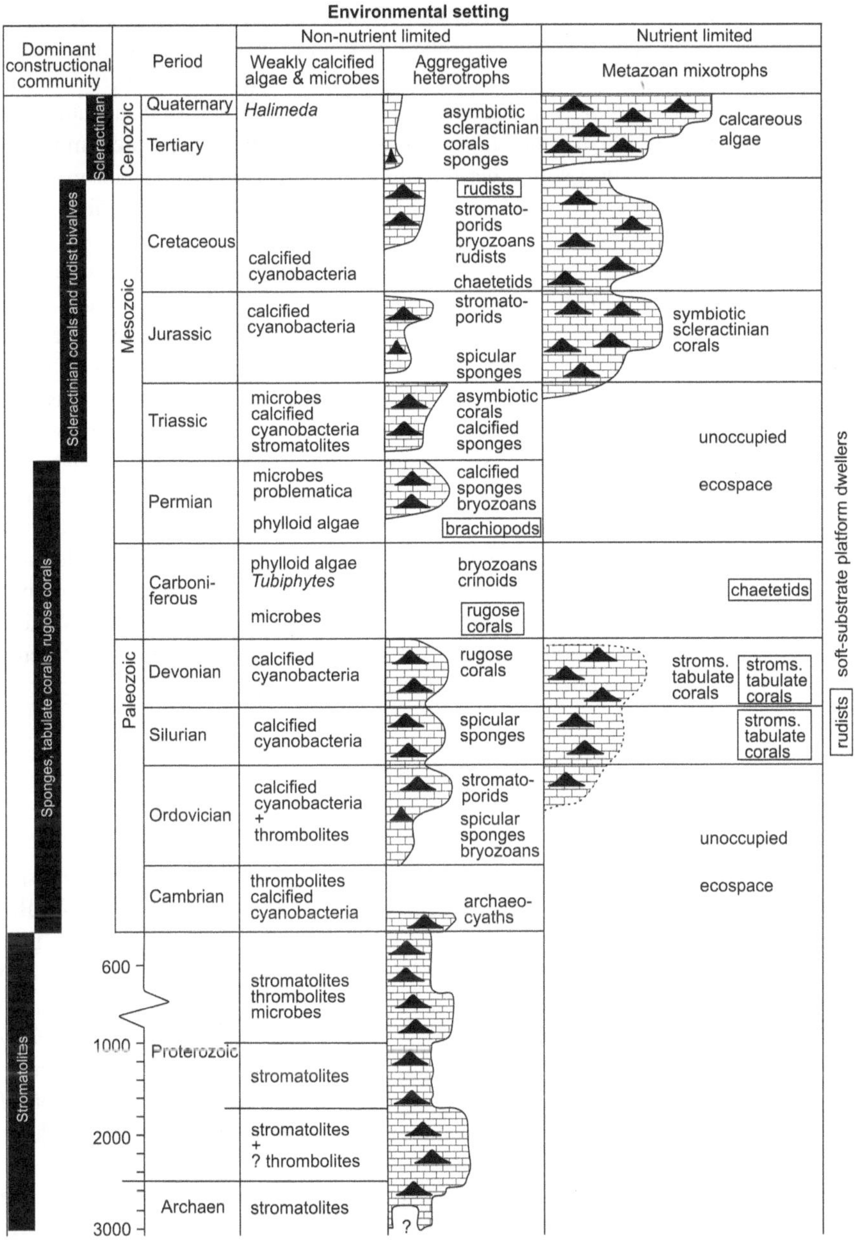

Fig. 10.9 The history of Reef habitat. There are two subhabitats—nutrient-rich and nutrient-poor. Until the Cambrian, Archeocythids, Stromatolites, and Thrombolites occupied the nutrient-poor environments whereas the nutrient-rich niche remained unoccupied. Serious reef-building started only in the Late Ordovician and dominated during Silurian-Devonian times. Carboniferous was essentially vacant, and by Permian, was occupied by calcified sponges and bryozoans. By Middle Triassic, the scleractinian corals occupied the long vacant nutrient-poor ecospace again and since have been major reef builders for the rest of the Mesozoic and Cenozoic times

construction allowed fast growth to keep ahead of burial. Moreover, corallites at the center of the colony could clean themselves and dump the material into the shafts between the interlocking palisades. Tabulates were unable, however, to cement themselves to their substrata—a disability that put them at a severe disadvantage in reef habitats (Fig. 10.9). In general, the Palaeozoic reefs were built predominantly by tabulates in conjunction with other reef-building, non-cnidarian organisms, such as stromatoporoids and bryozoans; rugose corals, only played a subordinate role (Fig. 10.9).

Most extant reefs have been growing for over 5000 years, spanning more than 100 countries and occupying more than 600,000 km^2 of tropical oceans. They require clear, warm waters with high intensity of light for growth and are thus limited to shallow waters, with maximum abundance between 10 and 30 m below the sea surface (Fig. 10.8). Reefs thrive in nutrient-poor environments and thus, any small changes in the nutrient content of the water adversely affects their survival (Fig. 10.9). Coral reefs also exert considerable control both on global climate and the marine environment, particularly in the recycling of carbon. The Great Barrier Reef, off the NE coast of Australia, is the largest (80 miles wide and 1200 miles long) and the most famous extant reef. The fossil reefs, prolific in the Silurian and Devonian and from Late Triassic to late Tertiary (Fig. 10.9), were largely confined to low latitudes. Best examples include those in the Alps (Triassic), Western Australia and Canada (Devonian). The Canadian reefs form important petroleum reservoirs because of porous limestones that result from reef growth.

10.5 Geological History

The Cambrian yields scarce tabulae-bearing coral-like fossils ("*skeletal organisms resembling tabulatomorph corals*"; Grotzinger et al. 2000; Wood et al. 2002). But the first real evidence of the tabulate corals is found in the Ordovician, when they first radiated (Fig. 10.10). The rugose corals also appeared in the Ordovician, and together they became common components of the Palaeozoic marine ecosystem (more so in Silurian-Devonian duration). The end-Devonian extinction severely affected both tabulate and rugose corals, although the latter recovered and became common in the Carboniferous (Fig. 10.10). But the next major mass extinction of the Permian, proved too much for them; they were completely wiped off as also a great many reef-dwelling organisms.

Most Tabulates are of no special stratigraphic value, though certain forms are useful markers, such as the *Pleurodictyum* which is restricted to the Early Devonian.

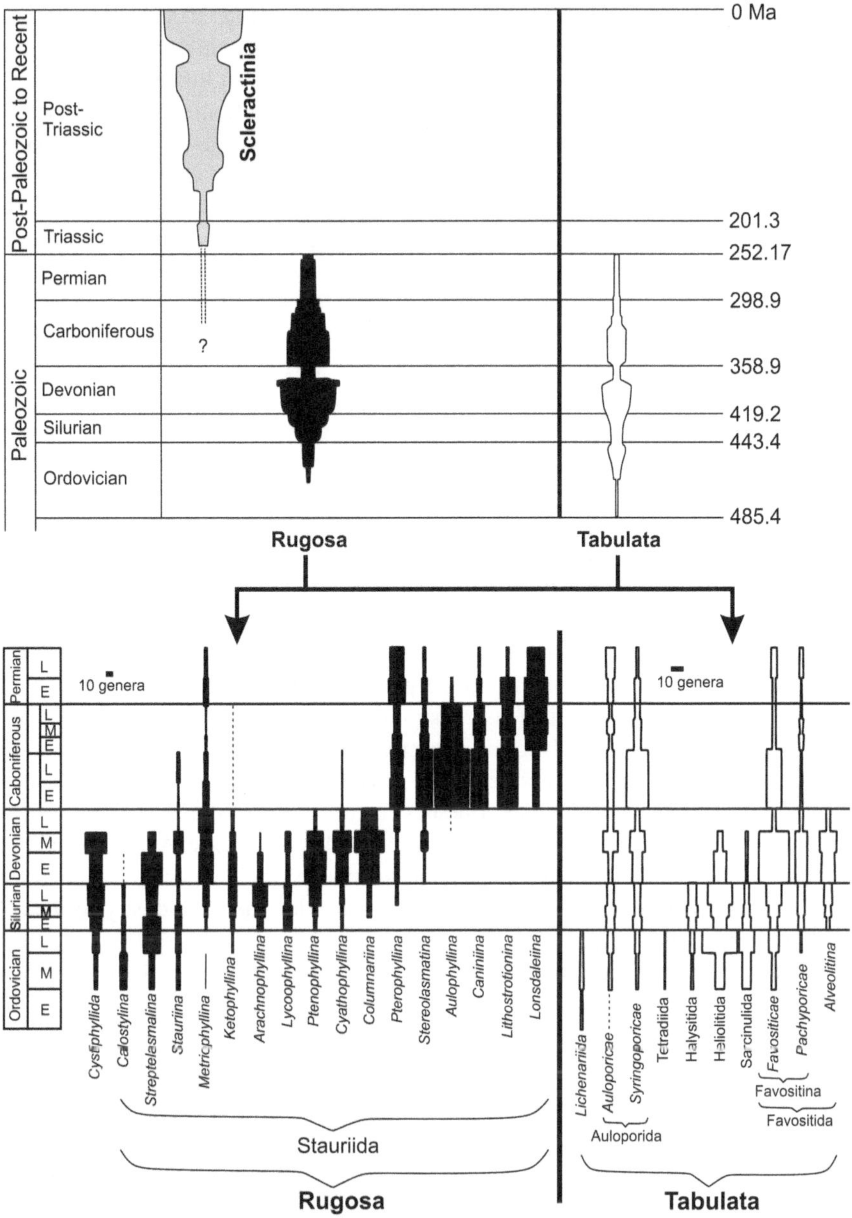

Fig. 10.10 Geological history of corals (modified from Clarkson 1998)

The Rugose corals, originating in the Ordovician (Fig. 10.10), also rose to prominence during the Silurian. They continued to diversify through the Devonian. It is during this time that slipper coral *Calceola* [Fig. 10.13(4, 5)] appeared, an excellent marker for the Middle Devonian. In the Carboniferous the rugose corals reached their acme, and are used as zonal indices. However they, like the tabulates, eventually died out—not, in this case, at the end of the Permian, but shortly after.

The first of the scleractinians appeared in the Triassic, replacing the rugose and tabulate corals, and since then remained the most important skeletal coral group. Their origin is still debated, and involves two possible routes. The first involves direct derivation from the rugose corals. However, there are two problems: no evidence of an intermediate between the two groups, and there is a considerable time gap from the last appearance of the rugosa in the late Permian to the first appearance of the scleractinians in the mid-Triassic. The second possibility is that the scleractinians were derived from a soft-bodied anthozoan ancestor, such as a sea anemone, via a route involving the creation of a mineralized skeleton. However, little evidence exists to substantiate either hypothesis, also. The Scleractinia, evolved possibly from the Rugosa, and dominated the Mesozoic and Cenozoic scene (Fig. 10.10). They developed along two lines, the reef-building hermatypic corals living in symbiosis with algae, the zooxanthellae; and the deeper water ahermatypic corals, without zooxanthellae. The hermatypic corals flourished in the later part of the Cenozoic, producing a great phase of reef development which has continued, with some interruption, to the present day (Fig. 10.9).

The scleractinians are not of much stratigraphic use, but they can build reefs ranging from very small atolls to massive structures on the scale of continents. The Great Barrier Reef, off the east coast of Australia, is even visible from space! Within these diverse and often enormous structures, the plasticity of coral growth and the distinct ecological zones into which they fall, are striking. An understanding of coral ecology is important in the interpretation of their environment, and the great reservoir potential of reefal build-ups and has led the oil industry to finance much research into this complex field.

Selected corals species characteristic of a particular time interval are illustrated in Figs. 10.11, 10.12, 10.13, 10.14, 10.15, 10.16, 10.17, 10.18, 10.19, 10.20, 10.21 and 10.22.

Appendix A gives the list of illustrated specimens mentioning the chapter number, species name, age, and locality along with its figure number within the said chapter.

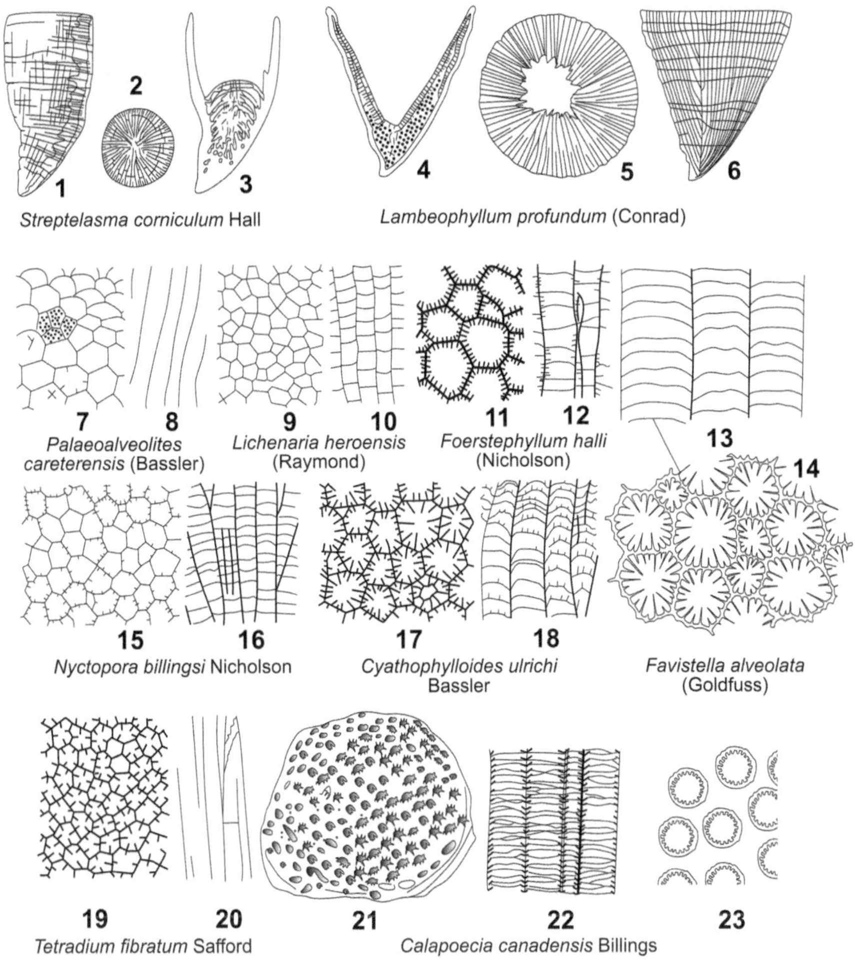

Fig. 10.11 Selected Ordovician corals and their major distinguishing characters

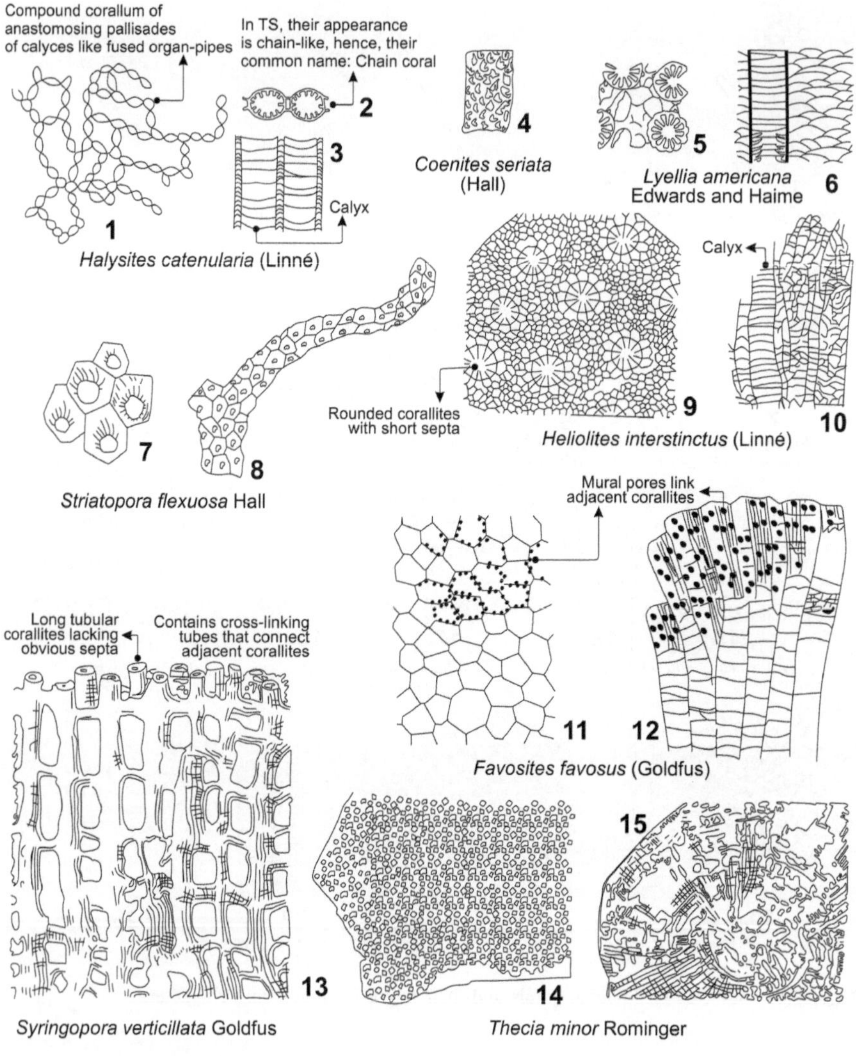

Fig. 10.12 Selected Middle Silurian corals and their major distinguishing characters

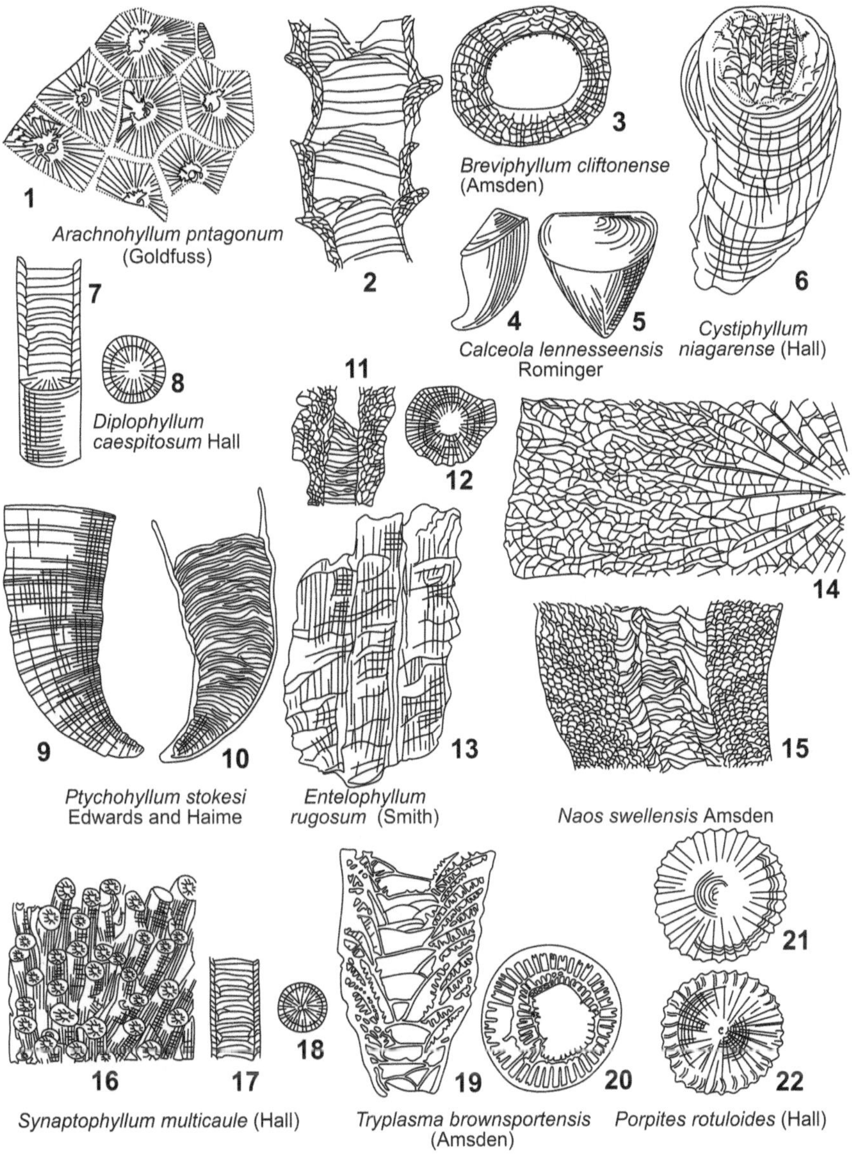

Fig. 10.13 Selected Middle Silurian corals and their major distinguishing characters

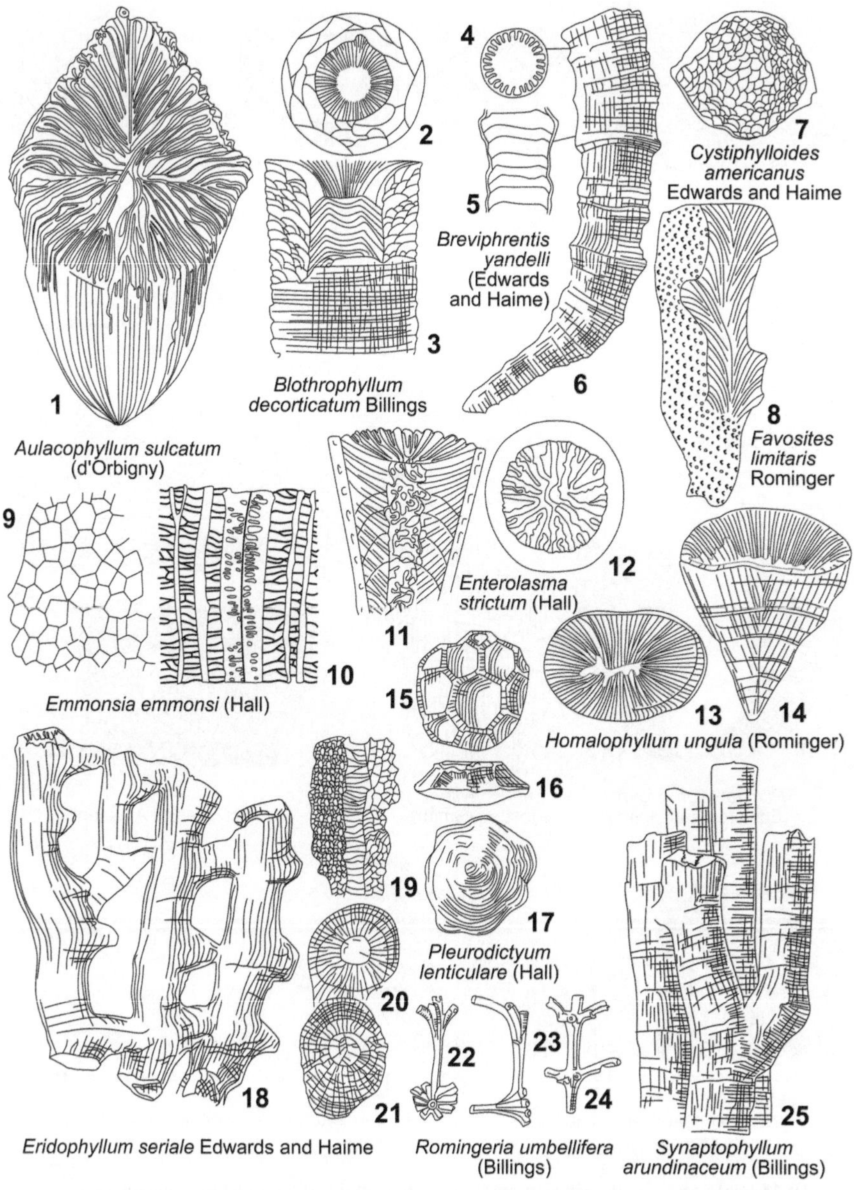

Fig. 10.14 Selected Early Devonian corals and their major distinguishing characters

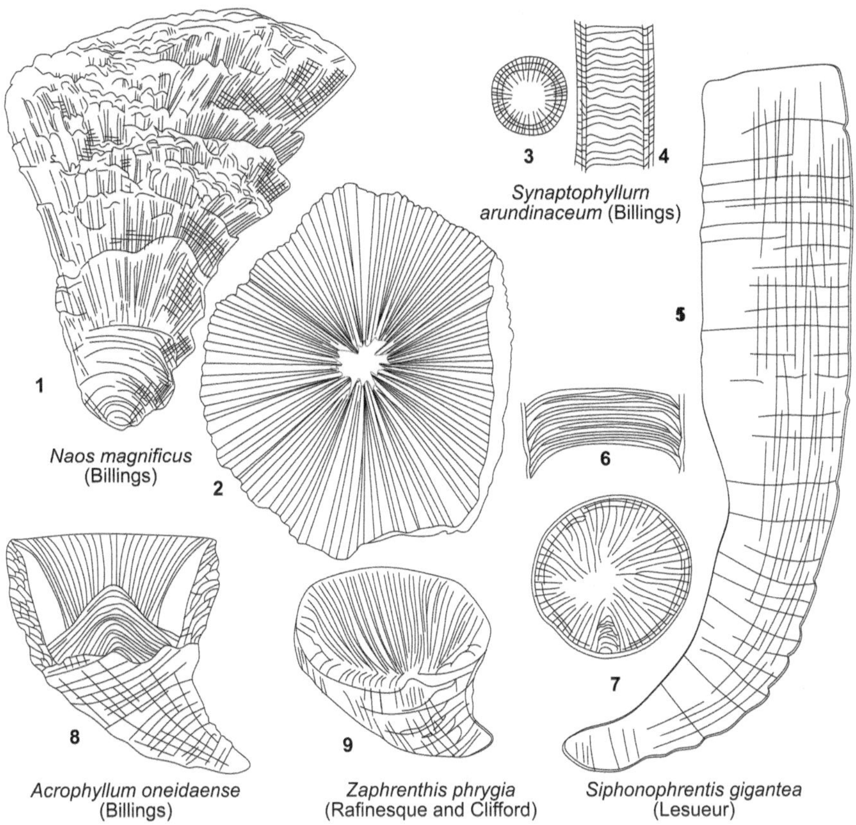

Fig. 10.15 Selected Early Devonian corals and their major distinguishing characters

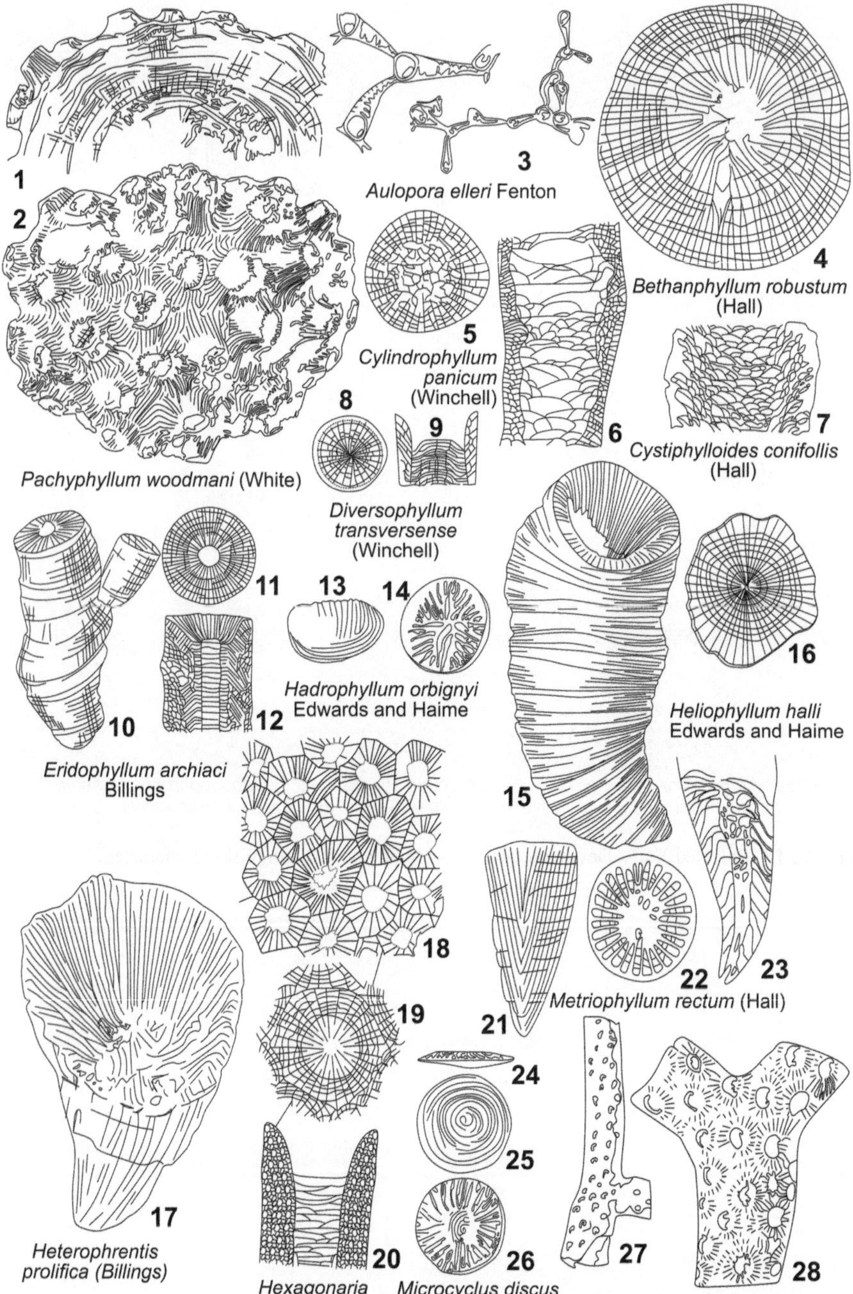

Fig. 10.16 Selected Late-Middle Devonian corals and their major distinguishing characters

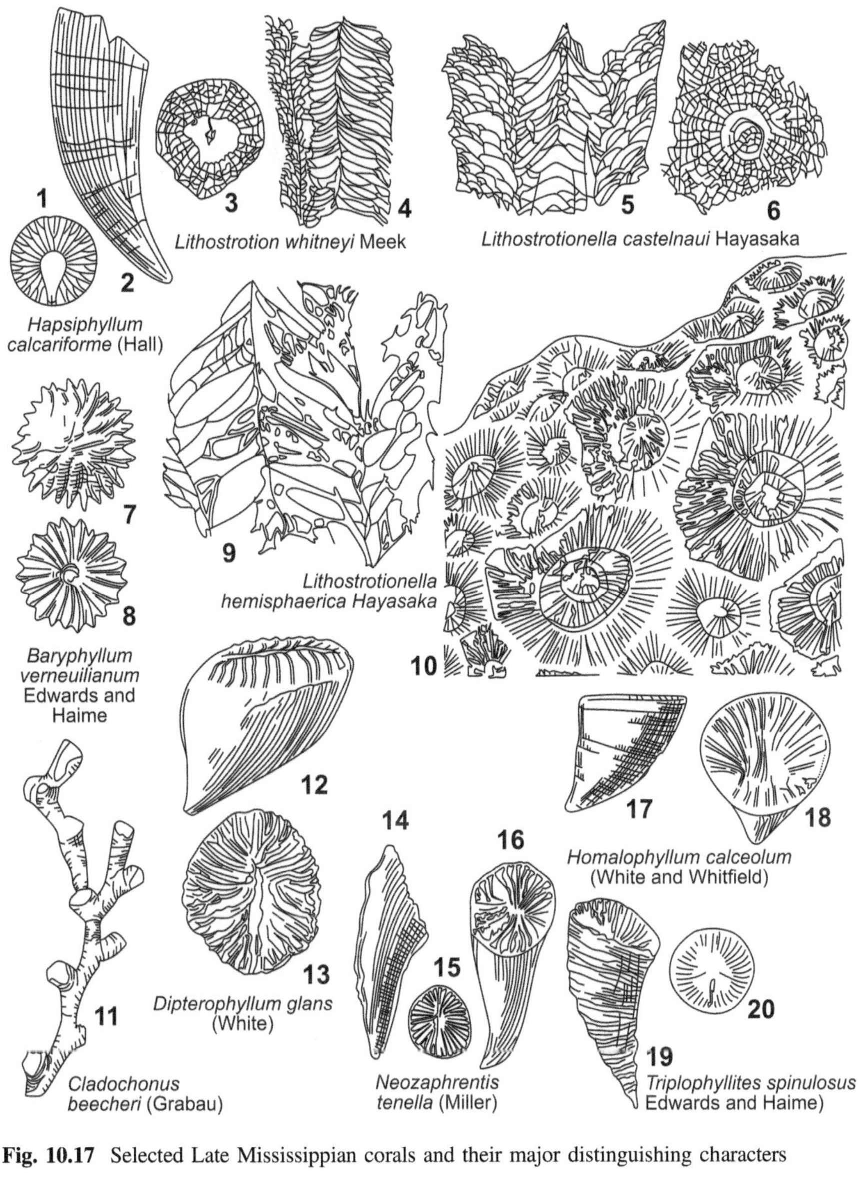

Fig. 10.17 Selected Late Mississippian corals and their major distinguishing characters

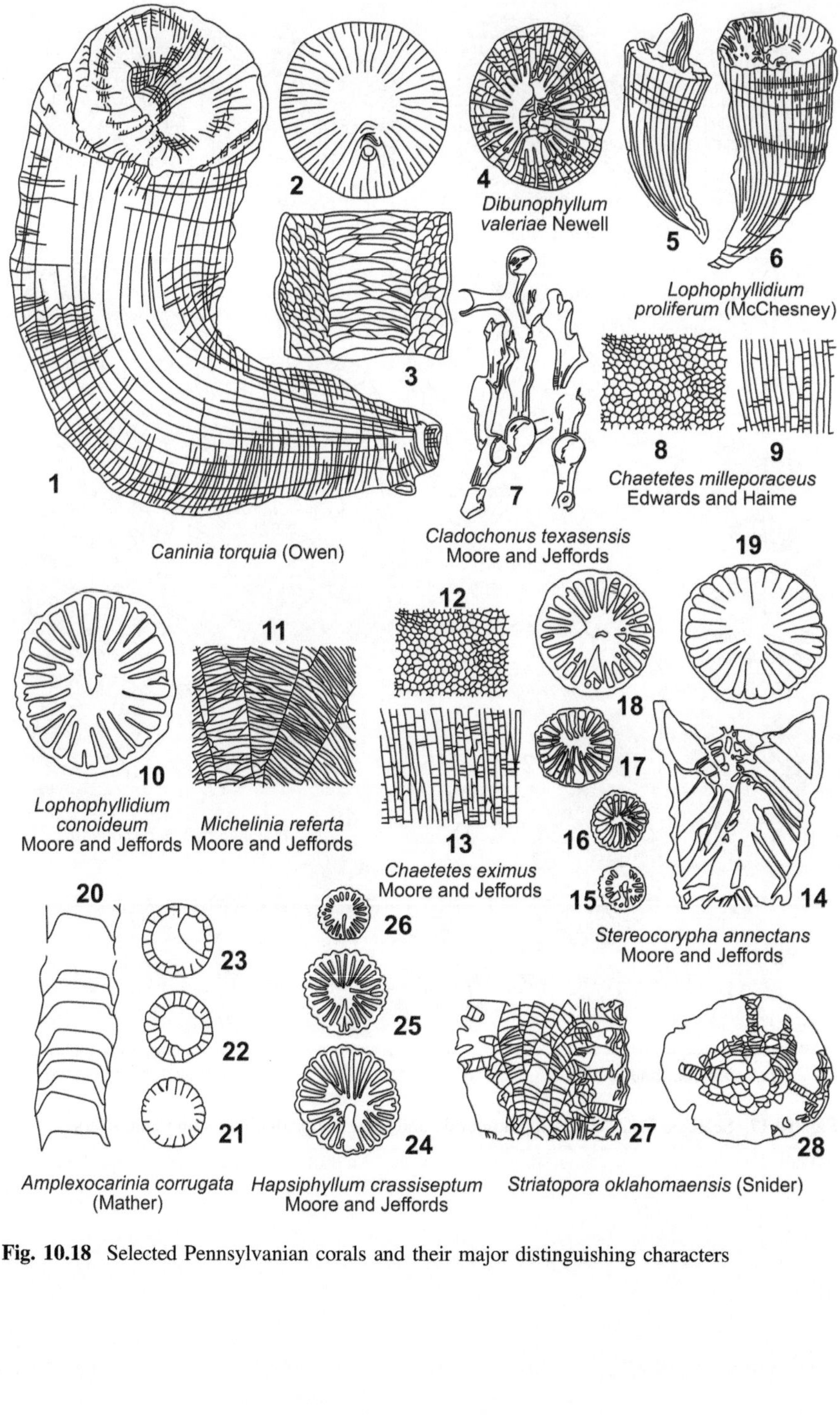

Fig. 10.18 Selected Pennsylvanian corals and their major distinguishing characters

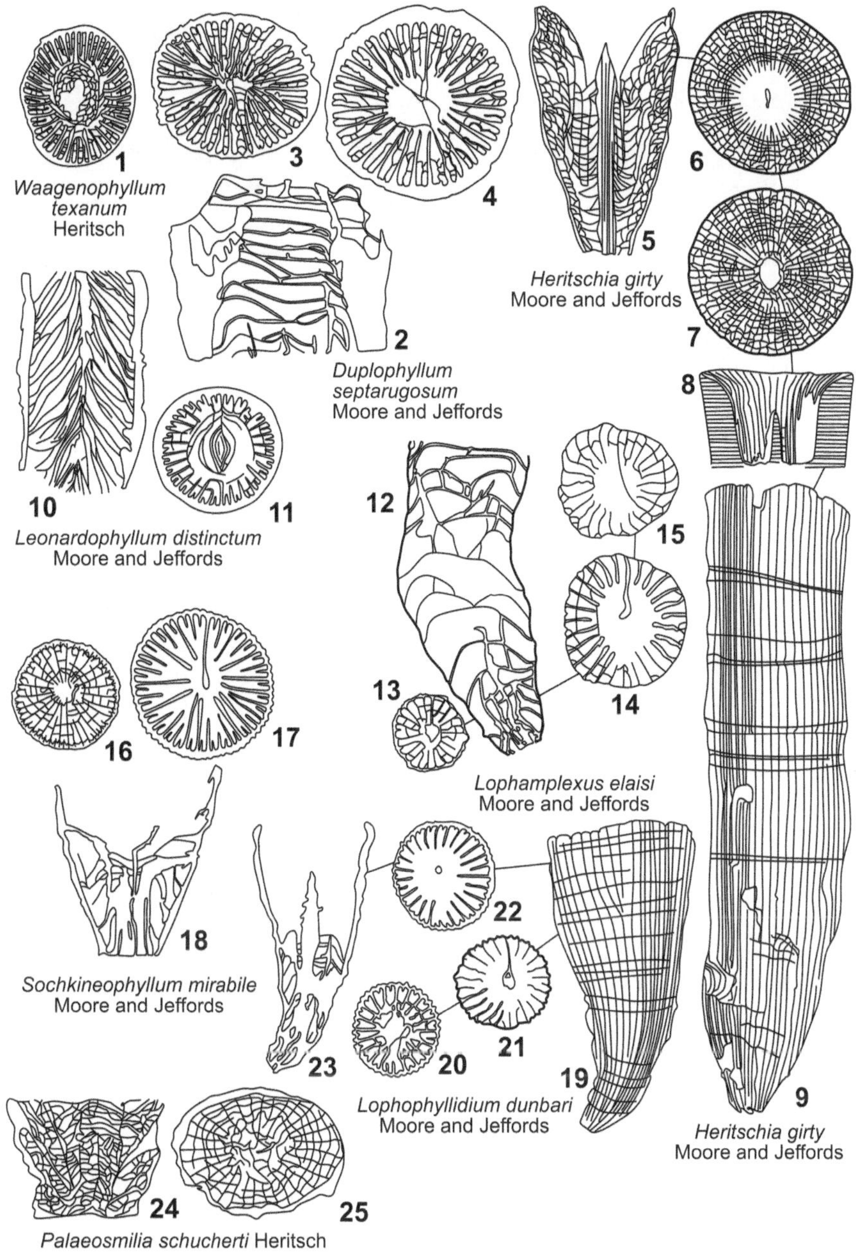

Fig. 10.19 Selected Early and Late Permian corals and their major distinguishing characters

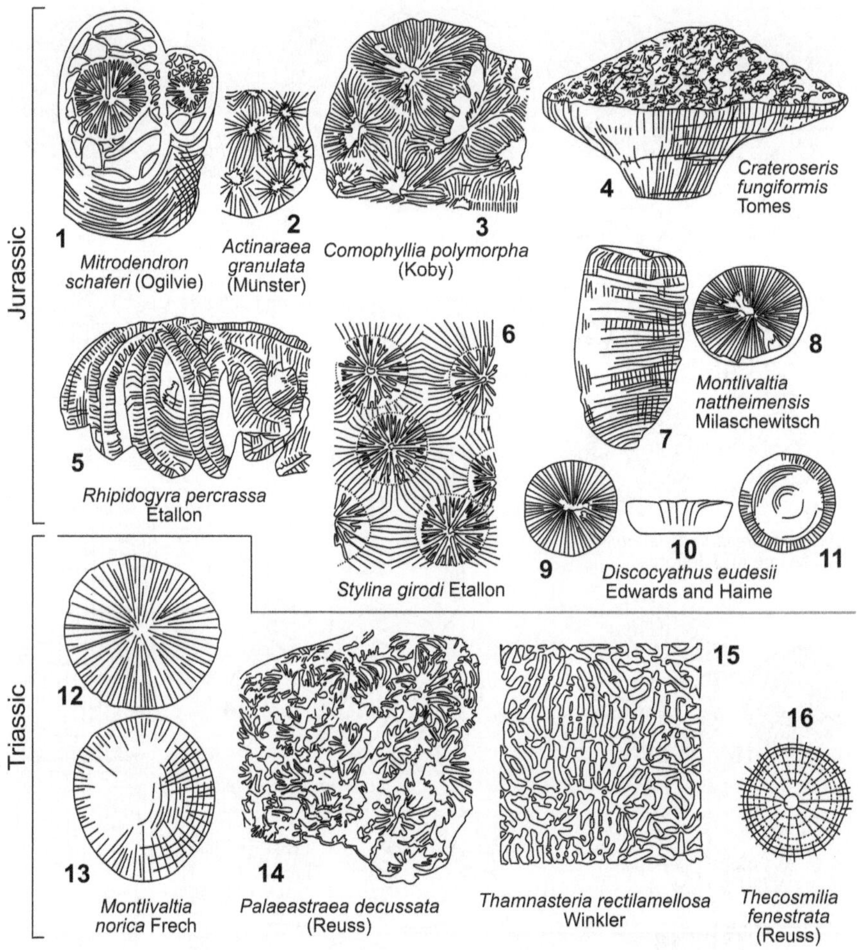

Fig. 10.20 Selected Triassic-Jurassic corals and their major distinguishing characters

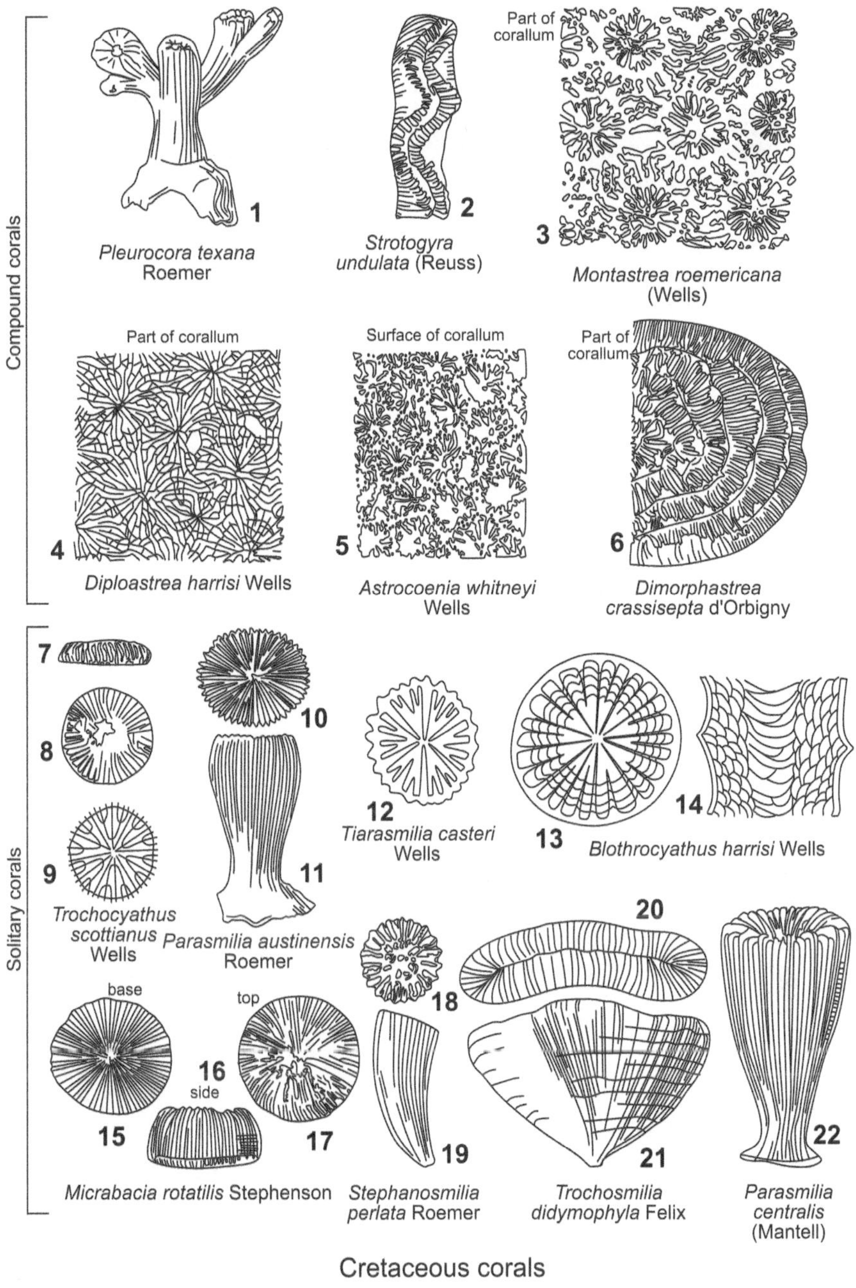

Fig. 10.21 Selected Cretaceous corals and their major distinguishing characters

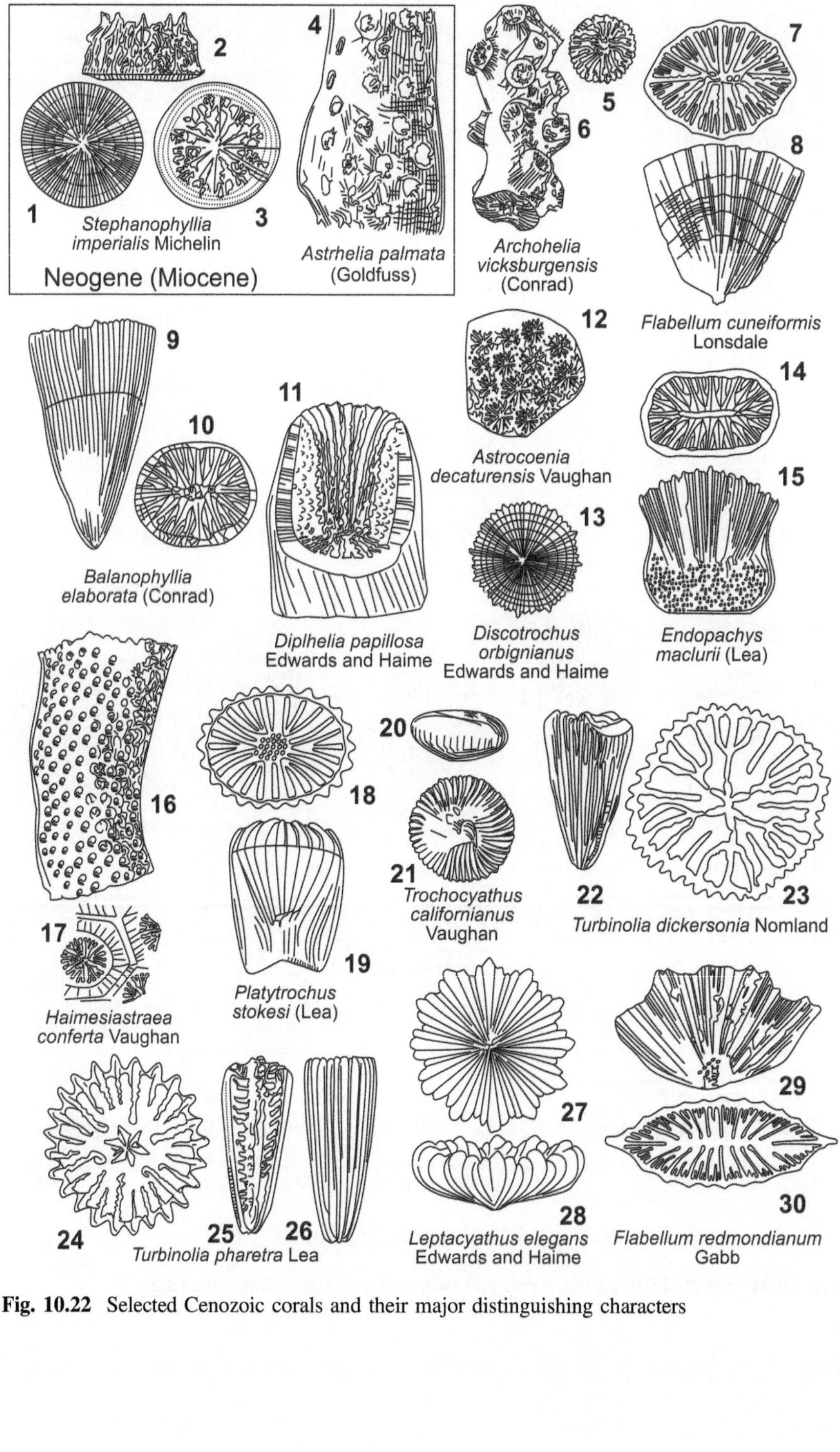

Fig. 10.22 Selected Cenozoic corals and their major distinguishing characters

References

Broadhurst, F.M., and I.M. Simpson. 1972. Bathymetry on a Carboniferous reef. *Lethaia* 6: 367–381.

Carruthers, R.G. 1910. On the evolution of *Zaphrentis delanouei* in Lower Carboniferous times. *Quarterly Journal of the Geological Society of London* 66: 523–536.

Clarkson, E.N.K. 1998. *Invertebrate palaeontology and evolution*, 4th ed, 452. Oxford: Blackwell Science.

Cousteau, J. Y. 1985. *The Ocean World.* 174–175. New York: Harry N. Abrams, Inc.

Doyle, P. 1996. *Understanding fossils: An introduction to invertebrate palaeontology*, 1st ed, 426. New York: Wiley.

Grotzinger, J.P., W.A. Watters, and A.H. Knoll. 2000. Calcified metazoans in thrombolite-stromatolite reefs of the terminal Proterozoic Nama Group, Namibia. *Paleobiology* 26: 334–359.

Hill, D. 1956. Rugosa. In *Treatise on invertebrate paleontology, part F, coelenterata*, edited by R. C. Moore, F233-324. Geological Society of America and University of Kansas.

Hill, D. 1981. Rugosa and Tabulata. In *Treatise on invertebrate paleontology, part F, coelenterata*, edited by C. Teichert, Vols. 2. Geological Society of America and University of Kansas.

Moore, R.C. ed. 1956. *Treatise on invertebrate paleontology, Part F, coelenterata.* Geological Society of America and University of Kansas Press, Lawrence, KS.

Nield, E.W., and V.C.T. Tucker. 1985. *Paleontology: An introduction*, 1st ed, 178. Oxford: Pergamon Press.

Oliver, W.A. 1980. The relationship of the scleractinian corals to the rugose corals. *Paleobiology* 6: 146–160.

Oliver Jr., W.A. 1996. Origins and relationships of Paleozoic coral groups and the origin of the Scleractinia. *Paleontological Society Papers* 1: 107–134.

Oliver, W.A., and A.G. Coates Jr. 1987. Phylum Cnidaria. In *Fossil invertebrates*, ed. R.S. Boardman, A.H. Cheetham, and A.J. Rowell, 140–193. Palo Alto, CA: Blackwell Scientific Publications.

Piper, R. 2007. *Extraordinary animals: an encyclopedia of curious and unusual animals*. Westport: Greenwood Press.

Ross, P. 2007. *Extraordinary Animals: An Encyclopedia of Curious and Unusual Animals*, 298. Greenwood Press.

Scrutton, C.T. 1997. The Palaeozoic corals, I: origins and relationships. *Proceedings of the Yorkshire Geological Society* 51: 177–208.

Wells, J.W. 1956. Scleractinia. In *Treatise on invertebrate paleontology, part F*, edited by R.C. Moore, F328-444. Geological Society of America and University of Kansas.

Wood, R.A., J.P. Grotzinger, and J.A.D. Dickson. 2002. Proterozoic modular biomineralized metazoan from the Nama Group, Namibia. *Science* 296: 2383–2386.

Appendix

S. Jain, *Fundamentals of Invertebrate Palaeontology*, Springer Geology,
DOI 10.1007/978-81-322-3658-0

Ch. no.	Species name	Age	Locality	Chapter	Fig. no.	Fig. no.
2	*Otavia antiqua* Brian et al.	Cryogenian-Ediacaran	Namibia, South Africa	Sponges	1	1
2	*Sycetta sagittifera* Haeckel	Recent	South India	Sponges	3	6
2	*Archaeocyaths atlanticus* Billings	Early Cambrian	West coast, USA	Sponges	7	3–4
2	*Ajacicyathus nevadensis* Oklulitch	Early Cambrian	West coast, USA	Sponges	7	5
2	*Ethmophyllum whitneyi* Meek	Cambrian	British Columbia, Canada	Sponges	7	6–7
2	*Pycnoidocyathus occidentalis* (Raymond)	Early Cambrian	British Columbia, Canada	Sponges	7	8
2	*Protospongia fenestrate* Salter	Middle Cambrian (Burgess)	British Columbia, Canada	Sponges	8	1
2	*Chancelloria eros* Walcott	Middle Cambrian (Burgess)	British Columbia, Canada	Sponges	8	2
2	*Eiffelia globosa* Walcott	Middle Cambrian (Burgess)	British Columbia, Canada	Sponges	8	3
2	*Choia carteri* Walcott	Middle Cambrian (Burgess)	British Columbia, Canada	Sponges	8	4
2	*Brachiospongia digitata* (Owen)	Middle Ordovician	Kentucky, USA	Sponges	9	1
2	*Hindia parva* Ulrich	Middle Ordovician	Minnesota, USA	Sponges	9	2–3
2	*Ischadites iowensis* (Owen)	Middle Ordovician	Kentucky, USA	Sponges	9	4
2	*Receptaculites oweni* Hall	Middle Ordovician	Kentucky, USA	Sponges	9	5
2	*Astaeospongia meniscus* (Roemer)	Middle Silurian (Niagaran)	Tennessee, USA	Sponges	10	1–2
2	*Prismodictya prismatica* (Hall)	Late Devonian	New York, USA	Sponges	10	3
2	*Prismodictya telum* (Hall)	Late Devonian	New York, USA	Sponges	10	4
2	*Astylospongia praemorsa* (Roemer)	Middle Silurian (Niagaran)	Tennessee, USA	Sponges	10	5–6
2	*Hydnoceras tuberosum* Conrad	Late Ordovician	New York, USA	Sponges	10	7
2	*Titusvillia drakei* Caster	Early Mississippian	NW Pennsylvania, USA	Sponges	11	1–2
2	*Girtyocoelia typica* King	Late Pennsylvanian (Missourian)	Texas, USA	Sponges	12	1–2
2	*Girtyocoelia beedei* (Girty)	Late Pennsylvanian (Missourian)	Kansas, USA	Sponges	12	3
2	*Maeandrostia kansasensis* Girty	Late Pennsylvanian (Missourian)	Texas, USA	Sponges	12	4–5
2	*Girtyocoelia dunbari* King	Permian (Leonardian)	Texas, USA	Sponges	12	6
2	*Cotykiscus ewersi* iKing				12	7–8
2	*Amblysiphonella prosseri* Clarke	Late Pennsylvanian (Virgilian)	Nebraska, USA	Sponges	12	9–10

(continued)

(continued)

Ch. no.	Species name	Age	Locality	Chapter	Fig. no.	Fig. no.
2	*Cystauletes mammilosus* King	Late Pennsylvanian (Desmoinesian)	Oklahoma, USA	Sponges	12	11–12
2	*Barroisia anastomans* (Mantell)	Early Cretaceous	England	Sponges	11	1–2
2	*Stellispongia glomerata* (Quenstedt)	Late Jurassic	Germany	Sponges	11	3
2	*Corynella quenstedi* Zittel	Late Jurassic	Germany	Sponges	11	4
2	*Coeleptychium agaricoides* Goldfuss	Late Cretaceous	Germany	Sponges	11	5–7
2	*Coscinopora infundibuliformis* Goldfuss	Late Cretaceous	Germany	Sponges	11	8–9
2	*Pachyteichisma carteri* Zittel	Late Jurassic	Germany	Sponges	11	10–11
2	*Ventriculites striatus* Smith	Late Cretaceous	Poland	Sponges	11	12–13
3	*Plectronoceras cambria* (Walcott)	Late Cambrian	China	Cephalopoda	1	1–3
3	*Palaeoceras mutabilis* Flower	Late Cambrian	North America	Cephalopoda	1	4–9
3	*Knightoconus antarticus* Yochelson, Flower and *Webers*	Late Cambrian	West Antarctica	Cephalopoda	1	10–11
3	*Hildoceras bifrons* (Bauguiere)	Early Jurassic (Early Toarcian)	Western Europe	Cephalopoda	5	3
3	*Dactylioceras directum* (Buckman)	Early Jurassic (Early Toarcian)	Western Europe	Cephalopoda	5	4
3	*Psiloceras planorbis* (Sowerby)	Early Jurassic (Hettangian)	United Kingdom	Cephalopoda	5	8
3	*Arcestes colonus* Mojsisovics	Late Triassic (Carnian-Rhaetian)	California, USA	Cephalopoda	5	9
3	*Ptycharcestes rugosus* Mojsisovics	Late Triassic (Carnian)	Western Europe	Cephalopoda	5	10
3	*Pararcestes sublabiatus* Rollier	Late Triassic (Carnian)	Northern Albaroz, Iran	Cephalopoda	5	11
3	*Arcestes pinacostomus* Diener	Late Triassic (Carnian)	California, USA	Cephalopoda	5	12
3	*Stenarcestes rolutaefirmis* Gammellaro	Late Triassic (Carnian-Norian)	Slovenia	Cephalopoda	5	13
3	*Proarcestes gibbus* Hauer	Late Triassic (Carnian-Norian)	Germany	Cephalopoda	5	14
3	*Hubertoceras omphalodes* (Waagen)	Middle Jurassic (Middle Callovian)	Kachchh, India	Cephalopoda	6	2
3	*Oecoptychius refractus* (Reinecke)	Middle Jurassic (Middle Callovian)	Germany	Cephalopoda	6	3
3	*Ebrayiceras pseudoanceps* (Waagen)	Middle Jurassic (Early-Middle Bathonian)	Germany	Cephalopoda	6	7
3	*Strenoceras bajocense* (Defrance)	Middle Jurassic (Late Bajocian)	France	Cephalopoda	6	8–9
3	*Cleistosphinctes cleistus* (Buckman)	Middle Jurassic (Late Bajocian)	France	Cephalopoda	6	10–11

(continued)

(continued)

Ch. no.	Species name	Age	Locality	Chapter	Fig. no.	Fig. no.
3	*Sphaeroceras brongniarti* (Sowerby)	Middle Jurassic (Middle-Late Bajocian)	France	Cephalopoda	6	12–13
3	*Dorsentensia edouardiana* (Orbigny)	Middle Jurassic (Middle Bajocian)	France	Cephalopoda	6	14–15
3	Cyrtosiceras *macrotelus* (Oppel)	Late Jurassic (Tithonian)	France	Cephalopoda	7	3
3	*Pectinatites pectinatus* (Pillips)	Late Jurassic (Late Kimmeridgian)	England	Cephalopoda	7	4-5
3	*Creniceras renggeri* (Oppel)	Late Jurassic (Oxfordian)	France	Cephalopoda	7	7
3	*Protophites christoli* (Beaudouin)	Late Jurassic (Oxfordian)	Germany	Cephalopoda	7	8–9
3	*Holcophylloceras ononense* (Stanton)	Early Cretaceous (Albian)	California, USA	Cephalopoda	8	1–2
3	*Angolaites gregoryi* Spath	Early Cretaceous (Albian)	Nigeria (Africa)	Cephalopoda	8	3
3	*Barremites difficilis* (Orbigny)	Early Cretaceous (Barremian)	France	Cephalopoda	8	4
3	*Kilianella pexiptycha* (Uhlig)	Early Cretaceous (Hautervian)	France	Cephalopoda	8	5
3	*Olcostephanus astierianus* (Orbigny)	Early Cretaceous (Valanginian)	France	Cephalopoda	8	6
3	*Kilianella pexiptycha* (Uhlig)	Early Cretaceous (Barriasian)	Madagascar	Cephalopoda	8	8
3	*Creniceras renggeri* (Oppel)	Late Jurassic (Oxfordian)	France	Cephalopoda	9	3
3	*Dactylioceras directum* (Buckman)	Pliensbachian (Eary Jurassic)	Hungary	Cephalopoda	9	8
3	*Protophites christoli* (Beaudouin)	Late Jurassic (Oxfordian)	France	Cephalopoda	9	9
3	*Arcestes colonus* Mojsisovics	Late Triassic (Norian)	Austria	Cephalopoda	9	11
3	*Pararcestes sublabiatus* Rollier	Late Triassic (Carnian)	Austria	Cephalopoda	9	12
3	*Psiloceras planorbis* (Sowerby)	Early Jurassic (Hettangian)	Hungary	Cephalopoda	9	14
3	*Olcostephanus astierianus* (Orbigny)	Early Cretaceous (Valanginian)	Mexico	Cephalopoda	9	15
3	*Ancyloceras matheroni* Orbigny	Early Cretaceous (Aptian)	France	Cephalopoda	16	9
3	*Cochloceras fischeri* Hauer	Late Triassic	Nevada, USA	Cephalopoda	16	10
3	*Hamites alternatus* Sowerby	Early Cretaceous	France	Cephalopoda	16	11
3	*Nipponites mirabilis* Yabe	Late Cretaceous	Japan	Cephalopoda	16	12
3	*Bactrites* Sandberger	Devonian	Michigan, USA	Nautiloids	26	1–2
3	*Lituites litmus* Montfort	Ordovician	Baltic region	Nautiloids	26	3

(continued)

(continued)

Ch. no.	Species name	Age	Locality	Chapter	Fig. no.	Fig. no.
3	*Ophioceras nackholmensis* Kjerulf	Ordovician	Norway	Nautiloids	26	4–5
3	*Augustoceras shideleri* Flower	Cincinnatian (Late Ordovician)	Kentucky, USA	Nautiloids	26	6–7
3	*Billingsites canadensis* (Billings)	Cincinnatian (Late Ordovician)	Anticosti Island, Canada	Nautiloids	26	8
3	*Shamattawaceras ascoceroides* Foerste and Savage	Late Ordovician	Manitoba, Canada	Nautiloids	26	9
3	*Cycloplectoceras miseri* Ulrich and Foerste	Early Ordovician	Arkansas, USA	Nautiloids	26	10–11
3	*Cyclostomiceras cassiense* (Whitefield)	Early Ordovician	Vermont, USA	Nautiloids	26	12
3	*Aphetoceras americanum* Hyatt	Early Ordovician	Newfoundland, USA	Nautiloids	26	13
3	*Cassinoceras grande* Ulrich and Foerste	Early Ordovician	Vermont, USA	Nautiloids	26	14
3	*Euophoiceras simplex* Barrande	Middle Silurian	Midwestern United States	Nautiloids	27	1
3	*Hexameroceras hertzeri* (Hall and Whitfield)	Middle Silurian	Ohio, USA	Nautiloids	27	2–3
3	*Phragmoceras broderipi* Barrande	Silurian	Czechoslovakia	Nautiloids	27	4–5
3	*Lechtricoceras desplainense* (McChesney)	Silurian	Wisconsin, USA	Nautiloids	27	6
3	*Ovoceras* oviforme (Hall)	Devonian	New York, USA	Nautiloids	28	1–2
3	*Goldringia* trivolve Flower	Middle Devonian	New York, USA	Nautiloids	28	3
3	*Lorieroceras* lorieri (Barrande)	Devonian	New York, USA	Nautiloids	28	4
3	*Centroceras* marcellense (Vanuxem)	Middle Devonian	New York, USA	Nautiloids	28	5
3	*Liroceras liratum* Girty	Pennsylvanian	Kansas, USA	Nautiloids	29	1–2
3	*Stenopoceras dumblei* (Hyatt)	Pennsylvanian-Permian	Kansas, USA	Nautiloids	29	3–4
3	*Solenocheilus springeri* (White and St. Jones)	Late Pennsylvanian	Iowa, USA	Nautiloids	29	5
3	*Domatoceras umbilicatum* Hyatt	Middle Pennsylvanian	Kansas, USA	Nautiloids	29	6–7
3	*Cooperoceras texanum* Miller	Permian	Texas, USA	Nautiloids	29	8–9
3	*Eutrephoceras dekayi* (Morton)	Late Cretaceous	North America	Nautiloids	30	1
3	*Proclydonautilus triadicus* Mojsisovics	Late Triassic	California, USA	Nautiloids	30	2–4
3	*Cosmonautilus dilleri* Hyatt and Smith	Late Triassic	California, USA	Nautiloids	30	5–7
3	*Cimomia* haughti (Olsson)	Eocene	Peru	Nautiloids	31	1–3

(continued)

(continued)

Ch. no.	Species name	Age	Locality	Chapter	Fig. no.	Fig. no.
3	*Hercoglossa harrisi* Miller and Thompson	Paleocene	Trinidad	Nautiloids	31	4–6
3	*Aturia vanuxemi* Conrad	Eocene	New Jersey, USA	Nautiloids	31	7–9
3	*Cameroceras* sp.	Middle Ordovician	Midwestern United States	Endocerida	32	1–3
3	*Endoceras* sp.	Early Ordovician	Midwestern United States	Endocerida	32	4–6
3	*Actinoceras beloitense* (Whitfield)	Middle Ordovician	Midwestern United States	Actinocerida	33	1–2
3	*Ordosoceras sphaeriforme* var. *otoktiense* Chang	Middle Ordovician	Myanmar	Actinocerida	33	3
3	*Bactrites budesheimensis* Roemer		England	Bactritida	34	1
3	*Bactrites ausavensis* (Steininger)	Middle Devonian	North America	Bactritida	34	2
3	*Timanites keyserlingi* Miller	Late Devonian	Alberta, Canada	Cephalopoda	35	1–3
3	*Acanthoclymenia neapolitana* (Clarke)	Late Devonian	New York, USA	Cephalopoda	35	4–6
3	*Manticoceras sinuosum* (Hall)	Late Devonian	New York, USA	Cephalopoda	35	7–9
3	*Parawocklumeria distorta* (Tietze)	Late Devonian	Germany	Cephalopoda	35	10–11
3	*Wocklumeria sphaeroides* (Richeter)				35	12–13
3	*Epiwocklumeria applatana* (Wedekind)	Late Devonian	Germany	Cephalopoda	35	14–15
3	*Agoniatites vanuxemi* (Hall)	Middle Devonian	New York, USA	Cephalopoda	35	16–18
3	*Soliclymenia paradoxa* (Muenster)	Late Devonian	Germany	Cephalopoda	35	19
3	*Prodromites gorbyi* Miller	Early Mississippian	Indiana, USA	Cephalopoda	36	1–3
3	*Muensteroceras parallelum* (Hall)	Early Mississippian	Indiana, USA	Cephalopoda	36	4–6
3	*Cravenoceras hesperium* Miller and Furnish	Late Mississippian	Navada, USA	Cephalopoda	36	7–9
3	*Prolecanites gurleyi* Smith	Late Mississippian	Missouri, USA	Cephalopoda	36	10–12
3	*Uddenites schucherti* Bose	Late Pennsylvanian	Texas, USA	Cephalopoda	37	1–3
3	*Gastrioceras listeri* (Martin)	Early Pennsylvanian	Texas, USA	Cephalopoda	37	4–6
3	*Shumardites uddeni* (Bose)	Late Pennsylvanian	Texas, USA	Cephalopoda	37	7–9
3	*Eothalassoceras kingorum* (Miller)	Late Pennsylvanian	Texas, USA	Cephalopoda	37	10–12
3	*Gonioloboceras goniolobus* Meek	Late Pennsylvanian	New Mexico, USA	Cephalopoda	37	13–15

(continued)

(continued)

Ch. no.	Species name	Age	Locality	Chapter	Fig. no.	Fig. no.
3	*Pronorites arkeansasensis* Smith	Early Pennsylvanian	Arkansas, USA	Cephalopoda	37	16–18
3	*Schistoceras missouriense* (Miller and Faber)	Late Pennsylvanian	Missouri, USA	Cephalopoda	37	19–21
3	*Eoasianites modestus* (Bose)	Early Permian	Texas, USA	Cephalopoda	37	22–24
3	*Properrinites mooreae* Miller and Furnish	Early Permian	Texas, USA	Cephalopoda	38	1–3
3	*Medlicottia whitneyi* Bose	Middle Permian	Texas, USA	Cephalopoda	38	4–6
3	*Pseudogastrioceras beedei* (Plummer and Scott)	Middle Permian	Texas, USA	Cephalopoda	38	7–9
3	*Perrinites hilli* (Smith)	Middle Permian	Texas, USA	Cephalopoda	38	10–12
3	*Stacheoceras toumanskyae* Miller and Furnish	Late Permian	Mexico	Cephalopoda	38	13–15
3	*Waagenoceras guadalupense* Girty	Late Permian	Texas, USA	Cephalopoda	38	16–18
3	*Timorites uddeni* Miller and Furnish	Late Permian	Texas, USA	Cephalopoda	38	19–21
3	*Meekoceras gracilitate* White	Early Triassic (Owenitian)	Idaho, USA	Cephalopoda	39	1–3
3	*Lecanites (Paralecanites) arnoldi* Hyatt and Smith	Early Triassic (Owenitian)	Idaho, USA	Cephalopoda	39	4–5
3	*Kymatites typus* Waagen	Early Triassic	Idaho, USA	Cephalopoda	39	6–8
3	*Ussuria waageni* Hyatt and Smith	Early Triassic	Eastern Siberia	Cephalopoda	39	9–11
3	*Columbites parisianus* Hyatt and Smith	Early Triassic	Idaho, USA	Cephalopoda	39	12–14
3	*Gyrophiceras gangeticum* Koninck	Early Triassic	Idaho, USA	Cephalopoda	39	15–17
3	*Tirolites pacificus* Hyatt and Smith	Middle Triassic	Navada, USA	Cephalopoda	40	1–3
3	*Ceratites (Gymotoceras) blakei* Gabb	Middle Triassic	Navada, USA	Cephalopoda	40	4–6
3	*Popanoceras (Parapopanoceras) haugi* Hyatt and Smith	Middle Triassic	California, USA	Cephalopoda	40	7–9
3	*Hungarites yatesi* Hyatt and Smith	Middle Triassic	California, USA	Cephalopoda	40	10–12
3	*Semiornites cordevolicus* (Mojsisovics)	Middle Triassic	California, USA	Cephalopoda	40	13–14
3	*Trachyceras (Anolcites) meeki* Mojsisovics	Middle Triassic	Navada, USA	Cephalopoda	40	15–17
3	*Tropites subbullatus* Hauer	Late Triassic	California, USA	Cephalopoda	41	1–3
3	*Cochloceras fischeri* Hauer	Late Triassic	Alps	Cephalopoda	41	4
3	*Leconteiceras californicus* Hyatt and Smith	Late Triassic	California, USA	Cephalopoda	41	5–7

(continued)

(continued)

Ch. no.	Species name	Age	Locality	Chapter	Fig. no.	Fig. no.
3	*Arcestes pacificus* Hyatt and Smith	Late Triassic	California, USA	Cephalopoda	41	8–10
3	*Discotropites laurae* Mojsisovics	Late Triassic	California, USA	Cephalopoda	41	11–13
3	*Schlotheimia angulata* Schluter	Hettangian (Eary Jurassic)	Germany	Cephalopoda	42	1–3
3	*Oxynoticeras oxynotum* (Quenstedt)	Sinemurian (Eary Jurassic)	Germany	Cephalopoda	42	4–6
3	*Rhacophyllites (Paradasyceras) vermosense* Herbiel	Eary Jurassic	England	Cephalopoda	42	7–9
3	*Liparoceras henleyi* Sowerby	Pliensbachian (Eary Jurassic)	England	Cephalopoda	42	10–11
3	*Vermiceras spiratissimum* (Quesntedt)	Sinemurian (Eary Jurassic)	Germany	Cephalopoda	42	12–14
3	*Hildoceras bifrons* Bruguiere	Toarcian (Eary Jurassic)	England	Cephalopoda	42	15–17
3	*Amaltheus margaritatus* Montfort	Late Pliensbachian (Eary Jurassic)	Western Europe	Cephalopoda	42	18–20
3	*Phylloceras heterophyllum* (Sowerby)	Toarcian (Eary Jurassic)	England	Cephalopoda	42	21–23
3	*Spiroceras bifurcatum* Quesntedt	Middle Jurassic	England	Cephalopoda	43	1–2
3	*Garantia garanti* (d'Orbigny)	Middle Jurassic	England	Cephalopoda	43	3–5
3	*Normannites* orbignyi Buckman	Middle Jurassic	England	Cephalopoda	43	6–8
3	*Oxycerites orbis* Oppel	Middle Jurassic	England	Cephalopoda	43	9–11
3	*Lytoceras (Hemilytoceras) immane* (Oppel)	Middle Jurassic	England	Cephalopoda	44	1–2
3	*Oecoptychius refractus* Reinecke	Middle Jurassic	England	Cephalopoda	44	3–5
3	*Reineckeia anceps* Bayle	Middle Jurassic	England	Cephalopoda	44	6–8
3	*Kheraiceras cosmopolita* Parona and Bonarelli	Middle Jurassic	England	Cephalopoda	44	9–11
3	*Phlycticeras pustulatum* Reinecke	Middle Jurassic	England	Cephalopoda	44	12–14
3	*Perisphinctes tiziani* Oppel	Middle Jurassic	England	Cephalopoda	44	15–17
3	*Phylloceras (Phyllopachyceras) infundibulum* d'Orbigny	Hauterivian to Berremian (Early Cretaceous)	Europe	Cephalopoda	45	1–3
3	*Hamulina astieri* d'Orbigny	Hauterivian to Berriasian (Early Cretaceous)	Europe	Cephalopoda	45	4–5
3	*Ptychoceras emerie* d'Orbigny	Berremian (Early Cretaceous)	Europe	Cephalopoda	45	6–7
3	*Streblites kraffti* Uhlig	Berriasian (Early Cretaceous)	Yemen	Cephalopoda	45	8–10
3	*Crioceras* duvali Leveille	Hauterivian to Berriasian (Early Cretaceous)	French Alps	Cephalopoda	45	11–12

(continued)

(continued)

Ch. no.	Species name	Age	Locality	Chapter	Fig. no.	Fig. no.
3	*Neolobites vibrayi* (d'Orbigny)	Cenomanian (Middle Cretaceous)	Mediterranean region	Cephalopoda	46	1–3
3	*Turrilites costatus* Lamark	Cenomanian (Middle Cretaceous)	Sultanate of Oman	Cephalopoda	46	4
3	*Engonoceras thomasi* Pervinquiere	Cenomanian (Middle Cretaceous)	North Africa	Cephalopoda	46	5–7
3	*Schloenbachia varians* Sowerby	Cenomanian (Middle Cretaceous)	Europe	Cephalopoda	46	8–10
3	*Sharpeiceras schluteri* Hyatt	Cenomanian (Middle Cretaceous)	Europe	Cephalopoda	46	11–13
3	*Dipoloceras (Oxytropidoceras) roissyi* d'Orbigny	Albian (Middle Cretaceous)	Europe	Cephalopoda	47	1–3
3	*Hamites alternatus* Sowerby	Early Cretaceous	Gault, England	Cephalopoda	47	4–5
3	*Pervunquieria inflata* Sowerby	Albian (Middle Cretaceous)	Europe	Cephalopoda	47	6–8
3	*Ancyloceras matheroni* d'Orbigny	Albian (Middle Cretaceous)	Europe	Cephalopoda	47	9–10
3	*Pictetia astieri* d'Orbigny	Aptian and Albian (Middle Cretaceous)	Gault, England	Cephalopoda	47	11–13
3	*Taxanites texanum* (Roemer)	Late Cretaceous	Texas, USA	Cephalopoda	48	1–3
3	*Scaphites hippocrepis* (Kay)	Late Cretaceous	Montana, USA	Cephalopoda	48	4–6
3	*Tissotia tissoti* (Bayle)	Coniacian (Late Cretaceous)	Mnorth Africa	Cephalopoda	48	7–8
3	*Baculites aquilaensis* Reeside	Late Cretaceous	Montana, USA	Cephalopoda	48	9
3	*Baculites ovatus* Say	Late Cretaceous	Montana, USA	Cephalopoda	48	10–11
3	*Sphenodiscus pleurisepius* (Conrad)	Late Cretaceous	Gulf coast, USA	Cephalopoda	48	12–14
3	*Placenticeras planum* Hyatt	Late Cretaceous	New mexico, USA	Cephalopoda	48	15–17
3	*Nipponites mirabilis* Yabe	Late Cretaceous	Japan	Cephalopoda	48	18
3	*Aptychus laevis* Mey	Late Jurassic (Tithonian)	Bulgaria	Aptychi	49	2
3	*Harpoceras lythense* Sowerby	Early Jurassic	Germany	Aptychi	49	3
3	*Oppelia steraspis* Oppel	Late Jurassic (Tithonian)	Germany	Aptychi	49	4
3	*Actinocamax granulatus* de Blainville	Late Cretaceous	Europe	Belemnoids	50	1
3	*Belemnitella mucronata* Scholenbach	Late Cretaceous (Maastrichtian)	Europe	Belemnoids	50	2
3	*Pseudobelus bipartitus* de Blainville	Early Cretaceous (Neocomian)	Europe	Belemnoids	50	3–4
3	*Pachyteuthis densus* (Meek)	Late Jurassic-Early Cretaceous	Rocky Mountains, USA	Belemnoids	50	5

(continued)

(continued)

Ch. no.	Species name	Age	Locality	Chapter	Fig. no.	Fig. no.
3	*Hibolites semisulcatus* (Munster)	Late Jurassic	Bavaria	Belemnoids	50	6–7
3	*Belemnopsis bessinus* (Orbigny)	Middle Jurassic	Europe	Belemnoids	50	8–9
3	*Salpingoteuthis acuarius* (Quesntedt)	Middle Jurassic	Europe	Belemnoids	50	10
3	*Hastites clavatus* (Schlotheim)	Early Triassic	England	Belemnoids	50	11–12
3	*Nannobelus acutus* (Bornn)	Early Triassic	Europe	Belemnoids	50	13
3	*Atractites haueri* Diener	Triassic-Jurassic	Alps	Belemnoids	50	14
3	*Aulacoceras timorense* Wanner	Late Triassic	Timor	Belemnoids	50	15–16
3	*Phragmoteuthis bisianuatus* (Bronn)	Late Triassic	Rocky Mountains, USA	Belemnoids	50	17
3	*Dictyoconites groenlandis* Rosenkrantz	Late Permian	Greenland	Belemnoids	50	18–19
4	*Fordilla troyensis* Barrande	Early Cambrian	New York, USA	Pelecypods	1	1
4	*Corbicula fluminea* (Müller)	Recent	Russia	Pelecypods	5	5
4	*Glycymeris subovata* (Say)	Pliocene	North Carolina, USA	Pelecypods	7	1
4	*Trigonia americana* Meek	Lower Jurassic	Utah, USA	Pelecypods	7	2
4	*Mytilus conradinus* d'Orbigny	Miocene	Maryland, USA	Pelecypods	7	3
4	*Spondylus regis* Linne	Middle Jurassic	Germany	Pelecypods	7	4
4	*Venericardia planicosta* Lamark	Eocene	Argentina	Pelecypods	7	5
4	*Cyprimeria alta* Conrad	Late Cretaceous (Maastrichtian)	Gulf coast, USA	Pelecypods	7	6
4	*Hippurites gosaviensis* Douville	Late Cretaceous (Campanian)	Romania	Pelecypods	7	7
4	*Nulcula percrassa* Conrad	Late Cretaceous (Maastrichtian)	Gulf coast, USA	Pelecypods	8	6
4	*Barbatia micronema* (Meek)	Late Cretaceous (Maastrichtian)	Germany	Pelecypods	8	7
4	*Ostrea cretacea* Morton	Late Cretaceous (Maastrichtian)	Alabama, USA	Pelecypods	8	8
4	*Cuspidaria ventricosa* Meek and Hayden	Late Cretaceous (Maastrichtian)	South Dakota, USA	Pelecypods	8	9
4	*Colpomya constricta* Ulrich	Late Ordovician (Trentonian)	New York, USA	Pelecypods	9	1
4	*Cleidophorus planulatus* (Conrad)	Late Ordovician (Cincinnatian)	Ohio, USA	Pelecypods	9	2
4	*Pterinea demissa* (Conrad)	Late Ordovician (Cincinnatian)	Ohio, USA	Pelecypods	9	3

(continued)

(continued)

Ch. no.	Species name	Age	Locality	Chapter	Fig. no.	Fig. no.
4	*Modiolopsis concentrica* Hall and Whitfeld	Late Ordovician (Richmondian)	Indiana, USA	Pelecypods	9	4
4	*Vanuxemia hayniana* (Safford)	Late Ordovician (Trentonian)	Tennessee, USA	Pelecypods	9	5–6
4	*Cyrtodonta grandis* (Ulrich)	Late Ordovician (Trentonian)	Wisconsin, USA	Pelecypods	9	7
4	*Whiteavesia cincinnatiensis* Hall and Whitfield	Late Ordovician (Trentonian)	Ohio, USA	Pelecypods	9	8
4	*Cypricardinia arata* Hall	Late Ordovician (Niagaran)	Indiana, USA	Pelecypods	9	9
4	*Orthodesma rectum* Hall and Whitfield	Late Ordovician (Cincinnatian)	Indiana, USA	Pelecypods	9	10
4	*Newsomella ulrchi* Foerste	Late Ordovician (Niagaran)	Tennessee, USA	Pelecypods	9	11
4	*Rhytimya producta* Ulrich	Late Ordovician (Cincinnatian)	Ohio, USA	Pelecypods	9	12
4	*Byssonychia radiata* (Hall)	Late Ordovician (Cincinnatian)	Ohio, USA	Pelecypods	9	13
4	*Lyrodesma major* (Ulrich)	Late Ordovician (Richmondian)	Minnesota, USA	Pelecypods	9	14–15
4	*Megalomus canadensis* Hall	Late Ordovician (Niagaran)	Ontario, Canada	Pelecypods	9	16
4	*Ambonychia bellistriata* Hall	Late Ordovician (Trentonian)	New York, USA	Pelecypods	9	17–18
4	*Ctenodonta gibberula* Salter	Middle Ordovician (Blackriveran)	Ontario, Canada	Pelecypods	9	19–20
4	*Goniophora bellula* Billings	Late Silurian	Nova Scotia, Canada	Pelecypods	9	21
4	*Conocardium cuneus trigonale* Hall	Early Devonian	New York, USA	Pelecypods	10	1–3
4	*Solemya vetusta* Meek	Early Devonian	New York, USA	Pelecypods	10	4–5
4	*Leiopteria rafinesquei* Hall	Middle Devonian	New York, USA	Pelecypods	10	6
4	*Cornellites flabella* (Conrad)	Middle Devonian	New York, USA	Pelecypods	10	7
4	*Limoptera macroptera* (Conrad)	Middle Devonian	New York, USA	Pelecypods	10	8
4	*Leiopteria dekayi* Miller	Middle Devonian	New York, USA	Pelecypods	10	9
4	*Orthonota undulata* Conrad	Middle Devonian	New York, USA	Pelecypods	10	10
4	*Paracyclas elliptica* Hall	Middle Devonian	New York, USA	Pelecypods	10	11
4	*Nuculoidea lirata* (Conrad)	Middle Devonian	New York, USA	Pelecypods	10	12
4	*Cypricardinia indenta* Conrad	Middle Devonian	New York, USA	Pelecypods	10	13
4	*Leptodesma rogersi* Hall	Middle Devonian	New York, USA	Pelecypods	10	14

(continued)

(continued)

Ch. no.	Species name	Age	Locality	Chapter	Fig. no.	Fig. no.
4	*Pholadella radiata* (Conrad)	Middle Devonian	New York, USA	Pelecypods	10	15
4	*Cypricardella bellastriata* (Conrad)	Middle Devonian	New York, USA	Pelecypods	10	16–17
4	*Palaeoneilo constricta* (Conrad)	Middle Devonian	New York, USA	Pelecypods	10	18
4	*Grammysia bisulcata* (Conrad)	Middle Devonian	New York, USA	Pelecypods	10	19
4	*Goniophora hamiltonensis* (Hall)	Middle Devonian	New York, USA	Pelecypods	10	20
4	*Amnigenia catskillensis* (Vanuxem)	Middle Devonian	New York, USA	Pelecypods	10	21
4	*Lunulicardium curtum* Hall	Middle Devonian	New York, USA	Pelecypods	10	22
4	*Sphenotus contractus* (Hall)	Late Devonian	New York, USA	Pelecypods	10	23
4	*Buchiola speciosa* (Hall)	Late Devonian	New York, USA	Pelecypods	10	24
4	*Lyriopecten tricostatum* (Vanuxem)	Late Devonian	New York, USA	Pelecypods	10	25
4	*Parallelodon chemungensis* (Hall)	Late Devonian	New York, USA	Pelecypods	10	26
4	*Nuculana diversa* (Hall)	Late Devonian	New York, USA	Pelecypods	10	27
4	*Pararca erecta* Hall	Late Devonian	New York, USA	Pelecypods	10	28
4	*Anthracomya elongata* (Dawson)	Early Pennsylvanian	Nova Scotia, Canada	Pelecypods	11	1
4	*Fasciculiconcha providencensis* (Cox)	Middle Pennsylvanian	Kentucky, USA	Pelecypods	11	2
4	*Nuculopsis girtyi* Schenck	Middle Pennsylvanian	Oklahoma, USA	Pelecypods	11	3
4	*Nadiadites carbonarius* Dawson	Middle Pennsylvanian	Oklahoma, USA	Pelecypods	11	4–5
4	*Lima retifera Shumard*	Late Pennsylvanian	Kansas, USA	Pelecypods	11	6
4	*Edmondia aspinwallensis* Meek	Late Pennsylvanian	Nebraska, USA	Pelecypods	11	7–8
4	*Parallelodon obsoletus* (Meek)	Late Pennsylvanian	Nebraska, USA	Pelecypods	11	9
4	*Pleurophorus oblongus* Meek	Late Pennsylvanian	Nebraska, USA	Pelecypods	11	10
4	*Myalina wyomingensis* (Lea)	Late Pennsylvanian	Illionos, USA	Pelecypods	11	11–12
4	*Clinopistha radiata* Hall	Late Pennsylvanian	Missouri, USA	Pelecypods	11	13–14
4	*Aviculopecten occidentalis* (Shumard)	Late Pennsylvanian	Missouri, USA	Pelecypods	11	15–16
4	*Chaenomya leavenworthensis* Meek and Hayden	Late Pennsylvanian	Kansas, USA	Pelecypods	11	17–18

(continued)

(continued)

Ch. no.	Species name	Age	Locality	Chapter	Fig. no.	Fig. no.
4	*Orthomyalina subquadrata* Shumard	Late Pennsylvanian	Kansas, USA	Pelecypods	11	19
4	*Pteria longa (Geinitz)*	Late Pennsylvanian	Kansas, USA	Pelecypods	11	20
4	*Annuliconcha interlineatus* (Meek and Worthen)	Late Pennsylvanian	Kansas, USA	Pelecypods	11	21
4	*Promytilus annosus senex* Newell	Late Pennsylvanian	Kansas, USA	Pelecypods	11	22
4	*Monopteria longispina* (Cox)	Late Pennsylvanian	Kansas, USA	Pelecypods	11	23
4	*Acanthopecten carboniferus* (Stevens)	Late Pennsylvanian	Kansas, USA	Pelecypods	11	24
4	*Mytilarca fibristriata* White and Whitfie d	Early Mississippian	Ohio, USA	Pelecypods	11	25
4	*Cypricardinia consimilis* Hall	Early Mississippian	Ohio, USA	Pelecypods	11	26
4	*Caneyella richardsoni* Girty	Late Mississippian	Oklahoma, USA	Pelecypods	11	27
4	*Palaeoneilo sulcatina* (Conrad)	Early Mississippian	Ohio, USA	Pelecypods	11	28
4	*Cypricardella oblongata* Hall	Late Mississippian (Meramecian)	Indiana, USA	Pelecypods	11	29
4	*Myalina keokuk* Worthen	Early Mississippian (Osagian)	Illionos, USA	Pelecypods	11	30
4	*Allorisma terminale* Hall	Early Permian	Kansas, USA	Pelecypods	12	1–2
4	*Dozierella gouldi* (Beede)	Early Permian	Oklahoma, USA	Pelecypods	12	3–4
4	*Nuculana bellistriata* (Stevens)	Early Permian	Kansas, USA	Pelecypods	12	5–6
4	*Nucula montplelierensis* Girty	Early Permian	Idaho, USA	Pelecypods	12	7–8
4	*Schizodus wheeleri* Swallow	Early Permian	Kansas, USA	Pelecypods	12	9
4	*Bakewellia parva* Meek and Hayden	Early Permian	Kansas, USA	Pelecypods	12	10
4	*Aviculopinna peracuta* (Shumard)	Early Permian	Kansas, USA	Pelecypods	12	11–12
4	*Myalina copie* Whitfield	Early Permian	Texas, USA	Pelecypods	12	13–14
4	*Pseudomonotis hawni* (Meek and Hayden)	Early Permian	Kansas, USA	Pelecypods	12	15
4	*Pleurophorus albequus* Beede	Early Permian	Oklahoma, USA	Pelecypods	12	16
4	*Aviculopecten vanvleeti* Beede	Early Permian	Oklahoma, USA	Pelecypods	12	17
4	*Grenvilla montanaensis* Meek	Late Jurassic	Montana, USA	Pelecypods	13	1
4	*Unio dockumensis* Simpson	Late Triassic	Texas, USA	Pelecypods	13	2

(continued)

(continued)

Ch. no.	Species name	Age	Locality	Chapter	Fig. no.	Fig. no.
4	*Camptonectes* bellistriatus Meek	Late Jurassic	Wyoming, USA	Pelecypods	13	3
4	*Unio felchi* White	Late Jurassic	Colorado, USA	Pelecypods	13	4
4	*Buchia piochei* (Gabb)	Late Jurassic	California, USA	Pelecypods	13	5
4	*Astarte packardi* White	Late Jurassic	Idaho, USA	Pelecypods	13	6–7
4	*Pleuromya subcompressa* Meek	Late Jurassic	Wyoming, USA	Pelecypods	13	8–9
4	*Trigonia americana* Meek	Late Jurassic	Montana, USA	Pelecypods	13	10–11
4	*Pteria submcconnelli* McLearn	Middle Jurassic	British Columbia, Canada	Pelecypods	13	12
4	*Myophoria alta* Gabb	Late Triassic	California, USA	Pelecypods	13	13
4	*Daonella americana* Smith	Middle Triassic	Nevada, USA	Pelecypods	13	14
4	*Monotis subcircularis* (Gabb)	Late Triassic	California, USA	Pelecypods	13	15
4	*Halobia superba* Mojsisovics	Late Triassić	Alps	Pelecypods	13	16
4	Pinna *laqueata* Conrad	Late Cretaceous	New Jersey, USA	Pelecypods	14	1
4	*Exogyra costata* Say	Late Cretaceous	Mississippi, USA	Pelecypods	14	2–3
4	*Volsella multilinigera* (Meek)	Late Cretaceous	Colorado, USA	Pelecypods	14	4
4	*Cuspidaria moreauensis* Meek and Hayden	Late Cretaceous	South Dakota, USA	Pelecypods	14	5
4	*Inoceramus labiatus* Schlotheim	Late Cretaceous	Kansas, USA	Pelecypods	14	6
4	*Cuspidaria ventricosa* Meek and Hayden	Late Cretaceous	Montana, USA	Pelecypods	14	7
4	*Anomia argenteria* Lamark	Late Cretaceous	Tennessee, USA	Pelecypods	14	8–9
4	*Pholadomya papyracea* Meek and Hayden	Late Cretaceous	Kansas, USA	Pelecypods	14	10–11
4	*Pteria petrosa* (Conrad)	Late Cretaceous	Tennessee, USA	Pelecypods	14	12
4	*Buchia terebratuloides* (Lahusen)	Early Cretaceous	California, USA	Pelecypods	14	13–14
4	*Exogyra arietina* Roemer	Early Cretaceous (Washitan)	Texas, USA	Pelecypods	14	15–16
4	*Paranomia scabra* (Morton)	Late Cretaceous	New Jersey, USA	Pelecypods	14	17
4	*Alectryonia quadriplicata* Shumard	Early Cretaceous (Washitan)	Texas, USA	Pelecypods	14	18
4	*Alectryonia carinata* Lamark	Early Cretaceous (Washitan)	Texas, USA	Pelecypods	14	19

(continued)

(continued)

Ch. no.	Species name	Age	Locality	Chapter	Fig. no.	Fig. no.
4	*Gryphaea corrugata* Say	Early Cretaceous (Washitan)	Texas, USA	Pelecypods	14	20–21
4	*Arctica ovata* (Meek and Hayden)	Late Cretaceous	North Dakota, USA	Pelecypods	15	1–2
4	*Barbatia micronema* (Meek)	Late Cretaceous	Wyoming, USA	Pelecypods	15	3–4
4	*Trigonia eufaulensis* Gabb	Late Cretaceous	North Carolina, USA	Pelecypods	15	5–6
4	*Corbula crassiplicata* Gabb	Late Cretaceous	Maryland, USA	Pelecypods	15	7–8
4	*Idonearca caroliensis* (Gabb)	Late Cretaceous	North Carolina, USA	Pelecypods	15	9–10
4	*Cardium eufaulensis* Gabb	Late Cretaceous	Maryland, USA	Pelecypods	15	11–12
4	*Ostrea cretacea* Morton	Late Cretaceous	North Carolina, USA	Pelecypods	15	13–14
4	*Breviarca haddonfieldensis* Stephenson	Late Cretaceous	New Jersey, USA	Pelecypods	15	15–16
4	*Nulcula percrassa* Conrad	Late Cretaceous	Tennessee, USA	Pelecypods	15	17–18
4	*Cyprimeria alta* Conrad	Late Cretaceous	Tennessee, USA	Pelecypods	15	19–20
4	*Veniella conradi* (Morton)	Late Cretaceous	Tennessee, USA	Pelecypods	15	21–22
4	*Crassatellites vadosus* (Morton)	Late Cretaceous	Maryland, USA	Pelecypods	15	23–24
4	*Lima reticulata* Forbes	Late Cretaceous	Maryland, USA	Pelecypods	15	25
4	*Nucula ovula* Lea	Eocene	North Carolina, USA	Pelecypods	16	1
4	*Nuculana parva* (Rogers)	Eocene	Virginia, USA	Pelecypods	16	2
4	*Glycimeris idonea* (Conrad)				16	3–4
4	*Taras hopkinsensis* (Clark)	Eocene	Virginia, USA	Pelecypods	16	5
4	*Lucina smithi* Meyer	Eocene	Virginia, USA	Pelecypods	16	6
4	*Volsella alabamensis* (Aldrich)	Eocene	California, USA	Pelecypods	16	7
4	*Pitar* uvasana (Conrad)	Eocene	California, USA	Pelecypods	16	8
4	*Acila schumardi* (Dall)	Oligocene	Washington, USA	Pelecypods	16	9
4	*Corbula aldrichi* Meyer	Eocene	Virginia, USA	Pelecypods	16	10–11
4	*Chlamys choctavensis* Aldrich	Eocene	Maryland, USA	Pelecypods	16	12
4	*Venericardia planicosta* Lamark	Eocene	Texas, USA	Pelecypods	16	13–14

(continued)

(continued)

Ch. no.	Species name	Age	Locality	Chapter	Fig. no.	Fig. no.
4	*Barbatia morsei* Gabb	Eocene	Washington, USA	Pelecypods	16	15–16
4	*Dosiniopsis lenticularis* (Rogers)	Eocene	Virginia, USA	Pelecypods	16	17–18
4	*Yoldia eborea* (Conrad)	Paleocene	Texas, USA	Pelecypods	16	19
4	*Cucullaea gigantea* Conrad	Paleocene	Maryland, USA	Pelecypods	16	20–21
4	*Venericardia smithi* Aldrich	Paleocene	Texas, USA	Pelecypods	16	22
4	*Crassatellites gabbi* (Safford)	Paleocene	Texas, USA	Pelecypods	16	23
4	*Tellina undulifera* Gabb	Paleocene	California, USA	Pelecypods	16	24
4	*Glycimeris subovata* (Say)	Miocene	Florida, USA	Pelecypods	17	1–2
4	*Plicatula densata* Conrad	Miocene	Florida, USA	Pelecypods	17	3–4
4	*Tellina declivis* Conrad	Miocene	New Jersey, USA	Pelecypods	17	5
4	*Chione latirata* Conrad	Miocene	Washington, USA	Pelecypods	17	6–8
4	*Clementia inoceriformis* (Wagner)	Miocene	Maryland, USA	Pelecypods	17	9
4	*Mactra clathrodon* Lea	Miocene	New Jersey, USA	Pelecypods	17	10–11
4	*Pallium swifti nutteri* Arnold	Miocene	California, USA	Pelecypods	17	12
4	*Panope generosa* Gould	Miocene	California, USA	Pelecypods	17	13
4	*Acila gettysburgensis* Reagen	Miocene	Washington, USA	Pelecypods	17	14
4	*Lyropecten estrellanus* (Conrad)	Miocene	California, USA	Pelecypods	17	15–16
4	*Mytilus conradinus* d'Orbigny	Miocene	South Carolina, USA	Pelecypods	17	17
4	*Taras acclinis* (Conrad)	Miocene	Alaska, USA	Pelecypods	17	18–19
4	*Mercenaria mercenaria* Linne	Miocene	New Jersey, USA	Pelecypods	17	20–21
4	*Hiatella artica* (Linne)	Miocene	New Jersey, USA	Pelecypods	18	1–2
4	*Semele subovata* (Say)	Miocene	Maryland, USA	Pelecypods	18	3
4	*Thracia trapezoides* Conrad	Miocene	Oregon, USA	Pelecypods	18	4
4	*Nuculana acuta* (Conrad)	Miocene	Florida, USA	Pelecypods	18	5–6
4	*Cumingia medialis* Conrad	Miocene	California, USA	Pelecypods	18	7–8

(continued)

(continued)

Ch. no.	Species name	Age	Locality	Chapter	Fig. no.	Fig. no.
4	*Arca trilineata* Conrad	Miocene	California, USA	Pelecypods	18	9
4	*Barbatia marylandica* Conrad	Miocene	California, USA	Pelecypods	18	10
4	*Ostrea trigonalis* Conrad	Miocene	Florida, USA	Pelecypods	18	11–12
4	*Glossus fraterna* Say	Miocene	California, USA	Pelecypods	18	13–14
4	*Macoma nasuta* (Conrad)	Miocene	Maryland, USA	Pelecypods	18	15
4	*Mya producta* Conrad	Miocene	New Jersey, USA	Pelecypods	18	16–18
4	*Ensis directus* (Conrad)	Miocene	Florida, USA	Pelecypods	18	19–20
5	*Isotelus brachycephalus* Foerste	Late Ordovician	Ohio, USA	Trilobites	4	1
5	*Acadoparadoxides briareus* (Geyer)	Middle Cambrian	Morocco	Trilobites	4	2
5	*Terataspis grandis* Hall	Early Devonian	New York, USA	Trilobites	4	3
5	*Uralichas hispanicus* tardus (Vela and Corbacho)	Late Ordovician	Morocco	Trilobites	4	4
5	*Isotelus* rex *Rudkin*	Late Ordovician	Manitoba, Canada	Trilobites	4	5
5	*Kjerulfia lata Kiaer*	Early Cambrian	Norway	Trilobites	6	1
5	*Asaphus raniceps* Dalman	Early Ordovician	Russia	Trilobites	7	4
5	*Reedops cephalotes* (Hawle and Corda)	Early Devonian	Bohemia	Trilobites	7	5
5	*Neocobboldia* chinlinica Lee	Early Cambrian	Henan, China	Trilobites	7	6–7
5	*Paedeumias transitans* Walcott	Early Cambrian	Vermont, USA	Trilobites	10	1
5	*Cryptolithus tesselatus* Green	Early Ordovician	Ohio, USA	Trilobites	10	2
5	*Phacops rana* Green	Middle Devonian	New York, USA	Trilobites	10	3
5	*Isotelus gigas* (Dekay)	Middle Ordovician	New York, USA	Trilobites	10	4
5	*Redlichia chinensis* Walcott	Early Cambrian	China	Trilobites	11	1
5	*Bathyurus extans* (Hall)	Middle Ordovician	Ontario, Canada	Trilobites	11	2
5	*Ogygiocaris saris* (Angelin)	Middle Ordovician	Russia	Trilobites	11	3
5	*Scutellum costatum* (Pusch)	Late Devonian	Belgium	Trilobites	11	4

(continued)

(continued)

Ch. no.	Species name	Age	Locality	Chapter	Fig. no.	Fig. no.
5	*Bathyuriscidella socialis Rasetti*	Middle Cambrian	Quebec, Canada	Trilobites	13	1
5	*Bonnia fieldensis* Walcott	Early Cambrian	British Columbia, Canada	Trilobites	13	2–3
5	*Callavia broggeri* Walcott	Early Cambrian	Newfoundland, USA	Trilobites	13	4
5	*Paedeumians transitans* Walcott	Early Cambrian	Vermont, USA	Trilobites	13	5
5	*Bonnia bubaris* (Walcott)	Early Cambrian	Quebec, Canada	Trilobites	13	6–8
5	*Kjerulfia lata* Kier	Early Cambrian	Norway	Trilobites	13	9
5	*Serrodiscus speciosus* (Ford)	Early Cambrian	New York, USA	Trilobites	13	10
5	*Redlichia chinensis* Walcott	Early Cambrian	Australia	Trilobites	13	11
5	*Neltneria jaquezi* (Neltner and Poctey)	Early Cambrian	Morocco	Trilobites	13	12
5	*Callavia broggeri* Walcott	Early Cambrian	Newfoundland, USA	Trilobites	13	13
5	Zacanthoides *romingeri* Walcott	Middle Cambrian	British Columbia, Canada	Trilobites	14	1
5	*Clappaspis typica* Deiss	Middle Cambrian	Monatana, USA	Trilobites	14	2
5	*Albertella helena* Mont.	Middle Cambrian	Monatana, USA	Trilobites	14	3
5	*Stephenaspis bispinosa* Rasetti	Middle Cambrian	British Columbia, Canada	Trilobites	14	4
5	*Ptychoparia striata* Emmrich	Middle Cambrian	Czechoslovakia	Trilobites	14	5
5	*Vanuxemella nortia* Walcott	Middle Cambrian	British Columbia, Canada	Trilobites	14	6
5	*Oryctocephalus primus* Walcott	Middle Cambrian	Nevada, USA	Trilobites	14	7
5	*Zocanthoides spinosus* Walcott	Middle Cambrian	Utah, USA	Trilobites	14	8
5	*Polypleuraspis insignis* Rasetti	Middle Cambrian	NW Greenland	Trilobites	14	9
5	*Alokistocare idahoense* Resser	Middle Cambrian	Utah, USA	Trilobites	14	10
5	*Dinesus ida* Etheridge	Early to Middle Cambrian	NW Queensland, Australia	Trilobites	14	11
5	*Tallaspis* pawlowskii (Schmidt)	Early to Middle Cambrian	Serbia	Trilobites	14	12
5	*Elrathina cordillerae* (Rominger)	Middle Cambrian	British Columbia, Canada	Trilobites	15	1
5	*Paradoxides pinus* Holm	Middle Cambrian	Sweden	Trilobites	15	2
5	*Conocoryphe sluzeri* (Schlotheim)	Middle Cambrian	Czechoslovakia	Trilobites	15	3

(continued)

(continued)

Ch. no.	Species name	Age	Locality	Chapter	Fig. no.	Fig. no.
5	*Paradoxides harlani* Green	Middle Cambrian	Massachusetts, USA	Trilobites	15	4
5	*Elrathia georgiensis* Resser	Middle Cambrian	Georgia, USA	Trilobites	15	5
5	*Ogygopsis klotzi* (Rominger)	Middle Cambrian	North America	Trilobites	15	6
5	*Bathyuriscus formosus* Deiss	Middle Cambrian	Montana, USA	Trilobites	15	7
5	*Olenoides curticei* Walcott	Middle Cambrian	Alabama, USA	Trilobites	15	8–9
5	*Asaphiscus wheeleri* Meek	Middle Cambrian	Utah, USA	Trilobites	15	10
5	*Olenus truncatus* (Brunnich)	Late Cambrian	Sweden	Trilobites	16	1
5	*Loganopeltoides kindlei* Rasetti	Late Cambrian	Newfoundland, USA	Trilobites	16	2
5	*Hausia canadensis* (Walcott)	Late Cambrian	British Columbia, Canada	Trilobites	16	3
5	*Litocephalus bilobatus* (Hall and Whitfield)	Late Cambrian	Nevada, USA	Trilobites	16	4
5	*Ctenopyge pecten* (Salter)	Late Cambrian	Eastern Canada	Trilobites	16	5
5	*Eurycare latum* (Boeck)	Late Cambrian	Sweden	Trilobites	16	6
5	*Bathyurus extans* (Hall)	Late Ordovician	Ontario, Canada	Trilobites	17	1
5	*Calliops callicephala* (Hall)	Late Ordovician (Katian)	New York, USA	Trilobites	17	2
5	*Achatella achates* (Billings)	Late Ordovician (Katian)	Minnesota, USA	Trilobites	17	3
5	*Eobronteus lunatus* (Billings)	Late Ordovician (Katian)	Ontario, Canada	Trilobites	17	4
5	*Basiliella barrandei* (Hall)	Late Ordovician (Katian)	New York, USA	Trilobites	17	5
5	*Ceraurus pleurexanthemus* Green	Late Ordovician (Katian)	New York, USA	Trilobites	17	6
5	*Illaenus davisii* Salter	Middle Ordovician	England	Trilobites	17	7
5	*Reedolithus carinatus* (Billings)	Middle Ordovician	Sweden	Trilobites	17	8
5	*Pliomerops canadensis* (Billings)	Middle Ordovician	New York, USA	Trilobites	17	9
5	*Asaphus expansus* Dalman	Middle Ordovician	Norway	Trilobites	17	10
5	*Kirkella vigilans* (Whittington)	Early Ordovician (Tremadocian)	Nevada, USA	Trilobites	17	11–14
5	*Highgatella gelasinassa* (Shaw)	Early Ordovician (Tremadocian)	NE Northern America	Trilobites	17	15
5	*Phacops rana* Green	Middle Devonian (Hamilton)	New York, USA	Trilobites	18	1–3

(continued)

(continued)

Ch. no.	Species name	Age	Locality	Chapter	Fig. no.	Fig. no.
5	*Dipleura dekayi* Green	Middle Devonian (Hamilton)	New York, USA	Trilobites	18	4
5	*Odontocephalus aegeria* Hall	Early Devonian (Onondaga)	New York, USA	Trilobites	18	5
5	*Coronura aspectans* (Conrad)	Early Devonian (Onondaga)	New York, USA	Trilobites	18	6
5	*Anchiopsis anchiops* (Green)	Early Devonian (Onondaga)	New York, USA	Trilobites	18	7
5	*Trimerus delphinocephalus* Green	Middle Silurian (Clintonian)	New York, USA	Trilobites	18	8
5	*Arctinurus boltoni* (Bigsby)	Middle Silurian (Clintonian)	New York, USA	Trilobites	18	9
5	*Calymene blumenbachi* Brongniart	Middle Silurian	England	Trilobites	18	10
5	*Encrinurus ornatus* Hall and Whitfield	Late Ordovician (Niagaran)	New York, USA	Trilobites	18	11
5	*Lichas speciosus* Beyrich	Silurian	Czechoslovakia	Trilobites	18	12
5	*Bumastus niagarensis* (Whitfield)	Late Ordovician (Niagaran)	Illinois, USA	Trilobites	18	13–14
5	*Terataspis grandis* (Hall)	Middle Devonian	New York, USA	Trilobites	19	1
5	*Akantharges gourdoni* (Barrois)	Middle Devonian	France	Trilobites	19	2
5	*Pennaia pauliana* Clarke	Early Devonian	Brazil	Trilobites	19	3
5	*Metacryphaeus australis* (Clarke)	Early Devonian	Brazil	Trilobites	19	4
5	*Greenops boothi* (Green)	Middle Devonian	Hamilton, New York	Trilobites	19	5
5	*Neogriffithides gemmellaroi* Toumansky	Middle Permian	Crimea	Trilobites	20	1
5	*Delaria antiqua* (Girty)	Early to Middle Permian	Texas, USA	Trilobites	20	2–3
5	*Paladin morrowensis* (Mather)	Early Permian	Arkansas, USA	Trilobites	20	4
5	*Phillipsia gemmulifera* (Phillips)	Mississippian	England	Trilobites	20	5
5	*Bollandia glibiceps* (Phillips)	Mississippian	England	Trilobites	20	6
5	*Eocyphinium clitheroense* Reed	Mississippian	England	Trilobites	20	7
5	*Excohops portlocki* (Meek and Worthen)	Mississippian	USA	Trilobites	20	8
5	*Griffithides longicepa* Vodges	Mississippian	England	Trilobites	20	9
5	*Weberides mucronatus* (M'Coy)	Pennsylvanian to Mississippian	England	Trilobites	20	10
5	*Griffithides seminiferus* (Phillips)	Pennsylvanian	England	Trilobites	20	11

(continued)

(continued)

Ch. no.	Species name	Age	Locality	Chapter	Fig. no.	Fig. no.
5	*Humilogriffithides divinopleurus* Inai	Pennsylvanian	England	Trilobites	20	12
6	*Stereocidaris tubifera* Mortensen	Recent	Philippine Islands	Echinoids	3	1–5
6	*Schizocidaris assimilis* Mortensen	Recent	Europe	Echinoids	5	4
6	*Goniocidaris sibogae* Mortensen	Recent	Europe	Echinoids	5	5
6	*Pseudosalenia zumoffenia*	Late Cretaceous (Cenomanian)	Lebanon	Echinoids	5	7
6	*Porocidaris schmidelii* (Munster)	Early Eocene	France	Echinoids	6	1
6	*Nortonechinus* welleri Thomas	Late Devonian	Iowa, USA	Echinoids	6	2
6	*Cyathocidaris erebus* Lambert	Late Cretaceous (Maastrichtian)	Seymour Island, Antarctica	Echinoids	6	3
6	*Balanocidaris glandifera* (Goldfuss)	Late Jurassic (Oxfordian)	Europe	Echinoids	6	4
6	*Balanocidaris pleracantha* (Agassiz)	Early Cretaceous (Albian)	USA	Echinoids	6	5
6	*Plegiocidaris cervicalis* (Agassiz)	Oxfordian	France	Echinoids	6	6
6	*Heterocentrotus* mammillatus (Linnaeus)	Recent	Indo–Pacific	Echinoids	6	7
6	*Hygrosoma petersii* (Agassiz)	Recent	West Indies	Echinoids	6	8–9
6	*Goniocidaris prunispinosa* (Chapman and Cudmore)	Recent	Indo-Pacific	Echinoids	6	10–11
6	*Allocentrotus fragilis* (Jackson)	Recent	California, USA	Echinoids	6	12
6	*Selenechinus armatus* (de Meijere)	Recent	Philippine Islands	Echinoids	6	13
6	*Aerosoma fenestratum* (Wyville-Thompson)	Recent	Illionis, USA	Echinoids	6	14
6	*Holectypus depressus* Leske	Jurassic	England	Echinoids	8	1
6	*Holectypus hemisphaericus* Agassiz	Jurassic	England	Echinoids	8	2
6	*Galeropygus agariciformis* (Forbes)	Late Jurassic (Oxfordian)	England	Echinoids	8	3–4
6	*Clypeus ploti Klein*	Jurassic	Europe	Echinoids	8	5–6
6	*Pygaster semisulcatus* Phillips	Jurassic	England	Echinoids	8	7–8
6	*Cassidulus eugeniae* (Agassiz)	Oligocene	USA	Echinoids	9	1
6	*Cassiduls subconicus* Clarke	Late Cretaceous	Mississippi, USA	Echinoids	9	2–6
6	*Encope tamiamensis* Mansfield	Pliocene (Neogene)	North America	Echinoids	10	1–5

(continued)

(continued)

Ch. no.	Species name	Age	Locality	Chapter	Fig. no.	Fig. no.
6	*Encope grandis* Agassiz	Pliocene (Neogene)	North America	Echinoids	10	6–8
6	*Eupatagus mooreanus* Pilsbry	Eocene	Florida, USA	Echinoids	11	1–3
6	*Pourtalesia miranda* Agassiz	Recent	North Atlantic	Echinoids	11	4
6	*Echinocardium cordatum* (Pennant)	Recent	Europe	Echinoids	11	5
6	*Linthia trechmanni* Hawkins	Eocene (Lutetian)		Echinoids	11	6–7
6	*Eothuria beggi* MacBirde and Spencer	Late Ordovician	England	Echinoids	12	1
6	*Aulechinus grayae* Bather and Spencer	Late Ordovician	England	Echinoids	12	2–3
6	*Palaeodiscus ferox* Salter	Silurian	England	Echinoids	12	4–5
6	*Triadocidaris subsimilis* (Munster)	Late Triassic (Cassian beds)	Alps	Echinoids	12	6–7
6	*Tiarechinus princeps* Neumayr	Late Triassic (Cassian beds)	Alps	Echinoids	12	8–11
6	*Hyattechinus rarispinus* Jackson	Early Carboniferous	Ohio, USA	Echinoids	13	1
6	*Melonechinus multiporus* (Norwood and Owen)	Mississippian (Carboniferous)	Missouri, USA	Echinoids	13	2–3
6	*Meekechinus elegans* Jackson	Early Permian	Kansas, USA	Echinoids	13	4
6	*Echinocrinus rossica* (van Buch)	Mississippian (Carboniferous)	Europe	Echinoids	13	5–6
6	*Plesiocidaris durandii* Peron and Gauthier	Jurassic	Algeria	Echinoids	14	16
6	*Pseudocidaris mammosa* Agassiz	Jurassic	France	Echinoids	14	17
6	*Hemicidaris crenularis* Lamark	Jurassic	France	Echinoids	14	10
6	*Hemicidaris jauberti* Cotteau	Jurassic	France	Echinoids	14	11
6	*Pedinothuria cidaroides* Gregory	Middle Jurassic (Bathonian)	Europe	Echinoids	14	1–4
6	*Palaeopedina globulus* Agassiz	Early Jurassic (Hettangian)	Europe	Echinoids	14	5–9
6	*Diademopsis heeri* Merian	Early Jurassic	Europe	Echinoids	14	12
6	*Pelanechinus corallinus* Keeping	Late Jurassic (Oxfordian)	England	Echinoids	14	13–15
6	*Micraster cortestudinarius* Goldfus	Late Cretaceous	Europe	Echinoids	15	1
6	*Pygurus oviformis* d'Orbigny	Cretaceous	France	Echinoids	15	2
6	*Hemiaster whitei* Clark	Middle Cretaceous (Fredericksburg)	Texas, USA	Echinoids	15	3–5

(continued)

(continued)

Ch. no.	Species name	Age	Locality	Chapter	Fig. no.	Fig. no.
6	*Fibularia subglobosa* Goldfus	Late Cretaceous (Senonian)	Europe	Echinoids	15	6–7
6	*Hyposalenia wrighti* (Cotteau)	Cretaceous	Europe	Echinoids	15	8
6	*Hyposalenia clathrata* (Cotteau)	Cretaceous	Europe	Echinoids	15	9
6	*Hyposalenia heliophora* (Cotteau)	Cretaceous	Europe	Echinoids	15	10
6	*Hyposalenia acanthoides* (Desmoulins)	Cretaceous	Europe	Echinoids	15	11
6	*Hyposalenia bunburyi* (Forbes)	Cretaceous	Europe	Echinoids	15	12
6	*Archiacia sandalina* (d'Archiac)	Late Cretaceous (Cenomanian)	Mediterranean	Echinoids	15	13–15
6	*Tylocidaris clavigera* Köing	Cretaceous	Europe	Echinoids	15	16
6	*Caenholectypus planatus* (Roemer)	Middle Cretaceous (Comanchean)	Texas, USA	Echinoids	15	17–19
6	*Dumblea symmetrica* Cragin	Middle Cretaceous (Washita)	Texas, USA	Echinoids	15	20–22
6	*Cyphosoma taxanum* Roemer	Middle Cretaceous (Fredericksburg)	Texas, USA	Echinoids	15	23–24
6	*Stereocidaris sceptrifera* (Mantell)	Cretaceous	France	Echinoids	15	25–26
6	*Schizaster arminger* Clark	Late Eocene (Paleogene)	SE USA	Echinoids	16	3
6	*Oligopygus wetherbyi* de Loriol	Eocene (Paleogene)	Florida, USA	Echinoids	16	4–5
6	*Fibularia vaughani* (Twitchell)	Eocene (Paleogene)	Florida, USA	Echinoids	16	6–7
6	*Periarchus lyelli* (Conrad)	Eocene (Paleogene)	SE USA	Echinoids	16	8–12
6	*Linthia tumidula* Clark	Paleocene (Paleogene)	New Jersey, USA	Echinoids	16	13–16
6	*Encope tamiamensis* Mansfield	Pliocene (Neogene)	North America	Echinoids	16	17
6	*Cassidulus gouldii* (Bouvre)	Oligocene (Paleogene)	SE USA	Echinoids	16	18–20
6	*Rotula orbiculus* (Linne)	Pliocene (Neogene)	West Africa	Echinoids	16	21
6	*Scutella leognanensis* Lambert	Miocene (Neogene)	France	Echinoids	16	22–23
6	*Dendraster gibbssi* (Remond)	Miocene (Neogene)	California, USA	Echinoids	16	24
7	*Dendrograptus communis* Kozlowski	Early Ordovician (Tremadocian)	England	Graptolites	4	5
7	*Dictyonema flabelliforme* (Eichwald)	Late Cambrian	England	Graptolites	4	7
7	*Dictyonema flabelliforme* (Eichwald)	Late Cambrian	England	Graptolites	7	1

(continued)

(continued)

Ch. no.	Species name	Age	Locality	Chapter	Fig. no.	Fig. no.
7	*Callograptus* sp.	Late Ordovician	England	Graptolites	7	2
7	*Acanthograptus* sp.	Late Cambrian – Early Ordovician	England	Graptolites	7	3
7	*Dendrograptus (Orthograptus) gracilis* Hall	Middle Ordovician	Germany	Graptolites	7	
7	*Staurograptus* sp.	Early Ordovician (Tremadocian)	England	Graptolites	8	1
7	*Didymograptus* sp.	Early Ordovician (Arenig)	England	Graptolites	8	2
7	*Phyllograptus angustifolius* Hall	Early Ordovician	England	Graptolites	8	3–4
7	*Goniograptus* sp.	Early Ordovician	New York, USA	Graptolites	8	5
7	*Tetragraptus sera* (Brongniart)	Early Ordovician (Arenig)	Norway	Graptolites	8	6
7	*Loganograptus* sp.	Early Ordovician	Quebec, Canada	Graptolites	8	7
7	*Tetragraptus fructocosus* Hall	Early Ordovician (Arenig)	Norway	Graptolites	8	8
7	*Leptograptus* sp.	Early Ordovician	England	Graptolites	8	9
7	*Tetragraptus phyllograptoides Linnarsson*	Early Ordovician	Quebec, Canada	Graptolites	8	10–11
7	*Tetragraptus bigsbyi* Hall	Early Ordovician	Sweden	Graptolites	8	12
7	*Tetragraptus headi* Hall	Early Ordovician	Quebec, Canada	Graptolites	8	13
7	*Ptilograptus plumosus* Hall	Early Ordovician	Quebec, Canada	Graptolites	8	14
7	*Dicellograptus* gurleyi Lapworth	Middle Ordovician	New York, USA	Graptolites	9	1
7	*Dicellograptus geniculatus* Bulman	Middle Ordovician	Sweden	Graptolites	9	2
7	*Nemagraptus gracilis* Hall	Middle Ordovician	New York, USA	Graptolites	9	3
7	*Retiograptus* sp.	Middle and Late Ordovician	New York, USA	Graptolites	9	4
7	*Nemagraptus* sp.	Middle and Late Ordovician	New York, USA	Graptolites	9	5
7	*Glossograptus ciliatus* Emmons	Middle and Late Ordovician	New York, USA	Graptolites	9	6–7
7	*Glossograptus quadrimucronatus approximatus* Ruedemann	Middle and Late Ordovician	New York, USA	Graptolites	9	8
7	*Climacograptus typicalis* Hall	Middle Ordovician	New York, USA	Graptolites	9	9–10
7	*Mastigograptus tenuiramus* Hall	Middle and Late Ordovician	New York, USA	Graptolites	9	11

(continued)

(continued)

Ch. no.	Species name	Age	Locality	Chapter	Fig. no.	Fig. no.
7	*Dicranograptus spinifer* arkansasensis Gurley	Middle and Late Ordovician	New York, USA	Graptolites	9	12
7	*Dicranograptus clingani* Hopkinson	Middle and Late Ordovician	New York, USA	Graptolites	9	13
7	*Coenograptus gracilis* Hall	Middle Ordovician	New York, USA	Graptolites	9	14
7	*Dicranograptus ramosus* Hall	Middle and Late Ordovician	Canada	Graptolites	9	15
7	*Diadymograptus pennatulus* Hall	Middle and Late Ordovician	New York, USA	Graptolites	9	16
7	*Tetragraptus bryonoides* Hall	Middle and Late Ordovician	New York, USA	Graptolites	9	17
7	*Lobograptus scanicus* (Tullberg)	Silurian	Germany	Graptolites	10	1–2
7	*Rastrites linnaei* Barrande	Silurian	Bohemia	Graptolites	10	3
7	*Dimorphograptus confertus* Nicholson	Silurian	Scotland	Graptolites	10	4
7	*Cyrtograptus kirki* Ruedemann	Silurian	Idaho, USA	Graptolites	10	5
7	*Monograptus bohemicus* Barrande	Silurian	Oklahoma, USA	Graptolites	10	6
7	*Monograptus lobiferus* McCoy	Silurian	Wales	Graptolites	10	7–8
7	*Monograptus dubius* Suess	Silurian	Oklahoma, USA	Graptolites	10	9
7	*Monograptus convolutus* Hisinger	Silurian	Wales	Graptolites	10	10
7	*Saetograptus chimaera* (Barrande)	Silurian (Ludlow)	Germany	Graptolites	10	11
7	*Gothograptus simplex* Eisenack	Late Silurian (Ludlow)	Baltic region	Graptolites	10	12
7	*Retiolites geinitzianus* Barrande	Silurian	Sweden	Graptolites	10	13
7	*Cyrtograptus murchisoni* Carruthers	Middle Silurian (Wenlock)	Idaho, USA	Graptolites	10	14
7	*Rastrites maximus* Carruthers	Early Silurian (Llandovery)	Scotland	Graptolites	10	15
7	*Spirograptus spiralis* (Geinitz)	Early Silurian (Llandovery)	Bohemia	Graptolites	10	16
7	*Monograptus lobiferus* McCoy	Silurian	Wales	Graptolites	10	17–18
7	*Saetograptus leintwardinensis* (Lapworth)	Late Silurian (Ludlow)	England	Graptolites	10	19
7	*Monograptus argenteus* (Nicholson)	Silurian	England	Graptolites	10	20–21
7	*Monograptus triangulatus* (Harkness)	Early Silurian (Llandovery)	Scotland	Graptolites	10	22
7	*Dimorphograptus elongatus* Elles and Wood	Early Silurian (Llandovery)	Scotland	Graptolites	10	23

(continued)

(continued)

Ch. no.	Species name	Age	Locality	Chapter	Fig. no.	Fig. no.
8	*Lingulella similis* Walcott	Late Cambrian	South Dakota, USA	Brachiopods	12	1–2
8	*Dicellomus politus* (Hall)	Late Cambrian	Wisconsin, USA	Brachiopods	12	3–5
8	Lingulelpis *pinnaformis (Owen)*	Late Cambrian	Wisconsin, USA	Brachiopods	12	6–7
8	*Siphonotreta tertia (Walcott)*	Late Cambrian	Alberta, Canada	Brachiopods	12	8–9
8	*Schizambon typicalis* Walcott	Late Cambrian	Nevada, USA	Brachiopods	12	10–11
8	*Apheoorthis linecosta* (Walcott)	Late Cambrian	Colorado, USA	Brachiopods	12	12–14
8	*Billingsella corrugata* Ulrich and Cooper	Late Cambrian	Oklahoma, USA	Brachiopods	12	15–16
8	*Billingsella perfecta* Ulrich and Cooper	Late Cambrian	Montana, USA	Brachiopods	12	17–20
8	*Acrrotreta idahoensis* Walcott	Late Cambrian	Idaho, USA	Brachiopods	12	21–23
8	*Nisusia montanaensis* Bell	Middle Cambrian	Montana, USA	Brachiopods	12	24–25
8	*Micromitra sculptilis* (Meek)	Middle Cambrian	Montana, USA	Brachiopods	12	26–28
8	*Prototreta trapeza* Bell	Middle Cambrian	Montana, USA	Brachiopods	12	29–32
8	*Dictyonina pannula* (White)	Middle Cambrian	Nevada, USA	Brachiopods	12	33–34
8	*Acrothele coriacea* Linnarsson	Middle Cambrian	Sweden	Brachiopods	12	35–36
8	*Obolella chromatica* Billings	Early Cambrian	Labrador, Canada	Brachiopods	12	37–38
8	*Rustella edsoni* Walcott	Early Cambrian	Vermont, USA	Brachiopods	12	39–40
8	*Kutorgina cingulata* (Billings)	Early Cambrian	Vermont, USA	Brachiopods	12	41–44
8	*Austinella kankakensis (McChesney)*	Late Ordovician	Illiois, USA	Brachiopods	13	1–3
8	*Catazyga headi* (Billings)	Late Ordovician	Indiana, USA	Brachiopods	13	4
8	*Hebertella sinuta* (Hall)	Late Ordovician	Kentucky, USA	Brachiopods	13	5–10
8	*Lepidocyclus* sp.	Late Ordovician	Kentucky, USA	Brachiopods	13	12
8	*Orthorhynchula linneyi* (James)	Late Ordovician	Ohio, USA	Brachiopods	13	13–17
8	*Retrorsirostra carleyi* (Hall)	Late Ordovician	Kentucky, USA	Brachiopods	13	18–19
8	*Plaesiomys subquadrata* (Hall)	Late Ordovician	Ohio, USA	Brachiopods	13	20–23
8	*Platystrophia crassa* James	Late Ordovician	Kentucky, USA	Brachiopods	13	24–26

(continued)

(continued)

Ch. no.	Species name	Age	Locality	Chapter	Fig. no.	Fig. no.
8	*Platystrophia cypha* James	Late Ordovician	Indiana, USA	Brachiopods	13	27–28
8	*Platystrophia ponderosa* Foerste	Late Ordovician	Ohio, USA	Brachiopods	13	29–32
8	*Rasserella meeki* (Miller)	Late Ordovician	Indiana, USA	Brachiopods	13	33–36
8	*Glyptorthis insculpta* (Hall)	Late Ordovician (Richmondian)	Ohio, USA	Brachiopods	14	1–4
8	*Rafinesquina loxorhytis* (Meek)	Late Ordovician (Richmondian)	Indiana, USA	Brachiopods	14	5–6
8	*Sowerbyella clarksvillensis* (Foerste)	Late Ordovician (Richmondian)	Ohio, USA	Brachiopods	14	7–8
8	*Rhynchotrema dentatum* Hall	Late Ordovician (Richmondian)	Indiana, USA	Brachiopods	14	9–11
8	Rhynchotrema *argenturbicum (White)*	Late Ordovician	New Mexico, USA	Brachiopods	14	12
8	*Strophomena neglecta James*	Late Ordovician	Ohio, USA	Brachiopods	14	13–15
8	*Strophomena planoconvexa* Hall	Late Ordovician	Ohio, USA	Brachiopods	14	16–18
8	*Zygospira modesta* (Say)	Late Ordovician	Kentucky, USA	Brachiopods	14	19–21
8	*Syntrophopsis magna* Ulrich and Cooper	Late Ordovician	Arkansas, USA	Brachiopods	14	22–25
8	*Dinorthis pectinella* Winchell and Pchuchert	Late Ordovician (Trentonian)	Kentucky, USA	Brachiopods	15	1
8	*Heterorthis clytie* (Hall)	Late Ordovician (Trentonian)	Kentucky, USA	Brachiopods	15	2–4
8	*Sowerbyella punctostriata* (Mather)	Late Ordovician (Trentonian)	New York, USA	Brachiopods	15	5–8
8	*Triplesia cuspidata* Clarke	Middle Ordovician	New York, USA	Brachiopods	15	9–13
8	*Dicaelosia biloba* (Linne)	Middle Ordovician	Indiana, USA	Brachiopods	15	14–15
8	*Finkelburgia virginica* Ulrich and Cooper	Late Cambrian to Early Ordovician	Virginia, USA	Brachiopods	15	16–19
8	*Strophomena nutans* (Meek)	Middle to Late Ordovician	New York, USA	Brachiopods	16	1
8	*Ancistrorhyncha costata* Ulrich and Cooper	Middle Ordovician	Tennessee, USA	Brachiopods	16	2
8	*Bimuria superba* Ulrich and Cooper	Middle Ordovician	Virginia, USA	Brachiopods	16	3–4
8	*Camerella plicata* (Schuchert and Cooper)	Middle Ordovician	Tennessee, USA	Brachiopods	16	5–6
8	*Christiania subquadrata* (Hall)	Middle Ordovician	Tennessee, USA	Brachiopods	16	7–10
8	*Doleroides gibbosus* (Billings)	Middle Ordovician	New York, USA	Brachiopods	16	11–12
8	*Hesperorthis tricenaria* (Conrad)	Middle Ordovician	New York, USA	Brachiopods	16	13–16

(continued)

(continued)

Ch. no.	Species name	Age	Locality	Chapter	Fig. no.	Fig. no.
8	*Opikina septata* Salmon	Middle Ordovician	Tennessee, USA	Brachiopods	16	17–18
8	*Valcourea strophomenoides* (Raymond)	Middle Ordovician	New York, USA	Brachiopods	16	19–20
8	*Pionodema subaequata* (Conrad)	Middle Ordovician	Minnesota, USA	Brachiopods	16	21–24
8	*Sowerbyites triseptatus* Willard	Middle Ordovician	Pennsylvania, USA	Brachiopods	16	25–27
8	*Leptellina tennesseensis* Ulrich and Cooper	Early to Middle Ordovician	Virginia, USA	Brachiopods	17	1–3
8	*Nanorthis hamburgensis* (Walcott)	Early Ordovician	Nevada, USA	Brachiopods	17	4–6
8	*Diparelasma elegantulum* (Butts)	Early Ordovician	Alabama, USA	Brachiopods	17	7–9
8	*Orthambonites eucharis* Ulrich and Cooper	Early Ordovician	Nevada, USA	Brachiopods	17	10–11
8	*Tetralobula delicatula* Ulrich and Cooper	Early Ordovician	Alabama, USA	Brachiopods	17	12–13
8	*Tritoechia typica* (Ulrich)	Early Ordovician	Oklahoma, USA	Brachiopods	17	14–15
8	*Dayia navicula* (Sowerby)	Late Silurian	England	Brachiopods	18	1
8	*Atrypella shrocki* Cooper	Middle Silurian (Niagaran)	Indiana, USA	Brachiopods	18	2–3
8	*Barrandella fornicata* (Hall)	Middle Silurian	New York, USA	Brachiopods	18	4–6
8	*Conchidium* sp.	Middle Silurian	Tennessee, USA	Brachiopods	18	7–8
8	*Cyrtia exporrecta* (Wahlenberg)	Middle Silurian	Indiana, USA	Brachiopods	18	9–11
8	*Trimerella ohioensis* Meek	Middle Silurian (Niagaran)	Ohio, USA	Brachiopods	18	12
8	*Dinobolus conradi* Hall	Middle Silurian	Iowa, USA	Brachiopods	18	13–14
8	*Eospirifer radiatus* (Sowerby)	Middle Silurian	New York, USA	Brachiopods	18	15–16
8	*Fardenia subplana* (Conrad)	Middle Silurian	Indiana, USA	Brachiopods	18	17–19
8	*Homeospira evax* (Hall)	Middle Silurian	Indiana, USA	Brachiopods	18	20
8	*Hyattidina congesta* (Conrad)	Middle Silurian	New York, USA	Brachiopods	18	21–23
8	*Meristina maria* (Hall)	Middle Silurian	Indiana, USA	Brachiopods	18	24–26
8	*Parmorthis waldronensis* (Foerste)	Middle Silurian	Indiana, USA	Brachiopods	18	27–29
8	*Pentamerus laevis* Sowerby	Middle Silurian	Illinois, USA	Brachiopods	18	30–31
8	*Rhipidium knappi* (Hall and Whitfield)	Middle Silurian	Tennessee, USA	Brachiopods	18	32–33

(continued)

(continued)

Ch. no.	Species name	Age	Locality	Chapter	Fig. no.	Fig. no.
8	*Plectodonta transversalis* (Dalman)	Middle Silurian	New York, USA	Brachiopods	18	34–36
8	*Rhynchotreta americana* Hall	Middle Silurian	Kentucky, USA	Brachiopods	19	1–3
8	*Stegerhynchus indianense* (Hall)	Middle Silurian	Indiana, USA	Brachiopods	19	3–4
8	*Stegerhynchus whitei* (Hall)	Middle Silurian	Indiana, USA	Brachiopods	19	5–7
8	*Whitfieldella nitida* (Hall)	Middle Silurian	Indiana, USA	Brachiopods	19	8–10
8	*Trimerella ohioensis* Meek	Middle Silurian (Niagaran)	Ohio, USA	Brachiopods	19	11–13
8	*Dolerorthis flabellites* (Foerste)	Middle Silurian (Niagaran)	Ohio, USA	Brachiopods	19	14–17
8	*Sieberella roemeri* Hall and Clarke	Middle Silurian	Oklahoma, USA	Brachiopods	19	21–22
8	*Stegerhynchus acinus* (Hall)	Middle Silurian	Indiana, USA	Brachiopods	19	18–20
8	*Cryptothyrella cylindrica* (Hall)	Early Silurian	Ohio, USA	Brachiopods	19	23–25
8	*Dalmanella edgewoodensis* Savage	Early Silurian	Illinois, USA	Brachiopods	19	26–28
8	*Triplesia ortoni* (Meek)	Early Silurian	Ohio, USA	Brachiopods	19	29–31
8	*Atrypa devoniana* Webster	Late Devonian	Iowa, USA	Brachiopods	20	1–3
8	*Cryptospirifer whtneyi* (Hall)	Late Devonian	Iowa, USA	Brachiopods	20	4–6
8	*Paurohyncha* sp.	Late Devonian	New Mexico, USA	Brachiopods	20	7–8
8	*Schizophoria australis* Kindle	Late Devonian	New Mexico, USA	Brachiopods	20	9–10
8	*Tenticospirifer cyrtiniformis* (Hall and Whitfield)	Late Devonian	Iowa, USA	Brachiopods	20	11–14
8	*Theodossia hungerfordi* (Hall)	Late Devonian	Iowa, USA	Brachiopods	20	15–16
8	*Ambocoelia umbonata* (Conrad)	Middle Devonian	New York, USA	Brachiopods	20	17–18
8	*Athyris spiriferoides* (Eaton)	Middle Devonian	New York, USA	Brachiopods	20	19–21
8	*Atrypa independensis* Webster	Middle Devonian	Iowa, USA	Brachiopods	20	22–24
8	*Atrypa reticularis* (Linne)	Middle Devonian	New York, USA	Brachiopods	21	1–5
8	*Brachyspirifer audaculus* (Conrad)	Middle Devonian	New York, USA	Brachiopods	21	6
8	*Camarotoechia congregata* (Conrad)	Middle Devonian	Pennsylvania, USA	Brachiopods	21	7–9
8	*Chonetes aurora* Hall	Middle Devonian	New York, USA	Brachiopods	21	10

(continued)

(continued)

Ch. no.	Species name	Age	Locality	Chapter	Fig. no.	Fig. no.
8	*Chonetes coronatus* (Conrad)	Middle Devonian	New York, USA	Brachiopods	21	11–14
8	*Cranaena romingeri* (Hall)	Middle Devonian	New York, USA	Brachiopods	21	15–16
8	*Cryptonella rectirostra* (Hall)	Middle Devonian	New York, USA	Brachiopods	21	17–18
8	*Douvillina inequistriata* (Conrad)	Middle Devonian	Michigan, USA	Brachiopods	21	19–20
8	*Cyrtina hamiltonensis* (Hall)	Middle Devonian	Michigan, USA	Brachiopods	21	21–22
8	*Fimbrispirifer venustus* (Hall)	Middle Devonian	Ohio, USA	Brachiopods	21	23
8	*Elytha fimbriata* (Conrad)	Middle Devonian	New York, USA	Brachiopods	21	24–26
8	*Gypidula comis* (Owen)	Middle Devonian	Iowa, USA	Brachiopods	21	27–28
8	*Gypidula romingeri* Hal and Clarke	Middle Devonian	Michigan, USA	Brachiopods	22	1–2
8	*Hypothyridina venustula* (Hall)	Middle Devonian	New York, USA	Brachiopods	22	3–5
8	*Leptaena rhomboidalis* Wilckens	Middle Devonian	Michigan, USA	Brachiopods	22	6–9
8	*Longispina emmetensis* (Winchell)	Middle Devonian	Michigan, USA	Brachiopods	22	10–11
8	*Megastrophia concava* (Hall)	Middle Devonian	Pennsylvania, USA	Brachiopods	22	12–13
8	*Mucrospirifer consobrinus* (d'Orbigny)	Middle Devonian	New York, USA	Brachiopods	22	14–15
8	*Mucrospirifer mucronatus* (Conrad)	Middle Devonian	New York, USA	Brachiopods	22	16–18
8	*Mucrospirifer thedfordensis* Shimmer and Grabau	Middle Devonian	Ontario, Canada	Brachiopods	22	19–20
8	*Pentagonia bisulcata* (Hall)	Middle Devonian	Ontario, Canada	Brachiopods	22	21–22
8	*Petrocrania hamiltoniae* (Hall)	Middle Devonian	Michigan, USA	Brachiopods	23	1
8	*Platyrachella oweni* (Hall)	Middle Devonian	Kentucky, USA	Brachiopods	23	2–3
8	*Pustulina pustulosa* (Hall)	Middle Devonian	New York, USA	Brachiopods	23	4
8	*Rhipidomella penelope* (Hall)	Middle Devonian	New York, USA	Brachiopods	23	5–6
8	*Spinocyrtia granulosa* (Conrad)	Middle Devonian	Pennsylvania, USA	Brachiopods	23	7–8
8	*Stropheodonta demissa* (Conrad)	Middle Devonian	New York, USA	Brachiopods	23	9–10
8	*Stropheodonta erratica* Winchell	Middle Devonian	Michigan, USA	Brachiopods	23	11–12
8	*Tropidoleptus carinatus* (Conrad)	Middle Devonian	New York, USA	Brachiopods	23	13–14

(continued)

(continued)

Ch. no.	Species name	Age	Locality	Chapter	Fig. no.	Fig. no.
8	*Trematospira gibbosa* (Hall)	Middle Devonian	New York, USA	Brachiopods	23	15–18
8	*Amphigenia elongata* (Vanuxem)	Early Devonian	New York, USA	Brachiopods	24	1–4
8	*Anastrophia verneuili* (Hall)	Early Devonian	Tennessee, USA	Brachiopods	24	5–8
8	*Atrypina imbricata* (Hall)	Early Devonian	Tennessee, USA	Brachiopods	24	9
8	*Beachia suessana* (Hall)	Early Devonian	New York, USA	Brachiopods	24	10-13
8	*Brevispirifer gregarius* (Clapp)	Early Devonian	Ohio, USA	Brachiopods	24	14—15
8	*Costellirostra tennesseensis* (Dunbar)	Early Devonian	Tennessee, USA	Brachiopods	24	16–20
8	*Costispirifer arenosus* (Conrad)	Early Devonian	Maryland, USA	Brachiopods	24	21–22
8	*Delthyris prelamellosus* (Hall)	Early Devonian	New York, USA	Brachiopods	24	23
8	*Eodevonaria arculata* (Hall)	Early Devonian	New York, USA	Brachiopods	24	24–25
8	*Eatonia medialis* (Hall)	Early Devonian	New York, USA	Brachiopods	24	26–28
8	*Ethymothyris gaspensis* (Clarke)	Early Devonian	Quebec, Canada	Brachiopods	24	29–30
8	*Gypidula pseudogaleata* (Hall)	Early Devonian	New York, USA	Brachiopods	25	1–2
8	*Gypidula coeymanensis* Schuchert	Early Devonian	New York, USA	Brachiopods	25	3–5
8	*Hipparionyx proximus* Vanuxem	Early Devonian	New York, USA	Brachiopods	25	6
8	*Leptocoelia flabellites* (Conrad)	Early Devonian	New York, USA	Brachiopods	25	7–9
8	*Meristella nasuta* (Conrad)	Early Devonian	New York, USA	Brachiopods	25	10-11
8	*Metaplasia pyxidata* (Hall)	Early Devonian	New York, USA	Brachiopods	25	12–13
8	*Paraspirifer acuminatus* (Conrad)	Early Devonian	Maryland, USA	Brachiopods	25	14–16
8	*Pentamerella arata* (Conrad)	Early Devonian	Ohio, USA	Brachiopods	25	17
8	*Platyorthis planaoconvexa* (Hall)	Early Devonian	New York, USA	Brachiopods	25	18–20
8	*Rensselaeria marylandica* (Hall)	Early Devonian	Maryland, USA	Brachiopods	25	21–22
8	*Rensselaeria elongata* (Conrad)	Early Devonian	Tennessee, USA	Brachiopods	25	23–24
8	*Rhynchospirina formosa* (Hall)	Early Devonian	New York, USA	Brachiopods	26	1–2
8	*Schizophoria nevadensis* Merriam	Early Devonian	Nevada, USA	Brachiopods	26	3

(continued)

(continued)

Ch. no.	Species name	Age	Locality	Chapter	Fig. no.	Fig. no.
8	*Schuchertella woolworthana* (Hall)	Early Devonian	New York, USA	Brachiopods	26	6–7
8	*Sphaerirhynchia nucleolatus* (Hall)	Early Devonian	Oklahoma, USA	Brachiopods	26	8–10
8	*Sphaerirhynchia ventricosus* (Hall)	Early Devonian	Oklahoma, USA	Brachiopods	26	11–14
8	*Trematospira multistriata* (Hall)	Early Devonian	Tennessee, USA	Brachiopods	26	15
8	*Athyris spiriferoides* (Eaton)	Middle Devonian	New York, USA	Brachiopods	26	16–17
8	*Strophonella ampla* (Hall)	Early Devonian	New York, USA	Brachiopods	26	18–19
8	*Avonia oklahomensis* Snider	Late Mississippian	Oklahoma, USA	Brachiopods	27	1-3
8	*Chonetes oklahomensis* Snider	Late Mississippian	Oklahoma, USA	Brachiopods	27	4–6
8	*Brachythyris subcardiformis* (Hall)	Late Mississippian	Illinois, USA	Brachiopods	27	7–8
8	*Buxtonia semicircularis* Sutton and Wagner	Late Mississippian (Chesteran)	Kentucky, USA	Brachiopods	27	9
8	*Diaphragmus cestriensis* (Worthen)	Late Mississippian (Chesteran)	Kentucky, USA	Brachiopods	27	10–11
8	*Spirifer arkansanus* Girty	Late Mississippian	Oklahoma, USA	Brachiopods	27	12–13
8	*Dictyoclostus inflatus* (McChesney)	Late Mississippian (Chesteran)	Illinois, USA	Brachiopods	27	14–17
8	*Dielasma illinoisense* Weller	Late Mississippian (Chesteran)	Illinois, USA	Brachiopods	27	18–19
8	*Nudirostra carbonifera* (Girty)	Late Mississippian	Arkansas, USA	Brachiopods	27	20–22
8	*Spirifer lateralis* Hall	Late Mississippian (Meramecian)	Indiana, USA	Brachiopods	27	23–25
8	*Shumardella missouriensis* (Shumard)	Early Mississippian	Missouri, USA	Brachiopods	27	26–28
8	*Linoproductus ovatus* (Hall)	Early Mississippian (Osagian)	Iowa, USA	Brachiopods	27	29–30
8	*Spirifer keokuk* Hall	Early Mississippian (Osagian)	Iowa, USA	Brachiopods	27	31–33
8	*Derbyia crassa* (Meek and Hayden)	Late Pennsylvanian	Texas, USA	Brachiopods	28	1–2
8	*Heteralosia slocomi* King	Pennsylvanian	Texas, USA	Brachiopods	28	3–4
8	*Spirifer rockymontanus* Marcou	Middle Pennsylvanian	Oklahoma, USA	Brachiopods	28	5
8	*Crurithyris planoconvexa* (Schumard)	Middle Pennsylvanian	Missouri, USA	Brachiopods	28	6–8
8	*Marginifera mauricatina* Dunbar and Condra	Middle Pennsylvanian	Missouri, USA	Brachiopods	28	9–10
8	*Mesolobus mesolobus* (Norwood and Pratten)	Middle Pennsylvanian	Texas, USA	Brachiopods	28	11–12

(continued)

(continued)

Ch. no.	Species name	Age	Locality	Chapter	Fig. no.	Fig. no.
8	*Neospirifer cameratus* (Morton)	Middle Pennsylvanian	Ohio, USA	Brachiopods	28	13–14
8	*Chonetina granulifer* Owen	Late Pennsylvanian	Kansas, USA	Brachiopods	29	1–3
8	*Dielasma bovidense* (Morton)	Late Pennsylvanian	Texas, USA	Brachiopods	29	4–6
8	*Dictyoclostus americanus* Dunbar and Condra	Late Pennsylvanian	Kansas, USA	Brachiopods	29	7-11
8	*Echinoconchus semipunctatus* (Shepard)	Late Pennsylvanian	Kansas, USA	Brachiopods	29	12–13
8	*Marginifera lasallensis* (Worthen)	Late Pennsylvanian	Texas, USA	Brachiopods	29	14–15
8	*Juresania nebrascensis* (Owen)	Late Pennsylvanian	Nebaraska, USA	Brachiopods	29	16–18
8	*Meekella striatocostata* (Cox)	Late Pennsylvanian	Missouri, USA	Brachiopods	29	19–21
8	*Juresania symmetrica* (McChesney)	Late Pennsylvanian	Kansas, USA	Brachiopods	29	22–23
8	*Spiriferella texanus* (Meek)	Late Pennsylvanian	Texas, USA	Brachiopods	29	24–25
8	*Chonetina flemingi* (Norwood and Pratten)	Late Pennsylvanian (Missourian)	Missouri, USA	Brachiopods	29	26–27
8	*Derbyia cymbula* Hall and Clarke	Late Pennsylvanian (Virgilian)	Kansas, USA	Brachiopods	29	28–31
8	*Nudirostra carbonifera* (Girty)	Late Pennsylvanian (Missourian)	Oklahoma, USA	Brachiopods	29	32–34
8	*Linoproductus oklahomae* Dunbar and Condra	Late Pennsylvanian (Missourian)	Oklahoma, USA	Brachiopods	29	35–36
8	*Enteletes hemiplicatus* (Hall)	Late Pennsylvanian (Missourian)	Kansas, USA	Brachiopods	29	37–43
8	*Buxtonia peruviana* (d'Orbigny)	Permian	Texas, USA	Brachiopods	30	1
8	*Heterelasma* sp.	Permian	Texas, USA	Brachiopods	30	2–4
8	*Leptodus americanus* (Girty)	Permian	Texas, USA	Brachiopods	30	5–6
8	*Prorichthofenia uddeni* (Bose)	Permian	Texas, USA	Brachiopods	30	7–9
8	*Stenoscisma venustum* (Girty)	Permian	Texas, USA	Brachiopods	30	10–15
8	*Neospirifer condor* (d'Orbigny)	Early Permian	Texas, USA	Brachiopods	30	16–18
9	*Latirus lynchi* (Basterot)	Miocene	France	Gastropods	1	1
9	*Dirhachopea normalis* Ulrich and Bridge	Late Cambrian	Missouri, USA	Gastropods	7	1–2
9	*Hypseloconus elongatus* Berkey	Late Cambrian (Dresbachian)	Wisconsin, USA	Gastropods	7	3
9	*Owenella antiquata* (Whitfield)	Late Cambrian (Trempealeauan)	Wisconsin, USA	Gastropods	7	4–5

(continued)

(continued)

Ch. no.	Species name	Age	Locality	Chapter	Fig. no.	Fig. no.
9	*Palaeacmaea typical* Hall and Whitfield	Cambrian	Wisconsin, USA	Gastropods	7	6
9	*Proplina cornutaformis* (Walcott)	Late Cambrian	Missouri, USA	Gastropods	7	7
9	*Sinuopea sweeti* (Whitfield)	Late Cambrian (Trempealeauan)	Wisconsin, USA	Gastropods	7	8
9	*Aldanella attleborensis* (Shaler and Foerste)	Early Cambrian (early Tommotian)	Newfoundland, USA	Gastropods	7	9–10
9	*Bagenovia sajanica* Radugin	Early Cambrian	Siberia	Gastropods	7	11
9	*Stenothecoides elongate* (Walcott)	Middle Cambrian	British Columbia	Gastropods	7	12
9	*Scenella reticulate* Billings	Early Cambrian	Newfoundland, USA	Gastropods	7	13
9	*Cyclonema bilix* (Conrad)	Late Ordovician (Richmondian)	Indiana, USA	Gastropods	8	1
9	*Liospira micula* (Hall)	Late Ordovician (Maquoketa)	Wisconsin, USA	Gastropods	8	2–3
9	*Ectomaria pagoda* (Salter)	Middle Ordovician (Blackriveran)	Quebec, Canada	Gastropods	8	4
9	*Eunema strigillatum* Salter	Middle Ordovician (Blackriveran)	Quebec, Canada	Gastropods	8	5
9	*Omospira laticincta* Ulrich and Scofield	Middle Ordovician (Blackriveran)	Tennessee, USA	Gastropods	8	6
9	*Raphistomina lapicida* (Salter)	Middle Ordovician (Blackriveran)	Quebec, Canada	Gastropods	8	7–8
9	*Raphistoma striatum* (Emmons)	Middle Ordovician (Chat' yan)	New York, USA	Gastropods	8	9–10
9	*Clathrospira subconica* (Hall)	Middle Ordovician (Trentonian)	New York, USA	Gastropods	8	11
9	*Holopea symmetrica* Hall	Middle Ordovician (Trentonian)	New York, USA	Gastropods	8	12
9	*Lophospira milleri* (Miller)	Middle Ordovician (Trentonian)	New York, USA	Gastropods	8	13
9	*Subulites regularis* Ulrich and Scofield	Middle Ordovician (Blackriveran)	Kentucky, USA	Gastropods	8	14
9	*Trochonema umbilicatum* (Hall)	Middle Ordovician (Trentonian)	New York, USA	Gastropods	8	15
9	*Tetranota wisconsinensis* (Whitfield)	Middle Ordovician	Wisconsin, USA	Gastropods	8	16
9	*Eotomaria supracingulata* (Billings)	Middle Ordovician	Eastern Canada	Gastropods	8	17
9	*Lytospira subrotunda* (Ulrich and Scofield)	Middle Ordovician	Minnesota, USA	Gastropods	8	18–19
9	*Phragmolites obliqua* (Ulrich and Scofield)	Middle Ordovician	Minnesota, USA	Gastropods	8	20–21
9	*Helicotoma tennesseensis* Safford	Middle Ordovician	Kentucky, USA	Gastropods	8	22–23
9	*Hormotoma trentonensis* Ulrich and Scofield	Middle Ordovician	Minnesota, USA	Gastropods	8	24

(continued)

(continued)

Ch. no.	Species name	Age	Locality	Chapter	Fig. no.	Fig. no.
9	*Ceratopea keithi* Ulrich	Early Ordovician	Virginia, USA	Gastropods	9	1–2
9	*Ecculiomphalus triangulus* Whitfield	Early Ordovician	Minnesota, USA	Gastropods	9	3–4
9	*Lecanospira compacta* Salter	Early Ordovician	Quebec, Canada	Gastropods	9	5
9	*Ophileta complanata* Vanuxem	Early Ordovician	New York, USA	Gastropods	9	6–8
9	*Ozarkispira leo* Walcott	Early Ordovician	Alberta, Canada	Gastropods	9	9–10
9	*Orospira bigranosa* Butts	Early Ordovician	Missouri, USA	Gastropods	9	11–12
9	*Plethospira cassina* (Whitfield)	Early Ordovician	Vermont, USA	Gastropods	9	13
9	*Schizopea typica* Ulrich and Bridge	Early Ordovician	Missouri, USA	Gastropods	9	14–15
9	*Auriptygma fortior* Barrande	Middle Silurian	Czechoslovakia	Gastropods	10	1
9	*Coelocaulus macrospira* (Hall)	Middle Silurian	New York, USA	Gastropods	10	2
9	*Cyclonemina delicatula* (Lindstrom)	Middle Silurian	Sweden	Gastropods	10	3
9	*Loxonema leda* Hall	Middle Silurian	Indiana, USA	Gastropods	10	4
9	*Mourlonia filitexta* (Foerste)	Middle Silurian (Clinton)	Ohio, USA	Gastropods	10	5
9	*Hormotoma whiteavesi* Clarke and Ruedemann	Middle Silurian	New York, USA	Gastropods	10	6
9	*Sinuspira tenera* Perner	Middle Silurian	Czechoslovakia	Gastropods	10	7
9	*Poleumita discors* (Sowerby)	Middle Silurian	England	Gastropods	10	8
9	*Tremanotus alpheus* Hall	Middle Silurian	New York, USA	Gastropods	10	9–10
9	*Bellerophon vasulites* Montfort	Middle Devonian	Germany	Gastropods	11	1
9	*Loxonema hamiltoniae* Hall	Middle Devonian	New York, USA	Gastropods	11	2
9	*Murchisonia turbinata* (Schlotheim)	Middle Devonian	Germany	Gastropods	11	3
9	*Philoxene laevis* (d'Archiac and Verneuil)	Middle Devonian	Germany	Gastropods	11	4
9	*Scalitina montana* Spriestersbach	Middle Devonian	Germany	Gastropods	11	5
9	*Straparolus cyclostomus* (Hall)	Middle Devonian	Iowa, USA	Gastropods	11	6
9	*Strobeus arculatus* (Schlotheim)	Middle Devonian	Germany	Gastropods	11	7
9	*Bembexia larteli* (Munier-Chalmas)	Early Devonian	France	Gastropods	11	8

(continued)

(continued)

Ch. no.	Species name	Age	Locality	Chapter	Fig. no.	Fig. no.
9	*Isonema depressum* Meek and Worthen	Early Devonian	Ohio, USA	Gastropods	11	9
9	*Murchisonia bachelieri* Rouault	Early Devonian	France	Gastropods	11	10
9	*Platyostoma ventricosum* Conrad	Early Devonian	New York, USA	Gastropods	11	11
9	*Ptychospirina varians* (Hall)	Early Devonian	New York, USA	Gastropods	11	12
9	*Strophostylus andrewsi* Hall	Early Devonian	New York, USA	Gastropods	11	13
9	*Tropidodiscus curvilineatus* (Conrad)	Early Devonian	New York, USA	Gastropods	11	14–15
9	*Turbonopsis shumardi* (Hall)	Early Devonian	Kentucky, USA	Gastropods	11	16
9	*Bulimorpha bulimiformis* (Hall)	Late Mississippian (Meramecian)	Indiana, USA	Gastropods	12	1
9	*Baylea yvanii* Léveillé	Late Mississippian (Meramecian)	Japan	Gastropods	12	2
9	*Eotrochus tenikimarginatus* (Miller)	Late Mississippian (Meramecian)	Indiana, USA	Gastropods	12	3
9	*Naticopsis carleyana* (Hall)	Late Mississippian (Meramecian)	Indiana, USA	Gastropods	12	4
9	*Rhineoderma wortheni* (Hall)	Late Mississippian (Meramecian)	Indiana, USA	Gastropods	12	5
9	*Gosseletina callosa* (De Koninck)	Late Mississippian (Visean)	Belgium	Gastropods	12	6
9	*Stegocoelia illinoiensis* (Weller)	Late Mississippian (Meramecian)	Illinois, USA	Gastropods	12	7
9	*Straparolus dionysii* Montfort	Mississippian	Belgium	Gastropods	12	8
9	*Mourlonia carinata* (Sowerby)	Mississippian	England	Gastropods	12	9
9	*Straparolus (Schizostoma) catillus* (Martin)	Late Mississippian (Visean)	Belgium	Gastropods	12	10–11
9	*Foordella hibernica* Longstaff	Mississippian	Ireland	Gastropods	12	12
9	*Platyceras vetustum* (Sowerby)	Mississippian	Ireland	Gastropods	12	13
9	*Rhineoderma radula* (De Koninck)	Early Mississippian (Tournaisian)	Belgium	Gastropods	12	14
9	*Angyomphalus radians* (De Koninck)	Early Mississippian	Belgium	Gastropods	12	15–16
9	*Euconospira turbiniformis* (Meek and Worthen)	Late Pennsylvanian (Missourian)	Illinois, USA	Gastropods	13	1
9	*Meekospira peracuta* Meek and Worthen	Late Pennsylvanian Missourian	Illinois, USA	Gastropods	13	2
9	*Phymatopleura brazoensis* (Shumard)	Late Pennsylvanian (Missourian and Virgilian)	North America	Gastropods	13	3–4

(continued)

(continued)

Ch. no.	Species name	Age	Locality	Chapter	Fig. no.	Fig. no.
9	*Soleniscus typicus* Meek and Worthen	Late Pennsylvanian (Missourian)	Illinois, USA	Gastropods	13	5
9	*Worthenia tabulata* (Conrad)	Late Pennsylvanian (Conemaugh)	Pennsylvania, USA	Gastropods	13	6–7
9	*Anthracopupa ohioensis* Whitfield	Late Pennsylvanian	Ohio, USA	Gastropods	13	8
9	*Anomphalus rotulus* Meek and Worthen	Middle Pennsylvanian (Desmoinesian)	Illinois, USA	Gastropods	13	9
9	*Cyclozyga mirabilis* Knight	Middle Pennsylvanian (Desmoinesian)	Missouri, USA	Gastropods	13	10
9	*Glabrocingulum grayvillense* (Norwood and Pratten)	Middle Pennsylvanian (Desmoinesian)	Oklahoma, USA	Gastropods	13	11–13
9	*Hemizyga elegans* Girty	Middle Pennsylvanian (Desmoinesian)	Missouri, USA	Gastropods	13	14
9	*Leptoptygma virgatum* (Knight)	Middle Pennsylvanian (Desmoinesian)	Missouri, USA	Gastropods	13	15
9	*Leptozyga minuta* (Knight)	Middle Pennsylvanian (Desmoinesian)	Missouri, USA	Gastropods	13	16
9	*Cymatospira montfortiana* Norwood and Pratten	Middle Pennsylvanian (Desmoinesian)	Oklahoma, USA	Gastropods	13	17–19
9	*Orthonema salteri* (Meek and Worthen)	Middle Pennsylvanian (Desmoinesian)	Illinois, USA	Gastropods	13	20
9	*Phymatopleura nodosa* (Girty)	Middle Pennsylvanian (Desmoinesian)	North America	Gastropods	13	21
9	*Straparolus (Amphiscapha) reedsi* Knight	Middle Pennsylvanian (Desmoinesian)	Missouri, USA	Gastropods	13	22–24
9	*Hypselentoma perhumerosa* (Meek)	Pennsylvanian	Nebraska, USA	Gastropods	14	1
9	*Pharkidonotus percarinatus* Conrad	Pennsylvanian	Pennsylvania, USA	Gastropods	14	2–3
9	*Plocezyga coronata* (Knight)	Middle Pennsylvanian (Desmoi–nesian)	Missouri, USA	Gastropods	14	4
9	*Straparolus (Euomphalus) pernodosus* Meek and Worthen	Middle Pennsylvanian (Desmoinesian)	Illinois, USA	Gastropods	14	5–6
9	*Trachydomia nodosa* Meek and Worthen	Middle Pennsylvanian (Desmoinesian)	Missouri, USA	Gastropods	14	7–8
9	*Trepospira sphaerulata* (Conrad)	Pennsylvanian	Pennsylvania, USA	Gastropods	14	9
9	*Bellerophon regularis* Waagen	Permian	India	Gastropods	14	10–12
9	*Plocostoma neumayri* Gemmellaro	Permian	Sicily	Gastropods	14	13
9	*Procerithiopsis ambigua* Mansuy	Permian	Indo-China	Gastropods	14	14
9	*Omphalotrochus whitneyi* Meek Permian California.	Permian	California, USA	Gastropods	14	15–16
9	*Warthia polita* Waagen	Permian	India	Gastropods	14	17–20
9	*Shansiella planicosta* Girty	Early Permian	Texas, USA	Gastropods	14	21

(continued)

(continued)

Ch. no.	Species name	Age	Locality	Chapter	Fig. no.	Fig. no.
9	*Amberleya berthel*oti d'Orbigny	Late Jurassic	France	Gastropods	15	1
9	*Eucyclomphalus cupido* d'Orbigny	Late Jurassic	Spain	Gastropods	15	2
9	*Pleurotomaria bitorquata* Deslongchamps	Late Jurassic	Spain	Gastropods	15	3
9	*Nerinea trinodosa* Voltz	Late Jurassic	France	Gastropods	15	4
9	*Ptychomphalus heliciformis* Deslongchamps	Late Jurassic	Spain	Gastropods	15	5
9	*Purpuroidea maureausea* Buvignier	Late Jurassic	France	Gastropods	15	6
9	*Amphitrochus duplicatus* Sowerby	Middle Jurassic	Spain	Gastropods	15	8
9	*Aptyxiella implicata* d'Orbigny	Middle Jurassic	France	Gastropods	15	7
9	*Eucyclus ornatus* Sowerby	Middle Jurassic	England	Gastropods	15	9
9	*Talantodiscus mirabilis* Deslongchamps	Early Jurassic	Spain	Gastropods	15	10–12
9	*Zygopleura periniana* d'Orbigny	Early Jurassic	France	Gastropods	15	13
9	*Tryoniella valida* (Stephenson)	Late Cretaceous	Tennessee, USA	Gastropods	16	1
9	*Volutotnorpha mutabilis* Wade	Late Cretaceous	Tennessee, USA	Gastropods	16	2
9	*Volutoderma appressa* Wade	Late Cretaceous	Tennessee, USA	Gastropods	16	3
9	*Actaeonella dolium* Roemer	Early Cretaceous (Comanchean)	Texas, USA	Gastropods	16	4
9	*Amberleya mudgeana* Meek	Early Cretaceous (Comanchean)	Kansas, USA	Gastropods	16	5
9	*Calliostoma cragini* Stanton	Early Cretaceous (Comanchean)	Kansas, USA	Gastropods	16	6
9	*Cerithium kickapooense* Stanton	Early Cretaceous (Comanchean)	Texas, USA	Gastropods	16	7
9	*Monodonta bartonensis* Stanton	Early Cretaceous (Comanchean)	Texas, USA	Gastropods	16	8
9	*Pseudonerinea proctori* Cragin	Early Cretaceous (Comanchean)	Texas, USA	Gastropods	16	9
9	*Turritella seriatim-granulata* Roemer	Early Cretaceous (Comanchean)	Texas, USA	Gastropods	16	10
9	*Lunatia cragini* Stanton	Early Cretaceous (Comanchean)	Kansas, USA	Gastropods	16	11
9	*Nerinea riograndensis* Stanton	Early Cretaceous (Washitan)	Texas, USA	Gastropods	16	12–13
9	*Cassiope branneri* Hill	Early Cretaceous (Trinitian)	Texas, USA	Gastropods	16	14
9	*Trochactaeon cumminsi* Stanton	Early Cretaceous (Washitan)	Texas, USA	Gastropods	16	15

(continued)

(continued)

Ch. no.	Species name	Age	Locality	Chapter	Fig. no.	Fig. no.
9	*Tylostoma elevatum* Shumard	Early Cretaceous (Comanchean)	Texas, USA	Gastropods	16	16
9	*Anchura mudgeana* White	Early Cretaceous (Washitan)	Texas, USA	Gastropods	16	17
9	*Bellifusus robustus* Stephenson	Late Cretaceous	Tennessee, USA	Gastropods	17	1
9	*Belliscala forsheyi* (Shumard)	Late Cretaceous	Tennessee, USA	Gastropods	17	2
9	*Beretra firma* Stephenson	Late Cretaceous	Tennessee, USA	Gastropods	17	3
9	*Boltenella excellens* Wade	Late Cretaceous	Tennessee, USA	Gastropods	17	4
9	*Ellipsoscapha striatella* Shumard	Late Cretaceous	Tennessee, USA	Gastropods	17	5
9	*Eoancilla acutula* Stephenson	Late Cretaceous	Tennessee, USA	Gastropods	17	6
9	*Epitonium bexarense* Stephenson	Late Cretaceous	Tennessee, USA	Gastropods	17	7
9	*Fulgerca venusta* Stephenson	Late Cretaceous	Tennessee, USA	Gastropods	17	8
9	*Graphidula terebriformis* Stephenson	Late Cretaceous	Tennessee, USA	Gastropods	17	9
9	*Liopeplum ldoderma longum* Stephenson	Late Cretaceous	Tennessee, USA	Gastropods	17	10
9	*Mataxa valida* Stephenson	Late Cretaceous	Tennessee, USA	Gastropods	17	11
9	*Medionapus elongatus* Stephenson	Late Cretaceous	Tennessee, USA	Gastropods	17	12
9	*Morea marylandica bella* Stephenson	Late Cretaceous	Tennessee, USA	Gastropods	17	13
9	*Napulus tuberculatus* Stephenson	Late Cretaceous	Tennessee, USA	Gastropods	17	14
9	*Odontofusus curvicostatus* Wade	Late Cretaceous	Tennessee, USA	Gastropods	17	15
9	*Ornopsis pulchra* Stephenson.	Late Cretaceous	Tennessee, USA	Gastropods	17	16
9	*Paladmete cancellaria* Conrad	Late Cretaceous	Tennessee, USA	Gastropods	17	17
9	*Piestochilus pergracilis* Wade	Late Cretaceous	Tennessee, USA	Gastropods	17	20
9	*Polinices rectilabrum* (Conrad)	Late Cretaceous	Tennessee, USA	Gastropods	17	19
9	*Pyropsis lanhami* Stephenson	Late Cretaceous	Tennessee, USA	Gastropods	17	18
9	*Remera microstriata* Stephenson	Late Cretaceous	Tennessee, USA	Gastropods	17	21
9	*Sargana stantoni* (Weller)	Late Cretaceous	Tennessee, USA	Gastropods	17	23
9	*Tornatellaea scatesi* Stephenson	Late Cretaceous	Tennessee, USA	Gastropods	17	22

(continued)

(continued)

Ch. no.	Species name	Age	Locality	Chapter	Fig. no.	Fig. no.
9	*Trochifusus perornatus* Wade	Late Cretaceous	Tennessee, USA	Gastropods	17	25
9	*Turritella bilira* Stephenson	Late Cretaceous	Tennessee, USA	Gastropods	17	24
9	*Actaeon idoneus* Conrad	Late Eocene	Mississippi, USA	Gastropods	18	1
9	*Architectonica trilirata* (Conrad)	Late Eocene	Mississippi, USA	Gastropods	18	2–4
9	*Athleta petrosa* (Conrad)	Late Eocene	Mississippi, USA	Gastropods	18	5
9	*Calyptraphorus velatus stamineus* (Conrad)	Late Eocene	Mississippi, USA	Gastropods	18	6–7
9	*Caricella polita* Conrad	Late Eocene	Mississippi, USA	Gastropods	18	8
9	*Cassidaria millsapsi* Sullivan and Gardner	Late Eocene	Mississippi, USA	Gastropods	18	9–10
9	*Cirsotrema nassulum creolum* Harris and Palmer	Late Eocene	Louisiana, USA	Gastropods	18	11
9	*Conus sauridens* Conrad	Late Eocene	Mississippi, USA	Gastropods	18	12
9	*Coronia genitiva* (Casey)	Late Eocene	Louisiana, USA	Gastropods	18	13
9	*Cypraedia fenestralis* Conrad	Late Eocene	Mississippi, USA	Gastropods	18	14
9	*Dentalium danvillense* Harris and Palmer	Late Eocene	Louisiana, USA	Gastropods	18	15
9	*Diodora tenebrosa veatchi* Harris and Palmer	Late Eocene	Louisiana, USA	Gastropods	18	16
9	*Ectinoehilus laqueatum* (Conrad)	Late Eocene	Alabama, USA	Gastropods	18	17–18
9	*Ficus filia* (Meyer)	Late Eocene	Mississippi, USA	Gastropods	18	19
9	*Globularia morgani* (Jordan)	Late Eocene	Mississippi, USA	Gastropods	18	19
9	*Harpa jacksonensis* Harris	Late Eocene	Mississippi, USA	Gastropods	19	1
9	*Lapparia dumosa exigua* Palmer	Late Eocene	Mississippi, USA	Gastropods	19	2
9	*Latirus leaensis* Harris	Late Eocene	Mississippi, USA	Gastropods	19	3
9	*Levifusus moodianus* Cooke	Late Eocene	Mississippi, USA	Gastropods	19	4
9	*Mazzalina inaurata oweni* (Dall)	Late Eocene	Arkansas, USA	Gastropods	19	5
9	*Murex vanuxemi* Conrad	Late Eocene	Mississippi, USA	Gastropods	19	6
9	*Polinices eminulus* (Conrad)	Late Eocene	Alabama, USA	Gastropods	19	7
9	*Fusimitra conquisita* (Conrad)	Late Eocene	Mississippi, USA	Gastropods	19	8

(continued)

(continued)

Ch. no.	Species name	Age	Locality	Chapter	Fig. no.	Fig. no.
9	*Pseudoliva vetusta prespectiva* Conrad	Late Eocene	Louisiana, USA	Gastropods	19	9
9	*Solariella cancellata jacksoniana* Harris and Palmer	Late Eocene	Mississippi, USA	Gastropods	19	10–12
9	*Sinum danvillense* Harris and Palmer	Late Eocene	Louisiana, USA	Gastropods	19	13–14
9	*Scobinella louisianae* Harris	Late Eocene	Louisiana, USA	Gastropods	19	15
9	*Terebra jacksonensis* Cooke	Late Eocene	Mississippi, USA	Gastropods	19	16
9	*Tornatellaea lata* Conrad	Late Eocene	Alabama, USA	Gastropods	19	17
9	*Tritonoatractus pearlensis montgomeryensis* (Vaughan)	Late Eocene	Mississippi, USA	Gastropods	19	18
9	*Turritella arenicola* Conrad	Late Eocene	Arkansas, USA	Gastropods	19	19
9	*Turritella clevelandia* Harris	Late Eocene	Arkansas, USA	Gastropods	19	20
9	*Ficus filia* (Meyer)	Late Eocene	Mississippi, USA	Gastropods	19	21
9	*Fossarus dalli* (Whitfield)	Miocene	Maryland, USA	Gastropods	20	1
9	*Lirosoma sulcosa* Conrad	Miocene	Maryland, USA	Gastropods	20	2
9	*Littorina irrorata* (Say)	Miocene	Maryland, USA	Gastropods	20	3
9	*Marginella calvertensis* Martin	Miocene	Maryland, USA	Gastropods	20	4
9	*Mangilia patuxentia* Martin	Miocene	Maryland, USA	Gastropods	20	5
9	*Nassa trivittata* Say	Miocene	Maryland, USA	Gastropods	20	6
9	*Oliva litterata* Lamarck	Miocene	Maryland, USA	Gastropods	20	7
9	*Melanella eborea* (Conrad)	Miocene	Maryland, USA	Gastropods	20	8
9	*Polinices heros* (Say)	Miocene	Maryland, USA	Gastropods	20	9
9	*Pterygia calvertensis* (Martin)	Miocene	Maryland, USA	Gastropods	20	10
9	*Ptychosalpinx altilis* (Conrad)	Miocene	Maryland, USA	Gastropods	20	11
9	*Pyramidella granulata* (Lea)	Miocene	Maryland, USA	Gastropods	20	12
9	*Scalaspira strumosa* Conrad	Miocene	Maryland, USA	Gastropods	20	13
9	*Surcula rotifera* Conrad	Miocene	Maryland, USA	Gastropods	20	14
9	*Terebra unilineata* Conrad	Miocene	Maryland, USA	Gastropods	20	15

(continued)

(continued)

Ch. no.	Species name	Age	Locality	Chapter	Fig. no.	Fig. no.
9	*Tritonalia topangensis* (Arnold)	Miocene	California, USA	Gastropods	20	16
9	*Scaphella typa* (Conrad)	Miocene	Maryland, USA	Gastropods	20	17
9	*Siphonalia marylandica* Martin	Miocene	Maryland, USA	Gastropods	20	18
9	*Tunis communis* (Conrad)	Miocene	Maryland, USA	Gastropods	20	19
9	*Volvula iota patuxentia* Martin	Miocene	Maryland, USA	Gastropods	20	20
9	*Urosalpinx rusticus* (Conrad)	Miocene	Maryland, USA	Gastropods	20	21
9	*Turritella variabilis* Conrad	Miocene	Maryland, USA	Gastropods	20	22
9	*Amphissa versicolor* Dall	Pliocene	California, USA	Gastropods	21	1
9	*Bittium asperum* (Gabb)	Pliocene	California, USA	Gastropods	21	2–3
9	*Cantharus fortis* (Carpenter)	Pliocene	California, USA	Gastropods	21	4
9	*Clathrodrillia mercedensis* Martin	Pliocene	California, USA	Gastropods	21	5
9	*Crepidula princeps* Conrad	Pliocene	California, USA	Gastropods	21	6
9	*Forreria magister* (Nomland)	Pliocene	California, USA	Gastropods	21	7
9	*Littorina mariana* Arnold	Pliocene	California, USA	Gastropods	21	8
9	*Nassarius californianus* Conrad	Pliocene	California, USA	Gastropods	21	9
9	*Olivella biplicata* Sowerby	Pliocene	California, USA	Gastropods	21	10
9	*Opalia varicostata* Stearns	Pliocene	California, USA	Gastropods	21	11
9	*Turritella cooperi* Carpenter	Pliocene	California, USA	Gastropods	21	12
9	*Barbarofusus arnoldi* (Cossmann)	Pleistocene	California, USA	Gastropods	21	13
9	*Calicantharus fortis* (Carpenter)	Pleistocene	California, USA	Gastropods	21	14
9	*Neverita reclusiana* (Deshayes)	Pleistocene	California, USA	Gastropods	21	15
9	*Nucella etchegoinensis* (Arnold)	Pleistocene	California, USA	Gastropods	21	16
9	*Neptunea tabulata* (Baird)	Pleistocene	California, USA	Gastropods	21	17
9	*Turcica caffea* brevis Stewart	Pleistocene	California, USA	Gastropods	21	18
9	*Turritella jewetti* Carpenter	Pleistocene	California, USA	Gastropods	21	19

(continued)

(continued)

Ch. no.	Species name	Age	Locality	Chapter	Fig. no.	Fig. no.
9	*Turritella pedroensis* Applin	Pleistocene	California, USA	Gastropods	21	20
9	*Tritonalia sqamulifera* (Carpenter)	Pleistocene	California, USA	Gastropods	21	21
9	*Calliostoma coalingaensis* Arnold	Pleistocene	California, USA	Gastropods	21	22
9	*Amnicola longiqua* Gould	Pleistocene	California, USA	Gastropods	22	1
9	*Calipyrgula carinifera* Pilsbry	Pleistocene	California, USA	Gastropods	22	2
9	*Discus cronkhitei* (Newcomb)	Pleistocene	Kansas, USA	Gastropods	22	3–4
9	*Gastrocopta cristata* (Pilsbry and Vanatta)	Pleistocene	Kansas, USA	Gastropods	22	5
9	*Goniobasis kettlemanensis* woodringi Pilsbry	Pleistocene	California, USA	Gastropods	22	6
9	*Gyraulus pattersoni* Baker	Pleistocene	Iowa, USA	Gastropods	22	7–8
9	*Helicodiscus parallelus* (Say)	Pleistocene	Iowa, USA	Gastropods	22	9–10
9	*Helisoma trivolvis* (Say)	Pleistocene	Texas, USA	Gastropods	22	11–12
9	*Hydrobia andersoni* (Arnold)	Pleistocene	California, USA	Gastropods	22	13
9	*Lymnaea caperata* Say	Pleistocene	Texas, USA	Gastropods	22	14
9	*Menetus pearlettei* Leonard	Pleistocene	Nebraska, USA	Gastropods	22	15–16
9	*Physa anatina* Lea	Pleistocene	Nebraska, USA	Gastropods	22	17
9	*Physa wattsi* Arnold	Pleistocene	Iowa, USA	Gastropods	22	18
9	*Planorbula vulcanata occidentalis* Leonard	Pleistocene	Texas, USA	Gastropods	22	19–20
9	*Polygyra texasiana* (Moricand)	Pleistocene	Iowa, USA	Gastropods	22	21–22
9	*Pomatiopsis cininnatiensis* (Lea)	Pleistocene	California, USA	Gastropods	22	23
9	*Pupilla muscorum* (Linne)	Pleistocene	Texas, USA	Gastropods	22	24
9	*Pupoides albilabris* (Adams)	Pleistocene	Nebraska, USA	Gastropods	22	25
9	*Succinea grosvenori* Lea	Pleistocene	Texas, USA	Gastropods	22	26
9	*Vallonia gracilicosta* Reinhardt	Pleistocene	Iowa, USA	Gastropods	22	27–28
9	*Valvata tricarinata* Say	Pleistocene	Nebraska, USA	Gastropods	22	29–30
9	*Vertigo ovata* Say	Pleistocene	Nebraska, USA	Gastropods	22	31

(continued)

(continued)

Ch. no.	Species name	Age	Locality	Chapter	Fig. no.	Fig. no.
10	*Heliophyllum halli* Edwards and Haime	Middle Devonian	Michigan, USA	Corals	2	1
10	*Heliolites interstinctus* (Linne)	Middle Silurian (Niagaran)	Tennessee, USA	Corals	2	2
10	*Montlivaltia nattheimensis* Milaschewitsch	Late Jurassic	Germany	Corals	2	3
10	*Astrhelia palmata* (Goldfuss)	Miocene	Maryland, USA	Corals	22	4
10	*Stephanophyllia imperialis* Michelin	Miocene	Austria	Corals	22	1–3
10	*Archohelia* vicksburgensis (Conrad)	Oligocene	Mississippi, USA	Corals	22	5–6
10	*Astrocoenia decaturensis* Vaughan	Oligocene	Georgia, USA	Corals	22	12
10	*Flabellum cuneiformis* Lonsdale	Late Eocene	Texas, USA	Corals	22	7–8
10	*Balanophyllia elaborata* (Conrad)	Eocene	Maryland, USA	Corals	22	9–10
10	*Diplhelia papillosa* Edwards and Haime	Eocene	England	Corals	22	11
10	*Discotrochus orbignianus* Edwards and Haime	Eocene	Louisiana, USA	Corals	22	13
10	*Endopachys maclurii* (Lea)	Eocene	Mississippi, USA	Corals	22	14–15
10	*Haimesiastraea conferta* Vaughan	Eocene	Alabama, USA	Corals	22	16–17
10	*Platytrochus stokesi* (Lea)	Eocene	South Carolina, USA	Corals	22	18–19
10	*Trochocyathus californianus* Vaughan	Eocene	California, USA	Corals	22	20–21
10	*Turbinolia dickersonia* Nomland	Eocene	California, USA	Corals	22	22–23
10	*Turbinolia pharetra* Lea	Eocene	Alabama, USA	Corals	22	24–26
10	*Leptacyathus elegans* Edwards and Haime	Early Eocene	England	Corals	22	27–28
10	*Flabellum redmondianum* Gabb	Paleocene	California, USA	Corals	22	29–30
10	*Streptelasma corniculum* Hall	Middle Ordovician	New York, USA	Corals	11	1–3
10	*Lambeophyllum profundum* (Conrad)	Middle Ordovician (Blackriveran)	New York, USA	Corals	11	4–6
10	*Palaeoalveolites* careterensis (Bassler)	Middle Ordovician (Blackriveran)	Tennessee, USA	Corals	11	7–8
10	*Lichenaria heroensis* (Raymond)	Middle Ordovician	Vermount, USA	Corals	11	9–10
10	*Foerstephyllum halli* (Nicholson)	Middle Ordovician (Blackriveran)	Tennessee, USA	Corals	11	11–12
10	*Favistella* alveolata (Goldfuss)	Middle Ordovician	Tennessee, USA	Corals	11	13–14

(continued)

(continued)

Ch. no.	Species name	Age	Locality	Chapter	Fig. no.	Fig. no.
10	*Nyctopora billingsi* Nicholson	Late Ordovician (Trentonian)	Kentucky, USA	Corals	11	15–16
10	*Cyathophylloides* ulrichi Bassler	Late Ordovician	Minnesota, USA	Corals	11	17–18
10	*Tetradium fibratum* Safford	Late Ordovician (Trentonian)	Tennessee, USA	Corals	11	19–20
10	*Calapoecia canadensis* Billings	Late Ordovician	Quebec, Canada	Corals	11	21–23
10	*Halysites catenularia* (Linne)	Middle Silurian (Niagaran)	Indiana, USA	Corals	12	1–3
10	*Coenites seriata* (Hall)	Middle Silurian (Niagaran)	New York, USA	Corals	12	4
10	*Lyellia americana* Edwards and Haime	Middle Silurian (Niagaran)	Indiana, USA	Corals	12	5–6
10	*Striatopora flexuosa* Hall	Middle Silurian (Niagaran)	New York, USA	Corals	12	7–8
10	*Heliolites interstinctus* (Linne)	Middle Silurian (Niagaran)	Tennessee, USA	Corals	12	9–10
10	*Favosites favosus* (Goldfus)	Middle Silurian (Niagaran)	Michigan, USA	Corals	12	11–12
10	*Syringopora verticillata* Goldfus	Middle Silurian (Niagaran)	New York, USA	Corals	12	13
10	*Thecia minor* Rominger	Middle Silurian (Niagaran)	Michigan, USA	Corals	12	14–15
10	*Arachnohyllum patagonum* (Goldfuss)	Middle Silurian (Niagaran)	Quebec, Canada	Corals	13	1–2
10	*Breviphyllum cliftonense* (Amsden)	Middle Silurian (Niagaran)	Tennessee, USA	Corals	13	3
10	*Calceola* lennesseensis Rominger	Middle Silurian (Niagaran)	Tennessee, USA	Corals	13	4–5
10	*Cystiphyllum niagarense* (Hall)	Middle Silurian (Niagaran)	Michigan, USA	Corals	13	6
10	*Diplophyllum* caespitosum Hall	Middle Silurian (Niagaran)	Ontario, Canada	Corals	13	7–8
10	*Ptychohyllum stokesi* Edwards and Haime	Middle Silurian (Niagaran)	Ontario, Canada	Corals	13	9–10
10	*Entelophyllum* rugosum (Smith)	Middle Silurian (Niagaran)	Ontario, Canada	Corals	13	11–13
10	Naos *swellensis* Amsden	Middle Silurian (Niagaran)	Tennessee, USA	Corals	13	14–15
10	*Synaptophyllum multicaule* (Hall)	Middle Silurian (Niagaran)	Michigan, USA	Corals	13	16–18
10	*Tryplasma brownsportensis* (Amsden)	Middle Silurian (Niagaran)	Tennessee, USA	Corals	13	19–20
10	*Porpites rotuloides* (Hall)	Middle Silurian (Niagaran)	Tennessee, USA	Corals	13	21–22
10	*Aulacophyllum sulcatum* (d'Orbigny)	Early Devonian	Indiana, USA	Corals	14	1
10	*Blothrophyllum decorticatum* Billings	Early Devonian	Ontario, Canada	Corals	14	2–3

(continued)

(continued)

Ch. no.	Species name	Age	Locality	Chapter	Fig. no.	Fig. no.
10	*Breviphrentis yandelli* (Edwards and Haime)	Early Devonian	Ohio, USA	Corals	14	4–6
10	*Cystiphylloides americanus* Edwards and Haime	Early Devonian	Ohio, USA	Corals	14	7
10	*Emmonsia emmonsi* (Hall)	Early Devonian	New York, USA	Corals	14	9–10
10	*Enterolasma strictum* (Hall)	Early Devonian	New York, USA	Corals	14	11–12
10	*Eridophyllum seriale* Edwards and Haime	Early Devonian	Kentucky, USA	Corals	14	18–21
10	*Favosites limitaris* Rominger	Early Devonian	Indiana, USA	Corals	14	8
10	*Homalophyllum ungula* (Rominger)	Early Devonian	Indiana, USA	Corals	14	13–14
10	*Pleurodictyum lenticulare* (Hall)	Early Devonian	New York, USA	Corals	14	15–17
10	*Romingeria umbellifera* (Billings)	Early Devonian	Michigan, USA	Corals	14	21–24
10	*Synaptophyllum arundinaceum* (Billings)	Early Devonian	Ontario, Canada	Corals	14	25
10	*Naos magnificus* (Billings)	Early Devonian	Michigan, USA	Corals	15	1–2
10	*Synaptophyllum arundinaceum* (Billings)	Early Devonian	Ontario, Canada	Corals	15	3–4
10	*Siphonophrentis gigantea* (Lesueur)	Early Devonian	Ohio, USA	Corals	15	5–7
10	*Acrophyllum oneidaense* (Billings)	Early Devonian	Ontario, Canada	Corals	15	8
10	*Zaphrenthis phrygia* (Rafinesque and Clifford)	Early Devonian	Kentucky, USA	Corals	15	9
10	*Pachyphyllum woodmani* (White)	Late Devonian	Iowa, USA	Corals	16	1–2
10	*Aulopora elleri* Fenton	Middle Devonian	New York, USA	Corals	16	3
10	*Bethanphyllum robustum* (Hall)	Middle Devonian	Ohio, USA	Corals	16	4
10	*Cylindrophyllum panicum* (Winchell)	Middle Devonian	Michigan, USA	Corals	16	5–6
10	*Cystiphylloides conifollis* (Hall)	Middle Devonian	New York, USA	Corals	16	7
10	*Diversophyllum transversense* (Winchell)	Middle Devonian	Michigan, USA	Corals	16	8–9
10	*Eridophyllum archiaci* Billings	Middle Devonian	Michigan, USA	Corals	16	10–12
10	*Hadrophyllum orbignyi* Edwards and Haime	Middle Devonian	Kentucky, USA	Corals	16	13–14
10	*Heliophyllum halli* Edwards and Haime	Middle Devonian	Michigan, USA	Corals	16	15–16
10	*Heterophrentis prolifica* (Billings)	Middle Devonian	Michigan, USA	Corals	16	17

(continued)

(continued)

Ch. no.	Species name	Age	Locality	Chapter	Fig. no.	Fig. no.
10	*Hexagonaria percarinata* Sloss	Middle Devonian	Michigan, USA	Corals	16	18–20
10	*Metriophyllum rectum* (Hall)	Middle Devonian	New York, USA	Corals	16	21–23
10	*Microcyclus discus* Meek and Worthen	Middle Devonian	Illinois, USA	Corals	16	24–26
10	*Trachypora ornata* Rominger	Middle Devonian	New York, USA	Corals	16	27–28
10	*Hapsiphyllum calcariforme* (Hall)	Late Mississippian	Indiana, USA	Corals	17	1–2
10	*Lithostrotion whitneyi* Meek	Late Mississippian	Montana, USA	Corals	17	3–4
10	*Lithostrotionella castelnaui* Hayasaka	Late Mississippian	Kentucky, USA	Corals	17	5–6
10	*Lithostrotionella hemisphaerica* Hayasaka	Late Mississippian	Missouri, USA	Corals	17	9–10
10	*Baryphyllum verneuilianum* Edwards and Haime	Early Mississippian	Tennessee, USA	Corals	17	7–8
10	*Cladochonus beecheri* (Grabau)	Early Mississippian	Missouri, USA	Corals	17	11
10	*Dipterophyllum glans* (White)	Early Mississippian	Missouri, USA	Corals	17	12–13
10	*Homalophyllum calceolum* (White and Whitfield)	Early Mississippian	Missouri, USA	Corals	17	17–18
10	*Neozaphrentis tenella* (Miller)	Early Mississippian	Missouri, USA	Corals	17	14–16
10	*Triplophyllites spinulosus* Edwards and Haime)	Early Mississippian	Illinois, USA	Corals	17	19–20
10	*Caninia torquia* (Owen)	Late Pennsylvanian	Nabaraska, USA	Corals	18	1–3
10	*Dibunophyllum valeriae* Newell	Late Pennsylvanian	Kansas, USA	Corals	18	4
10	*Lophophyllidium proliferum* (McChesney)	Late Pennsylvanian	Illinois, USA	Corals	18	5–6
10	*Cladochonus texasensis* Moore and Jeffords	Middle Pennsylvanian	Texas, USA	Corals	18	7
10	*Chaetetes milleporaceus* Edwards and Haime	Middle Pennsylvanian	Kansas, USA	Corals	18	8–9
10	*Lophophyllidium conoideum* Moore and Jeffords	Middle Pennsylvanian	Texas, USA	Corals	18	10
10	*Michelinia referta* Moore and Jeffords	Middle Pennsylvanian	Texas, USA	Corals	18	11
10	*Chaetetes eximus* Moore and Jeffords	Early Pennsylvanian	Oklahoma, USA	Corals	18	12–13
10	*Stereocorypha annectans* Moore and Jeffords	Middle Pennsylvanian	Texas, USA	Corals	18	14–19
10	*Amplexocarinia corrugata* (Mather)	Early Pennsylvanian	Oklahoma, USA	Corals	18	20–23
10	*Hapsiphyllum crassiseptum* Moore and Jeffords	Early Pennsylvanian	Oklahoma, USA	Corals	18	24–26

(continued)

(continued)

Ch. no.	Species name	Age	Locality	Chapter	Fig. no.	Fig. no.
10	*Striatopora oklahomaensis* (Snider)	Early Pennsylvanian	Arkansas, USA	Corals	18	27–28
10	*Waagenophyllum texanum* Heritsch	Late Permian	Texas, USA	Corals	19	1
10	*Duplophyllum septarugosum* Moore and Jeffords	Early Permian	Texas, USA	Corals	19	2–4
10	*Heritschia girty* Moore and Jeffords	Early Permian	Kansas, USA	Corals	19	5–9
10	*Leonardophyllum distinctum* Moore and Jeffords	Early Permian	Texas, USA	Corals	19	10–11
10	*Lophamplexus elaisi* Moore and Jeffords	Early Permian	Kansas, USA	Corals	19	12–15
10	*Sochkineophyllum mirabile* Moore and Jeffords	Early Permian	Kansas, USA	Corals	19	16–18
10	*Lophophyllidium dunbari* Moore and Jeffords	Early Permian	Kansas, USA	Corals	19	19–23
10	*Palaeosmilia schucherti* Heritsch	Early Permian	Kansas, USA	Corals	19	24–25
10	*Mitrodendron schaferi* (Ogilvie)	Late Jurassic (Portlandian)	Czechoslovakia	Corals	20	1
10	*Actinaraea granulata* (Munster)	Late Jurassic (Kimmeridgian)	Germany	Corals	20	2
10	*Comophyllia polymorpha* (Koby)	Late Jurassic	Portugal	Corals	20	3
10	*Crateroseris fungiformis* Tomes	Late Jurassic	England	Corals	20	4
10	*Rhipidogyra percrassa* Etallon	Late Jurassic	France	Corals	20	5
10	*Stylina girodi* Etallon	Late Jurassic	France	Corals	20	6
10	*Montlivaltia nattheimensis* Milaschewitsch	Late Jurassic	Germany	Corals	20	7–8
10	*Discocyathus eudesii* Edwards and Haime	Middle Jurassic (Bajocian)	France	Corals	20	9–11
10	*Montlivaltia norica* Frech	Late Triassic	Alps	Corals	20	12–13
10	*Palaeastraea decussata* (Reuss)	Late Triassic	California, USA	Corals	20	14
10	*Thamnasteria rectilamellosa* Winkler	Late Triassic	California, USA	Corals	20	15
10	*Thecosmilia fenestrata* (Reuss)	Middle Triassic	Alaska, USA	Corals	20	16
10	*Pleurocora texana* Roemer	Early Cretaceous	Texas, USA	Corals	21	1
10	*Strotogyra undulata* (Reuss)	Late Cretaceous	Austria	Corals	21	2
10	*Montastrea roemericana* (Wells)	Early Cretaceous	Texas, USA	Corals	21	3
10	*Diploastrea harrisi* Wells	Early Cretaceous	Texas, USA	Corals	21	4

(continued)

(continued)

Ch. no.	Species name	Age	Locality	Chapter	Fig. no.	Fig. no.
10	Astrocoenia *whitneyi* Wells	Early Cretaceous	Texas, USA	Corals	21	5
10	*Dimorphastrea crassisepta* d'Orbigny	Early Cretaceous	France	Corals	21	6
10	*Trochocyathus scottianus* Wells	Early Cretaceous	Texas, USA	Corals	21	7–9
10	*Parasmilia austinensis* Roemer	Early Cretaceous	Texas, USA	Corals	21	10–11
10	*Tiarasmilia casteri* Wells	Early Cretaceous	Texas, USA	Corals	21	12
10	*Blothrocyathus harrisi* Wells	Early Cretaceous	Texas, USA	Corals	21	13–14
10	*Micrabacia rotatilis* Stephenson	Late Cretaceous	Maryland, USA	Corals	21	15–17
10	*Stephanosmilia perlata* Roemer	Early Cretaceous	France	Corals	21	18–19
10	*Trochosmilia didymophyla* Felix	Late Cretaceous	Austria	Corals	21	20–21
10	*Parasmilia centralis* (Mantell)	Early Cretaceous	Texas, USA	Corals	21	22